Lecture Notes in Computer Science **13206**

Founding Editors

Gerhard Goos, Germany
Juris Hartmanis, USA

Editorial Board Members

Elisa Bertino, USA
Wen Gao, China
Bernhard Steffen, Germany
Gerhard Woeginger, Germany
Moti Yung, USA

FoLLI Publications on Logic, Language and Information

Subline of Lectures Notes in Computer Science

Subline Editors-in-Chief

Valentin Goranko, *Stockholm University, Sweden*
Michael Moortgat, *Utrecht University, The Netherlands*

Subline Area Editors

Nick Bezhanishvili, *University of Amsterdam, The Netherlands*
Anuj Dawar, *University of Cambridge, UK*
Philippe de Groote, *Inria Nancy, France*
Gerhard Jäger, *University of Tübingen, Germany*
Fenrong Liu, *Tsinghua University, Beijing, China*
Eric Pacuit, *University of Maryland, USA*
Ruy de Queiroz, *Universidade Federal de Pernambuco, Brazil*
Ram Ramanujam, *Institute of Mathematical Sciences, Chennai, India*

More information about this series at https://link.springer.com/bookseries/558

Aybüke Özgün · Yulia Zinova (Eds.)

Language, Logic, and Computation

13th International Tbilisi Symposium, TbiLLC 2019
Batumi, Georgia, September 16–20, 2019
Revised Selected Papers

Editors
Aybüke Özgün
University of Amsterdam
Amsterdam, The Netherlands

Yulia Zinova
Heinrich Heine University Düsseldorf
Düsseldorf, Germany

ISSN 0302-9743 ISSN 1611-3349 (electronic)
Lecture Notes in Computer Science
ISBN 978-3-030-98478-6 ISBN 978-3-030-98479-3 (eBook)
https://doi.org/10.1007/978-3-030-98479-3

© The Editor(s) (if applicable) and The Author(s), under exclusive license
to Springer Nature Switzerland AG 2022
This work is subject to copyright. All rights are reserved by the Publisher, whether the whole or part of the material is concerned, specifically the rights of translation, reprinting, reuse of illustrations, recitation, broadcasting, reproduction on microfilms or in any other physical way, and transmission or information storage and retrieval, electronic adaptation, computer software, or by similar or dissimilar methodology now known or hereafter developed.
The use of general descriptive names, registered names, trademarks, service marks, etc. in this publication does not imply, even in the absence of a specific statement, that such names are exempt from the relevant protective laws and regulations and therefore free for general use.
The publisher, the authors and the editors are safe to assume that the advice and information in this book are believed to be true and accurate at the date of publication. Neither the publisher nor the authors or the editors give a warranty, expressed or implied, with respect to the material contained herein or for any errors or omissions that may have been made. The publisher remains neutral with regard to jurisdictional claims in published maps and institutional affiliations.

This Springer imprint is published by the registered company Springer Nature Switzerland AG
The registered company address is: Gewerbestrasse 11, 6330 Cham, Switzerland

Preface

The Thirteenth International Tbilisi Symposium on Language, Logic and Computation (TbiLLC 2019) was held during September 16–20, 2019, in Kobuleti, Georgia. The symposium was organized by the Centre for Language, Logic and Speech at the Tbilisi State University, the Georgian Academy of Sciences, the Institute for Logic, Language and Computation (ILLC) of the University of Amsterdam, and the Collaborative Research Center 991 of Heinrich Heine University Düsseldorf. This biennial conference series and the proceedings are representative of the aims of the organizing institutes: to promote the integrated study of logic, information, and language. While the conference is open to contributions from any of the three fields, it aims to foster interaction among them by achieving stronger awareness of developments in the other fields, and of work that embraces more than one field or belongs to the interface between fields.

The scientific program of TbiLLC 2019 consisted of tutorials, invited lectures, contributed talks, and two workshops. The symposium offered two tutorials in language and logic and aimed at students as well as researchers working in the other areas: "Sign language linguistics. State of the art" by Fabian Bross (University of Stuttgart, Germany) and "Axiomatic Semantics" by Graham E. Leigh (University of Gothenburg, Sweden).

Seven invited lectures were delivered at the symposium: four on logic, by Gianluca Grilletti (Munich Center for Mathematical Philosophy, Germany), Philippe Balbiani (CNRS, University of Toulouse, France), Adam Bjorndahl (Carnegie Mellon University, USA), and Alexandru Baltag (University of Amsterdam, the Netherlands), two on language, by Thomas Ede Zimmermann (Goethe University Frankfurt, Germany) and Berit Gehrke (Humboldt University of Berlin, Germany), and one on computation by Libor Barto (Charles University in Prague, Czech Republic).

The workshop on Syntax, Semantics, and Pragmatics of Aspect Across Modalities (SSPAM), organized by Berit Gehrke and Fabian Bross, featured six contributed talks. The workshop on Topology and Modal Logic, organized by Adam Bjorndahl, featured five invited talks.

The contributed talks at the symposium were selected based on peer-reviewed extended abstract submissions. After the symposium, contributed talks, invited lecturers, and tutorial speakers were invited to submit full papers of their presented work for the post-proceedings. This volume contains a selection of papers that went through a rigorous two-stage, single-blind refereeing process during which each paper was reviewed by one to three anonymous referees. The post-symposium paper submissions and reviewing process were entirely organized by the editors of this volume. The papers are listed in alphabetical authorship order and divided into two main groups: Language and Logic, and Logic and Computation. Here we give a brief overview of their contribution.

Nino Amiridze investigates a relatively new development in modern spoken Georgian (Kartvelian) – the truncation of the final vowel in vocative forms of disyllabic nouns. The author considers a similar rule, operating both in some of the Georgian dialects and also in the former contact language Russian and argues between the language-family-internal vs. external contact scenarios, to find out the origin of the new pattern.

Fabian Bross offers a brief overview of linguistic research into sign languages. The target audiences are people with some background in linguistics of spoken languages. Bross briefly introduces sign languages, discusses some basics of phonological structure of these types of languages (including the use of space) as well as some new findings on the syntax of sign languages, and, finally, addresses some methodological issues. The majority of data comes from German Sign Language, although data from other sign languages is also included.

Stergios Chatzikyriakidis and Zhaohui Luo look at the issue of gradability within MTT-semantics. Specifically, they look at both gradable adjectives and nouns and show that the rich typing mechanisms afforded by MTT-semantics can provide us a natural account of gradability. Gradable adjectives take indexed nouns as their arguments, while gradable nouns are Σ-types where their first projection is a degree parameter. Chatzikyriakidis and Luo also look at multidimensional adjectives and use enumerated types to capture the multiple dimensions. They formalize their account in the Coq Proof Assistant and check its formal correctness. They also describe a recent proposal to model gradability by means of subtype universes in MTTs.

Oleg Kapanadze, Gideon Kotzé, and Thomas Hanneforth describe past and present work surrounding the development of treebank-related NLP resources for Georgian. In particular, they provide an overview of efforts made in the development of a morphologically and syntactically annotated treebank, as well as its application in the development of a syntactic parser. Building on this, the authors also report ongoing work in utilizing manual and automatic alignment solutions for the creation of a Georgian/German parallel treebank. The end goal is the development of resources and tools for improved computational processing and linguistic analysis of the Georgian language.

Ralf Naumann and Wiebke Petersen outline a formal framework that combines results from neurolinguistic research on two ERP components, the N400 and the LPP, with formal semantics. At the semantic level they combine de Groote's continuation-based version of Montague semantics with van Eijck's Incremental Dynamics enriched with frames. Naumann and Petersen provide an analysis in terms of complex properties that apply both to the semantic and the discourse level and which combine world knowledge with syntagmatic and paradigmatic relationships.

Sebastian Padó and Daniel Hole are concerned with the phenomenon of function word polysemy. They adopt the framework of distributional semantics, which characterizes word meaning by observing occurrence contexts in large corpora and which is in principle well situated to model polysemy. Although function words were traditionally considered as impossible to model reliably, due to their highly flexible use, Padó and Hole establish that contextualized word embeddings, the most recent generation of distributional methods, offer hope in this regard. Using the German reflexive pronoun *sich* as an example, they find that contextualized word embeddings capture theoretically motivated word senses for *sich* to the extent to which these senses are mirrored systematically in linguistic usage.

Swantje Tönnis takes a new perspective on *es*-clefts in German, focusing on how an *es*-cleft contributes to the discourse structure and how it does this differently than its canonical counterpart. The provided analysis combines an adapted version of Roberts' (2012) QUD stack and Velleman et al.'s (2012) approach to clefts. In particular, Tönnis presents a model that includes implicit and potential questions into the QUD stack and introduces the concept of expectedness, that she argues is crucial for the acceptability of clefts. Tönnis proposes that the cleft addresses a question that came up in the preceding context but that is not as urgent for the addressee to be answered at that point in the discourse compared to other questions. Those questions that are more urgent are answered with a canonical sentence. This approach is compatible with other functions that have been proposed for clefts, such as marking exhaustivity, maximality, or correction. However, it can also account for examples where the cleft serves to establish discourse coherence.

Thomas Ede Zimmermann scrutinizes the very notion of extension, which is central to many contemporary approaches to natural language semantics. The starting point is a puzzle about the connection between learnability and extensional compositionality, which is frequently made in semantics textbooks: given that extensions are not part of linguistic knowledge, how can their interaction serve as a basis for explaining it? Before the puzzle is resolved by recourse to the set-theoretic nature of intensions, a few clarifying observations on extensions are made, starting from their relation to (and the relation between) reference and truth. Extensions are then characterized as the result of applying a certain heuristic method for deriving contributions to referents and truth-values, which also gives rise to the familiar hierarchy of functional types.

Malte Zimmermann, Lea Fricke, and Edgar Onea present two novel diagnostics for gauging the exhaustivity level of German *wh*-interrogatives embedded under the predicates *wissen* 'know' and *überraschen* 'surprise'. The readings available in combination with the concessive particle combination *SCHON...aber* 'alright...but' and the Q-adverb *teilweise* 'partially' provide evidence that embedded *wh*-interrogatives under veridical and distributive *wissen* 'know' have a weakly exhaustive (WE) reading as their basic semantic interpretation. The logically stronger strongly exhaustive (SE) reading is a pragmatic enrichment that can be cancelled by *SCHON...aber*. Zimmermann, Fricke, and Onea provide an event-based analysis of *know+wh* as expressing the maximal plurality of sub-events of knowing the individual answers to the question. Under the cognitive-emotive attitude verb *überraschen* 'surprise', which is not obligatorily distributive, *wh*-interrogatives allow for two types of WE-interpretations, distributive and non-distributive. The *SCHON...aber*-diagnostic shows the logically stronger distributive WE-reading to be a pragmatic enrichment. In view of experimental evidence that *surprise+wh* allows for SE-interpretations, Zimmermann, Fricke, and Onea provide a tentative analysis of *surprise+wh* as expressing a psychological state caused by a complex situation, or subparts or missing parts thereof.

Bahareh Afshari and Graham E. Leigh prove Lyndon interpolation for the modal μ-calculus, a strengthening of Craig interpolation which is not implied by uniform interpolation. The proof utilises 'cyclic' sequent calculus and provides an algorithmic construction of interpolants from valid implications. This direct approach enables

Afshari and Leigh to derive a correspondence between the shape of interpolants and existence of sequent calculus proofs.

The core of the paper by Philippe Balbiani and Tinko Tinchev is constituted by Chagrova's Theorems about first-order definability of given modal formulas and modal definability of given elementary conditions. Balbiani and Tinchev consider classes of frames for which modal definability is decidable as well as classes of frames for which first-order definability is trivial, and provide a new proof of Chagrova's Theorem about modal definability as well as the sketches of the proofs of new variants of Chagrova's Theorem about modal definability.

Alexandru Baltag, Nick Bezhanishvili, and Saúl Fernández González introduce a multi-agent topological semantics for evidence-based belief and knowledge, which extends the dense interior semantics developed in Baltag, Bezhanishvili, Özgün, and Smets (2016). The authors provide the complete logic of this multi-agent framework together with generic models for a fragment of the language. They also define a new notion of group knowledge which differs conceptually from previous approaches.

In this paper, Dragan Doder, Zoran Ognjanović, Nenad Savić, and Thomas Studer present a logic for reasoning about higher-order upper and lower probabilities of justification formulas. They provide sound and strongly complete axiomatization for the logic and show that the introduced logic generalizes the existing probabilistic justification logic PPJ.

Besik Dundua, Temur Kutsia, and Mikheil Rukhaia define an unranked nominal language, an extension of the nominal language with sequence variables and term tuples. They define the unification problem for unranked nominal terms and present an algorithm solving the unranked nominal unification problem.

Gianluca Grilletti and Davide Emilio Quadrellaro focus on univariate formulae χ, that is, formulae containing at most one atomic proposition. For every such formula, they introduce a lattice of intermediate theories: the lattice of χ-logics. The key idea to define χ-logics is to interpret atomic propositions as fixpoints of the formula χ^2, which can be characterised syntactically using Ruitenburg's Theorem. Grilletti and Quadrellaro show that χ-logics form a lattice, dually isomorphic to a special class of varieties of Heyting algebras. This approach allows the authors to build five distinct lattices—corresponding to the possible fixpoints of univariate formulas—among which the lattice of negative variants of intermediate logics.

Temur Kutsia and Cleo Pau focus on proximity relations: fuzzy binary relations satisfying fuzzy reflexivity and symmetry properties. Tolerance, which is a reflexive and symmetric (and not necessarily transitive) relation, can be also seen as a crisp version of proximity. Kutsia and Pau discuss two fundamental symbolic computation problems for proximity and tolerance relations: matching and anti-unification, present algorithms for solving them, and study properties of those algorithms.

Graham E. Leigh presents a short introduction to the logical analysis of truth and related concepts. He examines which assumptions are implicit in the paradoxes of truth and self-reference, and presents some of the important formal theories of truth that have arisen out of these considerations.

We would like to thank all the authors for their contributions, and the anonymous reviewers for their high-quality reports. We would also like to express our gratitude to the organizers of the symposium, who made the event an unforgettable experience for all of its participants. The Tbilisi symposia are renowned not only for their high scientific standards, but also for their friendly atmosphere and heartwarming Georgian hospitality, and the 13th symposium was no exception. Finally, we thank the ILLC (University of Amsterdam) and the Department of Computational Linguistics at Heinrich Heine University Düsseldorf for their generous financial support for the symposium.

December 2021

Aybüke Özgün
Yulia Zinova

Organization

Organizing Institutions

Centre for Language, Logic and Speech at the Tbilisi State University, Georgia.
Georgian Academy of Sciences.
Institute for Logic, Language and Computation (ILLC) of the University of Amsterdam, The Netherlands.
Collaborative Research Center 991, Heinrich Heine University Düsseldorf, Germany.

Organization Committee

Rusiko Asatiani	Tbilisi State University, Gerogia
Anna Chutkerashvili	Tbilisi State University, Georgia
David Gabelaia	TSU Razmadze Mathematical Institute, Georgia
Kristina Gogoladze	ILLC, University of Amsterdam, The Netherlands
Marine Ivanishvili	Tbilisi State University, Georgia
Nino Javashvili	Tbilisi State University, Georgia
Mamuka Jibladze	TSU Razmadze Mathematical Institute, Georgia
Ramaz Liparteliani	Tbilisi State University, Georgia
Liana Lortkipanidze	Tbilisi State University, Georgia
Ana Kolkhidashvili	Tbilisi State University, Georgia
Evgeny Kuznetsov	Tbilisi State University, Georgia
Peter van Ormondt	ILLC, University of Amsterdam, The Netherlands
Khimuri Rukhaia	Tbilisi State University and Sokhumi State University, Georgia
Nutsa Tsereteli	Tbilisi State University, Georgia

Program Committee

Bahareh Afshari	University of Gothenburg, Sweden
Rusiko Asatiani	Tbilisi State University, Georgia
Guram Bezhanishvili	New Mexico State University, USA
Nick Bezhanishvili	ILLC, University of Amsterdam, The Netherlands
David Gabelaia	TSU Razmadze Mathematical Institute, Georgia
Katharina Hartmann	Goethe University Frankfurt, Germany
Jules Hedges	University of Oxford, UK
Daniel Hole (Co-chair)	University of Stuttgart, Germany
Sebastian Löbner	Heinrich Heine University Düsseldorf, Germany
Matteo Mio	CNRS, ENS Lyon, France
Sara Negri	University of Helsinki, Finland
Sebastian Padó	University of Stuttgart, Germany

Valeria de Paiva	Nuance Communications, USA
Alessandra Palmigiano	Technical University Delft, The Netherlands
Roland Pfau	University of Amsterdam, The Netherlands
Martin Schäfer	University of Tübingen, Germany
Lutz Schröder	University of Erlangen-Nürnberg, Germany
Kerstin Schwabe	Leibniz-ZAS Berlin, Germany
Alexandra da Silva	University College London, UK
Alex Simpson (Co-chair)	University of Ljubljana, Slovenia
Luca Spada	University of Salerno, Italy
Ronnie B. Wilbur	Purdue University, USA
Fan Yang	University of Helsinki, Finland

Standing Committee

Rusiko Asatiani	Tbilisi State University, Georgia
Matthias Baaz	TU Vienna, Austria
Guram Bezhanishvili	New Mexico State University, USA
George Chikoidze	Georgian Technical University, Georgia
Dick de Jongh (Chair)	ILLC, University of Amsterdam, The Netherlands
Paul Dekker	ILLC, University of Amsterdam, The Netherlands
Hans Kamp	University of Stuttgart, Germany
Manfred Krifka	Leibniz-ZAS Berlin, Germany
Temur Kutsia	Johannes Kepler University Linz, Austria
Sebastian Löbner	Heinrich Heine University Düsseldorf, Germany
Barbara Partee	University of Massachusetts Amherst, USA
Alexandra da Silva	University College London, UK

Tutorials

Fabian Bross	University of Stuttgart, Germany
Graham E. Leigh	University of Gothenburg, Sweden

Invited Speakers

Philippe Balbiani	CNRS, Toulouse University, France
Alexandru Baltag	ILLC, University of Amsterdam, The Netherlands
Libor Barto	Charles University, Czech Republic
Adam Bjorndahl	Carnegie Mellon University, USA
Berit Gehrke	Humboldt University of Berlin, Germany
Gianluca Grilletti	University of Stuttgart, Germany
Thomas Ede Zimmermann	Göthe University, Germany

Workshop Organizers

Fabian Bross	University of Stuttgart, Germany
Adam Bjorndahl	Carnegie Mellon University, USA
Berit Gehrke	Humboldt University of Berlin, Germany

Invited Speakers at Workshops

Alexandru Baltag	ILLC, University of Amsterdam, The Netherlands
Adam Bjorndahl	Carnegie Mellon University, USA
David Gabelaia	TSU Razmadze Mathematical Institute, Georgia
Tamar Lando	Columbia University, USA
Aybüke Özgün	ILLC, University of Amsterdam, The Netherlands

Additional Reviewers

Giovanna D'Agostino
Benjamin Anible
Mauricio Ayala-Rincón
Adam Bjorndahl
Harm Brouwer
Misha Daniel
Kilian Evang
Maribel Fernandez
David Fernández-Duque
Christopher Hampson
Pascual Julian-Iranzo
Natthapong Jungteerapanich
Aikaterini-Lida Kalouli
Stanislav Kikot
Ondrej Majer
Denis Paperno
Soroush Rafiee Rad
Lorenzo Rossi
Clemente Rubio-Manzano
Gil Sagi
Matteo Sammartino
Luigi Santocanale
Dmitri Sitchinava
Nadine Theiler
Ruben Van de Vijver
Matthijs Westera
Yiwen Zhang

Contents

Language and Logic

Final-Vowel Truncation in the Forms of Address in Modern Spoken Georgian

Nino Amiridze(✉)

Ivane Javakhishvili Tbilisi State University, Tbilisi, Georgia
nino.amiridze@gmail.com

Abstract. This paper studies a relatively new development in modern spoken Georgian (Kartvelian) – the truncation of the final vowel in vocative forms of disyllabic nouns. It considers a similar rule, operating both in some of the Georgian dialects and also in the former contact language Russian and argues between the language-family-internal vs. external contact scenarios, to find out the origin of the new pattern.

Keywords: Language contact · Borrowing · Vocative truncation · Georgian · Russian

1 Introduction

This paper deals with a relatively new development in modern spoken Georgian – vocative (VOC[1]) truncation of disyllabic proper nouns to express familiarity, close social relationship, affection, and endearment (cf. (1a) vs. (1b)). The truncation is optionally accompanied by lengthening of the remaining vowel (1c).[2]

(1) a. Standard Georgian
nana, ҫaiḳitxe es ҫign-i!
Nana you.SG.read.IMP.it this book-NOM
'Nana, read this book!' (simple address)

b. Spoken Georgian
nan, ҫaiḳitxe es ҫign-i!
Nan you.SG.read.IMP.it this book-NOM
'Nana, read this book!' (expressing familiarity)

The work was done within the project FR-19-18557, supported by the Shota Rustaveli National Science Foundation of Georgia.

[1] The following abbreviations are used in this paper: ADV = adverbial; DAT = dative; ERG = ergative; EV = epenthetic vowel; GEN = genitive; H = high; IMP = imperative; INST = instrumental; L = low; NARR = narrative; NEG = negation; NOM = nominative; PART = particle; PL = plural; SG = singular; TAM = Tense, aspect, mood; VOC = vocative.

[2] In Georgian, vowel quantity is not a phonological feature [55, p. 53], [7, p. 81].

© The Author(s), under exclusive license to Springer Nature Switzerland AG 2022
A. Özgün and Y. Zinova (Eds.): TbiLLC 2019, LNCS 13206, pp. 3–25, 2022.
https://doi.org/10.1007/978-3-030-98479-3_1

c. Spoken Georgian
naan, ҫaiḳitxe es ҫign-i![3]
Naan you.SG.read.IMP.it this book-NOM
'Nana, read this book!' (expressing familiarity)

VOC truncation has previously been known from several Georgian dialects [26], where it operates exclusively on nouns with more than two syllables (cf. (2a) vs. (2b)). In those dialects, it is not allowed to have lengthening of any of the remaining vowels of the truncated form (see (2c) and (2d)). Due to urbanization, the rule spread into the Tbilisi Georgian as well.

(2) Mtiulian-Gudamaqrian dialect, [26, p. 281]

a. manana!	b. manan!	c. *maanan!	d. *manaan!
Manana	Manan	Maanan	Manaan
'Manana!' (simple address)	'Manana!' (expressing familiarity)	'Manana!' (expressing familiarity)	'Manana!' (expressing familiarity)

The two groups of proper names, disyllabic ones vs. those with more than two syllables, when truncated in VOC, get a similar interpretation (cf. (1b) vs. (2b)). They are used informally, to express familiarity, close social relationship, affection, and endearment, as opposed to the non-truncated form having a neutral reading (see (1b) vs. (1a) and (2b) vs. (2a)). Then the following questions arise: Is it possible that the previously operating rule (originating from dialects) which truncated three-and-more-syllable names got analogically extended to truncate disyllabic names in Tbilisi Georgian? Or does it have a different origin, as patterns of vowel lengthening are different in the two groups of proper names?

To answer these questions and find out what is the origin of the more recent rule, I will briefly describe the vocative marking in standard Georgian (Sect. 2). Then, I will look into the VOC truncation in the diachrony of Georgian as well as in its modern dialects (Sect. 3). Section 4 overviews the VOC truncation in Russian, a former contact language for Georgian, in the search of a potential donor language. Section 5 will discuss whether the new pattern originates from Georgian dialects or is replicated from Russian. Section 6 concludes the paper.

Thus, this work addresses the following problems:

- What is the origin of the VOC truncation in disyllabic names in Georgian: does it come from Georgian dialects, or is it a change induced by contact with Russian, a genetically unrelated donor language?
- What type of change is it? What is its contribution to the theory of language contact?

[3] Posted on February 27, 2020 at https://forum.ge/?showtopic=35132192&view=findpost&p=55902429.

2 Vocative Marking in Georgian

According to the literature, there are seven cases in Georgian: nominative (NOM), narrative (NARR),[4] dative (DAT), genitive (GEN), instrumental (INST), adverbial (ADV), and vocative (VOC), see, e.g., [6,13,15,36,43,44,51,53]. These cases, except INST, ADV, and VOC, participate in the marking of arguments of verbs of different semantic classes in different TAM series. The INST and ADV cases are used to mark oblique arguments and adjuncts. As for the VOC, it is used in address forms only.

In general, in the standard variety, the vocative case ending depends on both formal and semantic properties of the stem to which it is applied to. In particular, it matters whether or not

- the address form consists of several components,
- the referent is human,
- it is a common noun,
- the stem is consonant-final,
- the stem is monosyllabic.

In this paper, I will only discuss the allomorphs of the vocative morpheme, when it attaches to a single item and will leave aside the syntax of applying VOC within a structure. The vocative ending for common nouns with consonant-final stems is *-o*, irrespective of the number of syllables or of the semantic class of the referent of the noun (e.g., human (3a), non-human animate (3b) or inanimate (3c)):

(3) Consonant-final stems

a. Human, animate	b. Non-human, animate	c. Inanimate
ḳac-o!	mgel-o!	mindor-o!
man-VOC	wolf-VOC	meadow-VOC
'[Hey,] man!'	'[Hey,] wolf!'	'[Oh,] the meadow'

Vowel-final common nouns, irrespective of the semantic class of the referent, group into two sub-groups, depending on the number of syllables. In VOC, those with monosyllabic stems, get the ending *-o* (4), while those with polysyllabic stems have *-v*:[5]

[4] The narrative case has also been referred to as *ergative* (ERG) in the Kartvelian literature [23]. There has been some debate, whether Georgian is an ergative language or not [3,19,21,22]. Here, I will be using the term *narrative* instead of *ergative* to avoid the bias towards the ergative alignment type.

[5] The VOC ending *-v* originates from the VOC ending *-o*. According to [44, p. 60], the VOC *-o*, when immediately following a vowel, historically lost syllabicity and went through the stages of impoverishment such as *-o* ≫ *-w*/*-v* ≫ -∅. In spoken language, both *v*- and -∅ are found and both are accepted by normative grammars as variants [17, p. 62].

(4) Vowel-final common monosyllabic nouns

a. Human, animate
ʒe-o!
son-VOC
'[Hey,] son!'

b. Non-human, animate
xbo-o!
calf-VOC
'[Hey,] the calf!'

c. Inanimate
mta-o!
mountain-VOC
'[Hey,] the mountain!'

(5) Vowel-final common polysyllabic nouns

a. Human, animate
gogo-v!
girl-VOC
'[Hey,] girl!'

b. Non-human, animate
lok̇ok̇ina-v!
snail-VOC
'[Hey,] the snail!'

c. Inanimate
samšoblo-v!
motherland-VOC
'[Oh,] the motherland!'

Proper nouns also illustrate differences in VOC marking. First of all, in VOC, consonant-final human first names (6a) and proper names of non-human animates (6b) are represented by the stem (and not by the NOM form (cf. (6a) vs. (6c) and (6b) vs. (6d))):

(6) Consonant-final proper names in Standard Georgian

a. Female first name
mariam, icek̇ve!
Mariam dance
'Mariam, dance!'

b. A name of a dog
julbas, iq̇epe!
Julbas bark
'Julbas, bark!'

c. Female first name
mariam-i cek̇vavs.
Mariam-NOM she.dances
'Mariam is dancing.'

d. A name of a dog
julbas-i q̇eps.
Julbas-NOM it.barks
'Julbas is barking.'

As for the consonant-final human last names (7a) and proper nouns with inanimate referents (7b), they are marked by *-o*.

(7) a. A family name
maγalašvil-o!
Maghalashvili-VOC
'[Hey,] Maghalashvili!'

b. A name of a mountain
mq̇invarc̣ver-o!
Mount.Kazbek-VOC
'[Oh,] Mount Kazbek!'

Proper nouns with vowel-final stems show a different grouping: human first (8a) and last names (8b) are represented by the stem in VOC (cf. (8a) vs. (9a) and (8b) vs. (9b)), while those with non-human animate (10a) and inanimate referents (10b) require *-v*[6] in VOC:

[6] See footnote 5.

(8) Vowel-final proper names in standard Georgian

a. First name
giorgi!
Giorgi
'[Hey,] Giorgi!'

b. Family name
beriʒe!
Beridze
'[Hey,] Beridze!'

(9) Vowel-final proper names in standard Georgian

a. First name
giorgi-m šeisc̣avla tekst-eb-i.
Giorgi-NARR he.studied.them text-PL-NOM
'Giorgi studied the texts.'

b. Family name
beriʒe-m gamoiḳvlia dialekṭeb-i.
Beridze-NARR (s)he.researched.them dialect-PL-NOM
'Beridze researched the dialects.'

(10) Proper nouns with non-human animate and inanimate referents

a. A name of a hamster
vanila-v!
Vanilla-VOC
'[Hey,] Vanilla!'

b. A name of a country
sakartvelo-v!
Georgia-VOC
'[Oh,] Georgia!'

Table 1 summarizes the vocative case marking in Georgian.

3 Truncation in Georgian, Its Diachrony and Synchronic Variation

3.1 No Vocative Truncation in Old or Middle Georgian

Old and middle Georgian do not illustrate truncation of any case endings. The only instance when a vowel gets deleted is *elision*, which refers to the deletion of the final vowel of vowel-final stems (cf. the stem *deda-* in NOM (11a) and VOC (11d) to the stem *ded-* in GEN (11b) and INST (11c)).

According to [45, pp. 37–38], elision characterized only a part of the old Georgian vowel-final common nouns, namely, those that end with the vowel *a* (see (11a)) or *e*, while the nouns with *o-* or *u*-final stems do not undergo elision [45, pp. 38–40]. However, the truncation is characteristic exclusively of genitive (cf. (11b) vs. (11a)) and instrumental case forms (cf. (11c) vs. (11a)) but not to vocative (11d):

Table 1. VOC marking of nouns in Georgian

Common vs. proper	Stem	Semantic category	Number of syllables	VOC marking	Example
Common	Consonant-final	Human	Any	*-o*	(3a)
		Non-human animate	Any	*-o*	(3b)
		Inanimate	Any	*-o*	(3c)
	Vowel-final	Human	Monosyll.	*-o*	(4a)
			Polysyll.	*-v*	(5a)
		Non-human animate	Monosyll.	*-o*	(4b)
			Polysyll.	*-v*	(5b)
		Inanimate	Monosyll.	*-o*	(4c)
			Polysyll.	*-v*	(5c)
Proper	Consonant-final	Human, first name	Polysyll.[a]	Represented by the stem	(6a)
		Human, last name	Polysyll.	*-o*	(7a)
		Non-human animate	Polysyll.	Represented by the stem	(6b)
		Inanimate	Polysyll.	*-o*	(7b)
	Vowel-final	Human, first name	Polysyll.	Represented by the stem	(8a)
		Human, last name	Polysyll.	Represented by the stem	(8b)
		Non-human animate	Polysyll.	*-v*	(10a)
		Inanimate	Polysyll.	*-v*	(10b)

[a]There are no monosyllabic proper names in Georgian.

(11) Vowel-final common noun

a. deda-j
mother-NOM
'mother'

b. ded-is(a)
mother-GEN
'mother's'

c. ded-it(a)
mother-INST
'by the mother'

d. deda-o! / *ded-o! / *ded!
mother-VOC mother-VOC mother.VOC
'[Hey,] mother!'

Proper nouns with vowel-final stem do not undergo elision (for GEN, cf. (12b) vs. (12a) or for INST, cf. (12c) vs. (12a)). Their vocative form is given as the stem and none of its parts get truncated either (12d):

(12) Vowel-final proper noun

a. iovane-j
Iovane-NOM
'Iovane'

b. iovane-js[7] / *iovan-js / *iovan-is
Iovane-GEN Iovane-GEN Iovane-GEN
'Iovane's'

c. iovane-jt / *iovan-jt / *iovan-it
Iovane-INST Iovane-INST Iovane-INST
'by Iovane'

d. iovane! / *iovan!
Iovane Iovane
'Iovane!'

Obviously, elision is not applicable to consonant-final stems (see (13), (14)), as there is no final vowel to be truncated there.[8]

(13) Consonant-final common noun

a. ḳac-i
man-NOM
'man'

b. ḳac-is(a)
man-GEN
'man's'

c. ḳac-it(a)
man-INST
'by a man'

d. ḳac-o!
Man-VOC
'[Hey], man!'

(14) Consonant-final proper noun

a. davit[9]
Davit
'Davit'

b. davit-is
Davit-GEN
'Davit's'

c. davit-it
Davit-INST
'by Davit'

d. davit!
Davit
'[Hey], Davit!'

However, note that the elision and the truncation of the final vowel of the VOC case-ending are essentially two different processes. While the former is morphophonological, the latter is phonetic.

Middle Georgian reflects practically the same situation in nominal case marking.[10] Namely, there is no truncation of VOC (cf. (15a) vs. (16) or (15b) vs. (16)) or any other case endings:

(15) Middle Georgian variation

a. From [38, p. 17], address form represented by the stem
ruka, did-ad miq̣varxar[...]
Ruka.VOC big-ADV I.love.you.SG
'Ruka, I love you much[...]'

---

[7] Consonant-final proper nouns do not get the epenthetic vowel (EV) *-a* in GEN (cf. (12b) vs. (11b)), INST (cf. (12c) vs. (11c)) (or DAT [45, p. 41], [41, p. 32]).

[8] Nouns with consonant-final stems get *syncopated* but I am not going to discuss the rules governing that operation in this paper. For more information see [14,41,45].

[9] Consonant-final proper nouns are represented by their stem in NOM (14a), NARR, and VOC (14d) forms [45, p. 41], [41, p. 32].

[10] The difference from the old Georgian is that proper nouns start resembling common nouns by developing case endings in NOM, NARR, and VOC.

b. From [38, p. 3]
ruka-v, šengan miḳvirs eget-i saubar-i!
Ruka-VOC from.you I.am.surprised that.kind-NOM talk-NOM
'Ruka, such talk from you surprises me!'

(16) Middle Georgian

*ruk!
Ruka.VOC

'Ruka!'

3.2 Modern Georgian Dialects and Vocative Truncation

Address Forms in Northeastern, Eastern, Central, and Southwestern Dialects. According to Jorbenadze [26], the Mtiulian-Gudamaqrian dialect in the Northeast, Kakhetian in the East, Kartlian in the Center, and the Meskhian dialect in the Southwest of Georgia show VOC truncation in trisyllabic (cf. (17a) vs. (17b)) and quadrisyllabic words (cf. (17c) vs. (17d)). The difference between the truncated and non-truncated forms is that the former express familiarity with the person referred to by the form and are used in informal contexts. As for the latter, non-truncated vocative, it represents a simple address and has a neutral reading:

(17) Georgian

a. Male	b. Male	c. Female	d. Female
mamuḳa!	mamuḳ!	duduxana!	duduxan![11]
Mamuka	Mamuk	Dudukhana	Dudukhan
'[Hey,] Mamuka!' (simple address)	'[Hey,] Mamuka!' (expressing familiarity)	'[Hey,] Dudukhana!' (simple address)	'[Hey,] Dudukhana!' (expressing familiarity)

Note, however, that in those dialects that allow truncation in nouns consisting of three and more syllables, there is no truncation in disyllabic ones [26] (cf. (18a) vs. (18b) and (18c) vs. (18d)):

[11] https://www.kvirispalitra.ge/2011-03-31-07-00-04/11031-qarthveli-msakhiobebi-dudukhana-tserodze.html.

(18) Mtiulian-Gudamaqrian dialect

a. Male	b. Male	c. Female	d. Female
mate!	*mat!	maro!	*mar!
Mate	Mat	Maro	Mar
'[Hey,] Mate!' (simple address)	'[Hey,] Mate!' (expressing familiarity)	'[Hey,] Maro!' (simple address)	'[Hey,] Maro!' (expressing familiarity)

Therefore, modern spoken Georgian could not have copied the new pattern of truncating the final vowel of voc forms of disyllabic names from these varieties of Georgian, as they do not allow such a truncation in disyllabic forms.

Address Forms in Western Dialects. Western varieties do not allow the truncation of the final vowel of address forms [26] (cf. (19a) vs. (19b)). However, it is common there to truncate the final CV sequence (cf. (19a) vs. (19c)):

(19) Western Georgian dialects

a. Full address form	b. Truncating final vowel	c. Truncating final CV
mamuḳa!	*mamuḳ!	mamu!
Mamuka	Mamuk	Mamu
'[Hey,] Mamuka!' (simple address)	'[Hey,] Mamuka! (expressing familiarity)	'[Hey,] Mamuka!' (expressing familiarity)

Thus, modern spoken Georgian could not have copied the new pattern of truncating the final vowel in disyllabic names to express familiarity, close social relationship, affection, and endearment from the Western varieties of Georgian simply because they do not possess such a pattern.

3.3 Tbilisi Georgian Between 1950–1980

Starting from around the 1950s,[12] the truncation of vocative forms of nouns consisting of three and more syllables spread in Tbilisi Georgian (20). Probably it can be attributed to the rapid urbanization of the previous decades, which started in the 1910s, since such a truncation is functional in Northeastern, Eastern, Central, and Southwestern dialects (see Sect. 3.2 and Table 2).

[12] The dates like 1950s for the start of the spread of truncated voc forms of tri- and quadrisyllabic names in the Tbilisi variety, as well as the 1980s for the start of the spread of truncated voc forms of disyllabic names are an approximation (see the Tables 2 and 3). In the absence of actual documented data of informal spontaneous speech of those decades, I exclusively rely on impressions and reports of a limited number of individuals as well as on address forms that sporadically occur in published interviews/recollections (see, for instance, the source interview for Example (17d)). A more thorough investigation will be needed to locate more precise dates.

(20) Tbilisi Georgian of the post-urbanization period

a. Female	b. Female	c. Male	d. Male
manana!	manan!	elguja!	elguj!
Manana	Manan	Elguja	Elguj
'Manana!' (simple address)	'Manana!' (expressing familiarity)	'Elguja!' (simple address)	'Elguja!' (expressing familiarity)

However, disyllabic names do not get truncated in those dialect varieties (Table 3). Nevertheless, the Tbilisi Georgian of around 1980s illustrates this use.

Table 2. Use of VOC truncation in names with three and more syllables in varieties of Georgian

	Western dialects	Northeastern, Eastern, Central and Southeastern dialects	Tbilisi variety
Before 1910s	No	Yes	No
1910s (start of urbanization) – 1950	No	Yes	No
1950–1980	No	Yes	Yes
Since 1980	No	Yes	Yes

Table 3. Use of VOC truncation in names with two syllables in varieties of Georgian

	Western dialects	Northeastern, Eastern, Central and Southeastern dialects	Tbilisi variety
Before 1910s	No	No	No
1910s (start of urbanization) – 1950	No	No	No
1950–1980	No	No	No
Since 1980	No	No	Yes

Starting around the 1980s, in Tbilisi Georgian it became possible to truncate the final vowel of the full vocative forms of disyllabic proper nouns (cf. (21a) vs. (21b) and (21c) vs. (21d)). The truncated forms were pragmatically heavily loaded and used to express familiarity, close social relationship, affection, and endearment:

(21) Georgian disyllabic names *giga* (male) and *nino* (female) truncated in VOC

a. giga!	b. gig!	c. nino!	d. nin!
Giga	Gig	Nino	Nin
'[Hey,] Giga!' (simple address)	'[Hey,] Giga!' (expressing endearment)	'[Hey,] Nino!' (simple address)	'[Hey,] Nino!' (expressing endearment)

We would expect there to be some other source besides Georgian dialect varieties from which the rule of truncating disyllabic nouns could have been replicated. A look at the sister languages of Georgian (Svan, Megrelian, and Laz) reveals that they do not employ the truncation of the final vowel in disyllabic nouns (or in nouns with more than two syllables) in address forms [8,27,29,37,42,52]. Thus, attributing the truncation of disyllabic nouns in VOC to Kartvelian origins and the influence from within the language family does not seem feasible.

4 Contact with Russian as a Possible Source

In the previous section, we have seen that for truncated vocatives of disyllabic nouns the inheritance scenario from the earlier stages of Georgian looks unlikely (Sect. 3.1). Modern spoken Georgian could not have directly copied the rule of truncating vocative of disyllabic nouns from dialects of Georgian or its sister languages either (Sect. 3.2).

Let us explore another possibility for how a new pattern could have been acquired. Namely, let us consider a contact hypothesis: the VOC truncation pattern for disyllabic names could be copied from a language in close contact with Georgian. A consideration of the time when the pattern became functional and operating in spoken Georgian, that is around the 1980s, leaves us with the only one possible option: the Russian language.

Russian is the most recent donor among the languages Georgian has been in direct contact with [9,18]. The two languages came into contact around the beginning of the 19th century [47,48]. Before the end of the active contact, in the 1990s, after the collapse of the former Soviet Union, Georgian-Russian language contact could be argued to have reached Stage 3 [4, p. 3] on the *borrowing scale* of Thomason and Kaufman [49, pp. 74–75].

The particular pattern of truncating the final vowel of vocative forms to express familiarity, close social relationship, affection, and endearment has been in Russian since the second half of the XIX century and ended up in the former Soviet intelligentsia speech around 1960s [12]. The Soviet Russian speech patterns have easily been copied into Georgian, as the Republic has been a part of the former USSR. The contact with Russian could be a plausible explanation for the spread of the new pattern of VOC truncation in Georgian.

In the next sections, I will overview VOC truncation in Russian (Sect. 4.1) and compare it to the use of the new pattern of VOC truncation in spoken Georgian (Sect. 4.2).

4.1 Vocative Truncation in Russian

According to [12], in Russian, the truncation of the final vowel in vocative forms was first used in the texts of the second half of the XIX century, mainly reflecting the speech of peasants (the actual usage must predate those texts). The use got spread to literature only later in the 1920s. Finally, in the 1960s, the pattern came into the speech of the intelligentsia, which was an upper social class in the

USSR. Today, the pattern is used in everyday spoken Russian, it has an informal usage, shows small social distance between interlocutors and is used to express familiarity, close social relationship, affection, and endearment.

In modern Russian, there is a use of nominative pro vocative (cf. (22a) vs. (22b) and (22d) vs. (22e)). Truncated vocative (or "new vocative" [12]) (see (22c) and (22f)) is formed on personal names (22) and kinship terms (23) having a penultimate-stressed nominative in *-a* [10,12]. Note that the truncation of the final vowel can operate both on disyllabic nouns (cf. (22b) vs. (22c), (23a) vs. (23b)) and on nouns consisting of three and more syllables (cf. (22e) vs. (22f), (23c) vs. (23d) and (24b) vs. (24c)):

(22) Russian personal names

a. Disyllabic reference form	b. Full vocative	c. Truncated vocative
Máša	Máša!	Máš!
Masha	Masha	Mash
'Masha' (female name)	'[Hey,] Masha!' (simple address)	'[Hey,] Masha!' (expressing affection)

d. Trisyllabic reference form	e. Full vocative	f. Truncated vocative
Nikíta	Nikíta!	Nikít!
Nikita	Nikita	Nikit
'Nikita' (male name)	'[Hey,] Nikita!' (simple address)	'[Hey,] Nikita!' (expressing affection)

(23) Russian kinship terms in vocative

a. máma!	b. mám!	c. dedúl'a!	d. dedúl'![13]
mom	mom	granddad	granddad
'[Hey,] mama!' (simple address)	'[Hey,] mama!' (expressing affection)	'[Hey,] grandad!' (simple address)	'[Hey,] grandad!' (expressing affection)

(24) Russian quadrisyllabic patronym

a. Reference form	b. Full VOC	c. Truncated VOC [24, p. 10]
Andréjevna	Andréjevna!	Andréjevn!
Andreevna	Andreevna	Andreevna
'Andreevna' (daughter of Andrej)	'[Hey,] Andreevna!' (simple address)	'[Hey,] Andreevna!' (expressing affection)

According to [39], the use of truncated vocative forms of nouns in Russian is optional, restricted to an informal setting with a relatively close interlocutor relationship, and expresses the following pragmatic meaning: familiarity, close social relationship, affection, and endearment. Note that in Russian truncated

[13] Examples (23b) and (23d) are taken from the Russian National Corpus, http://www.ruscorpora.ru/.

vocatives, one can get the remaining vowel lengthened[14] (cf. (23b) vs. (25) and (26b) vs. (26c)):

(25) Russian kinship term, truncated, lengthened

maam!
mom

'[Hey,] mom!' (expressing affection)

(26) Russian kinship term

a. Full form	b. Truncated	c. Truncated, lengthened
pápa!	páp!	paap!
dad	dad	dad
'[Hey,] dad!' (simple address)	'[Hey,] dad!' (expressing affection)	'[Hey,] dad!' (expressing affection)

In those nouns that consist of three or more syllables, only the vowel which gets an accent in the full form can be lengthened in truncated vocative forms. For instance, when the stress falls on the second *i* of the proper noun *Nikíta* (22e), the correct truncated and lengthened VOC form would be *nikiit!* (27a) but not **niikit!* (27b):

(27) Russian VOC forms

a. Truncated, lengthened	b. Truncated, lengthened
nikiit!	*niikit!
Nikiit	Niikit
'[Hey,] Nikita!' (expressing affection)	'[Hey,] Nikita!' (expressing affection)

The same is true of the truncation of three and more-syllable common nouns (here, kinship terms). It is the stressed vowel of the full form that becomes long in the truncated form (cf. (23c) vs. (28a) and (23c) vs. (28b)):

(28) Russian VOC forms

a. Truncated, lengthened	b. Truncated, lengthened
deduul'!	*deedul'!
granddad	granddad
'[Hey,] grandad!' (expressing affection)	'[Hey,] grandad!' (expressing affection)

As noted by [12], the lengthened vowel forms (see (25) and (26c)) can even be hyphenated in writing (see, accordingly, (29a) and (29b)), as data from the Russian National Corpus confirms:

[14] In Russian, vowel quantity is not a phonological feature [50, p. 41].

(29) Russian, hyphenated forms [12]

a. ma-am!
mom
'Mommy!'
(expressing affection)

b. pa-a-ap!
dad
'Daddy!'
(expressing affection)

4.2 Comparison of Vocative Truncation in Russian vs. Georgian

If we have a careful look at the Georgian truncated vocative forms of nouns, their use is similar to the use of Russian truncated vocatives. In both languages,

(i) the truncation is restricted to an informal setting with a relatively close interlocutor relationship;
(ii) the pragmatics of the uses in Georgian and Russian are similar: truncated forms show close social distance between interlocutors and express familiarity, close social relationship, affection, and endearment;
(iii) for both languages, in VOC truncation the lengthening of a vowel is optional (for Russian, see [12] and for Georgian cf. (30a) vs. (30b) and (30a) vs. (30c)).

(30) Georgian truncated vocative with an optionally lengthened vowel

a. Full
liḳa!
Lika
'Lika!'
(Simple address)

b. Truncated[15]
liḳ[,] ʒalian magariaaa[!]
Lik very coool.is
'Lika, it's very coool!'
(expressing affection)

c. Truncated, lengthened[16]
liiiḳ, madloba[!]
Liiik thank.you
'Lika, thanks[!]"
(expressing affection)

The features of being limited to informal setting (i) as well as similarity in the pragmatics (ii) are shared by address forms in many languages. The pragmatics of truncated forms could have also arisen in Georgian without any contact from an outside source as well. That makes (i) and (ii) less convincing arguments in favor of contact.

However, truncation accompanied by lengthening of a vowel might suggest the contact scenario. Truncation optionally coupled with lengthening of a vowel is not characteristic of any other variety of Georgian or its sister languages. Nor it is known from languages that have been in contact with Georgian before Russian. Thus, there is no possible source other than Russian to account for the vowel lengthening that optionally occurs in truncation in Georgian address forms. Therefore, (iii) might serve as an argument that the source language for the phenomenon is Russian.

Here are some other similarities and differences. In both languages, the truncation applies to the nouns ending on the vowel *a*. For Russian see (22b) vs.

[15] Taken from https://forum.ge/?showtopic=35090121&view=findpost&p=54687325.
[16] Taken from https://forum.ge/?showtopic=34490350&view=findpost&p=34939035.

(22c); (22e) vs. (22f); (23a) vs. (23b); (23c) vs. (23d); and for Georgian observe (21a) vs. (21b) and (30a) vs. (30b).

However, unlike Russian, Georgian also allows truncation of nouns that end on vowels other than *a* (cf. (21c) vs. (21d)). It would not be surprising if the truncated disyllabic vocatives started developing differently from those in the donor language: "once borrowed, a form or a pattern is likely to diverge from what it was in the source language, in terms of its formal adaptation, and also its semantics and function" [1, pp. 22].

Truncation in Georgian disyllabic names extends to nicknames (cf. (31a) vs. (31b) and (31a) vs. (31c)) and collective nouns (cf. (32a) vs. (32b) and (32a) vs. (32c)):

(31) Georgian truncated vocative with an optionally lengthened vowel

a. Full[17]
icocxle[,] dega-v[,]
live.IMP Dega-VOC
ʒma-o!
brother-VOC
'Live [long], Dega, brother!'
(Simple address)

b. Truncated[18]
au[,] deg[,] kargi ra[!]
oh Deg good PART
'Oh, Dega, come on!'
(expressing affection)

c. Truncated, lengthened[19]
ḳargi[,] ra[,] deeg[!]
good PART Deeg
'Come on, Dega[!]'
(expressing affection)

(32) Georgian truncated vocative with an optional vowel lengthening

a. Full VOC form of a collective noun
xalx-o!
people-VOC
'[Hey,] people!'

b. Truncated VOC[20]
daçq̣nardit[,] xalx...
calm.PL.down people
'Calm down, people...'

c. Truncated and lengthened VOC[21]
xaalx! ucxouri saiṭebi ar mušaobs[.]
people foreign sites NEG work
'[Hey,] people! Foreign sites aren't accessible[.]'

Note that Russian too allows VOC truncation of collective nouns (cf. (33a) vs. (33b) and (33c) vs. (33d), taken from [12, p. 226]) and nicknames[22]:

[17] Taken from https://forum.ge/?showtopic=34890421&view=findpost&p=48809259.
[18] Taken from https://forum.ge/?showtopic=34656179&view=findpost&p=41084605.
[19] Taken from https://forum.ge/?showtopic=34620382&view=findpost&p=39679738.
[20] Taken from https://forum.ge/?showtopic=35155595&view=findpost&p=56352277.
[21] Taken from http://karavi.ge/viewtopic.php?t=3228&start=15.
[22] One of the anonymous reviewers suggest that truncating non-conventional proper names is common in Russian as well.

(33)

a. Full	b. Truncated	c. Full	d. Truncated
rebjáta!	rebját!	devčáta!	devčát!
folks.VOC	folks.VOC	girls	girls
'Folks!'	'Folks!'	'Girls!'	'Girls!'

Note that Georgian truncated vocatives started to be used as reference forms as well. Example (34a) is a description of a young girl's photo posted by her friend on Facebook, illustrating the referential use of the phrase *čem-i niin* that has a truncated vocative form with a lengthened vowel as a head. Compare (34a) with the truncated form to (34b) with the regular NOM phrase *čem-i nino*:

(34) a. Posted on Facebook on 29.04.2018[23]

čem-i	niin	or-i	ç̣ika	šav-i	γvin-is	šemdeg
my-NOM	Niin	two-NOM	glass.NOM	black-NOM	wine-GEN	after

'My Niin after two glasses of red wine'

b.

čem-i	nino	or-i	ç̣ika	šav-i	γvin-is	šemdeg
my-NOM	Nino.NOM	two-NOM	glass.NOM	black-NOM	wine-GEN	after

'My Nino after two glasses of red wine'

This phenomenon of turning truncated vocatives into reference forms is attested in Russian as well. For instance, compare the truncated VOC form used as reference form in (35a) to the full form in (35b)):

(35) Russian

a. Truncated VOC as a reference form[24]	b. Non-truncated reference form[25]
Žal' Kat' ušla.	Žal' Kat'a ušla.
pity Kat' she.left	pity Kat'a she.left
'It's a pity [that] Kat' left.'	'It's a pity [that] Kat'a left.'

As known from the typological literature, it is common to have NOM forms used as forms of address, as many languages (including Russian) do not possess an actual VOC marker (see [11, pp. 631–632] and works discussed there). The truncated VOC forms used as reference forms (see (34a) for Georgian and (35a) for Russian) illustrate the reverse, namely, how address forms get reanalyzed as reference forms. Note that this too is a typologically standard development of reference forms out of address forms (*vocativus pro nominativo*), that is frequent in situations of language contact [11,46].

---

[23] https://www.facebook.com/Ekarochikashvilii/posts/10211427432799934

[24] Taken from https://eva.ru/forum/mobile/topic/3478053.htm#m94519819.

[25] Taken from https://www.labuhov.net/modules.php?name=Files&op=view_file&lid=1460#27862.

5 Discussion

5.1 Borrowing from Within or Outside of the Language Family?

Truncation of the final vowel to express familiarity, affection, and close social relationship has long been known in the Northeastern, Eastern, Central, and Southwestern dialects of Georgian. However, this use was restricted to nouns with three and more syllables. As for the new pattern of truncation that affects disyllabic nouns, there can be two possible alternative explanations:

1. The rule of truncation of trisyllabic words, characteristic of some Georgian dialects, which fed into Tbilisi Georgian, finally gets extended to the disyllabic nouns as well, or
2. the rule of truncation of disyllabic nouns gets copied from the contact language Russian which possesses such a rule.

The first of them would have been a sufficient explanation, if the disyllabic Georgian nouns, when being truncated, did not have lengthening. The lengthened remaining vowel of the truncated vocative clearly reflects the Russian rule, as Georgian has no long vowels[26] and no lengthening in native truncated trisyllabic nouns (cf. (20b) vs. (2c); (20b) vs. (2d); (20d) vs. (36a); (20d) vs. (36b)):

(36) Mtiulian-Gudamaqrian dialect

a.	Male	b.	Male
	*eelguj!		*elguuj!
	Elguj		Elguj
	'Elguja!' (expressing familiarity)		'Elguja!' (expressing familiarity)

Thus, the lengthening of the remaining vowel in disyllabic vocative forms under truncation cannot be explained via the extension of the native rule for tri- and other polysyllabic nouns to disyllabic ones.

What remains is to opt for the second alternative, namely, that truncation of disyllabic vocative forms in Georgian is a replication of the VOC truncation rule from Russian.

The arguments for the new pattern to be a borrowing from outside of the Kartvelian language family are the following:

- The pattern of truncating the vocative form of disyllabic nouns is not known in dialect varieties except Tbilisi Georgian;
- It is not known in the sister languages of Georgian;
- The lengthening of the vowel in truncated forms of Georgian is similar to the lengthening of the remaining vowel in Russian truncating vocatives;[27]

---

[26] See footnote 2.

[27] The fact that, apart from the simple truncation, the lengthening of the vowel (characteristic of the Russian pattern) got also copied, might reflect the frequent code switching and parallel use of Georgian and Russian, especially in the speech of the Georgian intelligentsia of the Soviet period.

- Russian has been the language in intensive contact with Georgian from the 1st half of the XIX century up to 1991. This time frame includes the period when the truncation of disyllabic nouns has started in Georgian;
- Russian has contributed to several contact phenomena in Georgian [5,32–34], which makes the language contact scenario worth considering for the new pattern.

Therefore, proper names can be divided into two main groups with regard to how they behave in VOC truncation:

N1. disyllabic names; and
N2. tri-, quadri-, etc.-syllabic names.

N1 get truncated in VOC and have the remaining vowel optionally lengthened, paralleling the Russian disyllabic names. As no dialect and no sister language of Georgian has VOC truncation in disyllabic names, N1 can be used to argue that Tbilisi Georgian replicated the rule from Russian.

N2 get truncated in VOC but have no lengthening at all. This does not parallel the rule of truncation and lengthening in Russian tri- and more-syllabic names. Rather it resembles the situation in several mountainous dialects of Georgian, where N2 names get truncated in VOC without lengthening of any of the remaining vowels.

5.2 Which Type of Borrowing Is It?

As known from the contact linguistics literature, there are two types of structural borrowing: *matter* (MAT) *borrowing*, when there is a direct replication of morphemes and phonological shapes from a source language and *pattern* (PAT) *borrowing*, when "only the patterns of the other language are replicated, i.e., the organization, distribution and mapping of grammatical or semantic meaning, while the form itself is not borrowed" [40, p. 15] (see also [16,30,31]).[28]

Usually, a borrowed matter in MAT borrowing and a borrowed rule in PAT borrowing will not carry all the functions characteristic of the source. Those that are taken over might further develop additional characteristics in the recipient language that are not known in the donor language. There are many factors that contribute to this, including sociolinguistic and structural factors (the structure and language internal development of the recipient language).

This particular case of copying a rule of VOC truncation from Russian into Georgian represents a case of PAT borrowing. This is because a deletion of a phonological matter in forms of address in Russian is a rule that gets replicated in the recipient language Georgian. There is no addition of some actual

[28] I am following here the terminology of [30,31,40]. Note that different terms have been used in the contact linguistics literature to describe essentially the same main distinction between replicating an actual matter and a rule/pattern (e.g., *importation* vs. substitution [20]; *transfer of elements* vs. *interference without outright transfer* [54]; *global copying* vs. *partial copying* [25], *material borrowing* vs. *loan-translation* [35], *diffusion of forms* vs. *diffusion of patterns* [2] among other terms).

phonological or morphological matter from the model language Russian to the recipient Georgian.

However, there is a question whether it is a PAT borrowing without any MAT or a PAT borrowing with some MAT. This question arises, as in Georgian disyllabic truncated forms the remaining vowel gets optionally lengthened (see, for instance, (30c) as opposed to (30b)), just as the stressed vowel of the truncated VOC forms does in Russian (see, for instance, (27a) as opposed to (22f)). The lengthening is optional and by no means obligatory. Still it replicates the rule of lengthening of the vowel of the truncated VOC form, functioning in Russian.

Can the optional lengthening of the vowel in VOC truncation in Georgian be considered as a MAT borrowing? The lengthening of a vowel is a replication of a rule of Russian truncated vocatives and can hardly be considered a MAT borrowing. However, if Georgian truncated disyllabic vocative forms get the same intonation contour as Russian truncated vocatives do, one could argue about borrowing prosodic material into Georgian.

According to the chapter on the prosody of address in Russian [28, pp. 161–174], the lengthened vowel of truncated VOC forms are characterized with a specific rising-falling contour [28, pp. 161–174]. The full address form *L'on'a* as a form of address is characterized with a rising-falling contour (37a). After truncation, it seems that there is a compensatory lengthening to maintain the intonation, the rising-falling one, and redistribute it on the two morae of the long vowel *o:* (37b):

(37) A Russian male name *Ljenja* as an address form, adopted from [28, p. 162]

a. Non-truncated form; two syllables
L'**o**. n'**a**!
H **L**
Ljonja.VOC
'Ljonja!'

b. Truncated form; one syllable with a long vowel consisting of two morae
L' **o:** n'!
HL
Ljonja.VOC
'Ljonja!'

If the contour was adopted for the Georgian truncated vocatives, we could have argued that such concrete matter as a specific contour has been adopted with a specific function. And thus, there would be a MAT borrowing of a prosodic contour, in addition to the PAT borrowing of truncation.

Up to the present, according to my observations, Georgian truncated vocatives do not have a rising-falling contour but only a lengthening of the vowel. However, this is not sufficient to resolve the issue whether the replication of the pattern of VOC truncation involves the borrowing of some prosodic material into Georgian or not. Rather resolution will depend on a detailed analysis of the prosodic structure of disyllabic truncated vocatives in Georgian and has to await further research.

6 Conclusion

In modern spoken Georgian, a VOC truncation rule for nouns with three and more syllables has been directly copied from the Northeastern, Eastern, Central, and Southwestern dialects of Georgian, where the rule does not operate on disyllabic nouns.

In modern spoken Georgian, a new pattern of VOC truncation has emerged that operates on disyllabic nouns, truncates the final vowel (with an optional lengthening of the vowel of the remaining syllable) to produce a pragmatically marked reading.

We could hypothesize that the rule of truncating trisyllabic and quadrisyllabic nouns, functioning in Northeastern, Eastern, Central, and Southwestern Georgian dialects was extended to other types of nouns (like disyllabic nouns) when it was copied into the standard variety. However, the pattern that operates on disyllabic nouns in the standard variety bears more resemblance to the VOC truncation rule, operating in Russian (a former contact language for Georgian).

In Russian, the rule of VOC truncation operates on proper names and kinship terms. It affects full vocative forms consisting of two and more syllables. The final vowel of the full vocative form gets deleted (with an optional lengthening of the vowel of the stressed syllable). As a result, a marked pragmatic reading is obtained.

What the modern spoken Georgian truncated vocatives of disyllabic nouns have in common with the Russian pattern is that in both languages, the remaining vowel gets optionally lengthened. However, it is not characteristic of the relevant dialects of Georgian, were we to argue for the new pattern to be a replication from a related dialect, within the Kartvelian language family.

Therefore, the developments in truncating forms in the vocative have been influenced from two different sources: (i) truncation in the tri- and more syllabic nouns came from some of the related Georgian dialects and (ii) truncation in the disyllabic nouns came from Russian. In both cases there is a contact – from within the language family and from outside of it.

Thus, as the functioning of VOC truncation in modern spoken Georgian illustrates, the same phenomenon in different sections of nominals (here, disyllabic vs. other polysyllabic nouns) can be a result of contact with different donors.

Acknowledgements. I am grateful to two anonymous reviewers for criticism and useful suggestions. Special thanks to Zurab Baratashvili, Winfried Boeder, Boyd H. Davis and Margaret Maclagan for reading and commenting on the paper. The editors, Aybüke Özgün and Julia Zinova, have been very supportive during the editing process. All possible shortcomings are mine. Additionally, I am forever grateful to my Kutsias for their love, to Ursula Pröll-List and Simone Greul for bringing peace and to the fellow Late-Discovery Adoptees for understanding and validation.

References

1. Aikhenvald, A.Y.: Grammars in contact: a cross-linguistic perspective. In: Aikhenvald, A.Y., Dixon, R.M.W. (eds.) Grammars in Contact. A Cross-Linguistic Typology. Explorations in Linguistic Typology 4, pp. 1–66. Oxford University Press, Oxford (2007)
2. Aikhenvald, A.Y., Dixon, R.M.W. (eds.): Grammars in Contact. A Cross-Linguistic Typology. Explorations in Linguistic Typology 4, Oxford University Press, Oxford (2007)
3. Amiridze, N.: Reflexivization Strategies in Georgian. LOT Dissertation Series 127, LOT, Utrecht (2006)
4. Amiridze, N.: Accommodating loan verbs in Georgian: observations and questions. Journal of Pragmatics **133**, 150–165 (2018)
5. Amiridze, N., Gurevich, O.: The sociolinguistics of borrowing: Georgian *moxdoma* and Russian *proizojti* 'happen'. In: Muhr, R. (ed.) Innovation and Continuity in Language and Communication of Different Language Cultures, pp. 215–234. Peter Lang Verlag, Frankfurt am Main (2006)
6. Boeder, W.: The South Caucasian languages. Lingua **115**, 5–89 (2005)
7. Butskhrikidze, M.: The Consonant Phonotactics of Georgian. LOT Dissertation Series 63. LOT, Leiden (2002)
8. Chikobava, A.: Grammatical Analysis of Laz with Texts. Georgian Branch of the Academy of Sciences of the USSR, Tbilisi (1936). (In Georgian)
9. Comrie, B.: The Languages of the Soviet Union. Cambridge University Press, Cambridge (1981)
10. Comrie, B., Stone, G., Polinsky, M.: The Russian Language in 20th Century. Clarendon Press, Oxford (1996)
11. Daniel, M., Spencer, A.: The vocative - an outlier case. In: Malchukov, A.L., Spencer, A. (eds.) The Oxford Handbook of Case, pp. 626–634. Oxford University Press, Oxford (2009)
12. Daniel, M.A.: "Novyj" russkij vokativ: Istorija formy usečennogo obraščenija skvoz' prizmu korpusa pis'mennykh tekstov. In: Kiseleva, K.L., Plungjan, V.A., Rakhilina, E.V., Tatevosov, S.G. (eds.) Korpusnye isslodovanija po russkoj grammatike, pp. 224–244. Probel, Moscow (2009)
13. Fähnrich, H.: Kurze Grammatik der georgischen Sprache. Verlag Enzyklopädie, Leipzig (1987)
14. Fähnrich, H.: Old Georgian. In: Harris, A.C. (ed.) The Indigenous Languages of the Caucasus. The Kartvelian Languages, vol. 1, pp. 129–217. Caravan Books, New York (1991)
15. Fähnrich, H.: Neugeorgisch. In: Fähnrich, H. (ed.) Kartwelsprachen, pp. 117–181. Reichert, Wiesbaden (2008)
16. Gardani, F., Arkadiev, P., Amiridze, N. (eds.): Borrowed Morphology. Language Contact and Bilingualism [LCB] 8. De Gruyter Mouton, Berlin/Boston/Munich (2015)
17. Gigineishvili, I. (ed.): Norms of the Modern Georgian Literary Language, vol. 1. Mecniereba, Tbilisi (1970)
18. Grenoble, L.A.: Language Policy in the Soviet Union. Kluwer Academic Publishers, Dordrecht (2003)
19. Harris, A.C.: Diachronic Syntax: The Kartvelian Case. Syntax and Semantics, vol. 18. Academic Press, Orlando (1985)
20. Haugen, E.: The analysis of linguistic borrowing. Language **26**, 210–231 (1950)

21. Hewitt, B.G.: Review-article of syntax and semantics 18: A.C. Harris' diachronic syntax: the Kartvelian case. Revue des Etudes Géorgiennes et Caucasiennes **3**, 173–213 (1989)
22. Hewitt, B.G.: Georgian–ergative, active or what? In: Bennet, D.C., Bynon, T., Hewitt, B.G. (eds.) Subject, Voice and Ergativity: Selected Essays, pp. 202–217. School of Oriental and African Studies, University of London, London (1995)
23. Hewitt, B.G.: Georgian: A Structural Reference Grammar. London Oriental and African Language Library, vol. 2. John Benjamins, Amsterdam/Philadelphia (1995)
24. Janda, L.A.: A Far North perspective on the "new" vocative in Russian. Slides of the presentation at the Slavic Cognitive Linguistics Conference 2015, Universities of Sheffield and Oxford, UK (2015)
25. Johanson, L.: Remodeling grammar: copying, conventionalization, grammaticalization. In: Siemund, P., Kintana, N. (eds.) Language Contact and Contact Languages. Empirical Approaches to Linguistic Typology (EALT); 35, pp. 61–79. John Benjamins, Amsterdam/Philadelphia (2008)
26. Jorbenadze, B.: kartuli dialekṭologia [Georgian Dialectology]. Mecniereba, Tbilisi (1989).(in Georgian)
27. Kipshidze, I.: Grammatika mingrel'skogo (iverskogo) jazyka s xrestomatiej i slovarem [Grammar of the Megrelian Language with reader and dictionary. VII. Typografija Imperatorskoj Akademii Nauk, St.-Petersburg, Materialy po jafeticheskomu Jazykoznaniju (1914). (in Russian)
28. Kodzasov, S.: Accent and Stress in Russian. Jazyki Slavjanskix Kul'tur, Moscow (2009). (in Russian)
29. Marr, N.: Grammatika chanskogo (lazskogo) jazyka s xrestomatiej i slovarem. Materialy po jafeticheskomu Jazykoznaniju, II. Typografija Imperatorskoj Akad. Nauk, St.-Petersburg (1910). (in Russian)
30. Matras, Y., Sakel, J.: Introduction. In: Matras, Y., Sakel, J. (eds.) Grammatical Borrowing in Cross-Linguistic Perspective, pp. 1–13. Mouton de Gruyter, Berlin/New York (2007)
31. Matras, Y., Sakel, J.: Investigating the mechanisms of pattern replication in language convergence. Studies in Language **31**(4), 829–865 (2007)
32. Mikiashvili, O.: On one phenomenon in Georgian after the Russian influence. Sazrisi **3**, 62–66 (2000). (in Georgian)
33. Mikiashvili, O.: On the results of the Russian influence on the Georgian dialect speech. Kartvelian Heritage 7, pp. 172–184, Kutaisi (2003)
34. Mikiashvili, O.: On the history of establishing modern Georgian literary language (Aspects of the Russian-Georgian language contact). Kartuli Ena, Tbilisi (2005). (in Georgian)
35. Nau, N.: Möglichkeiten und Mechanismen kontaktbewegten Sprachwandels unter besonderer Berücksichtigung des Finnischen. Lincom Europa, München/Newcastle (1995)
36. Oniani, A.: On the issue of vocative case in modern literary Georgian. In: Amaγlobeli, N., Khintibidze, E., Shanidze, M., Dzidziguri, S., Jibladze, G., Jorbenadze, B. (eds.) Akaki Shanidze - 100, pp. 143–156. Tbilisi State University Press, Tbilisi (1987). (in Georgian)
37. Oniani, A.: svanuri ena [The Svan Language]. Tbilisi State Pedagogical University Press, Tbilisi (1998). (in Georgian)
38. Orbeliani, S.-S.: A Book of Wisedom and Lies. Sabchota Sakartvelo, Tbilisi (1970). (In Georgian, written between 1686–1695)

39. Parrott, L.: Vocatives and other direct address forms: a contrastive study. In: Gr∅nn, A., Marijanovic, I. (eds.) Russian in Contrast, Oslo Studies in Language, vol. 2, no. 1, pp. 211–229. Oxford University Press, Oxford (2010)
40. Sakel, J.: Types of loan: matter and pattern. In: Matras, Y., Sakel, J. (eds.) Grammatical Borrowing in Cross-Linguistic Perspective, pp. 15–29. Mouton de Gruyter, Berlin/New York (2007)
41. Sarjveladze, Z.: The Old Georgian Language. Tbilisi State Pedagogical University Press, Tbilisi (1997). (in Georgian)
42. Schmidt, K.H.: Svan. In: Harris, A.C. (ed.) The Indigenous Languages of the Caucasus. The Kartvelian Languages, vol. 1, pp. 473–556. Caravan Books, New York (1992)
43. Shanidze, A.: The Fundamentals of Georgian Grammar I: Morphology. Tbilisi State University Press, Tbilisi (1953). (in Georgian)
44. Shanidze, A.: Foundations of Georgian Grammar, I, Morphology. Tbilisi University Press, Tbilisi (1973). (in Georgian)
45. Shanidze, A.: Grammar of the Old Georgian Language. Tbilisi University Press, Tbilisi (1976). (in Georgian)
46. Stifter, D.: Vocative for nominative. In: Sonnenhauser, B., Hanna, P.N.A. (eds.) Vocative! Addressing between System and Performance, Trends in Linguistics. Studies and Monographs [TiLSM] 261, pp. 43–85. De Gruyter Mouton, Berlin (2013)
47. Tabidze, M., Shavkhelishvili, B.: On some aspects of language contact between Georgian and Russian. Presentation at the International Conference "Conditions and Perspectives of Teaching Russian", Moscow (2008)
48. Tabidze, M., Shavkhelishvili, B.: The language of the capital. Russianisms in Georgian. Presentation at the International Conference, Perm, Russian Federation (2009)
49. Thomason, S.G., Kaufman, T.: Language Contact, Creolization, and Genetic Linguistics. University of California Press, Berkeley/Los Angeles/London (1988)
50. Timberlake, A.: A Reference Grammar of Russian. Cambridge University Press, Cambridge (2004)
51. Topuria, V.: ç̣odebiti brunvisatvis [On the vocative case]. In: Topuria, V. (ed.) On the History of Noun Declension in the Kartvelian Languages, vol. 1, pp. 36–47. Tbilisi State University Press, Tbilisi (1956). (in Georgian)
52. Tuite, K.: Svan. Language of the World/Materials 139. Lincom Europa, München-Newcastle (1997)
53. Vogt, H.: Grammaire de la langue géorgienne. Instituttet for sammenlignende kulturforskning, Serie B: Skrifter, LVII. Universitetsvorlaget, Oslo (1971)
54. Weinreich, U.: Languages in Contact. Findings and Problems. Linguistic Circle of New York, New York (1953). (Publications of the Linguistic Circle of New York; 1)
55. Zhgenti, S.: Georgian Phonetics. Tbilisi State University Press, Tbilisi (1956). (In Georgian)

Tutorial: Sign Language Linguistics

Fabian Bross(✉)

University of Stuttgart, Keplerstr. 17, 70174 Stuttgart, Germany
fabian.bross@ling.uni-stuttgart.de

Abstract. This tutorial offers a brief overview of linguistic research into sign languages. The tutorial's target audiences are people with some background in linguistics of spoken languages. For the sake of brevity, I will only concentrate on some major topics. I will briefly introduce sign languages, discuss some basics of phonological structure of these types of languages (including the use of space), discuss some new findings on the syntax of sign languages, and, finally, will briefly address some methodological issues. The majority of data will come from German Sign Language, although data from other sign languages is also included.

Keywords: Sign languages · Phonology · Syntax

1 Introduction

Sign languages are natural languages produced with the hands, arms, torso, head, and the face. World-wide there are 150 known sign languages with approximately 5.000.000 speakers (Eberhard et al. 2021). However, there may be up to 300 and 400 different sign languages used all over the world (Zeshan 2009).

Sign languages develop naturally when acoustic communication is blocked over a sufficiently long period of time (Kegl et al. 2014). The main reason for acoustic communication not being possible is deafness. Thus, most sign languages emerged as communicative devices for deaf people with most of the world's sign languages not being older than 200 to 300 years. This mainly holds true for western sign languages (note that due to a lack of written records it is often not possible to estimate the age of a sign language). The reason that most sign languages are relatively young is that a requirement for a language to emerge is that a sufficient number of potential language users are involved in social interaction with each other. The prerequisite for this kind of widespread interaction is the presence of large enough urban agglomerations, which only emerged with the age of industrialization. During this time, the first schools for the deaf were established (e.g., 1760 in Paris or 1817 in West Hartford, Connecticut). Thus, the reason for most (western) sign languages being young languages is that before industrialization one of the basic ingredients for a language, highlighted already by early Structuralist linguistics, was not met: "The existence of language is only possible within a society" (Baudouin de Courtenay 1904, p. 128) (translation from Adamska-Sałaciak 1998).

© The Author(s), under exclusive license to Springer Nature Switzerland AG 2022
A. Özgün and Y. Zinova (Eds.): TbiLLC 2019, LNCS 13206, pp. 26–37, 2022.
https://doi.org/10.1007/978-3-030-98479-3_2

Besides differences in modalities (visual-gestural versus auditory-vocal) there are surprisingly few differences between signed and spoken languages. Sign languages fulfill the same communicative and social functions as spoken languages, convey information at the same speed as spoken languages (Bellugi and Fischer 1972), are processed in the same brain regions as spoken languages (Emmorey 2002), and both are acquired without instruction, given normal exposure. The last point deserves some attention as sign languages are often acquired under special sociolinguistic circumstances. In most cases, children born deaf have hearing parents and only around 10% have deaf parents (the exact numbers are subject to variation depending on the general medical care situation in a given area). For children born deaf who do not have access to sign language input, the timing of their first language acquisition is often off-schedule (e.g., Mayberry 1993, 2002). However, being deaf and the use of a sign language, of course, cannot be equated. Additionally, it is worth mentioning that there is a widely adopted distinction between the audiological status of being deaf (written in lowercase) and the self-identification of individuals as members of the Deaf community which is spelled Deaf (with a capital letter) (Woodward 1972; see also the discussion in Phillips 1996).

Until the 1960s, sign languages were not regarded as natural human languages, but as underdeveloped gestural communication systems. This radically changed with the seminal publication of William Stokoe's book *Sign language structure* in 1960 (Stokoe 1960). Stokoe convincingly showed that individual signs are not holistic units, but can be segmented into smaller, meaningless units (see also Stokoe et al. 1965 and already Tervoort 1953).

In the following, I will briefly discuss some main topics from sign language research. The selection of topics is (by far) not exhaustive and follows the same order as in the tutorial held at the International Tbilisi Symposium on Logic, Language, and Computation 2019. In the next section, I will discuss the phonological make-up of manual signs. In Sect. 3, I briefly introduce how space is used in sign languages. Section 4 is devoted to the discussion of the expression of the three clausal layers (CP, TP/IP, VP/VoiceP), mainly building on my own work. In Sect. 5, I discuss some recent methodological issues. Finally, in Section 6 I conclude.

2 Phonology

For sign languages, the equivalent of spoken language phonology was originally referred to as 'chereology' (derived from the Greek word for hand). However, as it became clear that the underlying structural make-up of signed and spoken languages is extremely similar, linguists started to use the same terminology for both modalities (Bross 2015). The onset of sign language phonology is marked, as mentioned, by the discovery of the fact that individual signs are composed of smaller units.

One of the defining features of a natural language is taken to be that they are created using a limited set of meaningless units which can be combined (in a

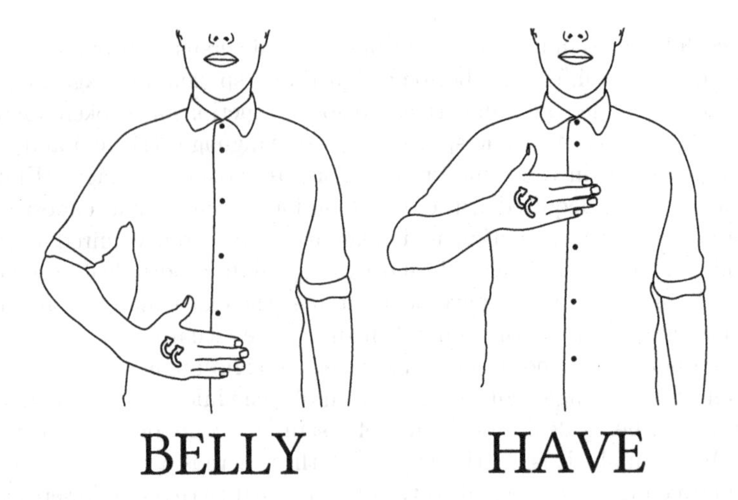

Fig. 1. An example of a minimal pair from German Sign Language. The noun sign BELLY and the verb sign HAVE only differ in location.

rule-governed way) to larger units with meaning. This way of creating meaning is called 'double articulation' (Hockett 1960; Martinet 1949). In spoken languages the building blocks of this process were first thought to be phonemes (Baudouin de Courtenay 1881) and later to be distinctive features such as [±voiced] (Jakobson et al. 1951).

The set of atomic building blocks a language uses can be identified by the formation of minimal pairs. The English words *car* and *par* are an example of a minimal pair only differing in the place of articulation. While the tongue creates an obstruction at the velum to produce the /k/ in *car*, there is an obstruction at the lips in the case of /p/.

Minimal-pair formation works in a very similar way in sign languages as uncovered by Stokoe (1960). The following parameters are traditionally distinguished (cf. Battison 1978; Stokoe 1960):

- place of articulation (also called location),
- movement direction,
- hand shape, and
- palm orientation.

Let us briefly examine how this works using the first of these parameters. Figure 1 shows two signs from German Sign Language. In both cases, the signs are produced by a flat hand, i.e., the same hand shape is used, the same tapping movements are performed, and the same palm orientation is used. The only thing that is different between the two signs is the location at which the sign is produced. The location itself does not have a meaning (at least not necessarily), but changing this parameter can lead to a difference in meaning.

In fact, there is not much difference between the basic phonological mechanisms between modalities. In both language types, language is produced via body movements. The movements produced with the speech organs in the case of spoken languages generate sound waves perceived with the ears where they are transformed into electrical signals. In the case of sign languages, the movements produced are perceived as light waves which are transformed into electrical signals within the retina.

Phonology in spoken languages is not merely a matter of concatenated material as there are additional suprasegmental features that play a role. Most prominently, intonation is produced "on top" of the linearly arranged segments. Again, the same mechanism is found in sign languages. "On top" of the concatenated manual signs we find non-manual articulation layered. A simple example from English would be a declarative sentence like *Paul drinks beer.* This sentence can be used as a statement or, with an appropriate rising intonation, also as a question (Gunlogson 2002). The very same is true for sign languages. A simple example from German Sign Language is given in (1). The sentence in (1a) is a declarative, while the sentence in (1b) is an interrogative. The only difference between the examples is that with the interrogative the manual signs are accompanied by non-manual markers abbreviated 'pol' (for polar interrogative). These markers consist of raised eyebrows and a slightly forwarded and tilted head.

(1) German Sign Language:

a. PAUL BEER DRINK
'Paul drinks a beer.'

b. $\overline{\text{PAUL BEER DRINK}}^{\text{pol}}$
'Does Paul drink a beer?'

Note that I follow the traditional convention that individual manual signs are written in capital letters. The starting and end points of non-manuals are indicated by a line and the non-manuals used are specified on top of this line (see Bross 2019b for a brief overview). Note that there are many more phonological topics (e.g., iconicity or word formation) which I cannot discuss here for reasons of space limitations.

3 The Use of Space

Sign languages are by definition spatial languages. While it is easy to refer to a present referent (e.g., by pointing), absent referents are assigned unique loci (i.e., locations) in space. This can be best illustrated by the use of personal pronouns which are realized by pointing with the index finger in many sign languages. The first person singular pronoun is realized by the signer pointing to herself, the second person singular is realized by the signer pointing to the addressee, and third person is realized by either pointing to a present referent or by pointing to a location where the referent in question was introduced before. A common way to gloss this is to write INDEX and indicate the person by an index (thus, INDEX_1 means first person and INDEX_2 means second person). Let's look at example (2). In (2a), Paul is introduced and assigned a locus in space (i.e., some point besides the signer). In (2b) this point is re-used and, again, refers to Paul for the rest of the conversation.

(2) German Sign Language:

a. PAUL INDEX3 BEER BUY
'Paul bought beer.'

b. INDEX3 BEER LIKE
'He likes beer.'

Now, let's look at verbal agreement. Following the seminal work of Padden (1983) on American Sign Language, three different classes of verbs are traditionally distinguished: agreement verbs, spatial verbs, and plain verbs. So far, we have seen that referents are assigned unique loci in space. Agreement verbs show overt agreement between the subject and the object, i.e., with the locations that were assigned to the subject and object. Spatial verbs also show agreement, but not between a subject and an object, but exhibit locative agreement (e.g., with verbs like *to put*). Plain verbs, finally, show no agreement at all. There are, however, also verbs which do not fit in the classification of verbs into these three classes. In many sign languages there are, for examples, verbs which only exhibit object, but no subject agreement. Additionally, there are so called "backward verbs" which do not exhibit subject-object, but object-subject agreement (see, for example Meir 1998). Although I do not know whether this holds true for the majority of sign languages, at least for German Sign Language it was claimed that the majority of verbs are plain (Pfau and Steinbach 2007, p. 310). In these cases, word order signals grammatical roles. Speaking of word order, we are now entering the realm of syntactic structures.

4 Syntax

The vast majority of spoken languages either follow an SVO or an SOV word order (Dryer 2013). The same is true for sign languages (Napoli and Sutton-Spence 2014). While American Sign Language, for example, is an SVO language (Fischer 1975), German Sign Language follows a rather strict SOV order (Keller 1998; Pfau and Glück 2000). Deviations in the basic word order are allowed for reasons of information-structural foregrounding. In German Sign Language, a neutral SOV sentence would look like (3a). The example (3b) shows an OSV structure resulting from a topicalization of the object (with t indicating its original position).

(3) a. PAUL BEER BUY

b. $\overline{\text{BEER}_i}^{\text{top}}$ PAUL t_i BUY

As in spoken languages, the topic is overtly moved into a structurally higher position in (3b). However, it additionally receives a non-manual topic marking consisting of a brow-raise. Syntactic structures are organized hierarchically. The position into which the topic moves is structurally rather high. To be more precise, it is located in a position in the structurally highest portion of the clause, the CP-layer (Rizzi 1997). At the same time, the eyebrows are the highest articulator available. Other CP-related functions are indeed also expressed with the face. We have already seen that polar interrogatives are encoded by raised eyebrows spreading over the whole clause. *Wh*-questions are expressed by brow-lowering and imperatives are accompanied by furrowed brows (and an increased signing speed). Thus clause-type marking is also marked via bodily high articulators. This is not a peculiarity of German Sign Language, but a cross-linguistically stable pattern in sign languages in general (cf., for example, Zeshan 2004, 1 for interrogative marking).

The structure below the CP-layer is the TP-layer, where T stands for tense. As you may have noticed with some examples (e.g., (2)) the English translations were given in the past tense, although nothing in the glosses indicated this. This is because the vast majority of sign languages does not have a tense system (similar to many spoken languages, like Mandarin). It is, of course, nevertheless possible to talk about time. This is achieved via temporal adverbs which, in German Sign Language, usually appear in a clause-initial position:

(4) YESTERDAY PAUL BEER BUY
'Paul bought beer yesterday.'

German Sign Language uses a topic time system. This means that once a topic time is set, the sentences to follow (4) will also be in the past until something else is specified. However, there is at least one known sign language with a documented tense system. Zucchi (2009) shows that tense marking is integrated into the verb signs in a variety of Italian Sign Language. Future tense is expressed by moving the shoulder forward while signing the verb, present tense is expressed by a neutral shoulder position, and past tense by putting the shoulder backwards.

As with the CP, the TP does not only host one category, but several. In German Sign Language (and all other sign languages I am aware of), these categories all find manual expression. Sometimes, the differences between a CP- and TP-category are not easy to see. Take the sentence *Paul must be at home early* as an example. The sentence can have different meanings. It can mean that it is necessary for Paul to be at home early given what the speaker knows about the world and Paul's behavior, or it can mean that it is necessary for Paul to be at home early given the power relations present. The reason for this ambiguity lies in the syntactic position of the modal verb *must*. The first interpretation is an instance of what is called epistemic modality. Epistemic modality is located in the CP-layer (e.g., Wurmbrand 2001). The second interpretation is an instance of deontic modality and deontic modals are located in the TP-layer. While the difference between epistemic and deontic modality is not visible on the surface in English, it is visible in German Sign Language. Examples are given in (5). While epistemic modality is expressed non-manually by furrowed brows spreading over the whole clause (often with additional clause-final head nods and closed eyes), as shown in (5a), the example in (5b) shows that deontic modality is expressed manually only by using the sign MUST.[1] Note that there are also sign languages in which there are also manual modal signs used to express epistemic modality. In these cases, however, epistemic modality is not expressed manually only, but by a combination of upper-face non-manuals a manual marker. This situation is found, for example, in Turkish Sign Language (Karabüklü et al. 2018).

(5) a. $\overline{\text{PAUL EARLY AT-HOME}}^{\text{epistemic}}$ **Epistemic modality (CP-layer)**
'Paul must be at home early (given what I know)'

b. PAUL EARLY AT-HOME MUST **Deontic modality (TP-layer)**
'Paul must be at home early (given what someone ordered).'

There is another shell structure below the TP which again hosts several categories. This structure, called VoiceP, also contains categories which have related meanings in a higher structure. A case in point is frequentative aspect I, being a TP-category, and frequentative aspect II, being a VoiceP-category. Again, the difference between these categories is often not easy to spot. The English sentence *Paul often insults Maria* can either mean that there are several different events of insulting (e.g., one on Monday, one on Tuesday, etc.) or that there is one single event of insulting with several subevents ('Paul insults Maria many times in a row'). The first case, called 'frequentative I' is an instance of a TP-category and the second case is an instance of a VoiceP-category. Again, the TP-category is expressed by using a manual sign. The VoiceP-category, in contrast, is expressed by manipulating the movement path of the verb sign. In this case, the verb sign is reduplicated, as indicated by the plus signs.[2] The claim that VoiceP-internal categories are expressed by manipulating the movement path of the verb sign is related to the idea that event structure in sign languages is made visible through the phonological shape of the manual verb sign, called "Event Visibility Hypothesis" (see, for example, Wilbur 2003, 2008).

(6) a. PAUL PAM MARIA OFTEN INSULT **Frequentative I (TP-layer)**
'Paul often insulted Maria.'

b. PAUL PAM MARIA INSULT++ **Frequentative II (VoiceP-layer)**
'Paul insulted Maria many times in row.'

The idea that there is a systematic mapping between the structural height of a category and the way it is expressed in sign languages is called the "Bodily Mapping Hypothesis" (Bross 2020b; Bross and Hole 2017). As I have shown by way of excursion, CP-categories are expressed with facial non-manuals, categories inside the TP are expressed by manual signs, and VoiceP-internal categories are expressed by manipulating the movement path of the verb sign. Thus, the hierarchical structure of a clause is mapped onto the body in German Sign Language, and perhaps in all sign languages of the world.

[1] A hyphen between two glosses, as in AT-HOME, indicates that we are dealing with one single manual sign that translates into several English words.

[2] The sign PAM is an abbreviation for "person agreement marker" (Rathmann 2003) and is sometimes thought to be used for differential object marking (Bross 2020a).

5 Methodological Considerations

There is a huge ongoing discussion about the standards of data in spoken language linguistics (e.g., Bross 2019a; Featherston 2007; Schütze 2016). This discussion is also important for sign language linguistics inter alia in relation to the use of non-manual markers discussed in the previous section. There are basically three approaches: the collection of judgment data can be informal, it can be carried out in a quasi-experimental way, or one can rely on corpus data. Informal data collection has been criticized, not only in spoken language linguistics, but also in the literature on sign languages (Kimmelman, to appear) for the reason that many previous studies have found conflicting data. One case Kimmelman (to appear) discusses are the judgments on American Sign Language presented in Neidle et al. (1998) and Petronio and Lillo-Martin (1997, 1). As the judgments presented by these authors contradict each other, he concludes that their methods must be questioned. It is indeed the case that these judgments differ. This, however, is no wonder. First, the studies mentioned are based on a very small amount of participants. Second, and more importantly, the grammar of sign languages is subject to large variations as they are usually used by a small, dispersed community. Thus, grammar may vary from community to community, from city to city, and sometimes even from signer to signer. Although I would agree that sign language linguistics is in need of more quasi-experimental judgment data, linguists first need to get a glimpse of where there might be (micro-)variation. Thus, I would argue, that it makes sense to collect informal judgments as a first step and then to proceed with collecting formal judgments on carefully selected phenomena.

Just to give one example, consider the long-extraction sentence in (7) including the *wh*-expletive WHAT. I informally surveyed eleven native signers of German Sign Language living in Southern Germany and asked them whether they would accept a structure like this.

(7) $\overline{\text{PAUL REPORT WHAT MARIA BUY WHICH COMPUTER}}^{\text{wh}}$
'Which computer did Paul say Maria bought?'

It turned out that some signers accepted the sentence, while other rejected it. The reason for this disagreement was that the signers came from different cities. If two linguists would have investigated the phenomenon in two different regions they would have come to opposing conclusions. But not because their methods are faulty, but because there is a lot of variation—which is often not random. Formal testing methods, however, will of course uncover such variation and are thus much needed.[3]

Other sign linguists stress the importance of corpus studies. Of course, corpus data is often natural data and, thus, very interesting for a lot of reasons. However,

[3] Note that there is also a lot of variation concerning the exact realization of WHICH in German Sign Language. Another reason for the variation found with this construction might also be an influence from spoken German. Thus, with formal testing it is also be important to collect data on linguistic competence.

corpus data has its limits, but not only because of its lack of negative evidence and the problem that it is virtually impossible to find well-balanced minimal pairs, i.e. two clauses only differing in one grammatical property. One important issue are non-manual markers: Above I have claimed that epistemic modality is expressed non-manually with the upper face in German Sign Language, while deontic modality is expressed manually only. However, a linguist looking at corpus data surely would have come to a different conclusion: sentences including deontic necessity modals are usually accompanied by non-manual markers of the upper face to indicate how strict an obligation is thought to be. These non-manuals, however, represent a facultative signer evaluation (located in the CP-layer). The non-manuals used for epistemic modality, in contrast, are obligatory. Additionally, there are cases where one non-manual marker overrides another. Reason clauses, for example, are marked by the manual sign REASON in German Sign Language which is obligatorily accompanied by a brow-raise. If the clause, however, contains an additional evaluation (e.g., *Unfortunately, Paul didn't come, because he is sick*) this brow-raise is sometimes absent as the evaluation is also marked non-manually with the eyebrows. This makes corpus studies sometimes extremely difficult—in this case because we do not know which non-manual markers are overridden in which cases. Taken together, I strongly defend the view that informal judgments, quasi-experimental judgments, and corpus studies form a triplet and should all be used to complement each other.

6 Conclusions

Much more could have been said about sign languages and their structures. Nevertheless, I hope to have shown that signed and spoken languages share many common features on all levels of linguistic description. If you want to know more about sign language linguistics, I recommend the very basic introduction in Baker et al. (2016), the handbook by Pfau et al. (2012), or the impressive guide to sign language grammar writing by Quer et al. (2017).

References

Adamska-Sałaciak, A.: Jan Baudouin de Courtenay's contribution to linguistic theory. Historiographia Linguistica **25**(1–2), 25–60 (2012). https://doi.org/10.1075/hl.25.1-2.05ada

Baker, A., van den Bogaerde, B., Pfau, R., Schermer, T. (eds.): The Linguistics of Sign Languages: An Introduction. John Benjamins Publishing Company, Amsterdam (2016)

Battison, R.: Lexical Borrowing in American Sign Language. Linstok, Silver Spring (1978)

Baudouin de Courtenay, J.: Nekotorye otdely 'sravniternoj grammatiki' slavjanskix jazykov [Some chapters of the 'comparative grammar' of the slavic languages]. Russkij Filologiceskij Vestnik **5**, 265–344 (1881)

Baudouin de Courtenay, J.: Szkice językoznawcze [Linguistic sketches]. Piotr Laskauer, Warsaw (1904)

Bellugi, U., Fischer, S.: A comparison of sign language and spoken language. Cognition **1**(2–3), 173–200 (1972)

Bross, F.: Chereme. In: Hall, T.A., Pompino-Marschall, B. (eds.) Dictionaries of Linguistics and Communication Science: Phonetics and Phonology. Mouton de Gruyter, Berlin; New York (2015)

Bross, F.: Acceptability ratings in linguistics: a practical guide to grammaticality judgments, data collection, and statistical analysis. version 1.02. Mimeo (2019a). www.fabianbross.de/acceptabilityratings.pdf. Accessed 19 Nov 2019

Bross, F.: Sign language glossing: using the sgloss package in LATEX. version 1.0. Mimeo (2019b). http://fabianbross.de/sign_language_glossing.pdf. Accessed 19 Feb 2020

Bross, F.: Object marking in German Sign Language (Deutsche Gebärdensprache): differential object marking and object shift in the visual modality. Glossa: J. General Linguist. **5**(1), 1–37 (2020a). https://doi.org/10.5334/gjgl.992

Bross, F.: The Clausal Syntax of German Sign Language. A Cartographic Approach. Language Science Press, Berlin (2020b)

Bross, F., Hole, D.: Scope-taking strategies and the order of clausal categories in German Sign Language. Glossa: J. General Linguist. **2**(1), 1–30 (2017)

Dryer, M.S.: Order of subject, object and verb. In: Dryer, M.S., Haspelmath, M. (eds.) The World Atlas of Language Structures Online. Max Planck Institute for Evolutionary Anthropology, Leipzig (2013). https://wals.info/chapter/81

Eberhard, D., Simons, G., Fennig, C.: Ethnologue: languages of the world (2021). Twenty-third edition. http://www.ethnologue.com. Twenty-fourth edition. http://www.ethnologue.com

Emmorey, K.: Language, Cognition, and the Brain: Insights from Sign Language Research. Lawrence Erlbaum Associates, Mahwah (2002)

Featherston, S.: Data in generative grammar: the stick and the carrot. Theor. Linguist. **33**(3), 269–318 (2007). https://doi.org/10.1515/TL.2007.020

Fischer, S.: Influences on word-order change in American Sign Language. In: Li, C. (ed.) Word Order and Word Order Change, pp. 1–25. University of Texas Press, Austin (1975)

Gunlogson, C.: Declarative questions. Semant. Linguist. Theory **12**, 124–143 (2002). https://doi.org/10.3765/salt.v12i0.2860

Hockett, C.F.: The origin of speech. Sci. Am. **203**(3), 88–97 (1960)

Jakobson, R., Fant, C.G., Halle, M.: Preliminaries to Speech Analysis: The Distinctive Features and their Correlates. MIT Press, Cambridge (1951)

Karabüklü, S., Bross, F., Wilbur, R., Hole, D.: Modal signs and scope relations in TİD. FEAST. Formal Exp. Adv. Sign Lang. Theory **2**, 82–92 (2018). https://doi.org/10.31009/FEAST.i2.07

Kegl, J.A., Senghas, A., Coppola, M.: Creation through contact: sign language emergence and sign language change in Nicaragua. In: DeGraff, M. (ed.) Language Creation and Language Change: Creolization, Diachrony, and Development, pp. 179–237. MIT Press, Cambridge (2014)

Keller, J.: Aspekte der Raumnutzung in der Deutschen Gebärdensprache. Signum, Seedorf (1998)

Kimmelman, V.: Acceptability judgments in sign linguistics. In: Goodall, G. (ed.) Cambridge Handbook of Experimental Syntax. Cambridge University Press, Cambridge (to appear)

Martinet, A.: La double articulation linguistique. Travaux du Cercle linguistique de Copenhague **5**, 30–37 (1949)

Mayberry, R.I.: First-language acquisition after childhood differs from second-language acquisition: the case of American Sign Language. J. Speech Lang. Hearing Res. **36**(6), 1258–1270 (1993). https://doi.org/10.1044/jshr.3606.1258

Mayberry, R.I.: Cognitive development in deaf children: the interface of language and perception in neuropsychology. In: Segalowitz, S., Rapin, I. (eds.) Handbook of Neuropsychology, vol. 8, Part II, pp. 71–107. Elsevier, Amsterdam (2002)

Meir, I.: Syntactic-semantic interaction in Israeli sign language verbs: the case of backwards verbs. Sign Lang. Linguist. **1**(1), 3–37 (1998). https://doi.org/10.1075/sll.1.1.03mei

Napoli, D.J., Sutton-Spence, R.: Order of the major constituents in sign languages: implications for all language. Front. Psychol. **5**, 376 (2014)

Neidle, C., MacLaughlin, D., Lee, R.G., Bahan, B., Kegl, J.A.: The rightward analysis of WH-movement in ASL: a reply to Petronio and Lillo-Martin. Language **74**(4), 819–831 (1998)

Padden, C.: Interaction of morphology and syntax in American Sign Language (Doctoral dissertation, University of California, San Diego, San Diego) (1983)

Petronio, K., Lillo-Martin, D.: WH-movement and the position of Spec-CP: evidence from American Sign Language. Language **73**, 18–57 (1997)

Pfau, R., Glück, S.: The pseudo-simultaneous nature of complex verb forms in German Sign Language. In: Proceedings of the 28th Western Conference on Linguistics, pp. 428–442 (2000)

Pfau, R., Steinbach, M.: Grammaticalization of auxiliaries in sign languages. In: Perniss, P., Pfau, R., Steinbach, M. (eds.) Visible Variation: Comparative Studies on Sign Language Structure, pp. 303–339. Mouton de Gruyter, Berlin; New York (2007)

Pfau, R., Steinbach, M., Woll, B.: Sign Language: An International Handbook. Walter de Gruyter, Berlin (2012)

Phillips, B.A.: Bringing culture to the forefront: formulating diagnostic impressions of deaf and hard-of-hearing people at times of medical crisis. Prof. Psychol. Res. Pract. **27**(2), 137–144 (1996). https://doi.org/10.1037/0735-7028.27.2.137

Quer, J., et al. (eds.): SignGram Blueprint: A Guide to Sign Language Grammar Writing. Mouton de Gruyter, Berlin; Boston (2017)

Rathmann, C.G.: The optionality of agreement phrase: evidence from German Sign Language (DGS). Texas Linguist. Forum **53**, 181–192 (2003)

Rizzi, L.: The fine structure of the left periphery. In: Haegeman, L. (ed.) Elements of Grammar. Handbook in Generative Syntax, pp. 281–337. Kluwer, Dordrecht (1997)

Schütze, C.T.: The empirical base of linguistics: grammaticality judgments and linguistic methodology (2016). https://doi.org/10.17169/langsci.b89.100

Stokoe, W.: Sign language structure: an outline of the visual communication system of the American deaf. Department of Anthropology and Linguistics, University of Buffalo, Buffalo (1960)

Stokoe, W., Casterline, D., Croneberg, C.: A Dictionary of ASL on Linguistic Principles. Gallaudet College Press, Washington, DC (1965)

Tervoort, B.T.: Structurele analyze van visueel taalgebruik binnen een groep dove kinderen (structural analysis of visual language use within a group of deaf children). North-Holland, Amsterdam (1953)

Wilbur, R.: Representations of telicity in ASL. In: Proceedings from the Annual Meeting of the Chicago Linguistic Society, vol. 39, no. 1, pp. 354–368. Chicago Linguistic Society (2003)

Wilbur, R.: Complex predicates involving events, time and aspect: is this why sign languages look so similar. In: Signs of the Time: Selected Papers from TISLR 2004, pp. 217–250 (2008)
Woodward, J.C.: Implications for sociolinguistic research among the deaf. Sign Lang. Stud. **1**, 1–7 (1972)
Wurmbrand, S.: Infinitives: Restructuring and Clause Structure. Mouton de Gruyter, Berlin; New York (2001)
Zeshan, U.: Interrogative constructions in signed languages: crosslinguistic perspectives. Language **80**, 7–39 (2004). https://doi.org/10.1353/lan.2004.0050
Zeshan, U.: Sign languages of the world. In: Brown, K., Ogilvie, S. (eds.) Concise Encyclopedia of Languages of the World, pp. 953–960. Elsevier, Amsterdam (2009)
Zucchi, S.: Along the time line: tense and time adverbs in Italian Sign Language. Nat. Lang. Seman. **17**(2), 99–139 (2009). https://doi.org/10.1007/s11050-008-9032-4

Gradability in MTT-Semantics

Stergios Chatzikyriakidis[1(✉)] and Zhaohui Luo[2]

[1] CLASP, Department of Philosophy, Linguistics and Theory of Science, University of Gothenburg, Gothenburg, Sweden
stergios.chatzikyriakidis@gu.se
[2] Royal Holloway, University of London, Egham, Surrey, UK

Abstract. In this paper, we look at the issue of gradability within formal semantics in modern type theories (MTT-semantics). Specifically, we look at both gradable adjectives and nouns, and show that the rich typing mechanisms afforded by MTT-semantics can give us a natural account of gradability. Gradable adjectives take indexed nouns as their arguments, while gradable nouns are Σ-types where their first projection is a degree parameter. Furthermore, we provide a standard polymorphic measure function applicable to all gradable adjectives and nouns. We also look at multidimensional adjectives and use enumerated types to capture multidimensionality. We formalize our account in the Coq proof assistant and check its formal correctness. Lastly, we briefly describe a recent proposal of model gradability by means of subtype universes in MTTs that can potentially give a unifying treatment of gradability for both regular gradable adjectives, but also multidimensional ones.

1 Introduction

The term gradable adjectives refers to the class of adjectives that involve some kind of grading property/parameter that allows them to be quantified according to it. For example, in the case of *small* and *large*, the grading parameter is size. Gradable adjectives have comparative and superlative forms and can be further modified by degree adverbs (e.g. *much*). Besides gradable adjectives, one also finds cases of gradable nouns, i.e. cases where the gradable element is not an adjective, but rather a noun:

(1) John is an enormous idiot/He is a big stamp collector.

In (1), the most natural reading is not one of large physical size, but rather of the nominal holding to a high degree.

S. Chatzikyriakidis—Supported by grant 2014-39 from the Swedish Research Council, which funds the Centre for Linguistic Theory and Studies in Probability (CLASP) in Department of Philosophy, Linguistics, and Theory of Science at the University of Gothenburg.
Z. Luo—Partially supported by EU COST Action EUTypes (CA15123, Research Network on Types).

© The Author(s), under exclusive license to Springer Nature Switzerland AG 2022
A. Özgün and Y. Zinova (Eds.): TbiLLC 2019, LNCS 13206, pp. 38–59, 2022.
https://doi.org/10.1007/978-3-030-98479-3_3

There are furthermore adjectives that can be quantified across more than one dimension. For example in the case of *tall*, there is only one dimension involved, tallness. This is not the case for adjectives like *healthy* and *sick*, which are called multidimensional. Following [47], two different classes of multidimensional adjectives are distinguished: positive and negative. The idea is that every positive adjective has a negative counterpart, i.e. its antonym (e.g. *healthy* and *sick*). What is different between the two is the form of quantification over dimensions in each case. Positive adjectives involve universal quantification over dimensions, while negative adjectives existential quantification. For example, for someone to be considered healthy, s/he must be healthy in all dimensions, whereas for someone to be considered sick, it suffices to be sick across one dimension only. In order for this intuition to be borne out more clearly, the exception phrase headed by *except* can be used. The interesting bit here is that this phrase is only compatible with universal quantification. As seen below, 'healthy' is compatible with an 'except' phrase, while 'sick' is not:

(2) Dan is healthy except with respect to blood pressure

(3) # Dan is sick except with respect to blood pressure

In this paper, we look at both gradable adjectives/nouns and multidimensional ones from the perspective of formal semantics in modern type theories (MTT-semantics) [10,35], arguing that MTT-semantics provides us with the mechanisms to give reasonable formal semantics accounts of these phenomena. The structure of the paper is as follows: in Sect. 2, we give a brief introduction to MTT-semantics, concentrating on the features that are mostly relevant to the analyses in this paper. In Sect. 3, we present our analysis of gradable and multidimensional adjectives/nouns. In Sect. 4, we formalize our account in the Coq proof assistant and check its correctness. In Sect. 5, we provide a brief investigation of an alternative way to deal with gradability by using recent work in Type Theory on subtype universes. Lastly, in Sect. 6, we conclude and discuss some future work.

2 Modern Type Theories: A Brief Introductioin

Formal semantics in modern type theories (MTT-semantics) [10,35] has been proposed as an alternative to Montague Semantics, and various semantic accounts have been given within this paradigm for a wide range of linguistic phenomena [8,11,32,35,50]. We use the term Modern Type Theories (MTTs) to refer to a class of type theories which have dependent types, inductive types and other powerful and expressive typing constructions. MTTs can be predicative, such as Martin-Löf's intensional type theory [39,44], and impredicative, such as the Unified Theory of dependent Types (UTT) [30]. In this paper, we shall employ UTT complemented with the coercive subtyping mechanism [31,36].

In this section we provide a brief introduction to MTTs, concentrating mostly on the features that are most relevant to this paper.

2.1 Many-Sortedness, Common Nouns as Types and Subtyping

A key difference between MTT-semantics and Montague semantics (MS) lies in the interpretation of common nouns (CNs). In [42], the underlying logic (Church's simple type theory [13]) is 'single-sorted' in the sense that there is only one type, e, of all entities. The other types such as the type of truth values, i.e. t, and the function types generated from types e and t do not stand for types of entities. Thus, no fine-grained distinctions between the elements of type e exist, and as such, all individuals are interpreted using the same type. For example, *John* and *Mary* have the same type in simple type theory, i.e. the type e of individuals. An MTT, on the other hand, can be regarded as a 'many-sorted' logical system in that it contains many types. In this respect, MTTs can make fine-grained distinctions between individuals and use those different types to interpret subclasses of individuals. For example, one can have $John : Man$ and $Mary : Woman$, where Man and $Woman$ are different types. Another very basic difference between MS and MTTs is that common nouns in MTTs (CNs) are usually interpreted as *types* [45] rather than sets or predicates (i.e., objects of type $e \to t$) as in MS. The CNs 'man, human, table' and 'book' are interpreted as types *Man, Human, Table* and *Book*, respectively. Then, individuals are interpreted as being of one of the types used to interpret CNs.

This many-sortedness has the welcome result that a number of semantically infelicitous sentences involving category mistakes, which are however syntactically well-formed, like e.g. 'he ham sandwich walks' can be explained easily. This is because a verb like 'walks' will be specified as being of type $Animal \to Prop$, while the type for 'ham sandwich' will be $Food$ or $Sandwich$:

(4) $the\ Ham\ sandwich : Food$

(5) $walk : Human \to Prop$

The idea that common nouns should be interpreted as types rather than predicates has been argued in [34] on philosophical grounds as well. There, it is claimed that the observation found in [20] according to which common nouns, in contrast to other linguistic categories, have criteria of identity that enable them to be compared, counted or quantified, has an interesting link with the constructive notion of set/type: in constructive mathematics, sets (types) are not constructed only by specifying their objects but they additionally involve an equality relation. The argument is then that the interpretation of CNs as types in MTTs is explained and justified to a certain extent. Extensions and further theoretical advances using the CNs as types approach can be found in [12] and an extension of the idea that CNs further specify their identity criteria with a case study on counting with numerical quantifiers under copredication cases is given in [9].

Interpreting CNs as types rather than predicates has also a significant methodological implication: compatibility with subtyping. For instance, one may introduce various subtyping relations by postulating a collection of subtypes (physical objects, informational objects, eventualities, etc.) of the type *Entity*

[2]. It is a well-known fact that if CNs are interpreted as predicates as in traditional Montagovian settings, introducing such subtyping relations would cause problems. This is because the contravariance of function types would predict that given the subtyping relation $A \leq B$, $B \rightarrow Prop \leq A \rightarrow Prop$ would be the case (the opposite relation than the one needed). Substituting A with type Man and B with type $Human$, we come to understand why interpreting CNs as predicates is not a good idea if we want to add a subtyping mechanism.

The subtyping mechanism used in the MTT endorsed in this paper is that of coercive subtyping [31,36]. Coercive subtyping can be seen as an abbreviation mechanism: A is a (proper) subtype of B ($A \leq B$) if there is a unique implicit coercion c from type A to type B and, if so, an object a of type A can be used in any context $\mathfrak{C}_B[\_]$ that expects an object of type B: $\mathfrak{C}_B[a]$ to be legal (well-typed) and equal to $\mathfrak{C}_B[c(a)]$.

To give an example: assume that both Man and $Human$ are base types. One may then introduce the following as a basic subtyping relation:

(6) $Man \leq Human$

2.2 Σ-types, Π-types, Indexed Types and Universes

Dependent Σ-types. One of the basic features of MTTs is the use of Dependent Types. A dependent type is a family of types that depend on some values. The constructor/operator Σ is a generalization of the Cartesian product of two sets that allows the second set to depend on values of the first. For instance, if $Human$ is a type and $Male$: $Human \rightarrow Prop$, then the Σ-type $\Sigma h\colon Human.\ Male(h)$ is intuitively the type of humans who are male.

More formally, if A is a type and B is an A-indexed family of types, then $\Sigma(A, B)$, or sometimes written as $\Sigma x\colon A.B(x)$, is a type, consisting of pairs (a, b) such that a is of type A and b is of type $B(a)$. When $B(x)$ is a constant type (i.e., always the same type no matter what x is), the Σ-type degenerates into product type $A \times B$ of non-dependent pairs. Σ-types (and product types) are associated projection operations π_1 and π_2 so that $\pi_1(a, b) = a$ and $\pi_2(a, b) = b$, for every (a, b) of type $\Sigma(A, B)$ or $A \times B$.

The linguistic relevance of Σ-types can be directly appreciated once we understand that, in its dependent case, Σ-types can be used to interpret linguistic phenomena of central importance, like adjectival modification (see above for interpretation of modified CNs) [45]. For example, *handsome Man* is interpreted as a Σ-type (7), the type of handsome men (or more precisely, of those men together with proofs that they are handsome):

(7) $\Sigma m\colon Man\ handsome(m)$

where $handsome(m)$ is a family of propositions/types that depends on the man m.

Dependent Π-Types. The other basic constructor for dependent types is Π. Π-types can be seen as a generalization of the normal function space where the second type is a family of types that might be dependent on the values of the first. A Π-type degenerates to the function type $A \rightarrow B$ in the non-dependent case. In more detail, when A is a type and P is a predicate over A, $\Pi x : A.P(x)$ is the dependent function type that, in the embedded logic, stands for the universally quantified proposition $\forall x : A.P(x)$. For example, the following sentence (8) is interpreted as (9):

(8) Every man walks.

(9) $\Pi x : Man.walk(x)$

Π-types are very useful in formulating the typings for a number of linguistic categories like VP adverbs or quantifiers. The idea is that adverbs and quantifiers range over the universe of (the interpretations of) CNs and as such we need a way to represent this fact. In this case, Π-types can be used, universally quantifying over the universe CN. (10) is the type for VP adverbs while (11) is the type for quantifiers:[1]

(10) $\Pi A : \mathsf{CN}.\ (A \rightarrow Prop) \rightarrow (A \rightarrow Prop)$

(11) $\Pi A : \mathsf{CN}.\ (A \rightarrow Prop) \rightarrow Prop$

Further explanations of the above types are given after we have introduced the concept of type universe below.

Indexed Types. An indexed type is a type of dependent type. They are families of types that are indexed by a parameter whose type is usually a simple one. Indexed types here will be used in the main analysis of gradable adjectives, as we will assume that gradable adjectives do not take simple CN types as their arguments but rather CN types indexed with a parameter. For example, we can think of the type representing humans along with their heights. We can do this using indexed types by considering the family of types $Human : Height \rightarrow Type$ indexed by heights: $Human(n)$ is the type of humans of height n.

Type Universes. An advanced feature of MTTs, which will be shown to be very relevant in interpreting NL semantics in general as well as adjectival modification specifically, is that of universes. Informally, a universe is a collection of (the names of) types put into a type [40].[2] For example, one may want to collect all the names of the types that interpret common nouns into a universe $\mathsf{CN} : Type$.

[1] The type for adverbs was proposed for the first time in [33].

[2] There is quite a long discussion on how these universes should be like. In particular, the debate is largely concentrated on whether a universe should be predicative or impredicative. A strongly impredicative universe U of all types (with $U : U$ and Π-types) is shown to be paradoxical [19,21] and as such logically inconsistent. The theory UTT we use here has only one impredicative universe $Prop$ (representing the world of logical formulas) together with an infinitely many predicative universes which as such avoids Girard's paradox (see [30] for more details).

The idea is that for each type A that interprets a common noun, there is a name $\overline{A}$ in CN. For example,

$$\overline{Man} : \mathsf{CN} \quad \text{and} \quad T_{\mathsf{CN}}(\overline{Man}) = Man.$$

In practice, we do not distinguish a type in CN and its name by omitting the overlines and the operator T_{CN} by simply writing, for instance, $Man : \mathsf{CN}$.

Having introduced the universe CN, it is now possible to explain (10) and (11). The type in (11) says that for all elements A of type CN, we get a function type $(A \to Prop) \to Prop$. The idea is that the element A is now the type used. To illustrate how this works let us imagine the case of quantifier *some* which has the typing in (11). The first argument we need, has to be of type CN. Thus *some human* is of type $(Human \to Prop) \to Prop$ given that the A here is $Human : \mathsf{CN}$ (A becomes the type $Human$ in $(Human \to Prop) \to Prop$). Then given a predicate like $walk : Human \to Prop$, we can apply *some human* to get $(some\ Human)(walk) : Prop$. Similar considerations apply for (10).

3 Gradability in MTT-Semantics

In this section, we present an MTT account of a number of aspects of gradable and multidimensional adjectives.

3.1 Gradable Adjectives

A standard assumption in the literature is that gradable adjectives involve some kind of measurement. Usually, this measurement is assumed to be a degree argument, whose presence or not, is then considered to be the main difference between gradable and non-gradable adjectives. This extra argument has been proposed to be formally encoded in the adjective's typing as in [3,23,49], or not as in [25,29,41,48].

The account we are going to pursue here is one where the arguments of gradable adjectives are not of simple types, but rather types indexed by degree parameters (dependent types). In MTT-semantics, the universe CN of common nouns are refined into subuniverses of CNs each of which is indexed by a degree. For example, the collection represented by the common noun *human* may be refined into the family of types indexed by heights: $Human : Height \to Type$ and $Human(n)$ is the type of humans of height n.[3] We can then define a function *height* that returns the value of the height-index of a human; i.e., $height(i, h)$ is the height of human h:

(12) $height : \Pi i : Height.\ Human(i) \to Height$

(13) $height(i, h) = i.$

[3] Informally, this family of types of humans are more refined than the type *Human* of all humans. Formally, we'd have $HHuman(i) \leq Human$.

With these assumptions in line, we may consider the semantic interpretation of *tall* to mean that the height of the human concerned is larger than some given standard n:

(14) $tall : \Pi i : Height.\ Human(i) \rightarrow Prop$

(15) $tall(i, h) = height(i, h) \geq n$

The above definition for *tall* specifies that for any i of type $Height$, *tall* takes a human argument indexed with i and returns the proposition saying that i, the height of the human, is greater than or equal to a natural number n, which stands for the contextually restricted parameter – humans taller than n are regarded as tall. In a similar fashion, we can define the comparatives, where the RHS of (17) is the same as $i > j$:

(16) $taller\_than : \Pi i, j : Height.\ Human(i) \rightarrow Human(j) \rightarrow Prop$

(17) $taller\_than(i, j, h_1, h_2) = height(i, h_1) > height(j, h_2)$.

From this definition, we can easily prove that, for example, if $height(i, h_1) \geq height(j, h_2)$ and $tall(j, h_2)$, then $tall(i, h_1)$.

The natural question to ask at this point is the following: where does this contextual parameter come from? In what we have provided so far, it is just a number that does not depend on anything. A better and more intuitive way to refer to this contextual parameter is to make its value dependent on the noun, the adjective, and sometimes even some other contextual information. These latter three parameters in MTT-semantics are represented as a type, a predicate and a context (in type theory), respectively. In order to fornalize this idea, we use polymorphism and type dependency. First, we introduce the universe of (totally ordered) degree types, *Degree*. As examples of degrees, one would find in *Degree* types such as $Height$, $Weight$ and $Width$, among many others. The inference rules of CN_G are given below, the second of which says that $\mathsf{CN}_G(D)$ is a subtype of CN and the third is an example of an introduction rule for CN_G:

$$\frac{D : Degree}{\mathsf{CN}_G(D) : Type} \qquad \frac{D : Degree \quad A : \mathsf{CN}_G(D)}{A : \mathsf{CN}} \qquad \frac{i : Height}{Human(i) : \mathsf{CN}_G(Height)}$$

We can now introduce the polymorphic standard, STND. First, for any common noun A, let $ADJ(A)$ be the type of syntactic forms of adjectives whose semantic domain is A. For instance, $TALL : ADJ(Human)$, where $TALL$ strands for the syntax of *tall*. Then, STND takes a degree D, a D-indexed common noun A and (the syntax of) an adjective whose domain is A, and returns the relevant standard for the adjective:

(18) $\mathrm{STND} : \Pi D : Degree.\ \Pi A : \mathsf{CN}_G(D).\ ADJ(A) \rightarrow D$

The next thing to consider in giving a more proper definition for *tall*, is a polymorphic type that is not restricted to $Human$ arguments with $Height$

parameters only. *Tall* can be used with types of non-humans: for example one can talk about a *tall building* or a *tall cat.* On the other hand, uses like *tall democracy* or *tall mind* do not seem to be felicitous, at least without some sort of contextual coercion. Using either $Human(i)$ as argument for *tall* or a polymorphic argument based over the universe CN will undergenerate and overgenerate respectively. One can try to use a subuniverse of CN, CN_{PHY}that basically includes all physical objects (types PHY and its subtypes). In this respect, we can introduce the universe $\mathsf{CN}_{PHY}$with the following introduction rule:

$$\frac{A : \mathsf{CN}, A < \text{PHY}}{A : \mathsf{CN}_{PHY}}$$

With this rule and assuming that every physical object has a height, we are now in a position to upgrade the definition for *tall* (we assume that the argument A is implicit in the definition):

(19) $tall : \Pi A : \mathsf{CN}_{PHY}.\Pi i : Height.A(i) \to Prop$

(20) $tall(i, h) = height(i, h) \geq \text{STND}(Height, Human, TALL).$

Note that indexing on the noun by means of a degree gives us for free the fact that we are not talking about tallness in general but tallness with respect to the relevant class (represented by the type $Human$ in the above example). In order to understand its importance, this indexing seems to be doing the work done by using the dot combinator of [26] to compose comparison classes with adjectives in the work of [24]. To give an example, let us say that one needs to compose *tall* with its comparison class, say basketball player (represented as BB). The typings we have are as follows: $BB : e \to t$ and $tall : e \to d$. However, we need functional application to return: $BB(tall) : e \to d$. As obvious, normal functional application will not work here. Thus, the dot combinator is used to remedy this. This additional and arguably not well-motivated extra machinery is not needed in our case. Furthermore, the polymorphic STND function can be seen as a more straightforward interpretation of Kennedy's context sensitive function from measure functions (adjectives basically) to degrees [24]. Lastly, one may consider standards that are dependent on other contextual information as well: for example, whether something is regarded as an expensive car might depend on where it is considered. In that case, the $STND$ function may take an additional parameter of locations that would take this into account.

Remark 1 (CN and Its Subuniverses). Type universes help us in MTT-semantic formalizations. For example, we have used the universe CN as the universe that makes polymorphism over all common nouns possible and allows adequate typing for phenomena like VP-adverbs and subsective adjectives to be provided:

(21) $VP_{ADV} : \Pi A : \mathsf{CN}.\ (A \to Prop) \to (A \to Prop)$

(22) $ADJ_{SUBS} : \Pi A : \mathsf{CN}.\ A \to Prop$

Furthermore, we have used the universe CN_{PHY} in our analysis of 'tall', in order to restrict the domain of polymorphism to the subuniverse that includes all physical objects and their subtypes. Other similar useful subuniverses can be constructed in order to help us in our semantic representations. Consider for example the subsective adjective 'skilful'. According to what we have been saying so far, it is of the type given in (22). Digging a bit deeper, one can see that 'skilful' is not really compatible with arguments that are not of type *Human*, or at least of type *Animal*. For example, one cannot talk about a skilful carpet or a skilful democracy. Thus, one could update the definition for 'skilful' taking these issues into consideration. On the assumption that 'skilful' is only relevant for human arguments, polymorphism is on the subuniverse CN_H, i.e. the universe including types *Human* and its subtypes:

(23) $skilful : \Pi A : \ \mathsf{CN}_H.\ A \rightarrow Prop$

An important question is, of course, how can we decide what the relevant universe is in each case? Well, one way to do it is by linguistic investigation as typically done in formal linguistics, i.e. getting judgments of native speakers that will help us decide the elements of the universe to be formed. Another way to do that is to use existing lexical-semantics resources that might contain such information. For example, in [7], the authors experiment with JeuxdeMots [27], a rich lexical-semantic network constructed using GWAPs [1], in order to extract information relevant for multi-typed systems, e.g. common noun types, subtyping relations, typings for predicates etc. We believe that such connections should be explored in future work combining lexical-semantic information drawn from linguistic resources with rich formal semantics formalisms like the one we are describing in this paper.

The other question one need to answer is whether such subuniverses are formally coherent in the sense that their introduction is logically okay. One has to be careful when constructing such universes. Some universes can be formally paradoxical even though they may seem justified from a linguistic perspective.[4] Thus, a better way to put what we have been saying is the following: we construct meaningful universes based either on linguistic intuitions and/or information from lexical/semantic networks, but only when we can formally justify it, i.e. to prove meta-theoretically that the incorporation of the new universe into the original type theory is OK (e.g., logically consistent, among other properties). The universes such as CN_{PHY} and CN_H are what we call *subtype universes* studied recently by Maclean and the second author [37], see Sect. 5. □

3.2 Gradable Nouns

As already discussed in the introduction, gradable nouns concern gradability cases where the relevant gradable element is not an adjective, but rather a noun, as (24) illustrates.

(24) John is an enormous idiot./He is a big stamp collector.

[4] For example, the type theory studied by Martin-Löf in [38] has a type U of all types (and hence, $U : U$) and has been proven to be logically inconsistent [19,21].

Indexed types, as already mentioned, are a type of dependent types, i.e. families of types indexed by a parameter whose type is usually a simple one. We have used indexed types so far in our treatment of gradable adjectives. The question is whether we can extend the usage of indexed types to gradable nouns as well. We will argue that this is indeed possible. What we want to propose here is that the distinction between nouns and adjectives is still clear: adjectives are taken to be predicates, nouns are taken to be types. At the same time, however, we assume that gradable nouns like *idiot* and gradable adjectives like *tall* both involve a degree parameter, albeit an abstract one in the former case. A natural way to capture this idea, i.e. abstract nouns being types but still involving a degree parameter, is to use Σ-types and assume that the first projection is actually the abstract parameter. To do this, we consider the type family $IHuman\colon Idiocy \to Type$ indexed by idiocy degrees of type $Idiocy\colon Degree$, where $Idiocy$ is a type whose objects form a total order and can be compared to each other by, for example, a $\geq$-relation. Then, *idiot* can be represented by means of (25):

(25) $Idiot = \Sigma i : Idiocy.IHuman(i) \times (i \geq \mathrm{STND}(Idiocy, Human, IDIOTIC))$

An idiot is thus a triple (i, h, p) where h is a human whose idiocy degree i is larger than or equal to the standard of being an idiot. Note that this account has not only similarities with the ideas proposed in [14] but also brings out a connection with gradable adjectives in the sense that both gradable adjectives and gradable nouns involve a degree parameter. However, these two are clearly different in terms of their formal status, adjectives being predicates while nouns types.

Let us now consider *enormous idiot*. The interpretation we want to get in this case is one where someone is an idiot to a very high degree. This means that this degree must be (much) higher than the degree of idiocy needed for someone to be considered an idiot (the standard $\mathrm{STND}(Idiocy, Human, IDIOTIC)$ in (25)). In order to capture that, we first propose that *enormous* can be interpreted as having the following type, where $PHY_D : \mathsf{CN}_{\mathsf{G}}(D)$ is the type of physical objects indexed by D:

(26) $enormous : \Pi D : Degree\ \Pi A : D \to \mathsf{CN}_{\mathsf{G}}(D)\ \Pi d : D.\ (A(D) \to Prop)$

Then we propose the following definition for 'enormous', for $D : Degree$, $A : D \to \mathsf{CN}_{\mathsf{G}}(D)$, $d : D$, and $a : A(D)$:

(27) $Enormous(D)(A)(d)(a) = d \geq \mathrm{STND}(D, PHY_D, ENORMOUS)$

We are now ready to interpret *enormous idiot* (D and A arguments are implicit):

(28) $Enormous\ Idiot = \Sigma h\colon Idiot.\ enormous((\pi_1(h), \pi_2(\pi_1(h))))$
where $\mathrm{STND}(D, PHY_D, ENORMOUS) \geq \mathrm{STND}(Idiocy, Human, IDIOTIC)$

Enormous idiot is thus a pair, where the first projection consists of a proof of being an idiot h (*Idiot* itself also a Σ-type, see (25)) and the second projection requires that the standard of idiocy associated with the first projection of the second projection of h is greater than the standard for *enormous*.

3.3 Multidimensional Adjectives

Multidimensional adjectives are adjectives that can be quantified across different dimensions. Adjectives like *sick* and *healthy* fall into this category. Following [47], two different classes of multidimensional adjectives are distinguished: positive and negative. The idea is that every positive adjective has a negative counterpart, i.e. its antonym (e.g. *healthy* and *sick*). What is different between the two is the form of quantification over dimensions in each case. Positive adjectives involve universal quantification over dimensions, while negative adjectives existential quantification. For example, for someone to be considered healthy, s/he must be healthy in all dimensions, whereas sick, it suffices to be sick across one dimension only. In order for this intuition to be borne out more clearly, the exception phrase *except* can be used. The interesting bit here is that this phrase is only compatible with universal quantification. As seen below, 'healthy' is compatible with 'except', but 'sick' is not:

(29) Dan is healthy except with respect to blood pressure

(30) # Dan is sick except with respect to blood pressure

This intuition can be implemented in an MTT setting using an inductive type for multiple dimensions. Consider an adjective like *healthy*. In order for someone to be considered healthy, one must be able to universally quantify over a number of "health" dimensions: *cholesterol, blood pressure* etc. To formalize this, one can introduce the inductive type $Health$ of type $Degree$ as follows:[5]

(31) $Health : Degree = heart \mid blood\_pressure \mid cholesterol$

We assume that the adjective *healthy* is of the following type (we use $Human$ as a simple type rather than a type-valued function as used earlier):

(32) $healthy : Health \rightarrow Human \rightarrow Prop$

We can now use this parameter as a primitive to define $Healthy$ and $Sick$ as follows:

(33) $Healthy = \lambda x : Human.\forall h : Health.\ healthy(h, x)$

(34) $Sick = \lambda x : Human.\neg(\forall h : Health.\ healthy(h, x))$

Note that, for multidimensional adjectives, each dimension may be gradable. For example, when we say that a healthy person is a person healthy in all dimensions, it basically means that each dimension surpasses a standard of healthiness. Take the dimension 'blood pressure' as an example: a child $x : Child \leq Human$ is healthy as far as blood pressure is concerned may mean that the blood pressure of x is less than some threshold with respect to children. (See Sect. 5 for formal details.)

[5] The inductive type $Health$ is a finite type (also called an enumeration type), sometimes written as $\{heart,\ blood\_pressure,\ cholesterol\}$.

Remark 2. With respect to multidimensional adjectives, there are a number of complications that need to be addressed. For example, the nature of the quantifier associated with positive adjectives does not seem to always be the universal quantifier. Sassoon and Fadlon [46] define quantificational multidimensional adjectives in the following sense:

> Quantificational adjectives like optimistic often involve counting of dimensions. As a default, entities fall under them iff they are classified under sufficiently many (e.g., some, most or all) dimensions.

Of course, this is not a problem in itself. One can modify the account w.r.t different adjectives as involving different quantificational force:

(35) $Healthy = \lambda x : Human.\ [\text{ALL,SOME,MOST}]\, h : Health.healthy(h, x)$

The choice of quantifier can be context dependent. One can assume that the quantifier quantifies over relevant dimensions in specific contexts. The definition of *Healthy* can be overloaded, picking the relevant dimensions in each case (relevant means available in that context). This is similar to the overloading technique as proposed by the second author to deal with homonymy [33].

□

3.4 Multidimensional Nouns

A further interesting discussion w.r.t multidimensionality concerns multidimensional nouns. For example, a noun like 'bird', at least according to theories like Prototype and Exemplar theories,[6] is argued to involve a rich couple of dimensions, i.e. in order for something to count as a bird, a couple of different dimensions (for example, dimensions like *winged, small, can breed,* etc.) have to be taken into consideration. Then, the idea is that the conceptual structure of a noun like 'bird' will involve an ideal value for each dimension. A similarity measure is mapping entities to degrees, representing how far from the ideal dimensions of the prototype the values for the respective entities are. This is represented as a weighted sum. The important thing, skipping formal details, is summarized in the following passage:

The distances of x from the prototypical values in the different bird dimensions integrate into a unique degree in the given noun by means of averaging operations, like weighted-sums... [47].

The above passage argues that dimensions integrate (another way of putting it is collapse) into a unique degree, and, thus, are not accessible for quantification as it is the case with multidimensional adjectives. Viewing common nouns as types seems to be compatible with this claim. The idea is as follows: in order for an object to be of a CN type, the standard of membership w.r.t the weighted sum of its similarity degrees to the ideal values in the dimensions of the noun has to be exceeded.[7] Actually, [46] revises later on her view, and talks about weighted

[6] See [47] for references to the relevant literature.

[7] See [43] for more details on this approach.

products in the case of these type of nouns. Somewhere in the middle between this two types of multidimensionality, i.e. multidimensional adjectives like *healthy* and multidimensional nouns like *bird*, we find social nouns like *linguist, artist.* These seem to behave like multidimensional adjectives, in that their dimensions seem to be accessible for quantification as witness the example below:

(36) He is an artist in many respects.

Such cases are then argued to represent intermediate cases, where the dimensions are integrated into a single degree, albeit the relevant operation is one of weighted sum and not product. The argument is that these dimensions are made easier available to quantification in these cases. This might then mean that the types become more elaborate in these cases. Consider the case for *artist*, and consider the inductive type for all its dimensions (we note them here as a_1, a_2, a_3 pending a more serious discussion of what these dimensions really are):

(37) Inductive $Art : D = a_1 \mid a_2 \mid a_3$

Now, one can think that in cases where social nouns make their dimensions accessible, what happens is that some sort of quantification is at play in the form of a Σ type, where the first projection is just a type $Human$, while the second projection specifies that all dimensions of artistry hold of this human above the relevant standard. Our definition for *artist* is given below:[8]

(38) $artist = \Sigma h : Human.\forall a : Art.DIM_{CN}(h, a)$

Notice, that the above is still a type and not a predicate. One can think that the creation of such types should be in general available, as even non-social nouns, e.g. natural-kind nouns like *duck*, can be sometimes, context allowing, used in a way that seems to make their dimensions available. For example, one can imagine a context where the following is true:

(39) My dog is a cat in most respects.

Thus, it seems that the operation to turn simple types into Σ types that make their dimensions available, is a more general one, and should be restricted w.r.t context and general world-knowledge considerations.

There are far more issues to consider when one looks at multidimensional adjectives (and nouns). However, we cannot go into detail into all these issues here. This topic deserves a separate paper in its own right. We direct the interested reader to [47] and [46] for literature review and a detailed exposition of the complexity of the phenomenon in question.

[8] With $DIM_{CN} : \Pi D : Degree.Human \rightarrow D \rightarrow Prop.$)

4 Coq Implementation

In this section, we present a Coq implementation for the different issues we have been discussing in this paper. But first things first. What is Coq? Simplifying things a bit, the main idea behind Coq can be roughly summarized as follows: you use Coq in order to see whether propositions based on statements previously pre-defined or user defined (definitions, parameters, variables) can be proven or not. Coq is a dependently typed proof assistant implementing the calculus of Inductive Constructions (CiC, see [18]). This means that the language used for expressing these various propositions is an MTT. This is a good start, at least for people using MTTs for NL semantics. Coq "speaks" so to say the language we use to interpret linguistic semantics. Given that Coq is in effect a reasoning engine, there are at least ways that can be used in studying linguistic semantics, to an extent overlapping with each other: a) as a formal checker for the semantic validity of proposed accounts in NL semantics and b) Natural Language Inference (NLI), i.e. reasoning with NL.

Remark 3 (interim note on installation). Coq can be installed easily for all platforms by visiting the system's website.[9] You can also get it using Macports, Homebrew or Nix. For mac and linux users, it is recommended to use Proof General,[10] a Coq interface for emacs that provides support for several proof assistants. □

Remark 4 (the type system implemented in Coq). The main difference between the type system that Coq implements [18] and the MTT we have been using so far (the type theory UTT [30]) is the use of coinductive types in Coq. Coinductive types are not used in any way in what we have been presenting so far, neither used in the Coq implementations. There are other minor differences between the two systems, but these are out of the scope of this paper, and play no important role in understanding the discussion in this section. □

Let us start with the formalization of gradable adjectives. We formulate the Degree universe Tarski-style in Coq:

```
(* Degree is type of names of degrees
d : Degree corresponds to type D(d) *)
(* So, Degree is a Tarski universe! *)
(* Here is an example with three degrees. *)
Require Import Omega.
Inductive Degree: Set:= HEIGHT | AGE | IDIOCY.
Definition D (d: Degree):= nat.
Definition Height := D(HEIGHT).
Definition Age:= D(AGE).
Definition Idiocy:= D(IDIOCY).
```

[9] http://coq.inria.fr/download.

[10] https://proofgeneral.github.io.

The code comments are enough to explain what is going on here: *Degree* is the type of names of degrees and $d : Degree$ corresponds to type $D(d)$. The next step is to formalize the universe CN_G, and then the context dependent standard, i.e. $STND$ function:

```
(* Universe CN_G of indexed CNs *)
Definition CN_G (_:Degree) := Set.
Parameter STND: forall d:Degree, forall A: CN_G(d), ADJ d A -> D(d).
```

Note that, in Coq, $forall$ stands for Π. With the previous parameter and definitions, *tall* can be defined:

```
Definition tall (h:Human):= ge (height h) (STND HEIGHT Human TALL).
```

With this at hand, one can define *taller*::

```
Definition taller_than (h1:Human) (h2:Human):= gt (height h2)
(height h1).
```

The next part involves formalizing gradable nouns, more specifically providing the type for *idiot* and the definitions for *enormous* and *enormous idiot.* The definitions follow closely the ones proposed in the paper. *Enormous idiot* is expressed as a Dependent Record Type:[11]

```
(**Definition for Idiot**)
Definition Idiot:=  sigT(fun x: Idiocy=> prod (IHuman x)
(ge x (STND IDIOCY Human IDIOTIC))).
Definition enormous  (d:Degree)(A:CN_G(d))(d1: D d) :=
    fun P: A =>  ge (d1)  (STND  d (PHY(d))(ENORMOUS d)).
Record enormousidiot: Set:= mkeidiot
    {h:> Idiot; EI: enormous IDIOCY
    (IHuman(projT1(h)))(projT1(h))(projT1(projT2(h)))
    /\  ge  (STND  IDIOCY (PHY(IDIOCY))(ENORMOUS IDIOCY))
    (STND IDIOCY Human IDIOTIC)}.
```

We continue with multidimensional adjectives. What we want to do in this case is implement the main idea we have been discussing in Sect. 3.3, namely the use of enumerated types in order to implement the many-dimensions aspect of multidimensional adjectives. Taking *healthy* as our example, we define the enumerated type Health that includes various health dimensions and then define adjectives *sick* and *healthy*, as involving universal quantification over the dimensions in *healthy*, and existential quantification in *sick*:

```
Definition  Degree:= Set.
Inductive Health: Degree:= Heart|Blood|Cholesterol.
Parameter Healthy: Health->Human->Prop.
Definition sick:= fun y: Human => ~ (forall x: Health, Healthy x y).
Definition healthy:= fun y: Human => forall x: Health, Healthy x y.
```

[11] Dependent Record Types in Coq are just syntactic sugar for Σ-types.

This suffices to give us the basic inferences with respect to multidimensional adjectives. For example one can prove that if John is healthy then he is healthy with respect to cholesterol, blood pressure and heart condition, if John is sick it suffices that he is not healthy across one dimension etc. These theorems, a number of other similarly relevant ones, as well as the formalization of the multidimensional noun *artist*, can be found in the Appendix A.2.

5 Modelling Gradability with Subtype Universes

Gradable adjectives and the related multidimensional cases provide challenging examples for MTT-semantics. This has led to further studies to develop type-theoretic mechanisms to formally deal with such phenomena. Recently, Maclean and the second author [37] have developed *subtype universes* for MTTs, which have interesting applications to programming and NL semantics. For the latter, they have pointed out that, employing subtype universes, one can obtain a nice semantics for gradable adjectives. We give a brief description here.

A subtype universe is a type that represents a collection of subtypes: for any type H, the universe $U(H)$ represents the collection of all subtypes of H. Such subtype universes can be specified formally by the following formation rule (U_F) and introduction rule (U_I), where $A \leq H$ is the shorthand for '$A \leq_c H$ for some coercion c' in the framework of coercive subtyping [36]:

$$(U_F)\quad \frac{\Gamma \vdash H : Type}{\Gamma \vdash U(H) : Type} \qquad\qquad (U_I)\quad \frac{\Gamma \vdash A \leq H : Type}{\Gamma \vdash A : U(H)}$$

Such type universes can be quantified over to form other propositions. For example, the proposition $\forall X : U(H).P(X)$ says that P holds for all subtypes of H. This, among other things, gives a nice treatment of *bounded quantification* of the form $\forall A \leq H.\ P(A)$ as proposed by Cardelli and Wegner [6], whilst avoiding the type checking issues traditionally associated with it. Also, Maclean and Luo have, for the first time, proved that extending MTTs with subtype universes preserves logical consistency [37], which is indispensable for a type theory to be used as a foundational semantic language.

Gradable adjectives such as 'tall' and 'healthy' can be modelled in MTT-semantics with the help of subtype universes. For example, let T be a type universe whose objects are base types H such as *Human* and *Building* for which the property $height\colon H \to Prop$ makes sense. Then, the type of *tall* can be given by means of subtype universes as in (40), which can be rewritten as (41) by means of bounded quantification as a notational abbreviation. So, *tall* is a predicate on subtypes of the base types. For instance, if $Human\colon T$ and $socrates\colon Man \leq Human$, then $tall(Human, Man, socrates)$ is a proposition. Given a threshold function $\xi\colon \Pi H : T.(U(H) \to Nat)$, one may define *tall* as $tall(H, A, x) = height(x) \geq \xi(H, A)$.

(40) $tall : \Pi H : T \Pi A : U(H).\ (A \to Prop)$

(41) $tall : \Pi H : T \Pi A \leq H.\ (A \to Prop)$

Note that, in modelling gradable adjectives as above, we have made use of the fact that applicability of gradable adjectives respects the usual subtyping relations (for example, if 'tall' can be applied to a type, it can be applied to any of its subtypes as well).

Multidimensional adjectives such as 'healthy' can also be modelled by means of subtype universes. For example, 'healthy' may be given the type (42) which can be rewritten as (43) in bounded quantification.

(42) $Healthy : \Pi A : U(Human).\ (A \rightarrow Prop)$

(43) $Healthy : \Pi A \leq Human.\ (A \rightarrow Prop)$

With healthy thresholds $\xi_i : \Pi A \leq Human.Nat$ with indexes i such as BP (for blood pressure), we have, for $A \leq Human$, $Healthy(A, x) = \bigwedge_i \chi_i(A, x)$, where χ_i's are the corresponding propositions: for instance, $\chi_{BP}(A, x) = BP(x) \leq \xi_{BP}(A)$, where A is a subtype of $Human$ examples of which include, for example, Boy and $Woman$.

Remark 5. As briefly described above, the approach to modelling gradability by means of subtype universes [37] results in simple semantic constructions and it is attractive and promising for modelling other linguistic features as well. It is worth remarking that most of the type constructions in the account in Sect. 3 are subtype universes to some extent. For example, CN_{PHY} is a subtype universe of those subtypes of PHY which are in CN as well. An in-depth comparative study would be interesting and may require further work. □

6 Conclusions and Future Work

In this paper, we have shown the use of MTT-semantics in the study of gradability. More specifically, we have shown that the rich typing mechanisms afforded by MTT-semantics can provide us with natural interpretations for both gradable and multidimensional adjectives/nouns. We have implemented the proposed accounts in the Coq proof-assistant and have checked their correctness. We have also briefly sketched an approach to modelling gradability by means of the recently studied notion of subtype universes. As mentioned, a comparative study of the two approaches to gradability is called for and left as future work.

One other issue that we have not looked at here and can be part of our future work is vagueness. In plain words, vagueness makes deciding what counts for something to be an X, where X is a gradable predicate (usually an adjective), difficult. There are three main problems associated with vagueness, the first one already mentioned and addressed in this paper: a) context dependency, b) the existence of borderline cases and c) the fact that vague adjectives (and predicates in general) give rise to the sorites Paradox. In the way our account stands, we cannot capture vagueness. We believe that this kind of problem needs to involve some kind of probabilistic reasoning. Indeed, a couple of researchers have pointed this out and have produced a body of research to this direction [4,5,22,28]. At the moment, the authors do not know of any successful work in combining probability with dependent types and some new idea would be needed in order to study probabilistic type theories.[12]

A Coq Code

A.1 Gradable Adjectives

```
(* Degree is type of names of degrees --*)
(*d: Degree corresponds to type D(d) *)
(* So, Degree is a Tarski universe! *)
(* Here is an example with three degrees. *)
Require Import Omega.
Inductive Degree: Set:= HEIGHT | AGE | IDIOCY |
Definition D (d: Degree):= nat.
Definition Height:= D(HEIGHT).
Definition Age:= D(AGE).
Definition Idiocy:= D(IDIOCY).

(* Universe CN_G of indexed CNs *)
Definition CN_G (_:Degree) := Set.
Parameter Human: CN_G(HEIGHT).
Parameter John Mary Kim : Human.
Parameter height: Human->Height.

(** Type of physical objects indexed with a degree**)
Parameter PHY : forall d: Degree, CN_G(d).

(* ADJ(D,A) of syntax of adjectives whose domain is A : CN_G(d) *)
Parameter ADJ: forall d:Degree, CN_G(d)->Set.
Parameter TALL SHORT: ADJ HEIGHT Human.
Parameter IDIOTIC: ADJ IDIOCY Human.
Parameter ENORMOUS: forall d: Degree,  ADJ d  (PHY(d)).

(* STND *)
Parameter STND: forall d:Degree, forall A:CN_G(d), ADJ d A -> D(d).

(* semantics of tall, taller_than *)
Definition tall (h:Human):= ge (height h) (STND HEIGHT Human TALL).
Definition taller_than (h1:Human) (h2:Human) := gt (height h2) (height h1).


Theorem TALLER:
taller_than Mary John /\ height Mary =
170 -> gt (height John) 170.
cbv. intro. omega. Qed.
Theorem trans:
taller_than Mary John /\ taller_than Kim Mary ->
taller_than Kim John.
cbv. intro. omega. Qed.
```

[12] The work on probability in TTR (see, for example, [17].) studies probability in a set-theoretical system, because TTR [15,16] is not a type theory, as the term is usually understood, but rather a set-theoretic notational system.

```
(**Definition for Idiot**)
Parameter IHuman : Idiocy -> CN_G(IDIOCY).
Definition Idiot:=
sigT(fun x: Idiocy=> (sigT (fun y: (IHuman x)
=> (ge x (STND IDIOCY Human IDIOTIC))))).
Definition enormous  (d:Degree)(A:CN_G(d))(d1: D d)
:= fun P: A =>  ge (d1)  (STND  d (PHY(d))(ENORMOUS d)).
Record enormousidiot: Set:= mkeidiot
{h1:> Idiot; EI1: enormous IDIOCY
(IHuman(projT1(h1)))(projT1(h1))(projT1(projT2(h1)))
/\ ge  (STND  IDIOCY (PHY(IDIOCY))(ENORMOUS IDIOCY))(STND IDIOCY Human IDIOTIC) }.

(*From enormous idiot it follows that there exists an idiot such
that their standard of idiocy is higher or equal to
the standard for  idiotic humans*)
Theorem ENORMOUS1:
enormousidiot -> exists H: Idiot,
projT1(H) >= STND IDIOCY (PHY IDIOCY) (ENORMOUS IDIOCY).
cbv. firstorder. Qed.

(*From enormous idiot it follows that there exists an idiot such
that their standard of idiocy is higher or equal to both
the standard for enormous idiots and the standard for idiotic
humans*)
Theorem ENORMOUS2:
enormousidiot -> exists H: Idiot,
projT1(H) >= STND IDIOCY (PHY IDIOCY) (ENORMOUS IDIOCY)
/\ projT1(H) >= (STND IDIOCY Human IDIOTIC).
cbv. firstorder. unfold Idiot in h2. exists h2. firstorder.
unfold enormous in H. firstorder. elim h2. intros. destruct p.
omega. Qed.

(*From enormous idiot it follows that there exists an idiot such
that their standard of idiocy is higher or equal to the standard
for enormous idiots and idiotic humans and also the standard for
enormous idiots is higher than that for idiotic humans*)
Theorem ENORMOUS3:
enormousidiot -> exists H: Idiot,
projT1(H) >= STND IDIOCY (PHY IDIOCY) (ENORMOUS IDIOCY)
/\ projT1(H) >= (STND IDIOCY Human IDIOTIC)
/\ STND IDIOCY (PHY IDIOCY) (ENORMOUS IDIOCY)
>= (STND IDIOCY Human IDIOTIC).
cbv. firstorder. unfold Idiot in h2. exists h2. firstorder.
unfold enormous in H. firstorder. elim h2. intros.
destruct p.
omega. Qed.
```

A.2 Multidimensional Adjectives

```
(*Dealing with multidimensional adjectives Health as an inductive
type where  the dimensions are enumerated. This is just an enumerated
type*)
Definition Degree:= Set.
Parameter Human: CN.
Parameter John: Human.
Inductive Health: Degree:= Heart|Blood|Cholesterol.
Parameter Healthy: Health -> Human -> Prop.
Definition sick:=fun y: Human => ~ (forall x : Health, Healthy x y).
Definition healthy:= fun y: Human => forall x: Health, Healthy x y.

Theorem HEALTHY:
healthy John -> Healthy Heart John /\ Healthy Blood John
/\ Healthy Cholesterol John.
cbv. intros. split. apply H.
split. apply H. apply H. Qed.

Theorem HEALTHY2:
healthy John -> not (sick John).
cbv. firstorder. Qed.

Theorem HEALTHY3:
(exists x: Health, Healthy x John) -> healthy John.
cbv. firstorder. Abort.

Theorem HEALTHY4:
(exists x: Health, not (Healthy x John)) -> healthy John.
cbv. firstorder. Abort.

Theorem HEALTHY5:
(exists x: Health, not (Healthy x John))  -> sick John.
cbv. firstorder.  Qed.

(*Multidimensional noun Artist*)
Inductive Art: Degree:= a1|a2|a3.
Set Implicit Arguments.
Parameter  DIM_CN : forall  D: Degree,   Human ->  D  -> Prop.
Record Artist: Set:= mkartist
{h:> Human; EI:  forall a: art,
(DIM_CN h a)}.
```

References

1. von Ahn, L., Dabbish, L.: Designing games with a purpose. Commun. ACM **51**(8), 58–67 (2008)
2. Asher, N.: Lexical Meaning in Context: A Web of Words. Cambridge University Press, Cambridge (2012)
3. Bartsch, R., Vennermann, T.: Semantic Structures. Athenaum, Frankfurt (1973)
4. Bernardy, J.P., Blanck, R., Chatzikyriakidis, S., Lappin, S.: A compositional Bayesian semantics for natural language. In: Proceedings of the First International Workshop on Language Cognition and Computational Models, pp. 1–10 (2018)

5. Bernardy, J.P., Blanck, R., Chatzikyriakidis, S., Lappin, S., Maskharashvili, A.: Bayesian inference semantics: a modelling system and a test suite. In: Proceedings of the Eighth Joint Conference on Lexical and Computational Semantics (*SEM 2019), pp. 263–272 (2019)
6. Cardelli, L., Wegner, P.: On understanding types, data abstraction, and polymorphism. ACM Comput. Surv. **17**(4), 471–523 (1985)
7. Chatzikyriakidis, S., Lafourcade, M., Ramadier, L., Zarrouk, M.: Type theories and lexical networks: using serious games as the basis for multi-sorted typed systems. J. Lang. Model. **5**(2), 229–272 (2017)
8. Chatzikyriakidis, S., Luo, Z.: Adjectival and adverbial modification: the view from modern type theories. J. Logic Lang. Inf. **26**(1), 45–88 (2017)
9. Chatzikyriakidis, S., Luo, Z.: Identity criteria of common nouns and dot-types for copredication. Oslo Stud. Lang. **10**(2), 21–141 (2018)
10. Chatzikyriakidis, S., Luo, Z.: Formal Semantics in Modern Type Theories. Wiley/ISTE, Hoboken (2020)
11. Duchier, D., Parmentier, Y. (eds.): CSLP 2012. LNCS, vol. 8114. Springer, Heidelberg (2013). https://doi.org/10.1007/978-3-642-41578-4
12. Chatzikyriakidis, S., Luo, Z., et al.: Modern Perspectives in Type-Theoretical Semantics, vol. 98. Springer, Cham (2017). https://doi.org/10.1007/978-3-319-50422-3
13. Church, A.: A formulation of the simple theory of types. J. Symbolic Logic **5**(1), 56–68 (1940)
14. Constantinescu, C.: Big eaters and real idiots: evidence for adnominal degree modification? In: Proceedings of the Sinn und Bedeutung. vol. 17, pp. 183–200 (2013)
15. Cooper, R.: Records and record types in semantic theory. J. Logic Comput. **15**(2), 99–112 (2005)
16. Cooper, R.: Context-passing and underspecification in dependent type semantics. In: Chatzikyriakidis, S., Luo, Z. (eds.) Modern Perspectives in Type-Theoretical Semantics. Springer, Cham (2017). https://doi.org/10.1007/978-3-319-50422-3
17. Cooper, R., Dobnik, S., Lappin, S., Larsson, S.: Probabilistic type theory and natural language semantics. Linguist. Issue Lang. Technol. **10**, 1–43 (2015)
18. The Coq Development Team: The Coq Proof Assistant Reference Manual (Version 8.1), INRIA (2007)
19. Coquand, T.: An analysis of Girard's paradox. In: Proceedings of the 31st Annual ACM/IEEE Symposium on Logic in Computer Science. IEEE (1986)
20. Geach, P.: Reference and Generality: An Examination of Some Medieval and Modern Theories. Cornell University Press, London (1962)
21. Girard, J.Y.: Interprétation fonctionelle et élimination des coupures de l'arithmétique d'ordre supérieur. Ph.D. thesis, Université Paris VII (1972)
22. Goodman, N., Lassiter, D.: Probabilistic semantics and pragmatics: uncertainty in language and thought. In: Lappin, S., Fox, C. (eds.) The Handbook of Contemporary Semantic Theory, 2nd edn., pp. 655–686. Wiley-Blackwell, Malden (2015)
23. Heim, I.: Degree operators and scope. In: Proceedings of the 31st Semantics and Linguistic Theory Conference, vol. 10, pp. 40–64 (2000)
24. Kennedy, C.: Vagueness and grammar: the semantics of relative and absolute gradable adjectives. Linguist. Philos. **30**(1), 1–45 (2007)
25. Klein, E.: A semantics for positive and comparative adjectives. Linguist. Philos. **4**(1), 1–45 (1980)
26. Kratzer, A., Heim, I.: Semantics in Generative Grammar, vol. 1185. Blackwell, Oxford (1998)

27. Lafourcade, M.: Making people play for lexical acquisition with the jeuxdemots prototype. In: SNLP2007, 7th International Symposium on Natural Language Processing, p. 7 (2007)
28. Lassiter, D., Goodman, N.D.: Adjectival vagueness in a Bayesian model of interpretation. Synthese **194**(10), 3801–3836 (2015). https://doi.org/10.1007/s11229-015-0786-1
29. Lewis, D.: General semantics. Synthese **22**(1), 18–67 (1970)
30. Luo, Z.: Computation and Reasoning: A Type Theory for Computer Science. Oxford University Press, Oxford (1994)
31. Luo, Z.: Coercive subtyping. J. Logic Comput. **9**(1), 105–130 (1999)
32. Luo, Z.: Type-theoretical semantics with coercive subtyping. In: SALT Proceedings, vol. 20 (2009)
33. Luo, Z.: Contextual analysis of word meanings in type-theoretical semantics. In: Pogodalla, S., Prost, J.-P. (eds.) LACL 2011. LNCS (LNAI), vol. 6736, pp. 159–174. Springer, Heidelberg (2011). https://doi.org/10.1007/978-3-642-22221-4_11
34. Luo, Z.: Common nouns as types. In: Béchet, D., Dikovsky, A. (eds.) LACL 2012. LNCS, vol. 7351, pp. 173–185. Springer, Heidelberg (2012). https://doi.org/10.1007/978-3-642-31262-5_12
35. Luo, Z.: Formal semantics in modern type theories with coercive subtyping. Linguist. Philos. **35**(6), 491–513 (2012)
36. Luo, Z., Soloviev, S., Xue, T.: Coercive subtyping: theory and implementation. Inf. Comput. **223**, 18–42 (2012)
37. Maclean, H., Luo, Z.: Subtype uiniverses. In: Post-Proceedings of the 26th International Conference on Types for Proofs and Programs (TYPES 2020). Leibniz International Proceedings in Informatics, vol. 188 (2021)
38. Martin-Löf, P.: An intuitionistic theory of types (1971), Manuscript
39. Martin-Löf, P.: An intuitionistic theory of types: predicative part. In: Rose, H., Shepherdson, J.C. (eds.) In: Proceedings of the Logic Colloquium 1973 (1975)
40. Martin-Löf, P.: Intuitionistic Type Theory. Bibliopolis, Naples (1984)
41. McConnell-Ginet, S.M.: Comparative Constructions in English: A Syntactic and Semantic Analysis (1973)
42. Montague, R.: Formal Philosophy. Yale University Press, New Haven. Collected papers edited by R. Thomason (1974)
43. Murphy, G.: The Big Book of Concepts, MIT Press, London (2004)
44. Nordström, B., Petersson, K., Smith, J.: Programming in Martin-Löf's Type Theory: An Introduction. Oxford University Press, Oxford (1990)
45. Ranta, A.: Type-Theoretical Grammar. Oxford University Press, Oxford (1994)
46. Sassoon, G.W., Fadlon, J.: The role of dimensions in classification under predicates predicts their status in degree constructions. Glossa: J. Gen. Linguist. **2**(1) (2017)
47. Sassoon, G.: A typology of multidimensional adjectives. J. Seman. **30**, 335–380 (2012)
48. Van Benthem, J.: The Logic of Time: a Model-Theoretic Investigation into the Varieties of Temporal Ontology and Temporal Discourse, vol. 156. Springer, Dordrecht (2012). https://doi.org/10.1007/978-94-010-9868-7
49. Von Stechow, A.: Comparing semantic theories of comparison. J. Seman. **3**(1), 1–77 (1984)
50. Xue, T., Luo, Z.: Dot-types and their implementation. In: Béchet, D., Dikovsky, A. (eds.) LACL 2012. LNCS, vol. 7351, pp. 234–249. Springer, Heidelberg (2012). https://doi.org/10.1007/978-3-642-31262-5_17

Building Resources for Georgian Treebanking-Based NLP

Oleg Kapanadze[1](✉), Gideon Kotzé[2], and Thomas Hanneforth[3]

[1] Ivane Javakhishvili Tbilisi State University, 1, Chavchavadze Avenue, 0179 Tbilisi, Georgia
okapanadze@uni-potsdam.de

[2] University of the Free State, 205 Nelson Mandela Dr, Bloemfontein 9301, South Africa

[3] University of Potsdam, Am Neuen Palais 10, House 9, 14469 Potsdam, Germany

Abstract. We describe past and present work surrounding the development of treebank related NLP resources for Georgian. In particular, we provide an overview of efforts made in the development of a morphologically and syntactically annotated treebank for this non-configurational language, as well as its application in the development of a syntactic parser. Building on this, we also report ongoing work in utilizing manual and automatic alignment solutions for the creation of a Georgian/German parallel treebank. The end goal is the development of resources and tools for improved computational processing and linguistic analysis of the Georgian language.

Keywords: Context-free grammars · Treebank · Parallel treebank · Syntactic parsing · Georgian · German

1 Introduction

The Georgian language has a multitude of language resources including grammars and dictionaries. However, these resources are not suited for NLP needs, where many of the developed tools and solutions exist for configurational languages such as English. Georgian, on the other hand, is a non-configurational language with rich derivational and inflectional morphology and very little fixed structure on the sentence level. These languages for morphologically rich and less-configurational features are referred to as MR&LC [1].

In this paper, we provide an overview of existing work done on the manual construction of a syntactically and morphologically annotated treebank for Georgian, and its application in the development of a syntactic parser. We also report ongoing work in utilizing manual and automatic alignment solutions for the creation of a Georgian/German parallel treebank. The end goal is the development of resources and tools for improved computational processing and linguistic analysis of the Georgian language.

© The Author(s), under exclusive license to Springer Nature Switzerland AG 2022
A. Özgün and Y. Zinova (Eds.): TbiLLC 2019, LNCS 13206, pp. 60–78, 2022.
https://doi.org/10.1007/978-3-030-98479-3_4

2 Background

2.1 Treebanks

Linguistic resource creation in a computational environment often consists of a series of steps that are applied in sequence. A text is pre-processed by ensuring that characters are consistently encoded, it may be filtered to create a corpus from a specific domain, or tokenization and sentence boundary detection tools may be applied so that words and sentences can be identified as units.

To create a useful linguistic resource for a language, often one or more of these steps is required. One level of representation for the study of language is the treebank, which is a corpus that is annotated with syntactic and/or semantic sentence structure in the form of skeletal parses – a *bank* of linguistic *trees*. Syntactic structure is commonly represented as a *tree structure* (in mathematical terms: an oriented graph) – hence, the name "treebank".

Treebanks are often created on top of a corpus that has already been annotated with part-of-speech tags. The annotation can vary from constituent to dependency or tecto-grammatical structures. Additionally, they are sometimes enhanced with semantic or other linguistic information. They can be created manually or semi-automatically, where a parser assigns some syntactic structure to a text that is checked and corrected, where necessary, by linguists.

Treebanks are now valuable resources as repositories for linguistic research, since corpus-based methods became useful in multilingual technology playing an important role in empirical language studies. They can be used in *contrastive studies* and *translation science*, in *corpus linguistics* for studying syntactic phenomena, in *computational linguistics* as evaluation corpora for different human language technology systems or for training and testing *parsers*, as well as functioning as databases for *translation memory* systems.

Many treebanks are annotated with syntactic structure that depends on a proper pre-analysis of the sentence, its part-of-speech tags and, in some cases, its morphology. As the Georgian language is assigned typologically to non-configurational languages, we believe that morphology plays an important role in the syntactic analysis of Georgian. This particular point of view will be discussed later in more detail.

Additionally, analysis is performed by a parser that is usually trained on a *grammar* that follows a given approach. Context-free grammars, as described in this paper, are one example. Grammars can be induced on treebanks or manually constructed.

Some treebanks follow a specific linguistic theory (for example, BulTreeBank for the Bulgarian language, which follows HPSG[1]), but most try to be less theory specific. However, two main groups can be distinguished: treebanks that annotate *phrase structure* (such as the Penn Treebank for Arabic [2], English [3] and Chinese [4]) and those that annotate *dependency structure* (such as the Prague Dependency Treebank for the Czech language [5]).

Phrase-structure annotation is a relatively dominant approach to modern computational linguistics, even though dependency structures are a much older idea and have

[1] http://bultreebank.org/en/btb/.

especially gained traction in the form of Universal Dependencies in recent years. Here, we describe our efforts to build a parser that annotate phrase structure in Georgian using a vanilla context-free grammar.

2.2 Parallel Treebanks and Alignment

Semantically equivalent or similar texts can be aligned on various levels in order to indicate equivalence or similarity. Apart from their obvious linguistic interest, these resources can also be used to train some NLP systems such as those used for machine translation [6, 7]. In practice, alignment can be indicated by two or more sections of texts sharing the same identification number or being put next to each other – aligned – within a text file.

A parallel corpus is a collection of *bitexts* that is aligned on one or more levels. A bitext can be defined as a document with one or more translations in other languages [8].

Historically, parallel corpora are aligned on sentence and word level for training statistical machine translation systems. Memory requirements at the time dictated the sentence boundary constraint, even though in practice, translations can overlap across these boundaries.

A *hierarchical iterative refinement strategy* [9] is often applied, where equivalent documents are aligned, after which, in these documents, paragraphs, sentences and then words are aligned. Each step creates an additional boundary constraint that limits the search space for which the specific alignment algorithm is to be applied, speeding up the process and limiting memory requirements.

For creating a *parallel treebank*, the sentence boundary constraint is often applied for the same reasons. A parallel treebank is a parallel corpus that has been analyzed and aligned both on and below the sentence level. In the case of phrase structures, words as well as so-called *subtrees* – smaller trees that have a constituent as its root node – can be aligned to indicate phrasal and structural equivalence. Figure 1 presents an example of a tree pair that is aligned on both the word and constituent level.

A phrase-structure parallel treebank is useful to indicate both structural and semantic equivalences of translated sentences and to provide additional data for syntax-based machine translation systems, since tree structure, not mere word-level annotation, is present. However, lexical and grammatical differences between the languages that are compared, as well as divergences between the tree structures, may lead to a lack of coverage. It is therefore important to note that phrase-structure alignment is but one level of representation of equivalence and that in the long term, additional layers and approaches will benefit further applications in future.

Um Gottes willen, erzählen Sie mir jetzt nicht, daß Schwarz Weiß sei.
თუ ღმერთი გწამთ, არ მითხრათ ახლა, რომ შავი თეთრია.

(Literally: “If you believe in God (=For God’s sake), do not tell me now that black is white”).

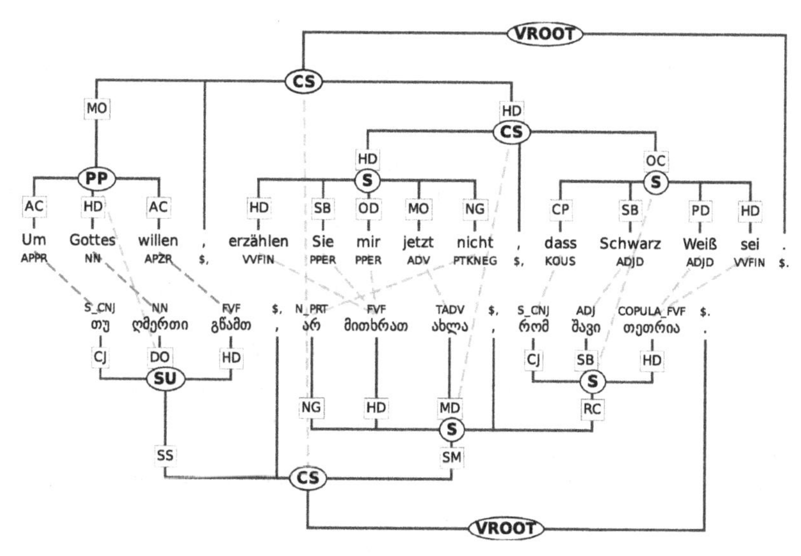

Fig. 1. Example of an aligned tree pair on sentence level, taken from the Stockholm TreeAligner [31]. Green lines indicate so-called "good" alignments that refer to an exact match, whereas red lines indicate "fuzzy" alignments that indicate a non-exact match. Note that words (leaves of the trees representing individual words) as well as constituents (non-terminal nodes representing multi-word phrases) are aligned. Some nodes remain unaligned, as the equivalent in the other tree is either not exactly represented by a constituent, or, mostly in the case of words, does not have an equivalent in the translation. (Color figure online)

In this paper, we present ongoing efforts in the tree alignment of Georgian and German for the creation of a parallel treebank. We have chosen German for the following reasons:

- The GRUG project [10][2] already involved the alignment of Georgian with German. Since the alignments in this resource are manually created, the natural next step would be to investigate the viability of automatic tree alignment.
- The Georgian approach is closely modelled after the TIGER treebank [11, 12]. See Sect. 3 for more detail.
- There is parallel data available for German/Georgian.
- Insights could be gained from comparing the agglutinative Georgian to the fusional German language.

3 Creating a Georgian Treebank

For the syntactic annotation of Georgian text, we draw on the experience of parallel treebanks for languages with different structures [13–17]. These studies provide useful information on processing the Turkish and Quechua languages. For instance, in a Quechua-Spanish parallel treebank, due to the strong agglutinative features of the

[2] http://fedora.clarin-d.uni-saarland.de/grug/.

Quechua language, the monolingual Quechua treebank had been annotated on morphemes rather than words. This allowed linking morpho-syntactic information precisely to its source. However, according to the authors, building phrase-structure trees over Quechua sentences does not capture the characteristics of the language. Therefore, a Role and Reference grammar has been implemented that uses nodes, edges and secondary edges to represent the most important aspects of Role and Reference syntax for Quechua sentences [18].

Georgian is also an agglutinating language. However, there is no need to annotate the Georgian treebank on morphemes. Its syntax can be reasonably well represented by functional relations on the clausal level. To paraphrase, morphological information for the head words of the constituent phrases is provided in feature characteristics of the corresponding wordforms. Therefore, there is no need to display morphological affixes on the syntactic layer as additional clues for syntactic chunking procedures and assigning them specific syntactic functions in the clause.

In general, for the manual syntactic annotation of GRUG monolingual trees, there were two options. The first is employed in the *INESS* project, an open system serving a range of research needs, offering an interactive, language independent platform for building, accessing, searching and visualizing treebanks [19].

For a Georgian sentence

ჯონს ლოდინის გარდა სხვა არაფერი შეეძლო
(English gloss: "John could do nothing, but to wait")

a simple XML code visualized by the INESS graphical viewer is depicted in Fig. 2. A German translation of the sentence in Fig. 3,

John konnte nichts anders tun als warten.

(English gloss: "John could do nothing, but to wait.")

can be viewed in Fig. 4.

As could be observed from the two tree structures, they lack morphological information presented in the XML. Moreover, the INESS graph is unable to display a clear linear order of the punctuation marks for the original input sentence, therefore visualizing it under the ROOT node at the top of the consequent graph.

An alternative for syntactic annotation is presented by *Synpathy*, a tool for the morphological and syntactic annotation of monolingual trees/oriented graphs [20] and used previously in the GRUG project [10] and subsequent work on Georgian data. It uses the so-called SyntaxViewer interface developed for the TIGER research project (Institut für Maschinelle Sprachverarbeitung, Universität Stuttgart[3]) and requires TIGER-XML format as input for syntactic tree visualization. This is especially useful since the TIGER-XML encoding format is compatible with multiple tools including those that we use for constituent alignment (Sect. 4).

[3] https://www.ims.uni-stuttgart.de/.

```
<s id="s69.36">
 <graph root="s69_502">
  <terminals>
   <t id="s69_1"word="ჯონს" pos="NE" morph="Dat.Sg." />
   <t id="s69_2"word="ლონდონის" pos="NN" morph="Gen.Sg." />
   <t id="s69_3" word="გარდა" pos="ENPOS" morph="--" />
   <t id="s69_4" word="სხვა" pos="IPRN" morph="Nom.Sg." />
   <t id="s69_5" word="არაფერი" pos="NPRN" morph="Nom.Sg." />
   <t id="s69_6" word="შეეძლო" pos="VMFIN" morph="3.Sg.Past.Ind" />
   <t id="s69_7" word="." pos="$." morph="--" />
 </terminals>
       <nonterminals>
         <nt id="s69_502" cat="S">
            <edge label="SB" idref="s69_510"/>
            <edge label="OO" idref="s69_500"/>
            <edge label="DO" idref="s69_4"/>
            <edge label="PD" idref="s69_501"/>
         </nt>
         <nt id="s69_500" cat="PP">
            <edge label="NK" idref="s69_2"/>
            <edge label="HD" idref="s69_3"/>
         </nt>
         <nt id="s69_501" cat="VP">
            <edge label="NG" idref="s69_5"/>
            <edge label="HD" idref="s69_6"/>
         </nt>
         <nt id="s69_510" cat="PN">
            <edge label="PNC" idref="s69_1"/>
         </nt>
       </nonterminals>
     </graph>
</s>
```

Fig. 2. An excerpt of the XML code for the Georgian sentence in the TIGER format.

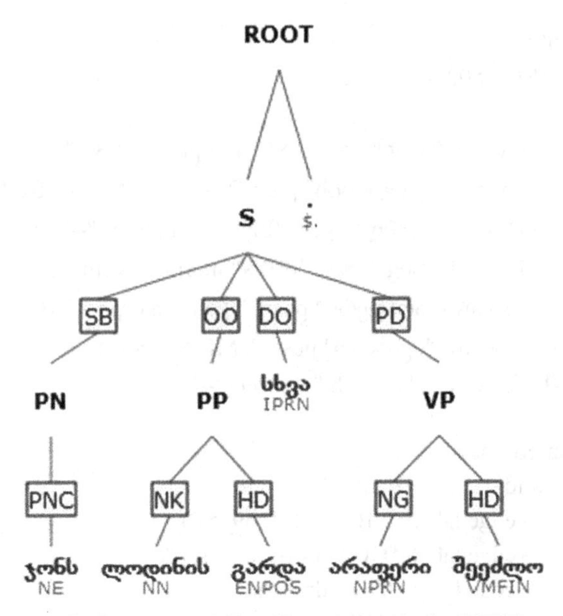

Fig. 3. A syntactically annotated Georgian sentence in the INESS output format.

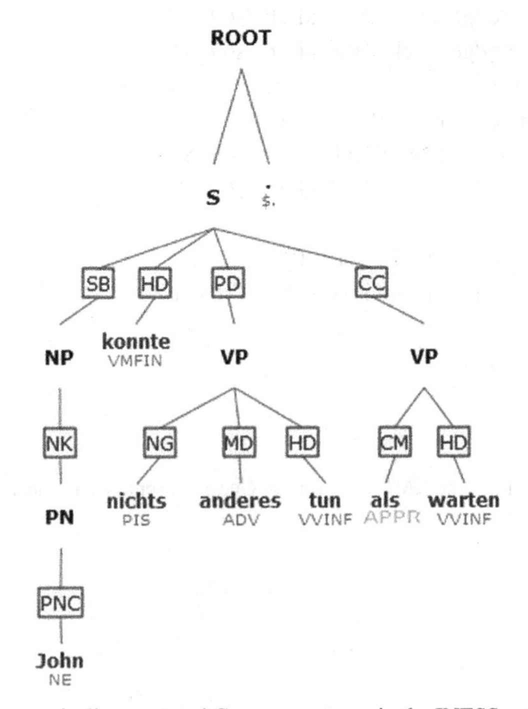

Fig. 4. A syntactically annotated German sentence in the INESS output format.

Synpathy is also equipped with a graphical shell for manual tree annotation. During this process, the shell is able to provide the terminal entities (tokens) of the plain input text with the necessary morphological features. In this way, morphological and syntactic annotation is combined into a single step, although the process can be time consuming. Nevertheless, both options are compatible with the TIGER annotation scheme implementation in the Synpathy environment.

For Georgian, the aforementioned two tasks were accomplished in separate steps. The Georgian text was tokenized, POS tagged and lemmatized by the FST morphological (morpho) parser [21, 22]. An output of the morphological transducer, before feeding to the Synpathy tool, is reformatted by a small Python script. It converts tagged and lemmatized tokens into a format required by the Synpathy processing engine. Manual annotation work is still ongoing; since TbiLLC 2019, the number of handcrafted syntactically annotated trees has reached 1,000 sentences.

For the German text, there are also two possible ways for building a syntactically annotated treebank. The first is an option outlined for the work with Synpathy—by hand, using the graphical shell, or compiling an XML file in TIGER format.

The **second possible scheme** for the tokenization, morphological analysis and automatic syntactic annotation of German text to create a treebank requires using *BitPar* [23], an efficient Bit-Vector-based CKY-style context-free parser that computes a compact parse forest representation of the complete set of possible analyses for large treebank grammars and long input sentences. The parser is particularly useful when all analyses are needed rather than just the most probable one. Its output must be converted from Penn format into TIGER-XML, requiring a relatively time-consuming manual post-editing step. Hence, for our data set – described in Sect. 4.1 – we have opted for using Synpathy.

4 Parallel Treebanking

As mentioned earlier, one way to validate the utility of a treebank is to apply it as a resource in one way or another. The same is valid of a parallel treebank, which can be invaluable to extract bilingual information, for use in comparative studies, NLP applications and the like.

As the Georgian treebank is in some respects modelled on the TIGER project and some parallel data exists, the decision was made to build a small constituent aligned parallel treebank comprising these languages. The intention is to use this effort as a testbed for gaining insights to help direct future work in this area.

The first step in obtaining a parallel treebank is to create a sentence aligned parallel corpus, the process of which is described in the next subsection. The words are also automatically aligned as this is a prerequisite for the constituent alignment solution that we applied.

We then create the tree aligned parallel treebank in three steps: (1) create a training and development test set by hand (2) train a model using a statistical tree aligner on the manual set (3) apply the model to the word aligned parallel corpus. We expand on these steps in the next few subsections.

4.1 Creating a Parallel Corpus

We follow convention, where documents are segmented into sentences and utilized as separate units within which further alignments with their equivalents in the other language are made below the sentence level on a 1:1 basis. By avoiding crossing sentence boundaries, the task is simplified, both for humans (in terms of effort) and computer systems (in terms of speed and memory requirements). Our data source is a selection of public domain texts from the legislative domain comprising 14,307 parallel sentences.[4,5] We also add the 308 sentence pairs from the set that we use as training data for our tree alignment experiments, the reason being that those sentence pairs have to be word aligned anyway and that we are using an unsupervised approach. See Sect. 4.3 for more information on the training data set.

Data preparation consists of sentence boundary detection, sentence alignment, and tokenization. The respective tokenizers described in Sect. 3 also performed the sentence boundary detection. Sentence alignment was performed manually,[6] after which the documents were converted and validated automatically. Validation consisted of steps such as checking for Georgian characters in expected German text and ensuring that each aligned document pair had the same number of lines. Conversion ensured that each sentence appears on a separate line in a format that is suitable for word alignment.

Next, we remove duplicate sentence pairs from the parallel corpus and randomize the set, after which we apply the `clean-corpus-n.perl` script[7] that is included with the Moses toolkit. This filters sentence alignments where there is a great difference in the word length of the source and target sides, as well as removing any sentence pairs where at least one contains fewer than or exceeds a certain number of words. Alignment performance can drop sharply as the length of a sentence pair increases. Hence, keeping to suggested parameter values, we set the maximum length at 80, noting that the longest sentence in the tree alignment set is 73. The final version of the parallel corpus used for word alignment is set at 13,425 sentence pairs.

4.2 Word Alignment

As previously mentioned, parallel treebanks are typically aligned on various levels in order to extract as much information as possible. In the case of phrase-based trees, a sentence-to-sentence alignment can contain sub-sentential alignments on the level of the constituent and word. The success of constituent alignment depends on features such as existing word alignments, POS and constituent tags, tree structure, as well as contextual and history features (depending on the algorithm, see, for example, [9]).

---

4 http://library.court.ge/login.php?geo&authorisation.

5 http://lawlibrary.info/ge/books/2020giz-ge-Georgische-Gesetze-auf-Deutsch.pdf.

6 Note that automatic sentence alignment using tools such as HunAlign usually speeds up the process, even with manual post-processing included. This approach will certainly be considered in future expansions of the parallel corpus. One notable advantage of the manual effort is that no valid parallel sentences are discarded, maximising the size of the – still rather small – data set.

7 https://github.com/moses-smt/mosesdecoder/blob/master/scripts/training/clean-corpus-n.perl.

Word alignments are best when trained on a large parallel corpus, typically in the millions of aligned sentence pairs. This, however, depends on factors such as sentence length, number of domains covered and the complexity of morphology, the latter of which can significantly increase the size of the vocabulary. As the quality of word alignment drops, so typically does that of constituent alignment (Chapter 5 in [32]).

Two approaches for word alignment are prevalent [25], depending on the nature of the application:

- The use of only alignments that have a high probability to be correct as deemed by the algorithm – typically the intersection of bi-directional alignment models – can be an option if the quality of the output needs to be high. This is generally seen as a viable option for tasks such as lexicon extraction. For tree alignment, the lesser number of word alignments may result in an insufficient weight value for a statistical aligner to align candidate constituent nodes. On the other hand, it may be less likely for wrong trees to be matched through their word alignments (see the well-formedness principle discussed in Sect. 4.3).
- Also allowing alignments with a lower probability is a prevalent option for statistical machine translation tasks. Some word alignment approaches, such as described hereafter, add alignments to the intersection according to specific heuristics. The union of bidirectional models can also be considered. The larger number of word alignments can be of sufficient weight for a statistical aligner to align candidate constituent nodes. However, it may be more likely that the wrong trees are matched in the process, as the higher number of incorrect word alignments may suggest more incorrect candidate constituent nodes.

In short, an important challenge is to find a suitable word alignment approach with just the right balance between high precision and high recall in order to optimize tree alignment quality.

For the experiments reported here, we utilize the well-established unsupervised word alignment approach implemented by the SyMGIZA++ tool [24]. SymGIZA++ is a tool based on GIZA++ [25] and MGIZA++ [26]. The original GIZA++ implements several different word alignment approaches, during which the EM [27] algorithm is applied to estimate the parameters of the models. MGIZA++ is a multi-threaded implementation of GIZA++, whereas SyMGIZA++ extends both programs by implementing a symmetrization mechanism that combines two directional models trained in parallel and using parameters of these models during training for computations of the next model.

Symmetrization of directed models allows for the application of heuristics, exploiting the fact that, for example, there is generally a high probability for the neighboring words of strongly aligned words to also be aligned.

Using the high-precision *intersection* alignment set as starting point, these refinement heuristics can be used to iteratively add these neighboring alignments according to specific conditions.[8] Traditionally implemented by the Moses statistical machine translation system [28], SyMGIZA++ also has the ability to do so.

[8] http://www.statmt.org/moses/?n=FactoredTraining.AlignWords.

We create word alignment models for a set of heuristics ranging from high precision to high recall, along with the intersection and union sets: *grow*, *grow-diag*, *grow-diag-final* and *grow-diag-final-and*, with the intention of analyzing their effects on tree alignment quality.

4.3 Tree Alignment

For our tree alignment experiments, we apply the toolbox Lingua-Align [29], which is a freely available and highly customizable statistical aligner using the maximum entropy learning method. It is able to process treebanks in TIGER-XML format and alignment files in Stockholm TreeAligner format (discussed next), and has been shown in previous work to perform well for language pairs such as Swedish/English [29] and Dutch/English [30].

The *Stockholm TreeAligner* (STA) [31] is a tool used to create, edit and view parallel treebanks in TIGER-XML format. Link types can be customized but defaults to "good" and "fuzzy", as described in the caption of Fig. 1, which is a screenshot from an aligned tree pair in STA. The tool generates an XML file containing node identification numbers and implicit alignments through grouping elements containing the IDs together.

We apply a data set that was manually built in STA for both words and constituents, partly containing data from the GRUG project and consisting of 178 sentence pairs from the legislative domain and 130 from general fiction-journalistic texts, adding up to 308. As [29] and [30] indicate that a training data size of this magnitude to be sufficient for producing balanced F-scores of over 70 and even 80 when applied to word aligned parallel corpora, at least for the language pairs of English/Dutch and English/Swedish, we decided to use this data set for our experiments.

After some feature engineering, we settled on a parameter configuration that was used in [32] to compare Lingua-Align with a transformation-based learning system discussed in the thesis. This uses all types of features described in [33], including those from the word alignment models that we created before. It uses the so-called *GreedyWellformed* alignment strategy, aligning candidate node pairs in a greedy fashion but only if one or more of the subtrees or their descendants do not share alignments with other subtrees. Ancestor nodes should also be aligned to each other. See [34] for more information on this so-called *well-formedness constraint*.

Next, we randomized the set and created folds for a ten-fold cross validation. 90% is used for training and the rest for testing. Each fold's last 31 sentence pairs are different, except in two cases where the count amounts to 30. In this way, the remainder of the division (308/10) is smoothed out over sets.

The next step was the implementation of a script to perform Lingua-Align training and testing using multiple cores in parallel, in order to speed up processing. To the output of each Lingua-Align model, we also apply a set of rule-based heuristics described in [30] that uses a more relaxed form of the well-formedness constraint and is applied in a greedy bottom-up fashion. In [30] and [32], it was shown that these heuristics increased both the recall as well as balanced F-scores of high-precision Lingua-Align models.

Application of the heuristics is centered around two concepts: the well-formedness constraint and subtree similarity. An unlinked subtree pair, with non-terminal nodes as their roots, can be viewed as a candidate node pair to be possibly linked. Each of these trees has a certain number of terminal nodes that it dominates, i.e. the phrases that it represents. If the terminal count difference between the trees is too great, it would seem less likely that they should be linked. On the other hand, if the terminals that they dominate share a great number of word alignments, it would seem more likely that they should be linked, but less so if some of those alignments link to words that are dominated by other subtrees. The regular well-formedness principle allows for none such "outside" alignments to happen, whereas with the heuristics, the rule can be relaxed by allowing, for example, one or two fuzzy word alignments.

With the subtree similarity measures, four features are important: the count difference between the source and target-side terminals, the ratio of the counts, the count difference of aligned and non-aligned terminals in both subtrees, and the ratio of those counts. If one assumes that the count differences and the ratios are equally important, we use the geometric mean of those scores. If one assumes that in the case of larger count differences, the ratio should be less important, we subtract a weighted difference from the ratio using a manually tuned normalization value.

For our experiments, we have applied both, which we call *geo* and *not-geo*. Similarly, we have also applied relaxed well-formedness where a single fuzzy word alignment is allowed to the "outside" (*fuzzy1*), as well as separate output sets where two are allowed (*fuzzy2*).

Since these heuristic alignments are performed greedily, they can be applied in two directions – from source tree to target tree or vice versa – of which we can calculate the intersection and union, with a greater focus on precision and recall, respectively. These different approaches can lead to a large number of different combinations of heuristic outputs for which we use identification names such as *LA+fuzzy2_not-geo_union* in order to track them (see Table 1).

4.4 Quantitative Evaluation

The model trained on each fold is applied to the aforementioned held-out section at the end of the fold. For evaluation against the gold standard – i.e. the original manually built data set – we use a script that is available with Lingua-Align. Our current intention is simply to optimize balanced F-scores for constituent alignment, with no direct application to tasks as of yet. Therefore, only non-terminal alignments are evaluated.

We run the ten-fold cross validation for the intersection and union of the directional models, as well as for each of the aforementioned word alignment heuristics: *grow*, *grow-diag*, *grow-diag-final* and *grow-diag-final-and*. For each, we use the same Lingua-Align parameters and heuristics, as well as the four different parameter values for the tree alignment heuristics to be applied afterwards. For other parameters, such as the aforementioned normalization factor, we use the default values. Table 1 presents the best and second-best averages across all ten-fold cross validation sets for precision, recall and balanced F-score.

Both the two best precision scores and two best F-scores are achieved by using the Lingua-Align model only. Heuristics applied afterwards increases recall by 3.029 (*gdf*) and 2.138 (*grow*) respectively, but precision drops by 4.829 (*gdf*) and 4.598 (*grow*). As the Lingua-Align models already have relatively high recall values, this is not too

Table 1. Best average precision, recall and F-score for different word and tree alignment approaches. The best scores are in bold and the second best in italics. *Word* refers to which word alignment heuristic was used, where *gdf* refers to *grow-diag-final*. *Tree* refers to whether the best score was achieved using Lingua-Align (LA) only, or also by applying a tree alignment heuristic to its output. *fuzzy2_not-geo_union* means allowing two fuzzy word alignments to the "outside", not using the geometric mean (i.e. the formula with the normalization factor) and the union of the heuristic applied bidirectionally.

Best	Word	Tree	Precision	Recall	F-score
Precision	gdf	LA	**72.627**	73.409	72.965
	grow	LA	*72.359*	73.653	72.939
Recall	gdf	LA+fuzzy2_not-geo_union	67.798	**76.438**	71.804
	grow	LA+fuzzy2_not-geo_union	67.761	*75.971*	71.576
F-score	gdf	LA	72.627	73.409	**72.965**
	grow	LA	72.359	73.653	*72.939*

unexpected. Also not surprising is the fact that the union of the heuristic with the most relaxed well-formedness setting (two fuzzy links) leads to the highest recall.

All the best scores are based on *gdf* alignments and all the second best on *grow* alignments. This shows that word alignment heuristics have a consistent effect no matter if we only run Lingua-Align or also apply tree alignment heuristics to the output. We also note that especially F-scores are very close to each other. This is consistent throughout the evaluation data, also of those not shown here. This suggests that word alignments have only a small effect and tree alignment heuristics almost no effect. We hypothesize that using a larger parallel corpus may lead to better word alignment and hence higher tree alignment precision, and a training curve may point to that trend, but this is left for future work.

Adding all the output alignments of the folds of the two "best" data sets together, we can extract some relevant statistics as found in Table 2.

The total number of tree alignment heuristic alignments added to the best recall set is 209, 11% of the total number of constituent alignments. In other data sets where the sole focus was not on maximizing recall, the number is lower. This reflects the relatively little effect that it has on the current set of alignments.

The number of non-terminals aligned in the gold standard and the other sets are relatively equal. However, note the large difference in aligned terminals count in the gold standard between the source and target trees (1241) as opposed to in the other two sets (4). This suggests an asymmetry between the two treebanks for which a unidirectional model may be a better fit.

The number of aligned terminal nodes in the automatic sets is much lower than in the gold standard, with a difference of 2,140. This suggests that many word alignment probabilities are not high enough to be used in the tree alignment data set, as Lingua-Align uses a default threshold for such alignments. Apart from the small size of the

Table 2. For each data set, comprising the combined output of each ten-fold cross validation, here we present some statistics concerning the number of terminal and non-terminal alignments in the sets (*Term alignments* and *NT alignments*), as well as the number of aligned terminals and aligned non-terminals in both the source-side tree (German; *stree*) and target-side tree (Georgian; *ttree*). Note that all three data sets are for the word alignment heuristic *grow-diag-final*, which explains the same number of terminal alignments and aligned terminals across different sets.

Data set	Term alignments	NT alignments	Term aligned stree	NTs aligned stree	Term aligned ttree	NTs aligned ttree
Gold	5,950	1,701	5,353	1,675	4,112	1,662
Precision+F-score	3,810	1,665	3,805	1,663	3,809	1,660
Recall	3,810	1,857	3,805	1,816	3,809	1,836

parallel corpus, the existence of a subset from the literature genre that is not represented by the parallel corpus is a possible factor here.

4.5 Qualitative Evaluation

Finally, we present two examples of tree pairs that illustrate some of the aforementioned issues. In Figs. 5 and 6, we show a small tree pair from the gold standard and its equivalent from the data set with the best F-score, respectively.

German gloss: "Sie unterhalten sich mit ihm über ihr Problem."
Georgian gloss: "ისინი მასთან თავიანთ პრობლემაზე საუბრობდნენ."
English gloss: "They talk to him about their problem."

In the gold standard version (Fig. 5), the Georgian PP is aligned with the second German PP. In Fig. 6, it is aligned with the first PP. Although Lingua-Align learns that PPs are more likely to align with other PPs, it is very likely that the alignments linking the wrong words have significantly contributed to the fact that the first PP has been chosen as the correct candidate.

Note also the relative lack of word alignments in the automatic output. There are only 1:1 alignments and the second PP dominates words that have no alignments at all. In general, candidate nodes cannot be aligned if there is no word alignment evidence to guide decision making.

In Fig. 7, we present an example of a subtree pair that was correctly aligned by a rule.

German gloss: "Er liefe wie ein richtiger Läufer."
Georgian gloss: "ის ნამდვილი მორბენალივით დარბოდა."
English gloss: "He ran like a real runner."

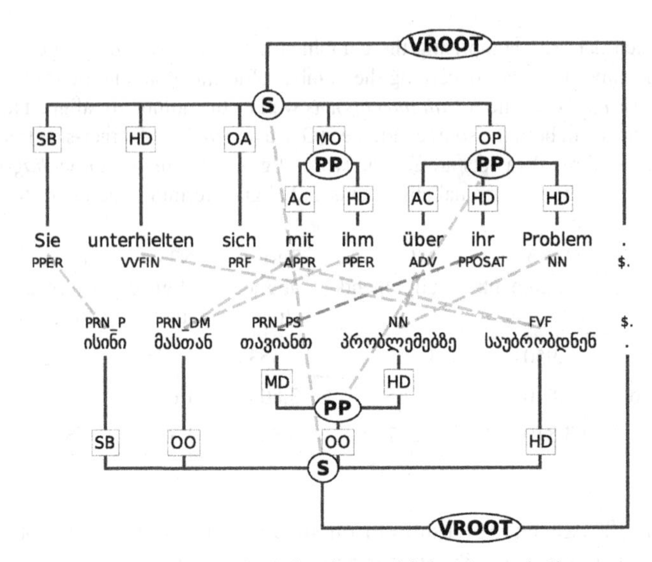

Fig. 5. A tree pair from the gold standard showing multiple n:1 word alignments. See Fig. 6 for the equivalent from the data set that achieved the best F-score.

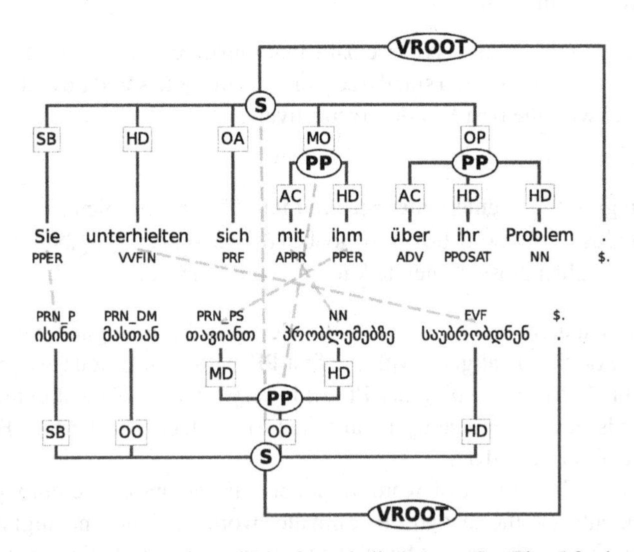

Fig. 6. An output tree pair showing a lack of word alignments. See Fig. 5 for its gold standard

The gold standard tree pair is very similar, but Lingua-Align did not align the German NP with the Georgian AVP, even though the trees are well-formed. A significant difference was probably the fact that they only share a fuzzy and a good word alignment. In the gold standard, "wie ein richtiger" all align with the same Georgian word, with "richtiger" also aligning with the same word as presented here. The well-formedness and perceived similarity of the subtrees led to the algorithm aligning the constituent nodes

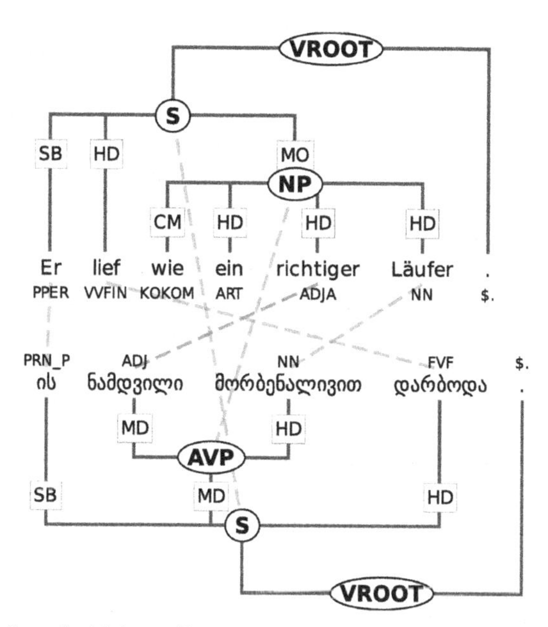

Fig. 7. A tree pair from the high-recall output data set showing a correct non-terminal alignment added by a rule (NP/AVP).

in question. With the relaxed rules, a fuzzy link may, for example, have gone from one of the two Georgian words to "lief", and the trees would still have been aligned.

Note also that Georgian words can sometimes equate to multi-word German phrases. This could be problematic since longer phrases tend to occur less frequently in corpora. It might therefore require a larger corpus for models to learn how to align certain Georgian words.

5 Discussion and Future Work

We have presented our efforts in creating treebank-related resources for the Georgian language. For building a full-scale Georgian syntactic parser, we intend to make use of the developed vanilla CFG that was extracted from the monolingual Georgian treebank. It will be utilized for finding optimal morphological features/preterminals for implementation in the probabilistic CFG parser. The reason for such a decision is the advantage of a deterministic part-of-speech tagger that can produce a morphologically annotated Georgian corpus achieving almost 100% accuracy after manual disambiguation [35]. Moreover, it is able to annotate the tokens with just POS tags, or also with morphological information using features such as case, number, person and tense.

At the first stage, the most successful supervised constituent parsers apply a probabilistic context-free grammar (PCFG) to extract possible parses. The n-best list parsers keep just the 50–100 best parses according to the PCFG. These feature templates exploit atomic morphological features and achieve improvements over the standard feature set.

These methods use a large feature set—usually a few million features—and are engineered for English, in some cases requiring distributed computing solutions (such as [36]) for proper application.

The innovative aspect of the proposed approach is a unique procedure for finding the optimal set of preterminals by merging morphological feature values. The main advantage of this methodology over previous undertakings is the performance speed—it operates inside a PCFG instead of using a parser as a black box with retraining for every evaluation of a feature combination—and it can investigate particular morphological feature values instead of removing a feature with all of its values.

The aforementioned GRUG project has provided some insights on how to achieve automatic tree alignment with German on a larger scale. Here, our present experiments highlighted some important issues. [30, 32] and [33] have shown that tree alignment F-scores approaching and surpassing 80 are possible. However, those use cases involved large parallel corpora with millions of sentence pairs. It is therefore desirable to increase the size of the Georgian/German parallel corpus both in terms of size as well as variety. There are also techniques to extract parallel sentences from comparable corpora, which are easier to obtain [37].

Apart from better feature engineering, more recent tools may also be used, or existing ones adapted. For example, the EFMARAL word alignment tool has been shown to be efficient for statistical machine translation [38]. Lingua-Align could be adapted to use the morpheme attribute as a feature. The transformation-based learning system presented by [32] could also be adapted and combined with existing solutions to improve tree alignment results. And of course, different machine learning approaches including deep learning methods may be used to improve current results.

In Figs. 1 and 5, as well as in Table 2, we have shown that n:1 alignments from German to Georgian may be common. It might be useful to consider a different segmentation approach, such as presented by [39], in order to improve alignment convergence, but this has to be guided by the approach taken with the parser.

Finally, a future parallel treebank can also be extended using other layers of data, such as dependency relations, valency frames, anaphoric resolutions and semantic knowledge, which in turn may provide the necessary clues to improve the parallel treebank as a knowledge source for linguistics and NLP applications.

Acknowledgments. This research was supported by **the Shota Rustaveli Georgian National Science Foundation Grant FR-18–15744**. We appreciate the project members Nunu Kapanadse, Nino Kimeridse, Ketevan Schekiladse, and Natia Putkaradze for their contribution to the presented research.

References

1. Fraser, A., Schmid, H., Farkas, R., Wang, R., Schütze, H.: Knowledge sources for constituent parsing of German, a morphologically rich and less-configurational language. In: Computational Linguistics, vol. 39, Issue 1. MIT Press Cambridge, Ma, USA (2013)
2. Maamouri, M., Bies, A., Buckwalter, T., Mekki, W.: The Penn Arabic Treebank: building a large-scale annotated Arabic corpus. In: The NEMLAR Conference on Arabic Language Resources and Tools, pp. 102–109 (2004)

3. Marcus, M., Santorini, B., Marcinkiewicz, M.A.: Building a large annotated corpus of English: the Penn Treebank. Computational Linguistics, vol. 19 (1993)
4. Xue, N., Xia, F., Chiou, F.-D., Palmer, M.: The Penn Chinese TreeBank: phrase structure annotation of a large corpus. Nat. Lang. Eng. **11**(2), 207–238 (2005). https://doi.org/10.1017/S135132490400364X
5. Bejček, E., et al.: Prague Dependency Treebank 3.0. Data/software. Univerzita Karlova v Praze, MFF, ÚFAL, Prague, Czech republic (2013). http://ufal.mff.cuni.cz/pdt3.0/
6. Tinsley, J.: Resourcing Machine Translation with Parallel Treebanks. A dissertation submitted in fulfilment of the requirements for the award of Doctor of Philosophy (Ph.D.) Dublin City University School of Computing. Supervisor: Prof. Andy Way (2009)
7. Vandeghinste, V., et al.: Parse and corpus-based machine translation. In: Spyns, P., Odijk, J. (eds.) Essential Speech and Language Technology for Dutch. TANLP, pp. 305–319. Springer, Heidelberg (2013). https://doi.org/10.1007/978-3-642-30910-6_17
8. Harris, B.: Bi-Text, a new concept in translation theory. Lang. Monthly **54**, 8–10 (1988)
9. Tiedemann, J.: Bitext alignment. In: Hirst, G. (ed.) Synthesis Lectures on Human Language Technologies. Morgan & Claypool Publishers (2011)
10. Kapanadze, O.: Multilingual GRUG Parallel TreeBank—Ideas and Methods, 52 p. LAMBERT Academic Publisher (2017). ISBN-13: 978-3-330-34810-3. EAN: 9783330348103
11. Smith, G.: A Brief Introduction to the TIGER Treebank, Version 1. Potsdam Universität (2003)
12. Brants, S., et al.: TIGER: linguistic interpretation of a German corpus. Res. Lang. Comput. **2**(4), 597–620 (2004)
13. Vincze, V., Szauter, D., Almási, A., Móra, G., Alexin, Z., Csirik, J.: Hungarian dependency treebank. In: Proceedings of the Seventh International Conference on Language Resources and Evaluation (LREC 2010). Valletta, Malta (2010)
14. Simov, K., Osenova, P., Simov, A., Kouylekov, M.: Design and implementation of the Bulgarian HPSG-based treebank. J. Res. Lang. Comput. **2**(4), 495–522 (2005)
15. Pretkalnina, L., Rituma, L.: Syntactic issues identified developing the Latvian treebank. In: Baltic HLT (2012)
16. Bengoetxea, K., Gojenola, K. Application of different techniques to dependency parsing of Basque. In: Proceedings of the NAACL/HLT Workshop on Statistical Parsing of Morphologically Rich Languages (2010)
17. Megyesi, B., Dahlqvist, B., Csató, E.A., Nivre, J.: The English-Swedish-Turkish Parallel Treebank. In: Proceedings of the International Conference on Language Resources and Evaluation, LREC2010. Valletta, Malta (2010)
18. Rios, A., Göhring, A., Volk, M.: Quechua- Spanish Parallel Treebank. In: 7th Conference on Treebanks and Linguistic Theories, Groningen. The Netherlands (2009)
19. INESS. http://clarino.uib.no/iness/
20. Synpathy: Syntax Editor. Manual – Nijmegen: Max Planck Institute for Psycholinguistics. The Netherlands (2006)
21. Kapanadze, O.: Finite state morphology for the low-density georgian language. FSMNLP 2009 Pre-proceedings of the Eighth International Workshop on Finite-State Methods and NLP. Pretoria, South Africa (2009)
22. Kapanadze, O.: Describing georgian morphology with a finite-state system. In: Yli-Jura, A., et al. (eds.): Finite-State Methods and Natural Language Processing 2009, Lecture Notes in Artificial Intelligence, vol. 6062, pp.114–122. Springer-Verlag, Berlin (2010). https://doi.org/10.1007/978-3-642-14684-8_12
23. Helmut, S.:. Efficient parsing of highly ambiguous context-free grammars with bit vectors. In: Proceedings COLING, pp. 162–168 (2004)

24. Junczys-Dowmunt, M., Szał, A.: SyMGiza++: Symmetrized word alignment models for machine translation. In: Bouvry, P., Klopotek, M.A., Leprévost, F., Marciniak, M., Mykowiecka, A., Rybinski, H., (eds.) Security and Intelligent Information Systems (SIIS), volume 7053 of Lecture Notes in Computer Science, pages 379–390. Springer, Warsaw, Poland (2012). http://emjotde.github.io/publications/pdf/mjd2011siis.pdf
25. Och, F.J., Ney, H.: A systematic comparison of various statistical alignment models. Comput. Linguist. **29**(1), 19–51 (2003)
26. Gao, Q., Vogel, S.: Parallel implementations of word alignment tool. In: Proceedings of SETQA-NLP, pp. 49–57 (2008)
27. Dempster, A.P., Laird, N.M., Rubin, D.B.: Maximum likelihood from incomplete data via the EM algorithm. J. Roy. Statist. Soc. Ser. B **39**(1), 1–38 (1977)
28. Koehn, P., et al.: Moses: Open Source Toolkit for Statistical Machine Translation. Annual Meeting of the Association for Computational Linguistics (ACL), demonstration session. Czech Republic, Prague (2007)
29. Tiedemann, J.: Lingua-align: an experimental toolbox for automatic tree-to-tree alignment. In: Proceedings of the 7th International Conference on Language Resources and Evaluation (LREC'2010) (2010). http://www.lrec-conf.org/proceedings/lrec2010/pdf/144_Paper.pdf
30. Kotzé, G., Vandeghinste, V., Martens, S., Tiedemann, J.: Large aligned treebanks for syntax-based machine translation. Lang. Resour. Eval. **51**(2), 249–282 (2016). https://doi.org/10.1007/s10579-016-9369-0
31. Lundborg, J., Marek, T., Mettler, M, Volk, M.: Using the Stockholm TreeAligner. In: 6th Workshop on Treebanks and Linguistic Theories. Bergen (2007)
32. Kotzé, G.: Complementary Approaches to Tree Alignment: Combining Statistical and Rule-Based Methods. PhD thesis. University of Groningen (2013). https://doi.org/10.13140/RG.2.2.31880.67841
33. Tiedemann, J., Kotzé, G.: A discriminative approach to tree alignment. In: NLP Methods and Corpora in Translation, Lexicography, and Language Learning 2009, pp. 33–39. Borovets, Bulgaria (2009)
34. Hearne, M., Tinsley, J., Zhechev, V., Way, A.: Capturing translational divergences with a statistical tree-to-tree aligner. In: Proceedings of the 11th International Conference on Theoretical and Methodological Issues in Machine Translation (TMI '07), pp. 85–94. Skövde, Sweden (2007)
35. May, P., Ehrlich, H.-C., Steinke, T.: ZIB structure prediction pipeline: composing a complex biological workflow through web services. In: Nagel, W.E., Walter, W.V., Lehner, W. (eds.) Euro-Par 2006. LNCS, vol. 4128, pp. 1148–1158. Springer, Heidelberg (2006). https://doi.org/10.1007/11823285_121
36. Foster, I., Kesselman, C.: The Grid: Blueprint for a New Computing Infrastructure. Morgan Kaufmann, San Francisco (1999)
37. Hangya, V., Braune, F., Kalasouskaya, Y., Fraser, A.: Unsupervised parallel sentence extraction from comparable corpora. In: Proceedings of IWSLT (2018)
38. Ostling, R., Tiedemann, J.: Efficient word alignment with Markov Chain Monte Carlo. Prague Bull. Math. Linguist. **106**, 125–146 (2016)
39. Kotzé, G., Wolff, F.: Syllabification and parameter optimisation in Zulu to English machine translation. S. Afr. Comput. J. **57**, 1–23 (2015). https://doi.org/10.18489/sacj.v0i57.323. (South African Institute of Computer Scientists and Information Technologists (SAICSIT))

Bridging the Gap Between Formal Semantics and Neurolinguistics: The Case of the N400 and the LPP

Ralf Naumann(✉) and Wiebke Petersen

Heinrich-Heine-Universität Düsseldorf, Düsseldorf, Germany
{naumann,petersen}@hhu.de

Abstract. We outline a formal framework that combines results from neurolinguistic research on two ERP components, the N400 and the LPP, with formal semantics. At the semantic level we combine de Groote's continuation-based version of Montague semantics with van Eijck's Incremental Dynamics enriched with frames. We analyze them in terms of complex properties that apply both to the semantic and the discourse level and which combine world knowledge with syntagmatic and paradigmatic relationships. DPs and common nouns are interpreted as sequences of update operations which are related to these properties. In turn, each ERP component is correlated to at least one update operation. Whereas the N400 is related to success of these operations and the way they reduce uncertainty about the situation described by a discourse, the LPP is correlated to the failure to execute these updates.

1 The N400 and the Late Posterior Positivity

An ERP-component is the summation of the post-synaptic potentials of large ensembles (in the order of thousands or millions) of neurons synchronized to an event. When measured from the scalp, continuous ERP waveforms manifest themselves as voltage fluctuations that can be divided into components. A component is taken to reflect the neural activity underlying a specific computational activity carried out in a given neuroanatomical module. The N400 component is a negative deflection in the ERP signal that starts around 200–300 ms post-word onset and peaks around 400 ms. Besides the N400 component, there is a set of later positive-going ERP components that is visible at the scalp surface between approximately 500 and 1000 ms. The most prominent element is the late posterior positivity (LPP, also known as semantic P600), which is maximal at parietal and occipital sites.

The two most prominent interpretations of the underlying neuro-cognitive function of the N400 are the integration and the retrieval view. On the integration account, the N400 amplitude 'indexes the effort involved in integrating the word

The research was supported by the German Science Foundation (DFG) funding the Collaborative Research Center 991.

© The Author(s), under exclusive license to Springer Nature Switzerland AG 2022
A. Özgün and Y. Zinova (Eds.): TbiLLC 2019, LNCS 13206, pp. 79–112, 2022.
https://doi.org/10.1007/978-3-030-98479-3_5

meaning of the eliciting word form with the preceding context, to produce an updated utterance interpretation', [DBC19]. On the retrieval/access account 'the N400 amplitude reflects the effort involved in retrieving from long-term memory conceptual knowledge associated with the eliciting word which is influenced by the extent to which this knowledge is cued (or primed) by the preceding context, [DBC19]. What is left open by the above characterization is which properties of words and the context underly the N400 amplitude. Five prominent properties that have been suggested are (i) semantic features, (ii) plausibility, (iii) semantic similarity, (iv) selectional restrictions and (v) schema-based knowledge. However, taken individually, none of the five features can explain the N400 amplitude.

Evidence for semantic features as being correlated with the N400 amplitude comes from the fact that the correlation between the N400 amplitude and the cloze probability, that is the probability of a target word to be the best completion in a cloze test, is not monotone.

(1) They wanted to make the hotel look more like a tropical resort. So along the driveway they planted rows of palms/pines/tulips. [FK99]

In (1) 'pine' but not 'tulips' comes from the same semantic category 'tree' as the best completion 'palms'. Though 'pines' and 'tulips' have the same low cloze probability (< 0.05), their N400 amplitudes differ. Within category violations (pines) elicit smaller N400 amplitudes than between category violations (tulips). Federmeier and Kutas argue that this result suggests that it is feature overlap like being tall or having a similar form that affords within category violations a processing benefit relative to between category violations, [FK99, p. 485].

However, feature overlap with the best completion is not without exceptions, as shown by the following example.

(2) The wreckage of the sunken ship was salvaged by the victims ... [PK12].

Though the critical word 'victims' shares few semantic features with the best completion 'divers', no N400 effect is observed.

A second candidate is plausibility which can be quantized by offline rating tasks using, e.g., a Likert scale. Plausibility is often related to the integration view of the N400. The less plausible a resulting interpretation is the more difficult it is to integrate the critical word in the preceding context. Evidence for the role of plausibility comes from the fact that in the Federmeier & Kutas study best completions elicited the smallest N400 amplitude and the highest plausibility ratings. Between category violations elicited the largest N400 amplitudes and got the lowest plausibility ratings. Within category violations were intermediate on both variables, [FK99, p. 486]. However, in semantic illusion data like that in (3) no N400 effect is observed although the sentence has an implausible interpretation.

(3) The fox that on the poacher hunted ....

A third candidate is semantic similarity. On this account the N400 amplitude is modulated by the degree to which a critical word in a target sentence is semantically related to the words preceding it in the context. One way of quantifying semantic similarity is to use Latent Semantic Analysis. On this account pairwise term-to-document semantic similarity values (SSVs) are extracted from corpora (see [KBW0] for an application). Semantic similarity underlies the Retrieval-Integration model of [VCB18]. One of its strengths is that it can explain semantic illusion data as given in (3). As there is a semantic relation between the arguments preceding the verb ('fox', 'poacher') and the verb itself ('hunted') no N400 effect is expected for the verb.

However, similar to both the notions of semantic feature overlap and plausibility, there are counterexamples to the thesis that the N400 amplitude is (monotonically) related to the corresponding LSA value. Kuperberg et al. [KPD11] showed that the degree of causal relationship in three-sentence scenarios with matched SSVs influences the N400 amplitude: highly related $<$ intermediately related $<$ causally unrelated. The authors conclude that it is the situation model constructed from the context (message-level meaning) that influences semantic processing of the critical word and not semantic relatedness. Similarly, [KBW0] could show an influence of high- versus low-constraint contexts on the N400 amplitude for controlled SSVs.

A fourth property is related to selection restrictions imposed by verbs. Each verb imposes constraints on its arguments that are independent of the context in which it is used. One prominent example of such a constraint is animacy. Violations of selectional restrictions (typically) evoke robust N400 effects that are larger than those for non-expected words that do not violate these restrictions. Furthermore, the amplitude of the N400 in case of such violations is not modulated by semantic similarity measured by LSA.

(4) The pianist played his music while the bass was strummed by the drum / coffin during the song.

In (4) taken from [PK12] both 'drum' and 'coffin' violate the animacy constraint imposed by 'strum' on its actor argument. Furthermore, though 'drum' is semantically more related to the preceding context than 'coffin' using LSA (0.18 vs. 0.01), the two N400 amplitudes did not differ. By contrast, the N400 amplitude evoked by words that do not violate selectional restrictions is modulated by semantic relatedness quantized by LSA.

(5) The pianist played his music while the bass was strummed by the drummer / gravedigger during the song.

Similar to the case of 'drum' and 'coffin', the semantic relatedness to the preceding context differs: 0.18 for 'drummer' vs. 0.00 for 'gravedigger'. However, in contrast to (4), in (5) the N400 amplitude for the semantically unrelated 'gravedigger' is larger than that for 'drummer'.

However, violations of selection restrictions need not always produce an N400 effect, which brings us to the fifth property that is related to schema-based knowledge.

(6) A huge blizzard swept through town last night. My kids ended up getting the day off from school. They spent the whole day outside building a big jacket in the front yard.

In (6) 'jacket' violates the selection restriction (animacy) imposed by the verb 'build'. Although a robust (large) N400 effect is expected due to the restriction violation only an attenuated N400 is measured compared to the expect 'snowman'. This data suggests that the N400 is also modulated by schema-based knowledge about a particular scenario that is depicted by a discourse (cf. [PK12], [KBW0]). This knowledge is based on a semantic network of interrelated concepts and goes beyond the information provided by words in a single sentence. For example, in (6) a winter scene involving children is described. The corresponding semantic network is related to the clothes of the children which are most likely such that they keep warm, a condition satisfied by jackets. Evidence for such a dependency of the N400 on schema-based knowledge comes from the fact that the attenuation of the N400 effect of such examples depends on the context in which the target sentence containing the critical word is embedded. Leaving this context out, e.g. the two sentences preceding the target sentence in (6), leads to a robust N400 effect on the critical word.

The second ERP component that we are considering here is the Late Posterior Positivity (LPP) which is usually associated with the impossibility of an interpretation. Evidence for this functional interpretation comes from examples like those in (7)

(7) a. He spread the warm bread with socks.
b. For breakfast, the eggs would eat ...
c. The lifeguards received a report of sharks right near the beach. Their immediate concern was to prevent any incidents in the sea. Hence, they cautioned the swimmers / trainees / drawer ....

In each case an LPP is elicited due to the violation of a selection restriction that blocks a direct interpretation. An LPP is not only elicited if there is a violation of selection restrictions, but also if direct interpretation is blocked differently. One example are so-called reversal anomalies that are a subset of the semantic illusion data.

(8) The restaurant owner forgot which waitress the customer had served.

In (8) no selection restriction is violated as both arguments satisfy the animacy constraint imposed by 'serve'. What is unexpected and explains the elicited LPP is the assignment of thematic roles. Instead of the waitress being the actor and the customer being the theme, the roles are reversed. An LPP can be also elicited on the discourse level:

(9) John left the restaurant. Before long, he opened the menu ...

In (9) it is the order of events which is unexpected. The first sentence triggers schema-based knowledge about a restaurant which includes particular kinds of actions that are partially ordered. This ordering requires the opening of the menu to occur before the leaving of the restaurant.

In summary, an LPP is elicited whenever a direct interpretation is impossible due to selection restrictions or world knowledge about thematic roles or schema-based knowledge.

2 The Functional Interpretation of the N400 and the LPP

Our main theses concerning the two ERP components are: (i) Two principle levels of representation must be distinguished: situation models (global) and event models (local); (ii) predictions are related to the level of situation models whereas integration operations are related to both levels; they are based on (a) syntagmatic relationships, (b) semantic features and (c) world knowledge; (iii) the N400 is directly related to predictions and, therefore, to the level of situation models; in addition, it is related to integration at the level of situation models but not to integration at the level of event models; its amplitude is modulated by a complex semantic-cognitive property and a pragmatic (discourse) property related to linking, i.e. the referent of the critical word needs to be linked to an object that is already part of the current situation model; and (iv) the LPP is related to failure at the level of integration at the situation model and at the event model.[1]

In this article we will pursue two aims that are closely related. On the one hand, we will combine functional interpretations of the N400 and the LPP that have been given in the neurolinguistic literature (access and integration) with concepts used in formal semantic theories (e.g. update operations). On the other hand, we will outline an extension of a dynamic semantics in which these functional interpretations can be incorporated. For example, we interpret access as the introduction of objects or features into the model and integration as an update operation. Predictions are modelled in terms of probability distributions on frames.

[1] We do not assume that there is a monophasic N400 activity. Rather, semantic processing in the brain is always biphasic with the first phase indexed by the N400 and the second phase indexed by late positivities like the LPP. Whereas N400 activity is always related to the situation model, late positivities are related to the situation model and the event model. Hence, integration at the (local) event model is not captured by N400 activity but only by the late positivities like the LPP. Thus, there will always be activity in the post N400 time window related to this kind of integration. We are indebted to one reviewer for stressing this relation between activity in the N400 time window and post-N400 time window.

2.1 Predictions and Situation Models

We follow growing evidence that predictions are based on scripts. A script is a standardized sequence of events that together make up a particular complex situation and that describes some stereotypical human activity such as going to a restaurant or visiting a doctor. Script knowledge is common knowledge that is shared between speakers of a community or culture. This knowledge comprises information about sorts of events and the sorts of objects typically involved in the realization of a script. In addition, it includes information about the temporal and causal relations between the events and which sorts of objects are related to which sorts of events. Consider the following example from [MTD+17].

(10) a. The waiter brought the ...
b. We got seated. The waiter brought the ...
c. We ordered. The waiter brought the ...

These examples are partial descriptions of a restaurant script. Knowledge about such a script includes knowledge about events like ordering, bringing and eating as well as objects participating in these events like waitresses, food and bills. One possible temporal ordering of the events is: enter, being seated, bring menu, order food, bring food, ask for bill, bring bill, pay bill, leave. Examples like (10) show that script knowledge not only constrains the sort of objects participating in an event relative to a particular thematic role but that the sort of the object also depends on the temporal placement of the event in the temporal order specified by the script. Theoretically, a bringing event as in (10-a) can be located at any of the three possibilities in the temporal order. Hence, expected objects are (instances of) food, the menu or the bill. By contrast, in the context of (10-b) the bringing is temporally located after the being seated so that the menu is the most expected object. Finally, in (10-c) the expected object is the food because the bringing event is temporally located after the ordering. The two above examples show that script knowledge can impose additional constraints on objects and events by constraining for particular events and objects participating in them. On the other hand, a context can constrain strongly for a particular situation model but not for a specific event or a specific object that is related to this event [KJ16]. For example in the blizzard example in (6), the jacket is not expected as a theme of the building event. Semantic processing of the critical word 'jacket' is facilitated because the semantic features associated with its interpretation are expected relative to a particular situation model (winter scene) and an object already introduced into this model (children) though these features are (highly) unexpected or even anomalous relative to the current event model (building). Objects of sort 'jacket' are expected as clothes of the children because the situation model 'winter scene' expects clothes that keep warm.

The important point about script knowledge is that upon its instantiation it activates a network of individuals and events as well as relations between these objects. Given such a network, predictions are not restricted to the current event (e.g. 'What is being brought?') or the next event (e.g. 'Which event is mentioned next and which objects participate in it?') but can target both objects that have already been introduced into the current situation model ('What were the children wearing?') as well as objects that are likely to be encountered in the continuation of the description of the situation (e.g. the bill and the leaving event). Two principle cases need to be distinguished: (a) to what degree are the features (properties) of a newly introduced object expected given the current partial description of a situation model?, and (b) can a newly introduced object be related to an object that has already been introduced into the current situation model?

More formally, suppose that a context specifies a situation model whose prototypical realization consists of the action sequence $e_1 \dots e_r$ with objects participating in them given by the set $\{o_1, \dots, o_t\}$ and that so far the initial sequence $e_1 \dots e_k$ has been introduced into the context. Predictions are possibly related to any of the events $e_{k+1} \dots e_r$ and objects participating in them as well as relative to participants that are related to objects involved in one of the events $e_1 \dots e_k$. Hence, script knowledge allows to capture 'long-range dependencies, [MTD+17].

2.2 The Functional Interpretation of the N400

Expectations are based on semantic features (or properties) of objects. Consider again example (1) repeated below for convenience.

(11) They wanted to make the hotel look more like a tropical resort. So along the driveway they planted rows of palms/pines/tulips.

Given the preceding context, expected features are the tropics as the natural geographical range, and tall trees as sort for visability. Objects that satisfy all of these features, like palms, are most expected, followed by objects like pines that satisfy a proper subset (being trees and being tall) and objects like tulips which satisfy none of these features being the least expected. The N400 amplitude is modulated in accordance with these expectations leading to our first thesis concerning the functional interpretation of the N400:

(12) One factor underlying the modulation of the N400 amplitude are paradigmatic relationships based on semantic features of objects that are related to a particular attribute in a situation model.

However, the following example shows that this thesis is too weak to fully capture the behavior of the N400.

(13) The pianist played the music while the bass was strummed by the guitarist / drummer / gravedigger / drum / coffin during the song.

(13) is a partial description of a concert scenario whereas 'the bass was strummed ...' is a partial description of an event in the concert. Each realization of such a scenario has attributes MUSICIANS, INSTRUMENTS and ACTIONS whose values are the set of musicians, instruments and actions, respectively. For example, for the partial description in (13) one has: MUSICIANS = $\{pianist\}$, INSTRUMENTS = $\{bass\}$ and ACTIONS $\{play, strum\}$. Predictions are related to

features of objects belonging to the values of these attributes: How likely is it that this concert also has a drummer or a guitarist, respectively and how likely is it that a drum is an instrument? For these objects, the respective probabilities are high. For example, guitarist and drummer are expected as extensions of the value of the MUSICIANS attribute whereas a drum is expected as an extension of the value of the INSTRUMENTS attribute. By contrast, neither a gravedigger nor a coffin are expected relative to these two attributes. Thus, according to thesis (12), one would expect a larger N400 amplitude for 'coffin' than for 'drum', contrary to the fact that both elicit amplitudes of the same magnitude. One may argue that the amplitude of the N400 in examples like (13) is due to a selection restriction violation which overrides any semantic relationships based on features and world knowledge. However, this strategy fails to explain the absence of an N400 effect for the critical words in (6) as well as for the critical words in the following semantic illusion data in which the thematic role assigned to the argument(s) clashes with the constraints imposed by the verb on these roles.

(14) a. For breakfast, the eggs would eat ...
b. De speer heeft de atleten geworpen. (The javelin threw the athletes)

The problem with (12) is that it ignores semantic relationships that exist between objects belonging to different attributes. It does not constrain how a newly introduced object is or can be related to objects that have already been introduced into the current situation model. We hypothesize that the difference between 'drummer' and 'guitarist' on the one hand and 'drum' and 'coffin' on the other lies in the way they can be anaphorically linked to the preceding context. Consider first the examples in (15) taken from [Bur06].

(15) a. Tobias besuchte einen Dirigenten in Berlin. Er erzählte, daß der Dirigent ...
(Tobias visited a conductor in Berlin. He said that the conductor ...)
b. Tobias besuchte ein Konzert in Berlin. Er erzählte, daß der Dirigent ...
(Tobias visited a concert in Berlin. He said that the conductor ...)
c. Tobias unterhielt sich mit Nina. Er erzählte, daß der Dirigent ...
(Tobias talked to Nina. He said that the conductor ...)

Burkhardt found an attenuated N400 effect for bridged DPs (Konzert - Dirigent) and an enhanced effect for new DPs (Nina - Dirigent) compared to the given DP (Dirigent - Dirigent). We follow Burkhardt and assume that this modulation of the N400 amplitude is related to discourse linking. In (15) this modulation cannot be related to the event model to which the object introduced by the interpretation of the critical word belongs because this object is the first to be introduced into this model. Rather, what is at stake in these contexts is a constraint to the effect that the newly introduced object needs to be linked to an object that has already been introduced. In (15-a) and (15-b) a concert script (scenario) is introduced in the first sentence. In (15-b) this situation model is explicitly introduced by the DP 'the concert'. In (15-a) the interpretation of the two occurrences of the DP 'the conductor' can be anaphorically linked by the relation of identity. In (15-b) the interpretation of 'the conductor' can be linked to the interpretation of 'the concert' in the preceding context. The conductor is the value of an attribute that is defined for the concert, e.g. the attribute CONDUCTOR. By contrast, in (15-c) no situation model to which an object of sort 'conductor' can be linked is explicitly introduced. As a result, 'the conductor' cannot be anaphorically linked to the preceding context.

We generalize discourse linking in the following way. Let $o_1 \ldots o_k$ be the objects already introduced into the current situation model and let o_{k+1} be the object related to the interpretation of the currently processed word, o_{k+1} has to be linked to an $o_i, 1 \leq i \leq k$. As a consequence, linking can also be done relative to the current event model. Linking defined in this way satisfies 'maximize anaphoricity' because each newly introduced object needs to be related to an object already introduced and is therefore a necessary condition to ensure discourse coherence. When taken together we arrive at our second hypothesis for the functional interpretation of the N400 component.

(16) A second factor underlying the modulation of the N400 amplitude is the possibility of linking the interpretation of the critical word to an object that has already been introduced into the situation model. Specifically, establishing such a linking relation consists in a bridging inference. The interpretation of the critical word is the value of an attribute associated with an object that has already been introduced.

On this approach, the effect of a selection restriction violation is to exclude *one* possibility of linking the critical word to the current situation model via a particular thematic role in the current event model. This violation alone is therefore not sufficient to block the establishment of a bridging inference. This is different if the situation model is reduced to a single event model, e.g. if the context is made up by of a single sentence.

(17) Dutch trains are sour.

In (17) 'sour' must be linked to the trains because no other objects have been introduced so far. Since there is no attribute of objects of sort 'train' for which 'sour' is an admissible value, linking fails. As expected 'sour' in (17) elicits an N400.

2.3 The Functional Interpretation of the LPP

According to the preceding section, the N400 is based on two factors: paradigmatic relationships based on semantic features and anaphoric linking, or, more generally, the establishing of a bridging inference. The linking operation fails, if no bridging inference can be established. We hypothesize that this failure triggers a revision-modification operation. At the ERP level, this operation is indexed by the LPP.

(18) One factor underlying the LPP is a revision-modification operation as a reaction to a failed linking operation.

An example of a revision operation is to question bottom-up information. Consider e.g. (15-c). A comprehender could countenance a reading or hearing error and assume 'a conductor' instead of 'the conductor'. Alternatively, bottom-up information already processed can be questioned in a similar way. Depending on which bottom-up information is questioned, situation models that have already been discarded can again become options. A third strategy is to extend the current situation model with additional information. One possibility is to introduce a concert as the subject or topic about which Tobias talked to Nina. This has the effect that some other situation models that are options according to bottom-up information become excluded, for example, models in which the topic is not a concert.

If the linking operation succeeds, the current situation model is updated with the information provided by the critical word. This success does not imply that a corresponding transition at the level of the current event model is possible as well. Two principle cases must be distinguished: For the first case consider example (19).

(19) The restaurant owner forgot which waitress the customer had <u>served</u>.

Although the critical word 'served' can be linked to an object already introduced into the current situation model ('restaurant') and no selection restriction violation occurs an LPP is elicited. Recall that predictions relative to arguments of a verb are dependent on the placement of the events in the temporal ordering if the sort of events denoted by the verb can occur more than once in this ordering. Generalizing this pattern, one has that each sort of objects that is admissible in a particular situation model is related to a particular set of action-role pairs that specify in which actions it can occur in which thematic roles in this situation model. For example, in a restaurant script an object of sort 'waitress' is at least assigned the set $\langle serve, actor\rangle, \langle ask, actor\rangle, \langle ask, theme\rangle$. If objects of a particular sort are assigned such a set, they are said to be *free* only for pairs in this set. We hypothesize that if the interpretation of the critical word is assigned an action and a thematic role for which it is not free, an LPP is elicited. This is the case if an object of sort waitress is assigned the theme role in a serving event in a restaurant scenario. The second principle case occurs if objects are not assigned action-role pairs that are relevant in the situation model.

(20) The pianist played the music while the bass was strummed by the <u>gravedigger</u> ...

In (20), none of the action role pairs like $\langle dig, actor\rangle$ associated with the object 'gravedigger' is licensed by the situation model 'playing music'. Such objects are free for any action-role assignment that respects the selection restrictions and no LPP is elicited. This accounts for the absence of an LPP for 'gravedigger' in a 'playing music' script. We hypothesize that freeness is a second factor underlying the LPP.

(21) A second factor underlying the LPP is a revision-modification operation as a reaction to a failure of the freeness constraint.

In response to a violation of a freeness constraint, one strategy open to a comprehender is to extend the set of possible situation models by changing the freeness constraint. For example, upon encountering 'The restaurant owner forgot which waitress the customer had served', a comprehender can extend his action-role assignments for restaurant scripts by adding the action role pair $\langle serve, actor\rangle$ to the sort 'customer' and the pair $\langle serve, theme\rangle$ to the sort 'waitress'. As a result, restaurant scripts now also allow serving events in which customers serve waitresses. Freeness is a special case of anaphoric linking that differs from it in the following two respects. First, in contrast to linking, freeness is restricted to the current event model and second, satisfaction of sortal constraints is not sufficient as shown by (19).

2.4 The Processing Model Underlying the N400 and the LPP

The processing model outlined in the last two sections based on particular functional interpretations of the N400 and the LPP consists of three steps. In the first step participants and actions must be linked to objects that have already been introduced into the current situation model. The leading question is: 'Does this information continue information already supplied in the context?' Success of this linking operation is a precondition for the next operation to be applied. This has the following consequences: (i) if the linking operation fails, neither paradigmatic relationships based on semantic features nor the freeness constraint play a role and (ii) as an effect, the N400 amplitude is therefore not modulated by this relationship and this constraint, in accordance with the empirical findings about this component. Failure of the linking operation triggers a revision-modification operation that is indexed by the LPP. Processing of the remaining text is continued on the basis of the result of this operation. In the case that the linking operation succeeds, the second step consists in integrating the new information into the current situation model. The leading question is: 'How probable is this information given the information in the context?'. This operation is related to paradigmatic relationships based on semantic features. As an effect, the modulation of the N400 amplitude is graded. Hence, the N400 is related to the linking operation in a double way. If it fails, an N400 effect is elicited and if it succeeds a graded N400 effect results with the limiting case that no N400 effect is elicited. The final step is related to integrating the new information in the current event model. This integration fails if the freeness constraint is violated. Similar to the failure of the linking operation, a revision-modification operation is triggered which is indexed by the LPP. Processing is continued with the result of this operation which, again, is a changed model. Since the N400 indexes integration at the situation model, no N400 is elicited if freeness is violated. If this constraint is not violated, the new information gets integrated into the current event model without eliciting an LPP.

The LPP indexes the impossibility of executing an integration operation, either at the level of situation models or at the level of event models. Both the N400 and the LPP are elicited at most once. If linking fails, a biphasic N400 - LPP is elicited. Since the other operations are not executed no second effect in relation to these two components is produced. If linking succeeds, no LPP in relation to this operation is elicited. Such an effect is still possible if freeness is violated. Similarly, an N400 effect can be produced in relation to the integration operation based on (successful) linking and the semantic-cognitive property. Since a violation of freeness does not elicit an N400 effect, this effect is produced at most once.

Empirical evidence for this model is based on two studies. First, a study by [DMK16] challenges the one-step model of language comprehension proposed in [HHBP04], who considered sentences like (22).

(22) Dutch trains are yellow / white / sour.

For each sentence, the N400 amplitude was measured relative to the critical word. They found that there was no difference in the N400 onset or peak latency between the semantic violation 'sour' and the world-knowledge violation 'white'. The authors concluded that semantic and world knowledge are processed in parallel during language comprehension. In a recent study this conclusion was challenged by [DMK16]. Similar to [HHBP04], the authors used correct sentences, semantically violated sentences and sentences violated by world knowledge. In contrast to [HHBP04], the critical word was kept constant. In addition to analyzing standard measures for component onset, i.e. the fractional area under the N400 curve and the relative-criterion-peak latency measure, they used a cluster-based permutation test that is sensitive to picking up differences by taking into account biophysical constraints in the testing procedure and which are able to deal with the multiple comparison problem. Specifically, this method allowed to determine the time point at which each of the conditions reached a fixed 2 μV criterion starting from the peak preceding the N400. When using this method, the authors found that the semantic violation condition differed significantly from the world-knowledge condition with regard to the time point when the 2 μV criterion was reached: the former crossed this criterion earlier than the latter.

The second study is [PK12] who found that the onset of the LPP to selection restriction violations in examples like 'The pianist played the music while the bass was strummed by the drum / coffin during the song' was somewhat later (approximately 100 ms) than the LPP evoked on verbs in semantic illusion data like 'The restaurant owner forgot which waitress the customer had served'. Applied to our approach, the results of the study in [DMK16] support a sequential execution of the operations associated with linking and the semantic-cognitive property. Success of the linking operation is a precondition for the execution of the operation associated with the semantic-cognitive property. The result by [PK12], on the other hand, is evidence for a temporal dissociation of the two conditions evoking an LPP: failure of linking and violation of a freeness constraint.

3 The Formal Framework

In order to account for the empirical neurophysiological findings in the previous sections in theoretical linguistics it is necessary develop a truth-theoretical formal semantics that reflects the empirical results. Our approach is based on Frame Theory and Incremental Dynamics extended by continuations.

3.1 Frame Theory

Frames are elements of a separate domain D_f of frames. Each frame is related to a particular object (an individual or a (complex) event) as its root and is a partial description of that object in a particular world. Being a partial description of an object, a frame is linked to a relational structure that is built by (finite) chains of attributes. This link is captured by a function θ which maps a frame f to a set of pairs $\theta(f) = \{\langle R_1, o_1\rangle, \ldots \langle R_n, o_n\rangle\}$; each pair consists of an attribute chain R_i and an object o_i that is related to the root of the frame by the chain. The R_i are 3-ary relations ($R_i \subseteq D_f \times D_o \times D_o$) that are functions in the sense that different objects cannot be related to the frame root by the same chain. Being partial descriptions, frames can be ordered by the information ordering $\sqsubseteq$. A frame f' is an *extension* of a frame f ($f \sqsubseteq f'$) iff (i) f and f' have the same root and (ii) $\theta(f) \subseteq \theta(f')$. Furthermore, f'' is said to be a *subframe* of f ($f'' \preceq f$) if it is embedded in f, that is in f there is a chain connecting the root of f with the root of f'' (e.g., a conductor frame is a subframe of a concert frame). For a given object, its associated frame stores information got during a discourse so far as well as world knowledge. Besides the domain D_f, there are the domains D_i of individuals and the domain D_e of events, which together make up the domain D_o of objects. We extend our approach in [NP19a] by set-valued frames for the current situation. Situation models sm are based on complex events. Their associated frames f_{sm} have an attribute ACTIONS whose value is the set of actions (events) occurring in this scenario together with an associated frame (denoted by $a(f_{sm})$). A second attribute is PARTICIPANTS whose value is a set of individuals together with an associated frame $p(f_{sm})$. Each element of this set is related to at least one action or one other participant, the set of these pairs $pr(f_{sm})$ is the value of the attribute PARTICIPANCY_RELATION. The value of the attribute ORDER is a set $o(f_{sm})$ of pairs of events that preorders the value of the ACTIONS attribute. Situation frames are sorted by SM which are sorts of complex events like 'wintery scenario' or 'restaurant scheme'.

Our frame theory is embedded into a particular type logic that combines de Groote's continuation-based framework with van Eijck's Incremental Dynamics, [DG06, vE01]. De Groote, [DG06], extends Montague's framework with a continuation-passing style technique. In addition to the two basic types e of entities and t of truth values, there is a third type γ, representing the type of contexts or environments. Terms of this type store the information from what has already been processed in the computation of the meaning of the whole discourse, [Leb12]. The type γ is taken as a parameter which can define any complex type. This has the effect that the context can easily be elaborated without affecting the core of the logical framework, [Leb12]. For example, in [DG06] the context is a list of objects (or discourse referents), whereas in [Leb12] it is taken as a list of propositions or a list of pairs consisting of an object and a proposition. The interpretation of a sentence can change the context, e.g. by adding a new object or by adding an anaphoric relationship between discourse referents. This updated context needs to be passed as an argument to the interpretation of the next sentence. In De Groote's approach this requirement is implemented by defining the meaning of a sentence not as a set of contexts or a relation between contexts but as a function of its (input) context and a continuation with respect to the computation of the meaning of the whole discourse. Specifically, continuations are of type $\langle \gamma, t \rangle$. Hence, a continuation denotes what is still to be processed in the computation of the meaning of the whole discourse, [Leb12]. As a result, the interpretation of a sentence is of type $\langle \gamma, \langle \langle \gamma, t \rangle, t \rangle \rangle = \Omega$. For example, the interpretation of (23-a) is (23-b).

(23) a. John loves Mary.
b. $\lambda c.\lambda \phi.love(j)(m) \wedge \phi(c^*)$.

In (23-b) c^* is the context obtained by updating the input context c. The conjunct $\phi(c^*)$ indicates that an updated context is passed as an argument to the continuation of the proposition expressed by (23-a). If the context c of type γ is interpreted as a list of objects (or discourse referents), both proper names in (23-a) contribute an object. For example, the interpretation of 'John' is (24).

(24) $\lambda P.\lambda c\phi.Pjc(\lambda c'.\phi(j :: c'))$.

In (24) P is a dynamic property of type $\langle e, \Omega \rangle$ and :: is an update function of type $\langle e, \langle \gamma, \gamma \rangle \rangle$, i.e. it maps an object and a context to a (new) context. Applied to (23-b), the updated context is $c^* = j{::}m{::}c$. When taken together, the interpretation is (25) and the updated context $j{::}m{::}c$ is accessible by future computations.

(25) $\lambda c.\lambda \phi.love(j)(m) \wedge \phi(j :: m :: c)$.

The update of a discourse interpreted as D with a sentence interpreted as S both of type Ω is defined by $\lambda c.\lambda \phi.D(c)(\lambda c'.S(c')(\phi))$.

We follow [NP19b, NP19a, NPG18], based on Incremental Dynamics, [vE01], and take a context as a stack. A stack can be thought of as a function from an initial segment $\{0, \ldots, n-1\}$ of the natural numbers $\mathbb{N}$ to entities of a domain D_o that are stored in the stack. Hence, a stack can equivalently be taken as a sequence of discourse objects $\{\langle 0, d_0 \rangle, \ldots, \langle n-1, d_{n-1} \rangle\}$ of length n. If c is a stack, $|c|$ is the length of c. By $c(i)$ we denote the object at position i at stack c. A link between stack positions and discourse objects that are stored at a position is established by two operations. First, there is a pushing operation:

(26) $c^\frown d := c \cup \{\langle |c|, d \rangle\}.$

Pushing an object d on the stack extends the stack by this element at position $|c|$. The second operation retrieves a discourse object from the stack.

(27) $ret := \lambda i.\lambda c.\iota d.c(i) = d.$

We write $c[i]$ for $ret(i)(c)$. In our application objects stored at a position i are pairs consisting of an object and an associated frame. Such objects are called *discourse objects.*

3.2 Adapting the Framework

The framework introduced in the last section still resembles standard semantic theories in one important respect. The interpretation of sentences is derived in parallel to its syntactic structure. This way of deriving the interpretation is not built on an incremental left-to-right processing strategy. For example, a sentence with a transitive verb is derived by first combing the verb with the direct object and only then is the resulting VP combined with the subject. By contrast, neuro- and psycholinguistic studies and experiments are based on an incremental left-to-right processing strategy. This makes it necessary to calculate semantic representations for non-constituents. For example, in the context of 'The cat chases ...' it is necessary to have a semantic representation of the combination of the NP and the verb before the second NP is encountered, [BS17]. This example also shows a second problem. 'The cat' can be interpreted e.g. as actor, as theme or as experiencer. This indeterminacy of a thematic role assignment must be modelled too in a formal framework.

Incremental Left-to-Right Processing. As our starting point for implementing an incremental left-to right processing strategy we choose [BS17], which presents an event semantics with continuations based on [DG06]. In this framework all expressions are translated as terms of type $\langle\langle t, t\rangle, \langle t, t\rangle\rangle$. For example, the general format for the interpretation of a verb is (28).

(28) $\lambda c.\lambda p.c(\exists e(verb(e) \wedge p)).$

In (28) c is of type $\langle t, t\rangle$ and ranges over continuations which take the existential quantifier in their scope; p is of type t and ranges over continuations within the scope of the quantifier and which provide additional information about the event. (28) maps two continuations to a truth value. This type is also used for the determiner 'a' and the interpretation of common nouns. This has the effect that the general rule of combination is functional composition: $[\![A + B]\!] := \lambda c.([\![A]\!]([\![B]\!](c)))$. A verb and its arguments are combined by thematic roles, which too are of type $\langle\langle t, t\rangle, \langle t, t\rangle\rangle$.

(29) a. $[\![a]\!] := \lambda c.\lambda p.\exists x_i.c(p)$.
b. $[\![boxer]\!] := \lambda c.\lambda p(boxer(x_i) \wedge c(p))$.
c. $[\![ag_i]\!] := \lambda c.\lambda p.c(p \wedge agent(e, x_i))$.
d. $[\![a\ boxer]\!] := \lambda c.\lambda p.\exists x_i.(boxer(x_i) \wedge c(p \wedge agent(e, x_i)))$.

Note that the interpretation of common nouns and the determiner 'a' contains a (possibly free) indexed object variable. When a determiner and a common noun are combined, it is supposed that both indices are the same. Furthermore, the interpretation of a thematic role contains a free event variable which is assumed to be the same variable as the event variable of the verb. This has the effect that constructions with more than one verb cannot be accounted for.

We will adapt this framework in the following way. First, instead of having contexts of type t and continuations of type $\langle t, t\rangle$, we follow de Groote and have contexts of type γ and continuations of type $\langle \gamma, t\rangle$. Second, in our approach objects that are associated with the interpretation of lexical elements are always related to the current situation model and/or the current event model, which are interpreted as discourse objects, i.e. pairs consisting of a (complex) event and an associated frame. Both kinds of models are not fixed once and for all but change during the processing of a discourse. For events, this is obvious because with each verb a new event is introduced. Empirical evidence for a fine-grained individuation of situation models comes from ERP-experiments using data like the following.

(30) Jörn ist mit dem Frühstück fertig. Er geht in die Küche, wo er Teller abwäscht. Dann beginnt er mit dem (a) Abtrocknen / (b) Joggen, ...
'Jörn has finished breakfast. He goes to the kitchen, where he washes plates. Then he starts to (a) dry / (b) jog, ...'

[DDC18] found an N400 effect at the critical word 'Joggen' compared to the critical word 'Abtrocknen'. This is taken by the authors as evidence that comprehenders expect the description of a situation model (or a complex event) to be continued in the next sentence or the subsequent discourse. Whenever this expectation is not satisfied because a new situation model (or complex event) is described an N400 effect is elicited. For example, in (30) a breakfast scenario is followed by a scenario describing an outdoor activity. Hence, two different situation models are involved. In our approach situation and event models are similar to indexical elements of a discourse like the speaker, the speech time and the reference time which, too, change during the processing of a discourse due to new bottom-up information. We therefore assume that the current situation model and the current event model are stored in particular stack position called sm and em, respectively. Specifically, we assume that they are stored at the positions 0 and 1, respectively. This has the effect that the current situation model and the current event model are always accessible if new bottom-up information is processed. In contrast to other elements like the speaker or the reference time situation models and event models are built up incrementally.

The Interpretation of DPs. We follow [Cha15] and [BS17] and assume that the interpretation of a verb in the lexicon does not (yet) provide information about thematic roles. Rather, thematic roles are introduced separately. Specifically, we assume the following structure for DPs: $[[DetN]_{DP_1}[TR]]_{DP_2}$. Whereas N provides sortal information, TR assigns a thematic role by which the object introduced by the interpretation of Det is related to the event introduced by the interpretation of the verb. On this interpretation the assignment of a thematic role can be taken as a non-deterministic operation that introduces branching.

Evidence for such a non-deterministic assignment is the fact that semantic processing in the brain is done in a left-to-right, incremental manner (see [BS17] for examples and further evidence). Further empirical evidence for such an analysis of thematic roles comes from studies involving languages like German in which the thematic role can at least sometimes be uniquely determined from the case of the determiner. Consider the following examples from [FS01].

(31) a. Paul fragt sich, welchen Angler der <u>Jäger</u> gelobt hat.
'Paul wonders which angler the hunter praised.'
b. Paul fragt sich, welchen Angler der <u>Zweig</u> gestreift hat.
'Paul wonders which angler the branch caught.'

The authors observed an N400 effect at the position of an inanimate subject (actor) following an animate object (theme) in German verb-final sentences, (31-b). No such effect was found for (31-a) where both arguments are animate. In our approach this is explained as follows. In (31-b) 'welchen Angler' is (deterministically) assigned the theme role because 'welchen' being accusative only allows for this role. As an effect, the actor argument is expected next. However, 'Zweig', being inanimate, cannot be assigned this role so that an N400 effect compared to 'der Jäger' is elicited (see [BSS08] for a similar analysis based on predictions). If in English or Dutch thematic roles were assigned on the basis of a thematic role hierarchy (actors outrank themes) or a syntactic analysis based on an NP VP structure, one would likewise expect an N400 to be elicited on the verb or the second noun in fragments like 'For breakfast, the eggs would <u>eat</u> ...' and 'De speer heeft de <u>atleten</u> ...'. However, no such effects are observed (see [BSS08] for further discussion).

The Interpretation of Common Nouns and Verbs. The interpretation of common nouns and verbs has to reflect the fact that each lexical element of one of these two syntactic categories can possibly modulate the amplitude of the N400 as well as that of the LPP. According to the analysis of these ERP-components given above, the N400 amplitude is modulated by a linking property and paradigmatic relationships based on features. By contrast, the

LPP is related to the failure of a constraint. Either linking fails or a freeness constraint is violated. These properties and constraints apply to the level of situation models and/or the level of event models. Following the considerations in Sect. 2.4, we further assume that there is a temporal dissociation between the two properties: the linking property applies before paradigmatic relationships are applied.

The relation between these constraints and our formal framework is the following. Consider the case of common nouns.[2] They are part of DPs with the structure $[[DetN]_{DP_1}[TR]]_{DP_2}$. Each component in this structure is related to a particular update operation, which, in turn, is correlated to a particular information ordering. Furthermore, the properties associated with the ERP components are related to these constituents and their update operations in a particular way. Similar to standard dynamic approaches, the interpretation of the determiners 'a' and 'the' is a domain expansion operation: a new object is pushed on the stack. Hence, this operation is directly related neither to the current situation model nor to the current event model. The interpretation of the nominal element (i.e. the head noun) is related to linking and paradigmatic relationships based on features and applies to the level of the current situation model. Linking is modelled as an update operation that targets the PARTICIPANCY_RELATION attribute in these models. This operation tests whether the frame component of a newly introduced discourse object o can be a subframe of an extension of the frame component of an object o' that has already been introduced into the current situation model. If this test is successful, the pair $\langle o', o\rangle$ is added to the value of the PARTICIPANCY_RELATION attribute. For example, in (15-a) above the conductor can be linked to the concert by extending the concert frame with the attribute CONDUCTOR whose value is the frame associated with the interpretation of 'conductor' in the second sentence. Linking fails, if no such relationship between o and some o' in the situation model can be established. In this case none of the remaining update operations are executed. The update operation associated with paradigmatic relationships is related to the PARTICIPANTS attribute. It adds the newly introduced object together with its associated frame to the value of this attribute. The precondition of this update operation is a successful execution of the linking update operation. This means that there must be an o' such that $\langle o', o\rangle$ is an element of the PARTICIPANCY_RELATION attribute. This operation has no side-effects, i.e. it always succeeds provided its precondition, the update operation associated with the linking property, is satisfied.

The operations associated with linking and paradigmatic relationships based on features together integrate a new object into the current situation model. However, success of these two operations does not guarantee that the newly introduced object can be successfully integrated into the current event model as well. Integration at the level of an event model is always related to the current event and a thematic role. This integration operation fails if a freeness constraint associated with the sort of o is violated. If successful, this update operation adds

[2] For verbs, we assume a decompositional structure that is strictly similar to that for DPs.

the pair $\langle R_{tr}, o\rangle$ to the value of θ for the current event. This operation is related to the TR element in a DP structure. The relationships between DP structure, update operations, and the levels and attributes they apply to is summarized in the Table 1 and formally defined in Sect. 3.3.

Table 1. Update operations and the level and attributes they apply to.

Operation	DP constituent	Level	Attribute
Pushing a stack object	Determiner	Global stack	n.a.
Linking	Head noun	SM	Participancy relation
SEM	Head noun	SM	Participants
Integration EM	TR element	EM	Thematic roles

3.3 Formal Definitions of the Update Operations

The update operations listed in Table 1 are uniformly of type $\langle\gamma, \langle\langle\gamma, t\rangle, t\rangle\rangle = \Omega$. The update operation interpreting the determiners 'a' and 'the' is defined in (32).

(32) a. $[\![det]\!] = \lambda c.\lambda\phi.\exists o.\exists f_o(\theta(f_o) = \{\langle R_o, o\rangle\} \wedge root(f_o) = o \wedge \phi(upd_{exp}(c, o, f_o)))$.

b. $upd_{exp}(c, o, f) = \iota c'.(c' = c^\frown\langle o, f\rangle)$.

The determiners 'a' and 'the' push a new discourse object on the stack, i.e. they add such an object to the input context. R_o is the lift of the domain D_o to the relational level with frames: $R_o(f)(o)(o') = 1$ iff $root(f) = o \wedge o = o' \wedge o \in D_o$. The frame component is the most general one which applies to any object in the domain because so far no sortal information is provided (see [NP19a] for details). The update operation correlated with linking is defined in (33).

(33) a. $[\![linking_\sigma]\!] = \lambda c.\lambda\phi.\exists o'.\exists f_{o'}.\exists o.\exists f_o.\exists f'_{o'}(\langle o', f_{o'}\rangle \in p(f_{c[sm]}) \wedge o \in D_\sigma \wedge c[|c|-1] = \langle o, f_o\rangle \wedge f_{o'} \sqsubseteq f'_{o'} \wedge f_o \preceq f'_{o'} \wedge \phi(upd_{link}(c, o, o')))$.

b. $upd_{link}(c, o, o') = \iota c'.(c' \approx_{\langle o', o\rangle} c)$.

c. $c' \approx_{\langle o', o\rangle} c := p(f_{c[sm]}) = p(f_{c'[sm]}) \wedge a(f_{c[sm]}) = a(f_{c'[sm]}) \wedge o(f_{c[sm]}) = o(f_{c'[sm]}) \wedge pr(f_{c'[sm]}) = pr(f_{c[sm]}) \cup \{\langle o', o\rangle\} \wedge \forall i(0 \leq i < |c| \wedge i \neq sm \rightarrow c'[i] = c[i]) \wedge |c| = |c'|)$.

The constraint $o \in D_\sigma$ is related to the sortal information of the head noun. For example, for 'dog', $\sigma = dog$ and D_σ is the set of dogs. The linking operation tests whether the frame f_o associated with the newly introduced object o is a subframe ($f_o \preceq f'_{o'}$) of an extension $f'_{o'}$ of the frame $f_{o'}$ ($f_{o'} \sqsubseteq f'_{o'}$) associated with an object o' already in the current situation model (see [NP19a] for definitions and further details). If this test succeeds, the pair $\langle o', o\rangle$ is added to the PARTICIPANCY_RELATION attribute of the current situation model. It is not required that o' be an element of the input context c. This is the case because objects can be added via a modification operation (accommodation) to the current situation model if linking fails. (see below Sect. 4.3 for details). The frame component is not added because this is accounted for in the update operation correlated with paradigmatic relationships based on features, which is defined in (34).

(34) a. $[\![sem_\sigma]\!] = \lambda c.\lambda\phi.\exists o.\exists f_o.\exists f'_o(c[|c|-1] = \langle o, f_o\rangle \wedge \theta(f'_o) = \theta(f_o) \cup \{\langle R_\sigma, o\rangle\} \wedge \phi(upd_{sem}(c, o)))$.

b. $upd_{sem}(c, o) = \iota c'.(c' \approx_{\langle o\rangle} c)$.

c. $c' \approx_{\langle o\rangle} c := p(f_{c'[sm]}) = p(f_{c[sm]}) \cup \{\langle o, f'_o\rangle\} \wedge a(f_{c[sm]}) = a(f_{c'[sm]}) \wedge o(f_{c[sm]}) = o(f_{c'[sm]}) \wedge pr(f_{c'[sm]}) = pr(f_{c[sm]}) \wedge \forall i(0 \leq i < |c| \wedge i \neq sm \rightarrow c'[i] = c[i]) \wedge |c| = |c'|)$.

This operation always succeeds provided the preceding update operation associated with linking succeeds. It adds the newly introduced object together with its associated frame o at position $c[|c|-1]$ to the PARTICIPANTS attribute of the current situation model. The associated frame is extended by the sortal information provided by the head noun. Similar to R_o, R_σ is the lift of the subdomain D_σ of objects of sort σ to the relational level: $R_\sigma(f)(o)(o') = 1$ iff $root(f) = o \wedge o = o' \wedge o \in D_\sigma$. The update operation correlated with a thematic role constituent is defined as follows.

(35) a. $[\![tr]\!] = \lambda c.\lambda\phi.\exists o.\exists f_o.\exists R_{tr}.\exists e_{em}.\exists f_{e_{em}}.\exists f'_{e_{em}}(c[|c|-1] = \langle o, f_o\rangle \wedge c[em] = \langle e_{em}, f_{e_{em}}\rangle \wedge \neg\exists o'.\langle R_{tr}, o'\rangle \in \theta(f_{e_{em}}) \rightarrow (\theta(f'_{e_{em}}) = \theta(f_{e_{em}}) \cup \{\langle R_{tr}, o\rangle\} \wedge \phi(upd_{tr}(c, o, R_{tr}, f'_{e_{em}}))))$.

b. $upd_{tr}(c, o, R_{tr}, f'_{e_{em}}) = \iota c'.(|c'| = |c| \wedge \forall i(0 \leq i < |c| \wedge i \neq em \rightarrow c'[i] = c[i]) \wedge c'[em] = \langle e_{em}, f'_{e_{em}}\rangle)$.

The update operation correlated with the thematic role constituent tests whether this role is already defined for the current event model. If this is not the case, this model is updated by adding the thematic role together with the object o to the value of θ yielding a new frame $f'_{e_{em}}$.

In contrast to the interpretation of DPs, the interpretation of verbs is related neither to a determiner nor to a thematic role. Therefore, the four update operations are related to the interpretation of a verb as a whole and are not distributed over several constituents. The first three update operations do not differ from those for DPs except for the fact that the newly introduced object is added to the ACTIONS attribute. The interpretation of a verb introduces a discourse object on the stack. This object needs to be linked to an object that is already an element of the current situation model. If this operation is successful, the pair relating the event to this object is added to the PARTICIPANCY_RELATION of the current situation model and next the object together with its associated frame is added to the ACTIONS attribute of the current situation model. The update operations differs w.r.t. the thematic role. In the case of a DP the thematic role constituent relates an object to the current event model by a thematic role. By contrast, the interpretation of a verb adds sortal information about the current event to this model. Hence, the contribution of tr is the relation R_σ and not a thematic role R_{tr}. Furthermore, the event e_{em} is updated by o because this event has now been introduced.

(36) a. $[\![tr_\sigma]\!] = \lambda c.\lambda\phi.\exists o.\exists f_o.\exists e_{em}.\exists f_{e_{em}}.\exists f'_{e_{em}}(c[|c|-1] = \langle o, f_o\rangle \wedge c[em] = \langle e_{em}, f_{e_{em}}\rangle \wedge \theta(f'_{e_{em}}) = \theta(f_{e_{em}}) \cup \{\langle R_\sigma, o\rangle\} \wedge \phi(upd_{tr}(c, o, R_\sigma, f'_{e_{em}}))$.

b. $upd_\sigma(c, o, R_\sigma, f'_{e_{em}}) = \iota c'.(|c'| = |c| \wedge \forall i(0 \leq i < |c| \wedge i \neq em \rightarrow c'[i] = c[i]) \wedge c'[em] = \langle o, f'_{e_{em}}\rangle)$.

Each update operation is correlated with a particular information ordering. For situation models, the most general ordering is defined in (37).

(37) $sm \sqsubseteq_{sm} sm'$ iff $o(f_{sm}) \subseteq o(f_{sm'}) \wedge pr(f_{sm}) \subseteq pr(f_{sm'}) \wedge \forall\langle o, f_o\rangle \in p(f_{sm}) \cup a(f_{sm}) : \exists\langle o, f'_o\rangle : f_o \sqsubseteq f'_o$.

According to (37), situation model sm' extends situation model sm if it contains at least the information about all objects in sm. Specifically, sm' extends sm by (i) possibly having information about more objects, (ii) by having more information about objects in sm or (iii) by having more information about relations between objects. The information ordering correlated with the update operation associated with linking is defined in (38).

(38) a. A situation model sm is related to a situation model sm' by a linking relation, denoted by $sm \sqsubseteq_{link} sm'$, iff $\exists o.\exists o' : o' \in p(f_{sm}) \cup a(f_{sm}) \wedge pr(f_{sm'}) = pr(f_{sm}) \cup \{\langle o', o\rangle\} \wedge p(f_{sm}) = p(f_{sm'}) \wedge a(f_{sm}) = a(f_{sm'}) \wedge o(f_{sm}) = o(f_{sm'})$.

b. $\sqsubseteq_{link} \subseteq \sqsubseteq_{sm}$.

The (possible) extension is related to the PARTICIPANCY_RELATION attribute of a situation model. $\sqsubseteq_{link}$ only reflects the changes of a (successful) linking operation to the value of the sm position. It does not reflect the test that is executed inside this operation. However, this test only checks whether the linking operation can be successfully executed. The information ordering correlated with the update operation associated with paradigmatic relationships based on features is defined in (39).

(39) a. A situation model sm is related to a situation model sm' by paradigmatic relationships based on features, denoted by $sm \sqsubseteq_{sem} sm'$, iff $\exists o.\exists f_o(p(f_{sm'}) = p(f_{sm}) \cup \{\langle o, f_o\rangle\} \wedge pr(f_{sm}) = pr(f_{sm'}) \wedge o(f_{sm}) = o(f_{sm'}) \wedge a(f_{sm}) = a(f_{sm'}))$.

b. $\sqsubseteq_{sem} \subseteq \sqsubseteq_{sm}$.

Both $\sqsubseteq_{link}$ and $\sqsubseteq_{sem}$ are subrelations of $\sqsubseteq_{sm}$. The information ordering correlated with the updated operation associated with thematic role assignment is defined in (40).

(40) An event model em is related to an event model em' by a thematic role, denoted by $em \sqsubseteq_{tr} em'$, iff $\exists e.\exists f_e.\exists f'_e(em = \langle e, f_e\rangle \wedge em' = \langle e, f'_e\rangle \wedge \theta(f_e) \subseteq \theta(f'_e))$.

The ordering $em \sqsubseteq_{em} em'$ on event models holds if the two models describe the same event and if f'_e contains at least the information about that event that f_e contains. The additional information is either sortal information about the event or information relating the event to an object by means of a thematic role.

4 Probability Distributions and Information Metrics

Having defined update operations together with their information orderings that are related to ERP components, we are interested in probabilities between a given context and its possible continuations relative to these update operations and orderings. The relation between probabilities and the ERP components, in particular the N400, is the following. The N400 amplitude on a word w in a context $c = w_1 \ldots w_t$ is typically inversely related to its conditional probability given this context: $P(w \mid c)$, [KJ16]. Underlying this relation is a model of online processing according to which at every step during this processing there exists a probability distribution over the words that could be encountered next. On this view, a prediction is simply the presence of such a probability distribution, (see [KJ16] for an overview). This conditional probability can be measured in at least two ways. The first way uses subjective human ratings and is based on the notion of cloze probability. Participants are presented the context plus the target sentence with the critical word missing. They are then asked to fill in the first word that comes to their mind. The cloze probability is the percentage of participants who provide this word as the filler. A second way of quantifying predictability is as the information-theoretic notion of surprisal. Given an initial sequence of words $w_1 \ldots w_{t-1}$, w_t can be viewed as a random variable. Its surprisal (or self-information) is defined as the negative logarithm of the conditional probability $P(w_t \mid w_1 \ldots w_{t-1})$ and is estimated by probabilistic language models trained on large text corpora. In contrast to these strategies, we define probabilities not at the level of word forms (or referring expressions) but at the semantic level. Interpreting lexical items as objects of type Ω has the effect that each input context is related to its set of possible continuations on which probability distributions can be defined. More specifically, one has the following. For a given context c, $\lambda\phi.T(c)(\phi)$ is of type $\langle\langle\gamma, t\rangle, t\rangle$ and, therefore, a set of continuations. Each continuation is a set of contexts. The contexts in a continuation can be ordered according to one of the information orderings defined in the preceding section. It is therefore necessary to lift the orderings on these models in a first step to the level of contexts. Since there are three orderings, we get a total of three lifts:

For the ordering correlated with the update operation associated with linking, the lifted ordering on contexts is defined in (41-a).

(41) a. $c \sqsubseteq_{link} c'$ iff $c[sm] \sqsubseteq_{link} c[sm']$.

b. $c \sqsubseteq_{sem} c'$ iff $c[sm] \sqsubseteq_{sem} c[sm']$.
c. $c \sqsubseteq_{tr} c'$ iff $c[em] \sqsubseteq_{tr} c'[em]$.

For example, the lifted ordering correlated with the update operation associated with linking $c[sm] \sqsubseteq_{link} c[sm']$ requires that the frame component of the discourse object stored in $c[sm']$ extends the corresponding component in $c[sm]$ according to (38). The other two lifted orderings are defined analogously.

4.1 Probability Distributions on Frames

Next we define properties of frames. We start with properties of events in event models. Let the current event be e of sort σ with a frame f_e and $\theta(f_e) = \{\langle R_1, o_1\rangle, \ldots \langle R_m, o_m\rangle\}$. Each R_i is a relation that maps f_e and its root to a (unique) object o. Hence, to each R_i corresponds the property of frames $Q_i = \{f \mid \exists o.R_i(f)(root(f))(o)\} = dom(R_i)$. The frame f_e is therefore related to the property $Q_1 \cap \ldots \cap Q_m$. If the next expression is a DP, it contributes the discourse object $\langle o, f_o\rangle$ with $\theta(f_o) = \{\langle R'_1, o'_1\rangle, \ldots, \langle R'_k, o'_k\rangle\}$. Relative to the current event model, this triggers a move along the information ordering $\sqsubseteq_{tr}$ based on the update operation tr defined in (35). Let $tr_1 \ldots tr_l$ be the thematic roles defined for events of sort σ. If in the given context $tr_1, \ldots, tr_j$ have already been discharged, information growth is possible only with respect to the thematic roles $tr_{j+1}, \ldots tr_l$. Hence, f_o is related to f_e by some thematic role $tr_k, j+1 \leq k \leq l$ with interpretation R_{tr_k}. One therefore has $\theta(f'_e) = \theta(f_e) \cup \{\langle R'_1, o'_1\rangle, \ldots, \langle R'_k, o'_k\rangle\}$ and $Q_{f'_e}$ the corresponding property of frames.

We define conditional probability functions on subsets of D_f. $P_{\sqsubseteq_{em}}(Q_{f'} \mid Q_f)$ is the probability that frame f can be extended to frame f' by a move along the information ordering $\sqsubseteq_{tr}$. $P_{\sqsubseteq_{tr}}(Q_{f'} \mid Q_f) > 0$ indicates that frame f' is accessible from frame f relative to $\sqsubseteq_{tr}$. The probability $P_{\sqsubseteq_{tr}}$ of a move along $\sqsubseteq_{tr}$ depends on the context. For example, given that the actor of an event has already been introduced, the probability of extending the frame of the current event by this relation is 0 because the corresponding update operation fails.

For situation models, properties of frames are defined in a way similar to that for event models. Given a situation model sm with associated frame f_{sm} and $\theta(f_{sm}) = \{\langle R_1, S_1\rangle \ldots, \langle R_n, S_n\rangle\}$, the corresponding property is $Q_1 \cap \ldots \cap Q_n$ where $Q_i = \{f_{sm} \mid \exists S.R_i(f_{sm})(root(f_{sm}))(S)\}$. Similar to the case of an event model, the contribution of the next word is based on the discourse object $\langle o, f_o\rangle$. For situation models, there are two update operations with corresponding information orderings $\sqsubseteq_{link}$ and $\sqsubseteq_{sem}$. Hence, one gets two conditional probability distributions: $P_{\sqsubseteq_{link}}$ and $P_{\sqsubseteq_{sem}}$. The constraints on these distributions are the same as in the case of $P_{\sqsubseteq_{em}}$. For example, $P_{\sqsubseteq_{link}}(Q_{f'} \mid Q_f) > 0$ means that frame f' is accessible from frame f along the ordering $\sqsubseteq_{link}$.

4.2 Information Metrics: Entropy and Entropy Reduction

The situation model sm stored at the stack of a context c can be taken as a partial description of a (complete) situation model sm_c, i.e. one has $sm \sqsubseteq_{sm} sm_c$. Given context c and sm, a comprehender wants to know which sm_c is described by the discourse. Each new word that is processed contributes additional information and therefore (possibly) decreases the uncertainty the comprehender has about which situation model is described. The comprehender expects the new information to comply with her expectations based on discourse principles (linking) and paradigmatic relationships as well as her world knowledge. The update operation associated with linking targets bridging inferences. The conductor example in (15) shows that at this level the N400 amplitude is smallest in the case of an identity relation as in (15-a). This kind of DP does not exclude any extensions that were possible before this DP was encountered because the information related to this DP was already known in the input information state. For bridged DPs, this will in general not be the case because some extensions are excluded by establishing a linking relation that was not known before. Take, for example, the case of the jackets in (6). This excludes situations in which the children were wearing coats or ski suits. If linking fails, no transition along $\sqsubseteq_{sem}$ is possible so that all continuations are discarded. This data suggests that the update operation associated with linking is related to the information metric of entropy reduction. Hence, we hypothesize the following relation to the modulation of the N400 amplitude:

(42) The modulation of the N400 amplitude is monotonically related to entropy reduction.

Let us make this idea formally precise. One way to proceed is to define entropy over maximal continuations relative to a particular situation model. However, the number of possible continuations in such contexts is in general far too large. We will therefore use another approach and define n-step entropy instead (see [Fra13] for further details). We start by defining conditional probabilities $P_{\sqsubseteq_{link}}(c_j \mid c_i)$ between contexts relative to the ordering $\sqsubseteq_{link}$ defined above for situation models. Let f_{sm_i} be the frame component of the discourse object stored at position sm in context c_i.

(43) $$P_{\sqsubseteq_{link}}(c_2 \mid c_1) = \begin{cases} P_{\sqsubseteq_{link}}(Q_{f_{sm_2}} \mid Q_{f_{sm_1}}) & \text{if } c_1 \sqsubseteq_{link} c_2 \\ 0 & \text{otherwise} \end{cases}$$

In the next step we define conditional probabilities for n-step transitions. This is done by using the chain rule from probability theory.

(44) $$P_{\sqsubseteq_{link}}(c_{t+1}^{t+n} \mid c_1^t) = \Pi_{i=1}^n P_{\sqsubseteq_{link}}(c_1^{t+i} \mid c_1^{t+i-1}).$$

In (44) c_1^{t+i-1} is the context got from c_1 by $t+i-2$ moves along the ordering $\sqsubseteq_{link}$. More generally, c_i^j is the context got from context i by $j-i$ moves along the ordering $\sqsubseteq_{link}$. The definition of n-step entropy is given in (45).

(45) $$H_{n_{\sqsubseteq_{link}}}(\Phi^n; c_1^t) = -\Sigma_{c_{t+1}^{t+n} \in \Phi^n} P_{\sqsubseteq_{link}}(c_{t+1}^{t+n} \mid c_1^t)\ log\ P_{\sqsubseteq_{link}}(c_{t+1}^{t+n} \mid c_1^t).$$

Φ^n is the set of n-step continuations. Processing word w_{t+1} leads to the new context c_1^{t+1} which drops out of the computation of uncertainty concerning the situation model described by the discourse. The relevant entropy at this point is over the probabilities of moves in Φ^{n-1} so that the simplified reduction in entropy due to w_{t+1} becomes (46).

(46) $$\Delta H_{n_{\sqsubseteq_{link}}}(\Phi^n; c_{t+1}) = H_{n_{\sqsubseteq_{link}}}(\Phi^n; c_1^t) - H_{n-1_{\sqsubseteq_{link}}}(\Phi^{n-1}; c_1^{t+1}).$$

However, using entropy reduction in this way is problematic for cases involving paradigmatical relationships as in the example of the holiday resort.

(47) They wanted to make the hotel look more like a tropical resort. So along the driveway they planted rows of palms / pines / tulips.

Recall that for 'pines' the N400 amplitude was less enhanced than that for 'tulips'. However, if for example pines and tulips have the same (low) conditional probability, they do not differ with respect to entropy reduction. As a result, the N400 amplitude for 'pine' and 'tulips' should be the same, contrary to the empirical findings. This shortcoming is similar to using cloze probabilities. Both 'pines' and 'tulips' have the same (low) cloze probability.

We suggest the following solution to this problem. One has to compare the actual (surviving) continuations with the continuations that have the highest conditional probability. Let us make this precise. Given a particular context c with $c[sm] = \langle e_{sm}, f_{sm} \rangle$, there is a maximal sortal constraint on elements of the values of f_{sm}. This constraint is determined by selectional restrictions, bottom-up information and world knowledge. Given these constraints, particular extensions are most expected, i.e. have the highest conditional probability relative to $\sqsubseteq_{sem}$. For example, in the case of (47) these are extensions which assign to the theme of the planting event objects that are tall trees whose geographical range are the tropics. Whereas palms satisfy all of these features, pines only satisfy two (they are trees and tall) and tulips satisfy none of these features. Hence, the question is: to what degree do the actual found features satisfy the most predicted ones? This idea can be made precise as follows.

Let the input context got after processing (47) up to but excluding the critical word be c_t and the next word be w_{t+1} with interpretation $\langle o, f_o \rangle$. In the case of (47) this is either 'palm', 'pine' or 'tulip'. In all three cases the plant o (i.e. the palms, the pines or the tulips) can be linked to the event of planting by the theme relation. One has that f_o is a subframe of an extension $f'_{o'}$ of the frame $f_{o'}$ associated with the planting event o' ($f_o \preceq f'_{o'}$ and $f_{o'} \sqsubseteq f'_{o'}$). Bottom-up information only yields the sortal information provided by the head noun. Enriching this information with world knowledge yields frames with the following values: $\theta(f_{palm}) = \{\text{SORT} = \textit{palm}, \text{RANGE} = \textit{tropics}, \text{SPECIES} = \textit{plant}, \text{SUBSPECIES} = \textit{tree}, \text{HEIGHT} = \textit{tall}\}$, $\theta(f_{pine}) = \{\text{SORT} = \textit{pine}, \text{RANGE} = \textit{moderate}, \text{SPECIES} = \textit{plant}, \text{SUBSPECIES} = \textit{tree}, \text{HEIGHT} = \textit{tall}\}$ and $\theta(f_{tulip}) = \{\text{SORT} = \textit{tulip}, \text{RANGE} = \textit{moderate}, \text{SPECIES} = \textit{plant}, \text{SUBSPECIES} = \textit{flower}, \text{HEIGHT} =$

small}.[3] These frames will be referred to by f_{found}. Predictions are calculated by extensions of c_t along the information ordering $\sqsubseteq_{sem}$. Instead of entropy reduction in the case of linking, we consider n-step conditional probabilities. Probabilities at the level of contexts relative to the ordering $\sqsubseteq_{sem}$ are defined in a way similar to those for $\sqsubseteq_{link}$.

(48) $$P_{\sqsubseteq_{sem}}(c_2 \,|\, c_1) = \begin{cases} P_{\sqsubseteq_{sem}}(Q_{f_{sm_2}} \,|\, Q_{f_{sm_1}}) & \text{if } c_1 \sqsubseteq_{sem} c_2 \\ 0 & \text{otherwise} \end{cases}$$

We are interested in those contexts got after n-steps that have the highest conditional probability given c_t relative to the ordering $\sqsubseteq_{sem}$. ϕ^n is the set of n-step continuations.

(49) For a given context c_t, let $S_{c_t} = \lambda\phi.T(c)(\phi)$. $max(c, \sqsubseteq_{sem}) = \{c_{t+1}^{t+n} \,|\, c_{t+1}^{t+n} \in \phi^n \wedge \phi^n \subseteq S_{c_t} \wedge \forall \hat{c}_{t+1}^{t+n} \in \phi^n : P_{\sqsubseteq_{sem}}(c_{t+1}^{t+n} \,|\, c_t) \geq P_{\sqsubseteq_{sem}}(\hat{c}_{t+1}^{t+n} \,|\, c_t)\}$.

Let's assume for the sake of simplicity that $max(c, \sqsubseteq_{sem})$ is a singleton, i.e. there is only one continuation of length n. Let c^* be the maximal element in this continuation relative to $\sqsubseteq_{sem}$ with $c^*[sm_{c^*}] = \langle e_{sm_{c^*}}, f_{e_{sm_{c^*}}} \rangle$ and $p(f_{e_{sm_{c^*}}}) = \{o_1, \ldots o_k\}$. Since w_{t+1} contributed the object o which is linked to the planting event o' by the theme relation R_{theme}, we need the object o_j in $p(sm_{c^*})$ for which one has $\langle o', o_j \rangle \in pr(f_{e_{sm_{c^*}}})$ and $\langle R_{theme}, o_j \rangle \in \theta(f_{o'})$. The frame associated with o_j is f_{o_j}. Recall that we are interested in the question: given f_{found} i.e. the frame for the palms, the pines or the tulips, what is the percentage of features that this frame has in common with f_{o_j}? The set of features common to both frames is given by $\theta^*(f_{found}) \cap \theta^*(f_{o_j})$ where $\theta^*(f)$ is the projection of $\theta(f)$ to its relational component. Finally, one calculates the percentage in (50).

(50) $$\frac{|\theta^*(f_{found}) \cap \theta^*(f_{o_j})|}{|\theta^*(f_{o_j})|}.$$

If $\theta^*(f_{o_j})$ is $\theta^*(f_{palm})$, one gets: For 'palm', (50) yields a value of 1. By contrast, for 'pine', $\theta^*(f_{found}) \cap \theta^*(f_{o_j})$ has three elements which yields a value of 0.60. For 'tulips', finally, one has $\theta^*(f_{tulip}) \cap \theta^*(f_{o_j}) = \{\text{SPECIES} = \textit{plant}\}$ and one gets 0.20. Tulips satisfy only the most general feature that is determined by 'plant' for its theme argument.

4.3 The LPP and Exception Handling

Due to lack of space, we can only sketch how the LPP component is related to our formal framework. By way of example, we will illustrate with the linking operation. Recall that the linking update operation is based on the establishment of a bridging inference. So far, there are only two possibilities: such an inference can be established or not. However, what is required is a threefold distinction

[3] To ease readability, we use a simplified notation. For a detailed analysis of this 'tropical resort' example refer to [NP17].

between true and false bridging inferences and the failure of such an inference. Recall that empirical evidence for such a distinction is twofold. First, N400 amplitudes that correspond to failure of linking in our approach are maximal and are independent of semantic similarity and paradigmatic relationships based on features. Second, cases of failure of linking in our approach elicit an LPP (usually associated with semantic violations) whereas this is not the case for cases in which linking succeeds but is false according to general world knowledge (usually associated with world knowledge violation).

We follow [dGL10] and [Leb12] and assume that update operations that have side-effects depend on a function *sel*. In our approach, for the linking update, *sel* takes a context c, an object o, a frame f and returns, if successful, another frame f'. It is defined in (51).

(51) $sel(c, o, f_o) = \iota o'.\exists f_{o'}.\exists f'_{o'}.\langle o', f_{o'}\rangle \in p(f_{c[sm]}) \wedge f_o \preceq f'_{o'} \wedge f_{o'} \sqsubseteq f'_{o'}.$

By itself, *sel* is a partial function: If it returns an object o', linking is successful. In this case the established bridging inference can either be true or false. If no object is returned, *sel* raises an exception to the effect that no object was found whose frame can be linked by a feature to the frame f_o. The exception is catched and the object will be returned to the exception handler. The handler introduces a new object into the context whose associated frame allows for a bridging inference with f_o. Hence, the linking update operation is called with an enriched context that makes a bridging inference possible. Formally, this can defined in terms of an exception handling mechanism (see [Leb12] for details).

(52) a. $D; S = \lambda\phi.D(\lambda c.S(c)(\phi))$ *handle* $(fail\ f_o)$ *with* $\lambda\phi.D(\lambda c'.\exists c.\exists o''.\exists f_{o''}.\exists e.\exists f_e.\exists f'_e$ $(c \sqsubseteq_{sm} c' \wedge c[sm] = \langle e, f_e\rangle \wedge c'[sm] = \langle e, f'_e\rangle \wedge \langle o'', f_{o''}\rangle \in p(f'_e) \wedge \phi(c')); S\ \lambda\phi.D(\lambda c'.\exists o''.\exists f_{o''}.\phi(update_{handle}(c, f_o, o'', f_{o''}))); S$

b. $update_{handle}(c, f_o, o'', f_{o''}) = \iota c' \exists e.\exists f_e.\exists f'_e(|c'| = |c| \wedge \forall i(0 \leq i < |c| \wedge i \neq sm \rightarrow c'[i] = c[i]) \wedge c[sm] = \langle e, f_e\rangle \wedge c'[sm] = \langle e, f'_e\rangle \wedge a(f'_e) = a(f_e) \wedge o(f'_e) = o(f_e) \wedge pr(f'_e) = pr(f_e) \wedge p(f'_e) = p(f_e) \cup \langle o'', f_{o''}\rangle) \wedge f_o \preceq f_{o''} \wedge \phi(c'))$

In (52) D is the discourse up to the linking operation. It is of type $\langle\langle\gamma, t\rangle, t\rangle$. S is the update operation associated with linking. *handle with* takes a set of continuations, an exception of type χ and a set of continuations and maps it to a set of continuations. The effect of the exception handling is to execute D with respect to continuations that are augmented by an addition object together with its associated frame so that a bridging inference relative to f_o becomes possible. Note that in this case the frame for o'' can directly be assumed to have the required attribute (feature) that links it to the frame f_o. The revised linking operation is given in (53).

(53) $[\![linking_\sigma]\!] = \lambda c.\lambda\phi.\exists o.\exists f_o(o \in D_\sigma \wedge c[|c| - 1] = \langle o, f_o\rangle \wedge \phi(upd_{link}(c, o, sel(c, o, f_o)))).$

The test for a bridging inference is now part of the *sel*-function which provides or fails to provide an argument of the update operation. Our hypothesis for the LPP is given in (54).

(54) The LPP is related to failure of the *sel* function to return an object. In this case the situation model in the input context is updated by a suitable discourse object. This is a case of accommodation.

For the update operation associated with thematic roles, other handling mechanisms are required that modify constraints like those imposed by freeness. We assume that such constraints are part of the common ground which, in turn, is part of the initial context of a discourse. Elaborating on this strategy must be left to another occasion. This account of the LPP may also shed some light on the fact that the evocation of an LPP is sometimes task-dependent. For example, the critical word in 'De bomen die in het park speelden ...' (The trees that in the park played ...') elicited an LPP effect compared to the expected 'stonden' ('stood') (and no N400 effect) when participants made explicit sentence acceptability judgments about these sentences, but when participants simply read the sentences for comprehension, the critical words only evoked an N400 effect and no LPP effect (see [Kup07] for references and further details). In our approach this difference is explained as follows. Participants execute an exception handling operation (accommodation) if they know that the discourse is continued or if they have to evaluate the coherence of the discourse so far. If they only have to read a particular discourse up to a particular point, there is no need to adapt the current context in order to continue or answer a question related to its coherence.

5 Comparison to Three Related Models

Three related models that have been proposed in the literature are the Retrieval-Integration model by Brouwer and colleagues, the MUC-model by Baggio and Hagoort and the approach by Rabovsky and colleagues that is based on a probabilistic representation of meanings.

The Retrieval-Integration model of Brouwer et al. [BFH12, BCVH17, DBC19], is based on the assumption that incremental, word-by-word language processing proceeds in retrieval-integration cycles where each cycle is modelled by a function *process* which maps a word form w_1 and a context to an updated context. The function *process*, in turn, is the composition of two functions *retrieve* and *integrate*. The former maps a word form and the prior context to the disambiguated meaning of the word form whereas the latter takes this meaning and the context and maps it to an updated context. The N400 component reflects the effort involved in retrieving from long-term memory conceptual knowledge associated with the eliciting word, which is influenced to the extent to which this information is cued (or primed) by the preceding context, [DBC19, p. 2]. The retrieval operation is viewed as a bottom-up process that does not involve integrative semantic processing or semantic composition, [BFH12, p. 134]. Top-down information, e.g. from the existing mental representation of the preceding sentence fragment, does play a role, but it *adds* to the activation pattern and does not *constrain* the pattern of activation. A reduced (attenuated) N400 amplitude reflects facilitated access, and hence retrieval, of lexical information, [DBC19, p. 2]. As an effect, the N400 amplitude for a critical word should be relatively insensitive to the plausibility of a sentence within which it is contained. For example, if one of two words makes a given sentence implausible, while the other does not, there will be no N400 effect if both are approximately equally primed by the preceding context. A by-product of this conception of the retrieval operation is that the language processing system is able to anticipate or predict upcoming words, [BFH12, p. 134]. In this approach, the absence of N400 effects in semantic illusion sentences results from contextually-cued retrieval mechanisms that are based on semantic similarity or semantic associations, [DBC19, p. 2]. An N400 effect is observed for critical words that are semantically weakly associated with the prior context. By contrast, if there is a strong semantic association, no N400 effect occurs.

According to the Retrieval-Integration model, late positivities to which the LPP belongs reflect the word-by-word construction, reorganization or updating of a mental representation of what is being communicated. It is functionally interpreted as the brain's natural electrophysiological reflection of updating a mental representation with new information. Each member of this family corresponds to a specific subprocess of this updating process. Subprocesses include: accommodating new discourse referents; establishing linking relations between discourse referents; assigning thematic roles to discourse referents; imposing constraints on discourse referents; revision of already established relations and resolving conflicts between different sources of information. Integration difficulty does not result from a conflict between two or more processing streams. Rather, it reflects the degree to which the current mental representation needs to be adapted to incorporate the current input, [BFH12, p. 138].

[BCVH17] use a neurocomputational model that is an extension of a Simple Recurrent Network to implement this approach. This network instantiates the *process* function with its two subprocesses *retrieve* and *integrate*. The N400 amplitude is an index of the amount of processing involved in activating the conceptual knowledge associated with an incoming word in memory. Specifically, the N400 amplitude for a word w is taken as the degree of change that w induces in the activity pattern of the RETRIEVAL layer that implements the 'retrieve' subprocess. Similarly, the LPP amplitude for a given word w is estimated as the degree of change that processing this words induces in the activity pattern of the INTEGRATE which implements the *integrate* subprocess.

The Retrieval-Integration model and our model have in common that language processing is taken as a biphasic process with the first phase indexed by the N400 and the second by the LPP. The difference is twofold. First, we distinguish between a global level of the situation model and a local level of the event model. The representation of an incoming word must be integrated at both levels, which is modelled by update operations. Integration at the level of the situation model is related to the N400. Adding the object is followed by an integration operation which adds the semantic representation of the incoming word to the situation model. Since this operation adds new information to this model, its associated probability distribution is changed. This change leads to a change in the expectations of the comprehender. Hence, the N400 is related to an operation that changes and, therefore, constrains the model.

In both models the LPP is indexed by integration. However, the set of operations modelling this integration operation is only a subset of those assumed in the Retrieval-Integration model. In the latter model integration captures all kinds of semantic update operations, whereas in our model these operations are restricted to those related to the current event model. For example, the LPP is related to establishing a linking operation.

Rabovsky et al. [RHM18], interpret N400 amplitudes as the change induced by an incoming word in a probabilistic representation of meaning. In this model each word in a sentence provides clues that constrain the formation of a probabilistic representation of the event described by the sentence, [RHM18, p. 693]. The context and each word is represented by a set of activation units which are modelled as probability distributions over features. Examples of such units are 'Agent', 'Action' and 'Patient'. Features for the 'Agent' unit include 'woman', 'man', 'boy' and 'girl' and capture semantic similarities among event participants. The magnitude of the activation update produced by each successive word of a sentence corresponds to the change in the model's probabilistic representation that is triggered by that word, [RHM18, p. 693]. The N400 amplitude of the n-th word is defined as the semantic update (SU) induced by this word. This update is defined as the sum of the absolute values of the change of each unit's activation (across the model) that the word triggers. For a given unit a_i the change is the difference between the unit's activation after processing the n-th word and the activation of this unit prior to processing it, i.e. after the (n-1)-th word.

(55) $N400_n = SU_n = \sum_i |a_i(w_n) - a_i(w_{n-1})|.$

Consider the sentence fragment 'I take my coffee with cream and ...'. The activation state associated with this fragment already implicitly represents a high subjective probability that in addition to cream the speaker takes her coffee with sugar. As an effect, the state will change very little if 'sugar' is in effect found as the next word and the N400 amplitude is small. If instead 'dog' is encountered, the activation state is changed to a much larger degree so that a larger N400 amplitude is elicited.

In contrast to most other accounts of N400 activity this model does not assume separate stages for lexical access and subsequent integration. It resembles an access view in that the change in activation state is fast, automatic and implicit. However, there is no separate step that consists of the isolated representation of the incoming word. Rather, the resulting activation state already is the updated activation state, i.e. the change that is triggered by this word. Hence, this activation state can be taken as representing the result of integrating the representation of the incoming word with the representation of the context. This model and our model have in common that the effect of processing a word is represented as a change in information state. However, in contrast to this model, in our model a static representation can be isolated for each word, which is the frame representation of the concept associated with this word. Furthermore, in our model, separate stages of processing are distinguished: the two stages of N400 activity and the stage indexed by the LPP. By contrast, the resulting activity state in the Rabovsky et al. approach represents all aspects of the event described by the sentence, [RHM18, p. 700].[4]

The approach by Baggio and Hagoort, [BH11], is based on the Memory-Unification-Control (MUC) model of language processing in the brain. The memory component is a lexicon that stores phonological, syntactic and semantic information about morphemes, words and other constructions. What gets stored are unification-ready structures which supply constraints across levels of description. The unification component combines stored lexical information to more complex units. This is done by solving (or unifying) sets of constraints given by the context and an input, say the next word in a sentence. This solving of constraints is done in a dynamic fashion. Memory supplies constraints for the Unification component, which retains a context for subsequent stages of memory retrieval and unification, [BH11, p. 1341f]. Finally, the Control component presides over executive functions in language like turn taking in conversations. Each component corresponds to a set of brain regions. The memory component is localized in temporal regions (superior temporal gyrus, STG; middle temporal gyrus, MTG; and inferior temporal gyrus, ITG). The unification component is subserved by the inferior frontal gyrus (IFG) and the Control component is localized in anterior cingulate and dorsolateral prefrontal cortices.

The N400 is explained as the result of the summation of currents injected by frontal into temporal areas (unification) with currents that are already circulating within temporal cortex due to the local spread of activation to neighbouring neuronal populations (pre-activation). More specifically, the N400 component reflect reverberating activity within the MTG/STG-IFG network, [BH11, p. 1358f]. Processing an initial fragment of a sentence or a discourse sets up a context, i.e. a set of unification-ready structures or constraints, in MTG/STG. This corresponds to the pre-activation component. Encountering the next word of the sentence/discourse similarly activates a unification-ready structure representing the meaning of this word. The next step is the unification component, i.e. the solution of the constraints representing the context and the new word, which amounts to calculating the unification of the unification-ready structures. If the constraints representing the context include features that are also part of the constraint associated with the new word, there will be some overlap between the populations in MTG/STG associated with the context and those associated with the word. The relation to the N400 is the following. The larger the overlap of features between the representations of the context and the new word, the smaller the amplitude of the N400. Consider the sentence 'The girl was writ-

[4] Though late positivities like the LPP are not captured in this model.

ing letters when her friend spilled coffee on the paper/tablecloth'. Processing the initial fragment up to but excluding the final word sets up a representation of the context that activates more features contained in the representation of 'paper' than in the representation of 'tablecloth'. This is due to features activated with the representations of the words 'write' and 'letters'. As a result, the N400 amplitude for 'paper' is smaller than that for 'tablecloth'.

Similar to this theoretical account of the N400 we assume that N400 activity is related to two components: prediction and integration at the level of situation models. However, whereas in the Baggio and Hagoort account the N400 amplitude is modulated only by the unification component, this amplitude is a function of both components in our model. Second, in the Baggio and Hagoort approach unification is an operation at the sentential or discourse level because the representation of the context and that of the incoming word are combined (unified) to a new (updated) context. By contrast, in our approach integration is related to two different levels: the situation model and the event model. The N400 activity is related to integration in the situation model, i.e. to the combination of the representation of the incoming word and the representation of the situation model. Integration at the event model is related to the LPP. Finally, in our approach stochastic frames are used as representations in the lexicon which results in a probabilistic framework that allows for a weighting of features.

6 Closing Outlook

We have outlined a formal framework in which results from neuro-linguistic research on the N400 and the LPP can be incorporated. Obviously, this framework needs to be extended in several directions. Two of the most important directions are: (i) besides the N400 and the LPP, data on the Left Anterior Positivity has to be accounted for as well as more data on the N400 and the LPP; (ii) our implementation of a left-to-right processing strategy only accounts for simple sentences. Extending it to include constructions like proper quantification and modification, e.g. in form of adjectives, adverbs or relative clauses, requires a more complex framework that has to use some kind of storing mechanism (see [BS17] for a similar argument).

References

[BCVH17] Brouwer, H., Crocker, M.W., Venhuizen, N.J., Hoeks, J.C.J.: A neurocomputational model of the N400 and the P600 in language processing. Cogn. Sci. **41**, 1318–1352 (2016)

[BFH12] Brouwer, H., Fitz, H., Hoeks, J.: Getting real about semantic illusions: rethinking the functional role of the P600 in language comprehension. Brain Res. **1446**, 127–143 (2012)

[BH11] Baggio, G., Hagoort, P.: The balance between memory and unification in semantics: a dynamic account of the N400. Lang. Cogn. Process. **26**(9), 1338–1367 (2011)

[BS17] Bott, O., Sternefeld, W.: An event semantics with continuations for incremental interpretation. J. Semant. **34**(2), 201–236 (2017)

[BSS08] Bornkessel-Schlesewsky, I., Schlesewsky, M.: An alternative perspective on "semantic P600" effects in language comprehension. Brain Res. Rev. **59**(1), 55–73 (2008)

[Bur06] Burkhardt, P.: Inferential bridging relations reveal distinct neural mechanisms: evidence from event-related brain potentials. Brain Lang. **98**(2), 159–168 (2006)

[Cha15] Champollion, L.: The interaction of compositional semantics and event semantics. Linguist. Philos. **38**(1), 31–66 (2014). https://doi.org/10.1007/s10988-014-9162-8

[DBC19] Delogu, F., Brouwer, H., Crocker, M.W.: Event-related potentials index lexical retrieval (N400) and integration (P600) during language comprehension. Brain Cogn. **135**, 103569 (2019)

[DDC18] Delogu, F., Drenhaus, H., Crocker, M.W.: On the predictability of event boundaries in discourse: an ERP investigation. Mem. Cognit. **46**(2), 315–325 (2017). https://doi.org/10.3758/s13421-017-0766-4

[DG06] De Groote, P.: Towards a montagovian account of dynamics. In: Gibson, M., Howell, J. (eds.) Proceedings of Semantics and Linguistic Theory XVI, pp. 1–16. Cornell University, Ithaca, NY (2006)

[dGL10] De Groote, P., Lebedeva, E.: Presupposition accommodation as exception handling. In: Fernández, R., Katagiri, Y., Komatani, K., Lemon, O., Nakano, M. (eds.) Proceedings of the SIGDIAL 2010 Conference, The 11th Annual Meeting of the Special Interest Group on Discourse and Dialogue, 24–15 September 2010, Tokyo, Japan, pp. 71–74. The Association for Computer Linguistics (2010)

[DMK16] Dudschig, C., Maienborn, C., Kaup, B.: Is there a difference between stripy journeys and stripy ladybirds? The N400 response to semantic and world-knowledge violations during sentence processing. Brain Cogn. **103**, 38–49 (2016)

[FK99] Federmeier, K.D., Kutas, M.: A rose by any other name: long-term memory structure and sentence processing. J. Mem. Lang. **41**, 469 (1999)

[Fra13] Frank, S.L.: Uncertainty reduction as a measure of cognitive load in sentence comprehension. Top. Cogn. Sci. **5**(3), 475–494 (2013)

[FS01] Frisch, S., Schlesewsky, M.: The N400 reflects problems of thematic hierarchizing. NeuroReport **12**(15), 3391–3394 (2001)

[HHBP04] Hagoort, P., Hald, L., Bastiaansen, M., Petersson, K.M.: Integration of word meaning and world knowledge in language comprehension. Science **304**(5669), 438–441 (2004)

[KBW0] Kuperberg, G.R., Brothers, T., Wlotko, E.W.: A tale of two positivities and the N400: distinct neural signatures are evoked by confirmed and violated predictions at different levels of representation. J. Cogn. Neurosci. **32**(1), 12–35 (2020)

[KJ16] Kuperberg, G.R., Jaeger, T.F.: What do we mean by prediction in language comprehension? Lang. Cogn. Neurosci. **31**(1), 32–59 (2016)

[KPD11] Kuperberg, G., Paczynski, M., Ditman, T.: Establishing causal coherence across sentences: an ERP study. J. Cogn. Neurosci. **23**(5), 1230–1246 (2011)

[Kup07] Kuperberg, G.R.: Neural mechanisms of language comprehension: challenges to syntax. Brain Res. **1146**, 23–49 (2007)

[Leb12] Lebedeva, E.: Expressing discourse dynamics through continuations. (Expression de la dynamique du discours à l'aide de continuations). Ph.D. thesis, University of Lorraine, Nancy, France (2012)

[MTD+17] Modi, A., Titov, I., Demberg, V., Sayeed, A., Pinkal, M.: Modeling semantic expectation: using script knowledge for referent prediction. Trans. Assoc. Comput. Linguist. **5**, 31–44 (2017)

[NP17] Naumann, R., Petersen, W.: Semantic predictions in natural language processing, default reasoning and belief revision. In: Hansen, H.H., Murray, S.E., Sadrzadeh, M., Zeevat, H. (eds.) TbiLLC 2015. LNCS, vol. 10148, pp. 118–145. Springer, Heidelberg (2017). https://doi.org/10.1007/978-3-662-54332-0_8

[NP19a] Naumann, R., Petersen, W.: Bridging inferences in a dynamic frame theory. In: Silva, A., Staton, S., Sutton, P., Umbach, C. (eds.) TbiLLC 2018. LNCS, vol. 11456, pp. 228–252. Springer, Heidelberg (2019). https://doi.org/10.1007/978-3-662-59565-7_12

[NP19b] Naumann, R., Petersen, W.: Combining neurophysiology and formal semantics and pragmatics: the case of the N400. In: Schlöder, J.J., McHugh, D., Roelofsen, F. (eds.) Proceedings of the 22nd Amsterdam Colloquium, pp. 309–318 (2019)

[NPG18] Naumann, R., Petersen, W., Gamerschlag, T.: Underspecified changes: a dynamic, probabilistic frame theory for verbs. In: Sauerland, U., Solt, S. (eds.) Proceedings of Sinn und Bedeutung 22, vol. 2 of ZASPiL 61, pp. 181–198. Leibniz-Centre General Linguistics (2018)

[PK12] Paczynski, M., Kuperberg, G.: Multiple influences of semantic memory on sentence processing: distinct effects of semantic relatedness on violations of real-world event/state knowledge and animacy selection restrictions. J. Mem. Lang. **67**(4), 426–448 (2012)

[RHM18] Rabovsky, M., Hansen, S.S., McClelland, J.L.: Modelling the N400 brain potential as change in a probabilistic representation of meaning. Nat. Hum. Behav. **2**, 693–705 (2018)

[VCB18] Venhuizen, N.J., Crocker, M.W., Brouwer, H.: Expectation-based comprehension: modeling the interaction of world knowledge and linguistic experience. Discourse Process. **56**(3), 229–255 (2018)

[vE01] van Eijck, J.: Incremental dynamics. J. Log. Lang. Inf. **10**(3), 319–351 (2001)

Distributional Analysis of Polysemous Function Words

Sebastian Padó[1(✉)] and Daniel Hole[2]

[1] Natural Language Processing (IMS), University of Stuttgart, Stuttgart, Germany
sebastian.pado@ims.uni-stuttgart.de
[2] Linguistics (IL), University of Stuttgart, Stuttgart, Germany
daniel.hole@ling.uni-stuttgart.de

Abstract. In this paper, we are concerned with the phenomenon of function word polysemy. We adopt the framework of distributional semantics, which characterizes word meaning by observing occurrence contexts in large corpora and which is in principle well situated to model polysemy. Nevertheless, function words were traditionally considered as impossible to analyze distributionally due to their highly flexible usage patterns.

We establish that *contextualized word embeddings*, the most recent generation of distributional methods, offer hope in this regard. Using the German reflexive pronoun *sich* as an example, we find that contextualized word embeddings capture theoretically motivated word senses for *sich* to the extent to which these senses are mirrored systematically in linguistic usage.

1 Introduction

Theoretical linguists observe with envy the way in which distributional semantics in computational linguistics renders research viable whose foundations were postulated by clear-sighted structuralists [10,13]. Their interest diminishes upon seeing that computational linguistics has dealt mainly with parts of speech dominated by content words (nouns, verbs, adjectives), whereas theoretical linguists firmly believe that function words and morphosyntax define the interesting backbone of natural language. In this respect, the focus of computational linguistics has broadened only in recent years.

This paper brings together the advanced computational tools of distributional semantics with the interest of formal linguistics in function words and in particular their disambiguation. We consider a multiply polysemous function word, the German reflexive pronoun *sich*, and investigate in which ways natural subclasses of this word which are known from the theoretical and typological literature map onto recent models from distributional semantics. Due to the differences between lexical and functional polysemy, our results are different from those of distributional studies of systematic polysemy in content words such as [5].

© The Author(s), under exclusive license to Springer Nature Switzerland AG 2022
A. Özgün and Y. Zinova (Eds.): TbiLLC 2019, LNCS 13206, pp. 113–127, 2022.
https://doi.org/10.1007/978-3-030-98479-3_6

We submit that our results open a window onto patterns of polysemy that may, in the long run, turn out at least as interesting and relevant to the computational study of natural languages as content words. What we find in our pilot is that some traditional subclasses of *sich* not only map neatly onto clusters produced by distributional methods, but that others which are predicted by theory to belong to constructional metaclasses with a wider distribution pervade the whole clustering space. What is more, the distribution of causative-transitive vis-à-vis anticausative verb types and of other verb classes partly reproduces the semantic map of the middle domain on a typological database [15]. We take these results to be a promising starting point for more in-depth studies of function morphemes in distributional semantics.

2 Background: Distributional Analysis

Today, distributional analysis is the dominant paradigm for semantic analysis in computational linguistics. Building on the distributional hypothesis, *"you shall know a word by the company it keeps"* [10], it typically represents words as high-dimensional vectors which summarize the words' occurrence contexts (see [19] for an introduction and overview). Traditionally, these vectors were obtained by counting: each dimension corresponded to one particular linguistic context (often, another word), and the value in the vector for this dimension was the co-occurrence frequency of the two words, or some function thereof.

This procedure was increasingly replaced by neural network-based methods, where the co-occurrence frequencies are not directly used as vectors. Instead, they form the "output" that the neural network is supposed to predict, and the vectors are given by the internal parameters of the neural network, now often called 'word embeddings' [3]. Crucially, traditional fundamental intuitions about distributional semantics mostly carry over to the new paradigm. In fact, some widely used types of word embeddings are mathematically equivalent to count vectors to which dimensionality reduction has been applied [17].

At the same time, the move to neural network-based vector learning has opened the door for innovative network architectures. Prominent among these are the recently introduced *contextualized embeddings*. These models concurrently learn (a) general vectors for word types (lemmas) and (b) specialized vectors for word tokens (instances) in their local context. In this manner, they overcome the traditional limitation of distributional semantics, which generally used to aggregate the contexts of all instances, and thus all senses, into one vector. The most successful model architecture to create contextualized embeddings are so-called *transformers* [20], a class of models which lets each context word directly influence the representation of the target word, and automatically learns to weigh these contributions using a mechanism called *self-attention*. In this process, which is carried out several times, transformers uncover (some degree of) implicit linguistic structure such as predicate-argument relations, coreference, or phrase structure [14].

As introduced above, the focus of this paper is the polysemy of function words such as *sich*. Traditionally, distributional analysis has concentrated mostly on

Table 1. Salient classes of *sich*, inspired by Kemmer (1991), plus feature representation (± indicates the possibility of both positive and negative cases depending on context)

Class/Example	Predictable	Agentive	Stressable	+*lassen*	Disposition
1. INHERENT REFLEXIVES: *Paul schämte sich*/ 'Paul felt ashamed'	+	±	–	–	±
2. ANTI-CAUSATIVES: *Die Erde dreht sich*/ 'The earth revolves'	+	–	–	–	±
3. CHANGE IN POSTURE: *Paul setzte sich hin*/ 'Paul sat down'	+	±	–	–	–
4. TYPICALLY SELF-DIRECTED: *Paul kämmte sich*/ 'Paul combed his hair'	–	+	–	–	–
5. TYPICALLY OTHER-DIRECTED: *Paul erschoss sich*/ 'Paul shot himself'	–	+	+	–	–
6. DISPOSITIONAL MIDDLE: *Die Dose lässt sich leicht öffnen*/ 'The can opens easily'	+	–	–	+	+
7. EPISODIC MIDDLE: *Paul lässt sich beraten*/ 'Paul takes advice'	+	+	–	+	–
8. RECIPROCALS: *Die Geraden schneiden sich im Unendlichen*/ 'The lines intersect in the infinite'	-	±	±	–	±

content words (common nouns, verbs and adjectives), following the intuition that these word classes refer to categories whose properties and relational structure can be learned from distributional analysis [6]. Exceptions notably include distributional studies of compositionality, which have modeled the semantic effects of quantifiers [4] and determiners [2] on sentence-level entailment.

Crucially, these studies do not consider polysemous function words. Indeed, the context of function words is typically so general that traditional methods of distributional analysis tended to fail in this domain, since any reflection of the function word meaning was likely to be masked by the topic of the surrounding linguistic material. Consequently, the only (partially functional) word category that has received more than cursory attention in distributional semantics with regard to senses and disambiguation are prepositions [1,18]. Our study takes benefit of the development that the contextualized embeddings created by transformers take a major step towards alleviating the generality problem: Even if the representation of the word type *sich* is still too general to be useful, the embeddings for each instance of *sich*, arising from the combination of word type meaning and context, is informative enough for analysis.

3 Phenomenon: The German Reflexive Pronoun *sich*

The reflexive pronoun in German is a notorious case of polysemy because prototypical instances such as *sich loben* 'praise oneself' are by far outnumbered by other uses. These other uses cover large portions of what has come to be known as the 'middle domain' in linguistic typology [15]. The classification in Table 1 provides our simplified overview of this domain in German including examples.

Class 1 is a metaclass, as it assembles historically fossilized combinations of verbs with reflexive pronouns (*sich benehmen* 'to behave oneself'). These verbs invariably occur with reflexive pronouns. This class includes fossilized combinations of *sich* with prepositions, such as Kant's *Ding an sich* 'thing in itself'. The anti-causatives of Class 2 derive non-agentive intransitive uses of transitive verbs (*sich drehen* 'to turn'), potentially expressing a disposition. Class 3 comprises constructions denoting changes in body posture with obligatory *sich*, such as *sich setzen* 'to sit down'. Class 4 consists of agentive predicates such as predicates of grooming (*sich kämmen* 'to comb one's hair') or predicates of assessment (*sich in der Lage sehen* 'to feel equal to doing sth.') which are typically, but not exclusively, used with *sich*. The 'prototypical' *sich* instances (*sich erschießen* 'to shoot oneself'), where *sich* is used to express the identity of subject and another argument, are concentrated in Class 5. Another diagnostic to distinguish classes 4 and 5 is that *sich* is typically unstressed in Class 4, whereas the reflexives of Class 5 may be stressed. The dispositional middles of Class 6 form a construction that encodes a disposition of the subject referent (*sich leicht öffnen lassen* 'to open easily'). Class 7 is similar, but an episodic event is referred to instead of the stative property of Class 6 (*sich beraten lassen* 'to get advice'). Class 8, finally, encompasses uses of *sich* that could be replaced by *einander* 'each other, one another' and are, hence, reciprocals (*sich kennen* 'to know each other').

One caveat is in order here. The classes are not completely mutually exclusive. If, for instance, *sich legen* 'lay down' is used as in *...der sich wie eine weiße Schimmelschicht auf die Kleidung legt...* '...which covers the clothes like a white layer of mold', either Class 2 or Class 3 (with a non-literal use) could host this example. We avoided multiple classifications and allotted examples of this kind on a 'best fit' basis (Class 2 for the example given).

As the right hand side of the table shows, these eight senses can be distinguished in terms of five properties:

- Is *sich* **predictable** in this context? Predictability is meant to describe the property that the reflexive pronoun in the relevant classes cannot be replaced by another 3rd person pronoun (* *Paul schämte ihn*).
- Is the event **agentive**?
- Is *sich* stressable in this context?
- Does the construction involve *lassen*?
- Does the construction describe a **disposition**?

In the table, the value ± indicates neutrality (both positive and negative values exist, depending on context). In our experience, these features can provide valuable criteria for choosing the right category in manual annotation.

4 Data and Annotation

As basis of our study, we use the 700M token SdeWAC web corpus [9]. We selected the first 335 out of more than 5.5 million instances of *sich* for manual annotation with the eight classes as defined above. The annotation was carried out by the two authors individually. We computed Cohen's kappa as a measure of inter-annotator agreement and obtained a value of 0.73, which indicates substantial agreement [16], despite the possible non-exclusivity of the classes.

The confusion matrix is shown in Table 2. The largest classes, according to both annotators, are Class 1, 2, and 4. There is essentially perfect agreement on the reciprocals and the middles and some disagreement on Classes 4 and 5 (typically self- vs. other-directed), but most diagreements involve Classes 1 through 3 – specifically Class 1 vs. Class 2 (31 cases – more than half of all disagreements), and Class 1 vs. Class 3 (8 cases).

Some of the disagreements were oversights by one of the two annotators. However, there were also cases of systematic differences in judgments. For instance, the Class 1 vs. Class 2 disagreements often concern instances where the main criterion for Class 2 (intransitive use of transitive verb) is debatable:

> Jedes Jahr wieder stauen *sich* zur Urlaubszeit die Blechlawinen auf den Autobahnen [...]
> 'Every year again, avalanches of metal back up Ø on the motorways during holiday time, [...]'

If one is willing to accept this as a reflexive analogue to transitive uses like *Blockaden stauen den Verkehr* 'blockades back up traffic', this is a case of Class 2, otherwise Class 1.

As for Class 1 vs. Class 3, a recurring problem is to delineate the verbs of change of posture (Class 3) – in particular with regard to nonliteral uses, which are frequent for motion verbs. For example,

> Die Revision wendet *sich* nur gegen die Ansicht des Berufungsgerichts [...].
> 'The revision only turns against/opposes Ø the view of the appellate court [...]'

We resolved these disagreements via joint adjudication. The resulting frequency distribution over classes is shown in Table 3. The final labeled dataset is available, together with the Jupyter notebook documenting the subsequent analysis, from https://www.ims.uni-stuttgart.de/forschung/ressourcen/korpora/sich20/.

5 Experimental Setup

The specific word embedding model we employ is BERT [8], a state-of-the-art transformer. We use the 'German BERT cased' model, which was trained on a variety of German corpora, including Wikipedia, OpenLegalData, and news

Table 2. Confusion matrix for *sich* categories by two annotators

	Class	Annotator 1							
		1	2	3	4	5	6	7	8
Annotator 2	1	143	6	6	1	0	0	0	0
	2	25	60	0	2	0	0	0	0
	3	2	1	11	0	0	0	0	0
	4	6	0	1	28	4	0	0	0
	5	2	0	1	3	18	0	0	0
	6	0	0	0	0	0	3	0	0
	7	0	0	0	0	0	0	8	0
	8	1	0	0	0	0	0	0	3

Table 3. Frequency distribution of *sich* senses in manually annotated *sich* dataset

Class	1	2	3	4	5	6	7	8	Sum
Frequency	161	84	11	42	22	3	8	4	335

articles [7]. In comparison to the 'BERT multilingual' model which can also be used to model the semantics of German text, the restriction of the training data to German leads in particular to better tokenization. The model provides 768-dimensional contextualized embeddings for all tokens presented to it as input.

We experiment with two conditions of presenting the *sich* instances in context to BERT. Recall that BERT learns contextualized word embeddings – that is, word embeddings that differ among instances of the same word, reflecting the influence of context on word meaning. In the first condition, we present *sich* instances in their local *phrasal* context, as approximated by punctuation. That is, the context is formed by all words surrounding *sich* up to the closest commas, (semi)colons, or other phrasal delimiters. The reason to use this oversimplification is that a proper syntactic identification of the current phrase would have involved full parsing of the sentences, which is still not possible at the near-perfect accuracy we would require as a starting point for our analysis.

In the second condition, we present them in their complete *sentential* context. To illustrate, the underlined part of the following sentence makes up the phrasal context for the italicized *sich* (the English gloss is designed so as to match German word order and punctuation):

> Unsere Universität hat exzellent abgeschnitten und war auch nur indirekt – aufgrund der landesweiten Unterauslastung – lediglich in den 3 Bereichen Chemie, Physik und Slawistik, tangiert, die für *sich* genommen allerdings ebenfalls exzellent dastehen: [...]

> 'Our university has performed excellently and was only indirectly affected – due to the countrywide underutilization – only in the three areas of

Chemistry, Physics and Slavic Studies, which, considered *on their own*, however also appear excellent: [...]'

Our hypothesis is that the phrasal context provides a better basis for distinguishing the senses of *sich*, since its contents are of higher average relevance. On the other hand, there is no guarantee that our shallow definition of phrasal context captures all relevant context cues. In the worst case, even the main verb may not be present in the phrasal context, as the next example illustrates:

Abschließend lässt sich sagen, **dass *sich* der Aufwand für diese Veranstaltung** (22 Stunden Zugfahrt an 2 Tagen für 2 Tage Seminar) insofern gelohnt hat, [...]

'In sum, we conclude, **that** *Ø the effort for this event* (22 hours of train ride on 2 days for 2 days of workshop) paid off *Ø* insofar as [...]'

This is why we also present *sich* in the full sentence context.

6 Exploratory Analysis

As a first step, we perform an exploratory analysis in which we assess to what extent we can recover the manually annotated senses in the contextualized word embeddings produced by BERT when presented in *phrasal context*. We do so visually, by performing principal components analysis (PCA), a dimensionality reduction method which constructs a two-dimensional approximation of a higher-dimensional space by capturing the directions of maximal variation (i.e., differences among instances). The result is a 2D representation of our 335 *sich* instances, as shown in Fig. 1 (above: all classes, below: without Class 1).

In our estimation, the overall picture is promising. Even though the classes are not completely separated, clear tendencies are visible. Our observations are:

- Inherently reflexive verbs (Class 1) are interspersed through all event types and do not form a cluster of their own, as can be expected given their predictable nature. This motivates our showing a figure with Class 1 removed.
- Typically other-directed reflexive events like 'shooting oneself' and typically self-directed reflexive events like 'defending oneself' or 'combing' (Classes 4, 5) form neighboring categories in the bottom and right sectors.
- The sectors at the bottom generally assemble agentive causative verb uses, whereas sectors in the top left corner assemble anticausative verb uses like 'diminishing' or 'revolving' (Class 2), which involve use of *sich* in German. Hence the gradient from top left to bottom right forms a path of growing agentivity, with traditional middle constructions (Classes 3, 6, 7) literally occupying the middle of the plot.
- Some of the classes show a 'core' surrounded by outlier clouds. For the change-of-posture verbs (Class 3), the outliers to the bottom and right are formed by the non-literal uses *sich aus dem Verderben erheben* 'to rise from doom' and *sich auf die Rechtsgrundlage stützen* 'to rest on the legal foundation').

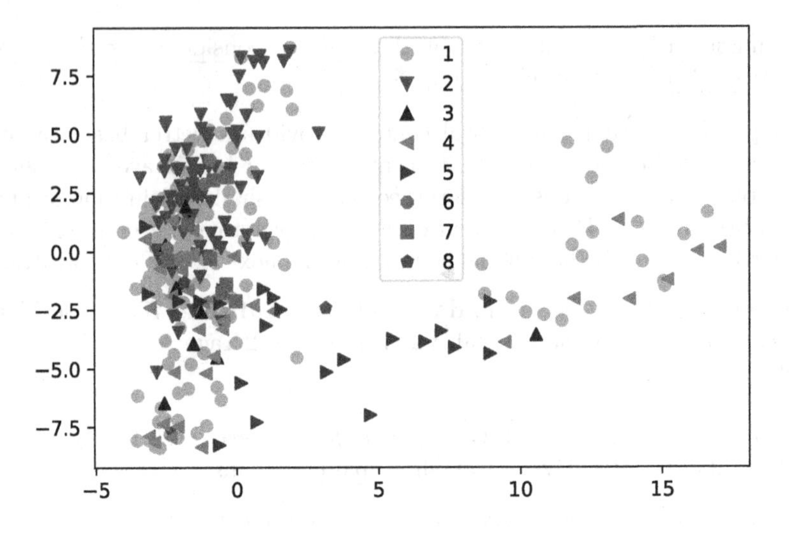

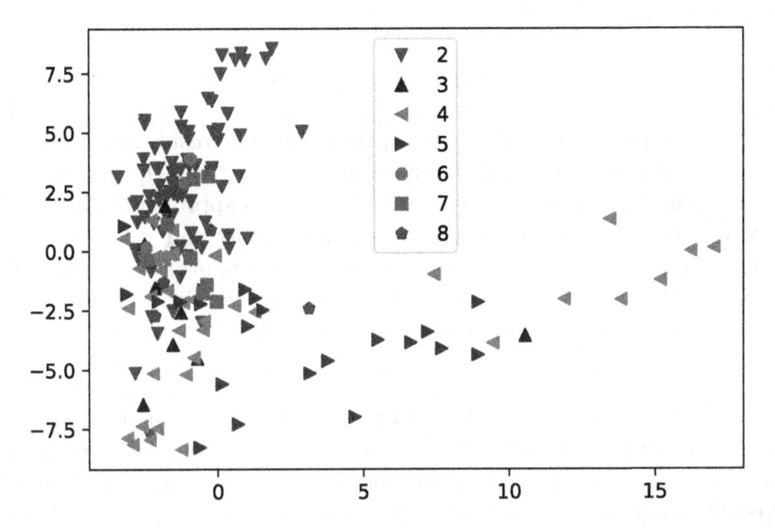

Fig. 1. Distributional representations of *sich* instances based on phrasal contexts. All classes (above), without inherent reflexives (below). Class labels according to Table 1.

- The most inhomogeneous class is the class of self-directed verbs (Class 4), with one cluster in the mid-left sector and another on the right hand side. This can be explained in terms of the distinction between PP-*sich* and DP-*sich* [11]: The mid-left 'core' of Class 4 consists of the DP cases, e.g. *sich unterziehen* 'to undergo'. In contrast, the outliers are made up of PP cases like *bei sich tragen* 'to carry'. The latter are clearly more causative, in line with the 'causation' gradient described above.

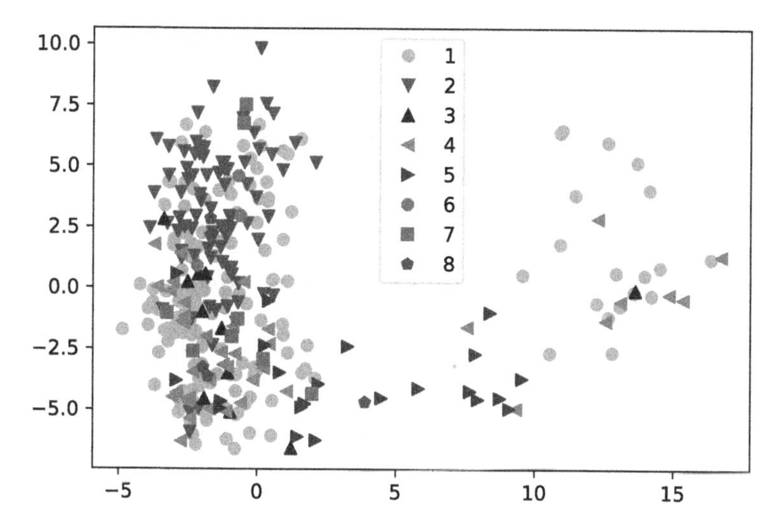

Fig. 2. Distributional representations of *sich* instances based on sentential contexts.

For comparison, Fig. 2 shows the instance embeddings for *sentential* contexts. The picture is overall similar to the phrasal contexts. However, the clusters for the classes tend to be even less tight than before, notably for Class 2 (for which we see instances also at the bottom) and Class 7 (which also occurs at the top). We interpret these observations as evidence for our hypotheses stated above: the phrasal contexts – which are on average 12 tokens long – are generally sufficient to disambiguate *sich*, while in the full sentential contexts – which are on average 77 tokens long – the contribution of *sich* is sometimes overwhelmed by the topic of the complete sentence, as was observed in pre-transformer distributional investigations. In the spirit of Occam's razor, we focus on the phrasal context condition in the remainder of this investigation.[1]

7 Classification Experiments

The analysis in the previous section took into account only the 335 instances that we annotated manually. Naturally, it would be desirable to scale up this analysis to large corpora and to automatically obtain a large number of disambiguated *sich* instances. In order to do so, we trained a classifier which takes the contextualized embeddings of *sich* instances as input and returns one of the eight senses as output. In essence, this classifier learns decision boundaries between regions in embedding space that map onto different classes.

To gauge the prospects for success in this procedure, we may inspect Fig. 1. Even though it is dangerous to draw strong conclusions from dimensionality

[1] We found a comparable, but slightly lower, performance for the sentential contexts in the classification experiments reported below. These experiments are part of the companion Jupyter notebook to this article.

reduced visualizations (since there is a loss of information compared to the original high-dimensional vectors), it appears clear that Class 1 (inherent reflexives) follows an essentially random distribution and will be hard to separate from the other classes. For this reason, we carry out two experiments: one where we consider all classes, and one where we leave Class 1 aside. Finally, we report on an experiment that attempts to predict the individual features of the classes.

7.1 Experiment 1: Classification with All Classes

For classification, we use a Support Vector Machine (SVM) with a linear kernel, a standard choice of classification model.[2] We perform 5-fold cross-validation, that is, we divide the dataset into five partitions of 20% each and run the model five times, training on four partitions and evaluating on the fifth.

For evaluation, we apply the standard classification evaluation measure, accuracy. As the percentage of correct predictions, accuracy ranges between 0% (all wrong) and 100% (all correct). As a point of comparison, we consider the *most frequent class baseline*, the accuracy achieved when always assigning the predominant class. According to Table 3, Class 1 is the most frequent class, with a relative frequency of 48.1% – that is, simply assigning Class 1 to each datapoint would lead to an overall accuracy of 48.1%. Clearly, an informed model should outperform this baseline.

The SVM model, using the phrasal context, achieves an accuracy of 63.8%. This result is some 15 points accuracy above the baseline, but not even two out of three model's predictions are correct. This indicates that the classification is relatively hard to make based on the information present in the word embeddings. Table 4 shows a simplified confusion matrix for Class 1 vs. all other classes, where correct predictions are shown on the diagonal and incorrect predictions off-diagonal. Indeed, this distinction is the main problem of the classification. Most Class 1 instances are classified as such, but more than one third of the instances of other classes are also classified as Class 1. This is consistent with the classifier's attempt to model the largest class (Class 1) as well as possible. Unfortunately, this also means that the smaller classes are not modeled appropriately.

However, the blame should probably not fall entirely on the classifier: As we saw in Sect. 4, the human annotators also ran into problems to agree on some of the borderline Class 1–Class 2 and Class 1–Class 3 cases, pointing towards the inherent difficulty of these distinctions.

[2] We also experimented with fine-tuning the embeddings, but did not obtain competitive results, presumably due to the small size of the training set.

Table 4. Experiment 1: Confusion matrix. Aggregated version: shows only Class 1 vs. all other classes. Overall accuracy of model: 63.8%.

		Predicted	
		Class 1	Other
Actual	Class 1	129	32
	Other	72	102

Table 5. Experiment 2: confusion matrix for classification among all classes except Class 1. Full version. Overall model accuracy: 78.7%)

		Predicted						
		Class 2	Class 3	Class 4	Class 5	Class 6	Class 7	Class 8
Actual	Class 2	77	1	5	1	0	0	0
	Class 3	3	2	5	1	0	0	0
	Class 4	4	1	34	3	0	0	0
	Class 5	1	0	5	16	0	0	0
	Class 6	2	0	0	0	0	1	0
	Class 7	0	0	0	0	0	8	0
	Class 8	2	0	2	0	0	0	0

7.2 Experiment 2: Classification Without Inherent Reflexives

Motivated by this finding, we tested in a second experiment how well the other classes can be distinguished from one another. We adopted the same setup as in Experiment 1 (SVMs with cross-validation), but used only the 174 instances that were labeled as not Class 1 in the gold standard.

This time, the classifier achieved an accuracy of 78.7%, whereas the most frequent class baseline is almost unchanged at 48.3% (now the most frequent class is Class 2). This is a clear improvement over the accuracy shown in Experiment 1 – the model outperforms the baseline by 30 points accuracy. Clearly, the model is not perfect – however, its performance appears fair given the presence of ambiguous cases, as discussed above.

The confusion matrix in Table 5 shows that the highest numbers are indeed on the diagonal. In this setup, the hardest part of the problem appears to be to distinguish Class 4 from Classes 2 and 3. This corresponds to our observations in Sect. 6, where we found Class 4 to be represented in a relatively scattered manner due to its internal heterogeneity (NP-*sich* vs. PP-*sich*, nonliteral cases).

Table 6. Experiment 3: prediction of individual semantic features

Feature	Predictable	Agentive	Stressable	+*Lassen*	Disposition
# Instances	335	159	331	335	86
Accuracy	80.0%	88.6%	95.4%	99.4%	96.5%

7.3 Experiment 3: Prediction of Semantic Features

A different approach towards distinguishing among the senses of *sich* is to consider these senses as bundles of features, as defined in Table 1. Concretely, this means that we can predict the presence (or absence) of the five features from the embeddings by phrasing them as binary classification tasks, again with contextualized word embeddings as input. This approach enables us to investigate whether any of these features are particularly easy or difficult to predict.

We carried out this experiment for each of the features, using the same experimental setup, model, and evaluation measure as in Experiments 1 and 2. For each feature, we removed the instances for classes which are neutral with regard to this feature (± in Table 1) from consideration.

The results are shown in Table 6, including the number of remaining instances. Overall, the numbers look positive, with even the hardest feature showing an accuracy of more than 80% correct predictions.

The easiest feature to predict is '+*lassen*', which is not altogether surprising, given the obligatory presence of (an inflected form of) *lassen* in the context. In fact, the only error of this classifier is an instance where *lassen* was over ten words away from *sich*. The features 'stressable' and 'disposition' are also relatively easy to predict (>90% accuracy). In the case of 'disposition', this may be an effect of correlation with '+*lassen*', since, excluding the classes that are neutral for this feature, the 'disposition' instances are a strict subset of the '+*lassen*' instances. This interpretation is bolstered by the observation that two of the three errors again involve large distances between *sich* and *lassen*, as above. It is interesting that 'stressable' belongs to this category, since stressability is a prosodic property that might not be reflected directly in word embeddings, and arguably a property of the construction rather than the individual instance.

The two features that are more difficult to predict are 'agentive' and 'predictable'. Again, it is not surprising that 'predictable' is a hard feature, since this feature captures idiosyncratic, historically fossilized properties of the predicate which, as we found over the course of this article, are hard to capture for the embedding-based methods we employed. There are also some borderline cases such as the following:

> [...] wenn sie *sich* redlich informiert haben und vom geschichtlichen Hintergrund der Chilbi wissen [...]
>
> 'If they have informed *themselves* honestly and know about the historical background of the Chilbi'

We analysed this instance of *sich informieren* 'to inform oneself' as an inherent reflexive (and thus 'predictable') despite the existence of the transitive *jmd. informieren* 'to inform someone'. The reasons for our analysis are that *sich* is always unstressed in this collocation, and that **er hat sich und andere informiert* '*he informed himself and others' is not possible, further evidence for the independence of the two constructions. The classifier, however, did not reproduce our analysis. At any rate, these results tie in well with our observation in Experiment 2 about the difficulty of distinguishing Class 4 from Classes 2 and 3, which differ exactly with regard to these two features.

Unfortunately, 'recomposing' predictions for the individual features into predictions about classes is not straightforward. The reason is the partial neutrality of the classes with respect to the features, which makes the mapping from features onto classes underspecified. For example, an instance which is predictable, not agentive, not stressable, without *lassen*, and not dispositional, could belong to either Class 1 or Class 2.

8 Discussion and Conclusion

In this study, we have investigated the use of distributional meaning representations to characterize the senses of a function word, the German reflexive pronoun *sich*. The main outcome of our study is a positive one: the recent advances in distributional modeling of lexical semantics, namely transformer-based contextualized embeddings, have substantially increased the 'resolution' of distributional analysis: we can now characterize the meaning of function words not only at the lemma level, but also at the level of individual instances. In turn, this enables us to computationally model function word polysemy and use the associated tools, such as visualization and quantitative evaluation, to develop a better understanding of the senses at hand.

An important limitation which we encountered in this study was that one of the senses – (meta-)Class 1, 'inherent reflexives' – turned out to be rather difficult to distinguish from the other Classes, due to the idiosyncratic behavior of its instances. This is an important take-home message regarding the generalizability of our approach to other function words or other phenomena in general: distributional approaches, at the least in the incarnation we considered in this study, are apt at capturing distinctions that can be grounded in linguistic patterns, but they cannot account well for patterns that are the result of historical fossilization.

This means that the classification setup that we used in the present study, does not scale up directly to large corpora, as the results for the other classes would be polluted by instances of Class 1, and vice versa. Note that this negative result hinges on the fact that we used the standard formulation of classification, where we force the model to assign a class to each and every instance. In view of the very large number of attested *sich* instances, which number 5.5 million in the SdeWAC corpus alone, this may not be the best approach. A promising avenue for future work appears to be experimenting with classifiers that only assign a class to instances that they are very confident about. These 'high-precision, low-recall' classifiers would stand a better chance at identifying 'prototypical' instances of the various classes (maybe with the exception of Class 1) and should still be able to collect substantial numbers for each class. Evaluating such an approach would however require annotating another sample of *sich* instances, based on the confidence estimates of the classifiers for the various classes.

Our present study can be compared and contrasted to another recent study which investigated to what extent word embeddings encode world knowledge attributes such as countries' areas, economic strengths, or olympic gold medals [12]. The findings of that study were remarkably similar to the present one in that the result was also overall positive, but the difficulty of individual attributes was directly related to the extent to which these attributes correlate with salient patterns of linguistic usage in the underlying newswire corpus – high for area, low for olympic gold medals. Taken together, these observations reaffirm the tight interactions between linguistic and referential considerations in forming language, and the difficulty of distinguishing between them in distributional analysis.

References

1. Bannard, C., Baldwin, T.: Distributional models of preposition semantics. In: Proceedings of the ACL-SIGSEM Workshop on the Linguistic Dimensions of Prepositions and Their Use in Computational Linguistics Formalisms and Applications, Toulouse, France, pp. 169–180 (2003)
2. Baroni, M., Bernardi, R., Do, N.Q., Shan, C.C.: Entailment above the word level in distributional semantics. In: Proceedings of EACL, Avignon, France, pp. 23–32 (2012)
3. Baroni, M., Dinu, G., Kruszewski, G.: Don't count, predict! A systematic comparison of context-counting vs. context-predicting semantic vectors. In: Proceedings of ACL, Baltimore, Maryland, pp. 238–247 (2014)
4. Bernardi, R., Dinu, G., Marelli, M., Baroni, M.: A relatedness benchmark to test the role of determiners in compositional distributional semantics. In: Proceedings of ACL, Sofia, Bulgaria, pp. 53–57 (2013)
5. Boleda, G., Schulte im Walde, S., Badia, T.: Modeling regular polysemy: a study on the semantic classification of Catalan adjectives. Comput. Linguist. **38**(3), 575–616 (2012)
6. Cimiano, P., Hotho, A., Staab, S.: Learning concept hierarchies from text corpora using formal concept analysis. J. Artif. Intell. Res. **24**, 305–339 (2005)
7. Deepset.AI: German BERT (2019). https://deepset.ai/german-bert
8. Devlin, J., Chang, M., Lee, K., Toutanova, K.: BERT: pre-training of deep bidirectional transformers for language understanding. In: Proceedings of NAACL, Minneapolis, pp. 4171–4186 (2019)
9. Faaß, G., Eckart, K.: SdeWaC – a corpus of parsable sentences from the web. In: Gurevych, I., Biemann, C., Zesch, T. (eds.) GSCL 2013. LNCS (LNAI), vol. 8105, pp. 61–68. Springer, Heidelberg (2013). https://doi.org/10.1007/978-3-642-40722-2_6
10. Firth, J.R.: Papers in linguistics 1934–1951. Oxford University Press, Oxford (1957)

11. Gast, V., Haas, F.: On reciprocal and reflexive uses of anaphors in German and other European languages. In: König, E., Gast, V. (eds.) Reciprocals and Reflexives: Theoretical and Typological Explorations, pp. 307–346. Mouton de Gruyter, Hague (2008)
12. Gupta, A., Boleda, G., Baroni, M., Padó, S.: Distributional vectors encode referential attributes. In: Proceedings of EMNLP. Lisbon, Portugal (2015)
13. Harris, Z.S.: Distributional structure. Word **10**(2–3), 146–162 (1954)
14. Jawahar, G., Sagot, B., Seddah, D.: What does BERT learn about the structure of language? In: Proceedings of ACL. Florence, Italy, pp. 3651–3657 (2019)
15. Kemmer, S.: The Middle Voice, Typological Studies in Language, vol. 23. John Benjamins, Amsterdam and Philadelphia (1991)
16. Landis, J.R., Koch, G.G.: The measurement of observer agreement for categorical data. Biometrics **33**(1), 159–174 (1977). http://www.ncbi.nlm.nih.gov/pubmed/843571
17. Levy, O., Goldberg, Y.: Neural word embedding as implicit matrix factorization. In: Proceedings of NeurIPS. Montréal, QC, pp. 2177–2185. (2014)
18. Schneider, N., et al.: Comprehensive supersense disambiguation of English prepositions and possessives. In: Proceedings of ACL, Melbourne, Australia, pp. 185–196 (2018)
19. Turney, P.D., Pantel, P.: From frequency to meaning: vector space models of semantics. J. Artif. Intell. Res. **37**(1), 141–188 (2010)
20. Vaswani, A., et al.: Attention is all you need. In: Proceedings of NeurIPS, Long Beach, CA, pp. 5998–6008 (2017)

It is not the Obvious Question that a Cleft Addresses

Swantje Tönnis(✉)

Stuttgart University, Stuttgart, Germany
swantje.toennis@ling.uni-stuttgart.de

Abstract. I take a new perspective on *es*-clefts in German, that focuses on how an *es*-cleft contributes to the discourse structure and how it does this differently than its canonical counterpart. My analysis is inspired by naturally occurring examples from German novels. It combines an adapted version of Roberts' (2012) QUD stack and Velleman et al.'s (2012) approach to clefts. In particular, I present a model that includes implicit and potential questions into the QUD stack and I introduce the concept of expectedness, that I argue is crucial for the acceptability of clefts. I propose that the cleft addresses a question that came up in the preceding context but that is not as expected for the addressee to be answered at that point in the discourse compared to other questions. Those question that are more expected are answered with a canonical sentence. This approach is compatible with other functions that have been proposed for clefts, such as marking exhaustivity, maximality, or contrast. However, it can also account for examples where the cleft serves to establish discourse coherence.

Keywords: German *es*-clefts · QUD · Discourse expectations

1 Introduction

In German, cleft structures are not very frequent (especially in spoken German) and one could wonder why an author would even use an *es*-cleft, such as (1-a)[1], instead of the the much less complex canonical equivalent in (1-b).

(1) a. Es ist die Ungewissheit, die mir keine Ruhe lässt.
It is the uncertainty that me no quiet let.
'It is the uncertainty that bothers me.'

I thank Edgar Onea, Lea Fricke, Maya Cortez Espinoza, the reviewers, and the audience of TbiLLC 2019 for valuable feedback and comments. The analysis of German *es*-clefts presented in this paper is based on my dissertation, Tönnis (2021), which includes a much more detailed version of this analysis.

[1] Taken from the novel *Herzenhören* (Sendker (2012). *Herzenhören.* Heyne, München, p. 21.)

© The Author(s), under exclusive license to Springer Nature Switzerland AG 2022
A. Özgün and Y. Zinova (Eds.): TbiLLC 2019, LNCS 13206, pp. 128–147, 2022.
https://doi.org/10.1007/978-3-030-98479-3_7

b. Die Ungewissheit lässt mir keine Ruhe.
The uncertainty let me no quiet.
'The uncertainty bothers me.'

A preliminary corpus search, based on data from German novels, revealed some interesting occurrences of *es*-clefts, that seemed much more natural than those examples found in a corpus study that mainly investigated newspaper texts, such as Tönnis et al. (2016). As Wedgwood et al. (2006) pointed out, clefts have often been analyzed in unnatural contexts (or even without a context), which failed to capture the effect of clefts as part of a discourse. The preliminary corpus search showed that it does make a difference when the cleft is replaced with its canonical equivalent, which was not so obvious in the newspaper texts. The difference has to do with discourse coherence, but is not easily explicated precisely. Intuitively, the *es*-cleft in German seems to pick up a question that is less prominent at the moment it is uttered. However, in many occurrences of clefts in novels, it was still only a small degradation of acceptability when it was replaced with a canonical sentence.

In order to get to the core of the cleft's discourse function, I constructed an example, inspired by several examples from novels, which clearly favors the cleft over the canonical sentence. Example (2) presents a context in which the cleft is more appropriate than its canonical equivalent.[2]

(2) Lena hat gestern auf der Party mit einem Typen_1 gesprochen. Die beiden haben viel gelacht und sich direkt für den nächsten Abend verabredet. Dann ist Lena glücklich nach Hause gefahren.
'Yesterday at the party, Lena talked to some $\textit{guy}_1$*. The two of them laughed a lot and they agreed to meet again the next evening. Then, Lena went home happily.'*

a. Es war Peter_1, mit dem sie gesprochen hat.
it was Peter_1 with whom she talked has
'It was $\textit{Peter}_1$ *she talked to.'*

b. ?Sie hat mit Peter_1 gesprochen.
She has with Peter_1 talked
'She talked to $\textit{Peter}_1$*.'*

The *es*-cleft in (2-a) can easily be interpreted as referring back to the discourse referent introduced by *einem Typen* ('*some guy*'), which establishes discourse coherence and which makes it an appropriate continuation. The canonical sentence in (2-b), in contrast, seems to be incapable of referring back to that discourse referent and leaves the reader a bit puzzled. The canonical sentence in (2-b) does not seem relevant in this context, which makes it an inappropriate discourse continuation.

In this paper, I will be concerned with the question that a cleft addresses in comparison to the question a canonical sentence addresses. Example (2) suggests

[2] The judgments for this example, as well as for examples (3)–(5), are confirmed by a couple of speakers but still have to be tested empirically.

that the cleft addresses a less expected question than the canonical sentence, here *Which guy did Lena talk to?*. The canonical sentence can only address more expected questions and can, thus, not refer back to the discourse referent introduced in the very first sentence by *einem Typen* (*'some guy'*).

A cleft can, however, not address a question that is extremely unexpected. More precisely, the appropriateness of the cleft decreases when a pressing question arises between the cleft and the question it addresses[3], as in (3).

(3) Lena hat gestern auf der Party mit einem Typen_1 gesprochen. Die beiden haben viel gelacht und sich direkt für den nächsten Abend verabredet. **Lena hat ihm sogar ein Geheimnis verraten.**
'Yesterday at the party, Lena talked to some guy_1. The two of them laughed a lot and they agreed to meet again the next evening. ***Lena even told him a secret.****'*

a. ?Es war Peter_1, mit dem sie gesprochen hat.
it was Peter_1 with whom she talked has
'It was Peter_1 she talked to.'

b. ?Sie hat mit Peter_1 gesprochen.
She has with Peter_1 talked
'She talked to Peter_1.'

In example (3), the sentence in bold evokes the question *'What was the secret?'*. This seems to make the cleft and the canonical sentence equally inappropriate.

However, if a sentence interferes between the second pressing question and the cleft/canonical sentence, such as the bold sentence in (4), the situation is similar to (2). The cleft is again better than the canonical sentence.

(4) Lena hat gestern auf der Party mit einem Typen_1 gesprochen. Die beiden haben viel gelacht und sich direkt für den nächsten Abend verabredet. Lena hat ihm sogar ein Geheimnis verraten. **Dann ist Lena glücklich nach Hause gefahren.**
'Yesterday at the party, Lena talked to some guy_1. The two of them laughed a lot and they agreed to meet again the next evening. Lena even told him a secret. ***Then, Lena went home happily.****'*

a. Es war Peter_1, mit dem sie gesprochen hat.
it was Peter_1 with whom she talked has
'It was Peter_1 she talked to.'

b. ?Sie hat mit Peter_1 gesprochen.
She has with Peter_1 talked
'She talked to Peter_1.'

[3] Thanks to Edgar Onea (p.c.) for raising this issue.

Example (5) provides another context in which the cleft and the canonical sentence are equally degraded.[4]

(5) **Lena ist gestern Abend bei der Party angekommen und hat erstmal einen leckeren Cocktail getrunken.** Danach hat sie mit ihrer Freundin Andrea getanzt und die beiden hatten sehr viel Spaß. Dann ist Lena glücklich nach Hause gefahren.
'Lena arrived at the party yesterday and first of all she had a tasty cocktail. *Thereafter, she danced with her friend Andrea and the two of them had a lot of fun. Then, Lena went home happily.'*

a. ?Es war ein Bloody Mary, den sie getrunken hat.
it was a Bloody Mary that she drunk has
'It was a Bloody Mary she drank.'

b. ?Sie hat einen Bloody Mary getrunken.
She has a Bloody Mary drunk
'She drank a Bloody Mary.'

Here, the sentence in bold raises the question Q: *'Which cocktail did Lena drink?'*. Moreover, the subsequent context does not give rise to any other pressing question. Nevertheless, the cleft is again as inappropriate as the canonical sentence, just like in (3). It seems that the cleft, even without a pressing question interfering, still cannot address a question that is too unexpected to be addressed at that point in the discourse.

I approach these puzzles by providing a discourse model that makes different predictions about the discourse behavior of clefts and canonical sentences. I argue that existing approaches, which focus on the exhaustivity/maximality or contrastivity of clefts, cannot account for the effect of discourse (in-)coherence of clefts or canonical sentences, respectively. My approach takes a new perspective on clefts, while still being compatible with those cases of clefts that mark exhaustivity/maximality or contrast.

In Sect. 2, I briefly present previous approaches to different features of clefts. In Sect. 3, I develop an adapted discourse model, which introduces a QUD set, based on Roberts' (2012) QUD stack, including implicit questions and discourse expectations. In Sect. 4, I present an application of the proposed model to the examples presented above. Finally, Sect. 5 concludes.

2 Background

Several features of clefts are discussed in the literature, and different approaches tend to focus on one feature that determines the function of the cleft. In this section, I will briefly introduce those features and the most important approaches to analyzing the semantics and pragmatics of cleft structures.

[4] An anonymous reviewer questions the proposed judgments for examples (3) and (5), suggesting that the cleft is still more acceptable than the canonical sentence, as in the other examples. If this was the case, it would still need to be explained why the clefts in (3) and (5) are less acceptable than the clefts in (2) and (4).

The cleft in (6) has the EXISTENCE PRESUPPOSITION in (6-a), which is rather uncontroversial.

(6) It is Peter Lena talked to.
  a. **Existence presupposition:** Lena talked to somebody.
  b. **Exhaustivity inference:** Lena talked to nobody else than Peter.
  c. **Canonical inference:** Lena talked to Peter.

Furthermore, it has the EXHAUSTIVITY INFERENCE in (6-b). It is still a point of debate what status this inference has (see Onea (2019) for an overview of exhaustivity in *it*-clefts). Furthermore, the cleft in (6) is assumed to have the at-issue content in (6-c), which is called the CANONICAL INFERENCE or the PREJACENT.

In this paper, I address the question of what function the cleft has, especially when compared to its canonical form. Why would the author of a text use a cleft instead of the structurally much simpler canonical form in German?

Horn (1981), Büring and Križ (2013) and others argued that clefts are used to mark exhaustivity. Others, e.g., Destruel and Velleman (2014), consider clefts to mark contrast or to correct a previous statement. A third approach analyzes clefts as marking focus unambiguously or marking prominence of the clefted element (cf. De Veaugh-Geiss et al. 2015; Tönnis et al. 2016).

My analysis, to be presented in Sect. 3 and Sect. 4, is closely related to a fourth approach by Velleman et al. (2012). They take a discourse-oriented approach by treating clefts as inquiry terminating (IT) constructions. They show that the exhaustivity inference of clefts is focus sensitive. Compare (7-a) and (7-b), where the inference changes depending on the focus in the cleft pivot.

(7) a. It was PETER's$_F$ eldest daughter that Lena talked to.
  $\rightarrow$ Lena did not talk to anybody else's eldest daughter.
  b. It was Peter's ELDEST$_F$ daughter that Lena talked to.
  $\rightarrow$ Lena talked to no other daughter of Peter's.

Hence, Velleman et al. (2012) argue that clefts contain a focus sensitive operator in the sense of Beaver and Clark (2008). This operator refers to the CURRENT QUESTION (CQ).[5] More precisely, Velleman et al. argue that clefts provide a maximal answer to the CQ and, thereby, terminate the ongoing inquiry about CQ. They predict the cleft in (6) to maximally answer the question *Who did Lena talk to?*. The focus sensitive cleft operator is composed of $\text{MIN}_S(p)$ and $\text{MAX}_S(p)$, where p is the prejacent and S is the context, that contains the current question CQ: "$\text{MIN}_S(p)$, which ensures that there is a true answer to the CQ which is at least as strong as the prejacent p, and $\text{MAX}_S(p)$, which ensures that no true answer is strictly stronger than p." (Velleman et al. 2012:450) While the cleft marks $\text{MIN}_S(p)$ as at-issue content, it presupposes $\text{MAX}_S(p)$.

[5] I will later adopt a different version of CQ based on Simons et al. (2017).

Velleman et al. (2012) point out a problematic example for their own approach, which is illustrated in (8) and (9) (a slightly adapted version of Velleman et al. 2012:449).

(8) A: What did Mary eat?
B: ?It was a PIZZA that Mary ate.

(9) A: What did Mary eat?
B: I thought she said she was gonna get a pasta dish, but I might be wrong.
A: And did she also order a salad?
C: Guys, I was there and actually paid attention. It was a PIZZA that Mary ate.

While the cleft is odd as a direct answer to a question as in (8), it is felicitous once other material intervenes between the question and the answer, as in (9). The approach of Velleman et al., however, incorrectly predicts B's answer to be felicitous as long as it is a maximal answer to A's question, for both (8) and (9).

Velleman et al. (2012) argue that it is not necessary to mark the end of the inquiry in (8), since it is not an extended inquiry. Accordingly, no special cleft marking is needed and that is why the cleft is infelicitous. This argument does not seem very convincing. Strictly speaking, one could even argue that an extended inquiry is more likely to terminate than a short one and, therefore, the latter needs more marking than the former. This would predict the cleft in (8) to be felicitous. Thus, the extent of the inquiry per se cannot be the crucial factor, neither can maximality.

Destruel et al. (2019) also take an approach that is related to the discourse function of the cleft, but from a different perspective. They provide empirical evidence for the hypothesis that the acceptability of French *c'est*-clefts improves the more they indicate that an utterance runs contrary to a doxastic commitment of the interlocutor. Destruel et al. focus on examples of clefts that express a contrast or, more precisely, correction. However, as the data in Sect. 1 shows, not all clefts are contrastive. The referent of *Peter* in example (2) to (5) does not contrast with another alternative in the presented context.

In my analysis, I will adopt the approach of Velleman et al. (2012) by analyzing what kind of question the cleft addresses and how it addresses this question. I will argue that not maximality is crucial, but the expectedness of the question addressed by the cleft (see Definition 1 in the next section).

3 Analysis

I focus on how an *es*-cleft in German contributes to the discourse structure and, in particular, how it does this differently than its canonical counterpart. In a nutshell, I propose that an *es*-cleft addresses a question that came up in the preceding context, but that is not as expected for the addressee to be answered at that point in the discourse as other questions. Those questions that are more expected are answered with a canonical sentence.

In order to provide a formal analysis of *es*-clefts, I first describe the discourse model that I am using. I further develop Roberts' (2012) Question Under Discussion stack. She assumes that, in the course of a conversation, several questions under discussion (QUDs) are piled up on a QUD stack one after the other. Whenever a question is accepted by the interlocutors, it is added to the stack as the top-most element. The top-most question is the one that is supposed to be addressed first. In the following, I will spell out what exactly needs to be on the QUD stack in order to explain the discourse function of a cleft and other sentences. Furthermore, I will argue that the discourse is better represented by assuming a QUD set instead of a QUD stack. I, first, discuss different kinds of questions, that will play a role for my approach. Based on Roberts, Simons et al. (2017) define the notion of the CURRENT QUESTION as opposed to the DISCOURSE QUESTION.

CURRENT QUESTION
The current question (CQ) is directly associated with a corresponding utterance, in particular, it is dependent on the focus of that utterance. Accordingly, the CQ of (10-a) is (10-b).

(10) a. LENA$_F$ talked to Peter.
b. **CQ:** Who talked to Peter?

Simons et al. (2017) analyze the CQ as the domain-restricted subset of the focus alternatives of the utterance. I will adopt this definition of CQ (which differs from Velleman et al.'s (2012) version). For the cleft, I take the CQ to be associated with the cleft pivot, as (11) indicates.[6]

(11) a. It was Lena who talked to Peter.
b. **CQ:** Who talked to Peter?[7]

This definition of CQ is independent of the discourse function of its corresponding utterance. This means that the identification of the CQ of an utterance is independent of whether that utterance is later accepted or not. In order to model

[6] Even with narrow focus inside of the pivot, as in (i), I assume that the CQ is associated with the entire pivot. This is still a point of debate that I will not be able to solve in this paper (see Velleman et al. (2012) and É. Kiss (1998) among others for discussion).

(i) a. It is LENA'S$_F$ boyfriend who talked to Peter.
b. **CQ:** Who talked to Peter?

[7] I intentionally do not assume the cleft question *Who is it who talked to Peter?* as the CQ of the cleft. It is possible that the cleft question would be more adequate as the CQ of the cleft. However, the semantics and pragmatics of cleft questions are even less understood than of cleft assertions. Therefore, the predictions made on the basis of a clefted CQ would be unclear. For reasons of feasibility, I assume an unclefted CQ, which is well understood. Hopefully, the insights about cleft assertions from this paper can contribute to the investigation of cleft questions in future research.

the acceptability of an utterance in a discourse, Simons et al. (2017) introduce the notion of a DISCOURSE QUESTION.

DISCOURSE QUESTION
The discourse question (DQ) is a discourse-segment relative notion and can intuitively be interpreted as the topic of a (sub-)inquiry (*inquiry* in the sense of Roberts 2012). This concept can also be found in van Kuppevelt (1995) who calls it the question that corresponds to the discourse topic. Admittedly, the DQ is not always unambiguously identifiable, which is also pointed out by Tonhauser et al. (2018:footnote 7). I make the simplifying assumption that it is given by the linguistic context. Importantly, it does not depend on the continuation, i.e. cleft versus canonical sentence.

According to Roberts (2012) and Simons et al. (2017), a discourse move realized by utterance *U* is accepted if the CQ of *U* is identical to the DQ or if its CQ contributes to the ongoing inquiry about DQ. If either of these conditions is fulfilled for *U*, it is accepted as a valid discourse move and the CQ of *U* is added to the QUD stack as the top-most element. Whether a CQ contributes to an inquiry, is subject to relevance constraints. The more relevant the CQ is, given the preceding context, the higher its probability to be accepted.

Example (2) from the introduction, however, cannot be explained relying only on the CQ and the DQ. The *es*-cleft and the canonical sentence in (2) have the same CQ: *Who did Lena talk to?*, and whatever the DQ is, it is also identical for both sentences (because they occur in the same linguistic context). Hence, they are either both acceptable or both unacceptable, based on the relation between CQ and DQ. More precisely, it seems that both sentences do not contribute to the DQ, assuming that the DQ is *What did Lena do after the party.* Given that the cleft is still acceptable, I assume that other kinds of questions must be relevant for the acceptability of an *es*-cleft in German, such as sub-questions and implicit questions.

IMPLICIT/POTENTIAL QUESTIONS
My analysis refers to van Kuppevelt (1995) and Onea (2016), who use the concept of IMPLICIT QUESTIONS (van Kuppevelt 1995) or POTENTIAL QUESTIONS (PQs) (Onea 2016). Van Kuppevelt notices that discourses contain many questions that are not formulated explicitly, but arise implicitly, especially in monologues. He characterizes implicit questions as questions "which the speaker anticipates to have arisen with the addressee as the result of the preceding context." (van Kuppevelt 1995:110) He also includes sub-questions of explicit or implicit questions and follow-up questions, if the answer to a preceding question was unsatisfactory, into his discourse model. He concludes that many implicit questions arise due to unsatisfactoriness of the provided answer to a previous question. I will make use of van Kuppevelt's concept of satisfactoriness.

Onea (2016) focuses on potential questions, which are evoked by the immediately preceding utterance, and which are explicitly not sub-questions of preceding questions. An example of a PQ is given in (12-b).

(12) a. My boss called me an idiot today.
b. **PQ:** Why did he call you an idiot?

Most of the examples I analyze in this paper are narratives which do not include explicit questions. Accordingly, I use a model that focuses on the questions that the author anticipates the addressee to have, based on the previous text. I assume that those questions are not all equally expected to be addressed from the perspective of the addressee. Consider example (13).

(13) Lena told Andrea a secret.

The author of this example would anticipate the addressee to wonder what the secret was, but also why Lena told the secret to Andrea. However, intuitively, the first question has a higher expectedness than the second. Whether a question arises and how expected it is, depends on many factors such as the addressee, the situation, and the common ground. The latter is illustrated by (14-a) and (14-b). In (14-a), the potential question *Why did he call you an idiot?* arises. In (14-b), in contrast, the answer to that question is already in the common ground and the question does not arise.

(14) a. Yesterday, I came to the meeting. My boss called me an idiot.
b. Yesterday, I forgot about the meeting. My boss called me an idiot.

In order to account for this difference, I need to include the common ground into my analysis. My definition of the common ground follows Cohen and Krifka (2014) and Krifka (2015), who introduce *Commitment Space Semantics*. Intuitively, a *commitment space* C subsumes all sets of propositions that contain the shared information and any number of continuations that are consistent with that shared information. Furthermore, Krifka (2015:329) defines the *update* of a commitment space C, $C + p$, where p is an assertion. Moreover, in Kamali and Krifka (2020), the update of C with a question is defined. My approach does not rely on this specific definition of common ground. Any other system that incorporates the concept of an update with a proposition or a question would serve the same purpose.

Having introduced all the preliminaries, I will now adapt Roberts' QUD stack. First of all, I introduce the concept of EXPECTEDNESS. I assume that the author of a text can anticipate how strongly the addressee expects a certain question to be addressed, given the provided information (recall example (13)). This understanding of expectedness is based on Zimmermann (2011), who assumes that unexpected discourse developments need extra linguistic marking. I interpret these discourse expectations as expectations with respect to questions. I assume that the addressee has expectations about whether a question will be asked or answered in the next discourse move, and those expectations differ depending on the kind of question and on the previous discourse, hence on the commitment space.[8]

[8] Note that expectedness is formulated from the addressee's perspective. The speaker comes into play when s/he anticipates the expectations of the addressee and chooses her/his next discourse move accordingly.

In order to formalize expectedness, I define the expectedness function f_e, which takes a commitment space C and a possible question q as its input and yields the respective expectedness value e between 0 and 1. Expectedness values (EVs) can also be seen as probabilities for the addressee to expect q to be addressed in a commitment space C. It is a recursive function that evaluates expectedness of a question by describing the difference between the current commitment space and the previous commitment space. This way, each update of the commitment space incrementally changes the EVs of the questions. The formal definition is provided in the following, and will be explained step by step below.

Definition 1 (Expectedness function). *The expectedness function* $f_e : \mathcal{CS} \times \mathcal{Q} \rightarrow \mathcal{E}$ *is a recursive function, where*

- $\mathcal{CS}$ *is the set of all possible commitment spaces,*
- $\mathcal{Q}$ *is the set of all possible questions,*[9]
- $\mathcal{E}$ *is the set of expectedness values from 0 to 1,*

such that for a given $C \in \mathcal{CS}$

$$\sum_{x \in \mathcal{Q}} f_e(C)(x) = 1.$$

f_e *is defined recursively as follows:*

i. For any $q \in \mathcal{Q}$ *and for* $C_0 \in \mathcal{CS}$ *such that* C_0 *is the commitment space at the beginning of a conversation,* $f_e(C_0)(q)$ *assigns a prior EV to* q.[10]

ii. For any $q \in \mathcal{Q}$*, any* $C \in \mathcal{CS}$*, and any update* p*:*

$$f_e(C+p)(q) \propto \begin{cases} f_e(C)(q) + \alpha \quad \textit{iff } q \textit{ is a subquestion of } p & \text{SUB.Q} \\ f_e(C)(q) + \beta \quad \textit{iff } q \textit{ is a PQ of } p & \text{PQ} \\ max((f_e(C)(q) - \gamma), 0) \textit{ iff } q \textit{ is the CQ of } p. \gamma \textit{ differs} & \text{CQ} \\ \quad \textit{with respect to completeness of } p \textit{ as answer to CQ.} & \\ f_e(C)(q) + \alpha + \beta + \delta \quad \textit{iff } q = p. & \text{EXPL.Q} \\ f_e(C)(q) \quad \textit{otherwise.} & \text{OTHER} \end{cases}$$

(For all conditions, it holds that q *has not been answered yet. If* q *has already been answered in* $C + p$*, then* $f_e(C + p)(q) = 0$*.)*

The condition that the EVs have to add up to 1 (for the range of f_e) expresses the similarity of EVs and probabilities. Hence, it implies that once an EV is raised, others must decrease.[11] The function f_e is defined recursively, which means that not only the current update but each former update has an effect on the expectedness of all the questions.

[9] The term *question* refers to the discourse move of asking a question.

[10] I am not concerned with those priors here and will just assume them to be well-defined.

[11] Empirically, the EVs would probably not exactly add up to 1. This needs to be considered if this model is tested empirically.

Step *i.* of the recursion provides the prior EVs at the beginning of a conversation, given that we never start a conversation without any shared information. Based on this common ground, some questions are a priori more expected than others. Step *ii.* defines the effect of an update on the expectedness of different kinds of questions in relation to the previous commitment space.[12] The effect is modeled by adding α, β, or δ, or subtracting γ to/from the EV of q in the commitment space before the update.[13] Those variables are context-dependent, hence, no constants. As we will see in the application of the model, some of the cases must be further subcategorized leading to α_1, ..., α_n, β_1, ..., β_n, etc. The conditions of SUB.Q, PQ, CQ, and EXPL.Q are not at all an exhaustive list of effects on expectedness, but rather a first approach to provide an idea of what can affect expectedness. I will now explain each of the conditions of step *ii.* of f_e.

Condition SUB.Q: This condition describes the strategy of asking a sub-question when one cannot answer the super-question, as in example (15).

(15) $\underbrace{\text{Peter celebrated his birthday.}}_{\text{C}}$ $\underbrace{\text{I wonder who had a present for him?}}_{C+p}$

Here, a sub-question of the broader super-question *Who had a present for him?* would be *Did Nina have a present for him?*, which would have a higher expectedness in $C + p$ as compared to C, if we do not know anything specific about Nina's and Peter's relationship.

Condition PQ: If a commitment space C is updated with a proposition p and q is a PQ of p, then the expectedness is higher after the update (in the commitment space $C + p$) than before in C. An example for this situation would be (14-a), repeated in (16-a).

(16) a. $\underbrace{\text{Yesterday, I came to the meeting.}}_{\text{C}}$ $\underbrace{\text{My boss called me an idiot.}}_{\text{C+p}}$

b. $\underbrace{\text{Yesterday, I forgot about the meeting.}}_{\text{C}}$ $\underbrace{\text{My boss called me an idiot.}}_{\text{C+p}}$

After the first sentence in (16-a), the question of why the speaker's boss called him/her an idiot is not very expected. After the second sentence, however, it is expected, given that it is a potential question of the second sentence. The addition in brackets in Definition 1 explains why, in (16-b), $f_e(C + p)(PQ)$ is not higher than $f_e(C)(PQ)$ for the PQ *Why did he call you an idiot?*, namely because that PQ is already answered.

[12] The definition uses '+' for two different operations: an update as in $C + p$ and for adding a variable to an EV.

[13] Strictly speaking, those variables should not be added or subtracted, but α, β and $(\alpha + \beta + \delta)$ should be increasing functions and γ should be a decreasing function, that take $f_e(C)(q)$ as their argument.

Condition CQ: If a commitment space C is updated with a proposition p and q is the CQ of p, the expectedness of q is lower in $C+p$ than in C. The CQ of p in (17) is *Who did Lena talk to?*, which is very expected at C. At $C+p$, however, this question is addressed and has, thus, a reduced expectedness.

(17) $\underbrace{\text{Who did Lena talk to?}}_{\text{C}}\ \underbrace{\text{Lena talked to PETER.}}_{\text{C+p}}$

This example also shows how γ could differ depending on whether we take p to be an exhaustive answer, a partial answer or a mention-some answer. If we had expected that Lena would talk to many people, the expectedness would be reduced less than if we had expected her to talk to just one person. In the terminology of van Kuppevelt (1995), this describes that the more satisfactorily a question is addressed, the more its EV is reduced.

The function *max* in CQ makes sure that there will be no negative EVs and will yield 0 in case $f_e(C)(q) - \gamma$ would be negative. Otherwise it yields $f_e(C)(q) - \gamma$.

Condition EXPL.Q: For an explicit question q, the expectedness is forced to be high. By taking the sum of α and β plus an additional constant δ, it is guaranteed that an explicit question will always have the highest EV. This means that an explicit question will reduce the expectedness of all questions in $C+q$ compared to C, except for q itself of course.

The actual prior EVs, as well as the actual values for α, β, γ and δ must be determined empirically. One preliminary approach to do this is presented in Westera and Rohde (2019). They presented snippets of texts to participants and asked them which questions are evoked. The EV of a question could be calculated from the relative frequency of that question.

Based on the expectedness function, I now define an adapted version of Roberts' (2012) QUD stack. It is actually not a stack anymore but a set of pairs of questions and their EVs, hence it is called the QUD SET. This set depends on the commitment space, since each update changes the EVs of questions.

Definition 2 (QUD set). *For a $C \in \mathcal{CS}$, the QUD set is defined as the set $\mathcal{S}_C = \{\langle q, e_i\rangle \mid q \in \mathcal{Q}\}$, such that $e_i = f_e(C, q)$.*

My definition of the QUD set differs from earlier versions, like Roberts' (2012), with respect to including also implicit questions. Actually, it includes all possible questions paired with their EVs. The consequence is that accepting a discourse move never implies adding a new question since all questions are already included in the set. Furthermore, the top-most question in the stack looses importance. The model I am proposing allows, in principle, to address any question in the QUD set. Acceptance conditions are modeled via expectedness. Accepting a discourse move means identifying its CQ with a question in the QUD set which has a sufficiently high EV. This is defined as follows for the default case.

Definition 3 (Accepting a sentence – default case). *A sentence p_n is acceptable iff for a given $C \in CS$ and cq_n being the CQ of p_n*

$$f_e(C, cq_n) > e_{def}$$

where e_{def} is the EV that is necessary for a sentence to be accepted by default.

The value e_{def} is most likely not a constant value, but depends on the context and must again be determined empirically. Turning back to the difference between clefts and canonical sentences, I assume the canonical sentence to be a default case which is covered by Definition 3. The cleft, however, is a non-default case resulting in additional requirements for acceptability, as defined below.

Definition 4 (Accepting a cleft sentence – non-default case). *A cleft sentence p_{cl} is acceptable iff for a given $C \in CS$ and cq_{cl} being the CQ of p_{cl}*

$$e_{def} > f_e(C, cq_{cl}) > e_{cl}$$

where e_{cl} is the EV that is necessary for a cleft sentence to be accepted.

What Definition 4 says, is that the EV for an acceptable cleft has to exceed the threshold e_{cl} for an acceptable cleft. Furthermore, the EV of an acceptable cleft has to fall below the threshold for the EV of an acceptable sentence in the default case. In other words, clefts address less expected questions than other sentences do in most of the default cases.[14] Still, also the cleft needs a minimum value of expectedness e_{cl} in order to be acceptable. It cannot address just any question with an EV below e_{def}.[15]

4 Applying the Model

I will now apply the proposed model to the examples mentioned before, which showed a difference in acceptability of the canonical sentence and the cleft. First of all, I will provide a different explanation for the example of Velleman et al. (2012), repeated in (18).

(18) A: What did Mary eat?
B: ?It was a PIZZA that Mary ate.

[14] An anonymous reviewer pointed out that there are other non-default cases, besides clefts, that impose additional restrictions on acceptability, and that could be grouped with clefts. One such example is a sentence including the phrase *by the way*. I argue that we still need to assume different thresholds for each of these non-default cases, since *by the way*-sentences can address even less expected questions than clefts. Even in examples (3) and (5), in which the cleft in unacceptable, a *by the way*-sentence would be acceptable.

[15] In order to account for those cases where both a cleft and canonical sentence are acceptable, one would have to introduce a variable m, that is added to e_{def} in Definition 4. This would make sure that there is an interval of EVs $(e_{def}, e_{def} + m)$ where both the cleft and the canonical sentence are acceptable. For presentational purposes in Sect. 4, I will use the simpler definition in this paper.

According to Definition 1:EXCL.Q, the CQ of the cleft will have the highest EV in the QUD set in the commitment space C_0+q: *What did Mary eat?*. The highest EV will exceed the threshold e_{def} for default cases and, therefore, the cleft is not acceptable in (18) (Definition 4), but a canonical sentence would be acceptable (Definition 3). My approach can also explain why the acceptability of the cleft improves in (9), repeated in (19).

(19) A: What did Mary eat?
B: I thought she said she was gonna get a pasta dish, but I might be wrong.
A: And did she also order a salad?
C: Guys, I was there and actually paid attention. It was a PIZZA that Mary ate.

The difference between (18) and (19) is not only the extendedness of the inquiry but also the amount and the kind of questions that are evoked. B's answer evokes the question *Did Mary get a pasta dish?*, and A's second statement evokes the question *Did Mary order a salad?*. The EVs of both questions increase after the updates, the former is a sub-question of the first question and the latter is an explicit question. This means that they push down the EV of the CQ of the cleft, which is the more general question *What did Mary eat?*. It is plausible that the value is pushed below or at least close to e_{def}. If it is pushed below e_{def}, the cleft is predicted to be acceptable in (19) (Definition 4). Example (19) could also be a case for which both the canonical sentence and the cleft are acceptable.

Also example (2), repeated in (20), can now be explained by analyzing the anticipated questions and their EVs. I indicated the commitment spaces C_0 – C_3 in the example.

(20) (C_0)Lena hat gestern auf der Party mit einem Typen$_1$ gesprochen. (C_1) Die beiden haben viel gelacht und sich direkt für den nächsten Abend verabredet. (C_2) Dann ist Lena glücklich nach Hause gefahren.(C_3)
'(C_0)Yesterday at the party, Lena talked to some guy$_1$. (C_1)The two of them laughed a lot and they agreed to meet again the next evening. (C_2)Then, Lena went home happily.(C_3)'

a. Es war Peter$_1$, mit dem sie gesprochen hat.
it was Peter$_1$ with whom she talked has
'It was Peter$_1$ she talked to.'

b. ?Sie hat mit Peter$_1$ gesprochen.
She has with Peter$_1$ talked
'She talked to Peter$_1$.'

I will discuss this example in a bit more detail, also in order to illustrate how the proposed model works. For simplicity, I assume a very reduced set of possible questions $\mathcal{Q} = \{q_1, q_2, q_3, q_4, q_5\}$. The questions are explicated in (21).

(21) $S_{C_0} = \{$
$\langle q_1$: *What did Lena do after the party?*, $0\rangle$,

$\langle q_2$:*What happened to Lena and the guy after the party?*, $0\rangle$,
$\langle q_3$:*Which guy did Lena talk to?*, $0\rangle$,
$\langle q_4$:*What happened to Lena at the party?*, $1\rangle$,
$\langle q_5$:*How was the conversation?*, $0\rangle\}$

If (20) took place in a context where it is common knowledge that we are talking about Lena and yesterday's party, the QUD set S_{C_0} at C_0 would look like (21). Accordingly, only q_4 will be an a priori expected question in C_0.

Table 1 presents how $f_e(C)(q)$ changes for each question progressing from C_0 to C_3. For the variables from Definition 1, I stipulate values that will make the correct predictions for the purpose of illustration. As mentioned before, the real values would have to be determined empirically. For a strong PQ, I take $\beta_1 = 0.5$ and, for a weaker PQ, I take $\beta_2 = 0.3$. Furthermore, I assume $\gamma = 0.8$ for a rather satisfactory answer (in van Kuppevelt's (1995) sense of satisfactoriness).

Table 1. Application of the model for example (20) for commitment spaces C_0–C_3 and $\beta_1 = 0.5$, $\beta_2 = 0.3$, and $\gamma = 0.8$.

q	$f_e(C_0)(q)$	$f_e(C_1)(q)$	$f_e(C_2)(q)$	normalized
q_1	0	$f_e(C_0)(q_1) = 0$	$f_e(C_1)(q_1) + \beta_2 = 0.3$	0.23
q_2	0	$f_e(C_0)(q_2) = 0$	$f_e(C_1)(q_2) + \beta_1 = 0.5$	0.38
q_3	0	$f_e(C_0)(q_3) + \beta_1 = 0.5$	$f_e(C_1)(q_3) = 0.5$	0.38
q_4	1	$max(f_e(C_0)(q_4) - \gamma) = 0.2$	$max(f_e(C_1)(q_4) - \gamma) = 0$	0
q_5	0	$f_e(C_0)(q_5) + \beta_2 = 0.3$	0 (answered)	0

q	$f_e(C_3)(q)$	normalized
q_1	$f_e(C_2)(q_1) + \beta_1 = 0.73$	0.41
q_2	$f_e(C_2)(q_2) + \beta_2 = 0.68$	0.38
q_3	$f_e(C_2)(q_3) = 0.38$	0.21
q_4	$f_e(C_2)(q_4) = 0$	0
q_5	$f_e(C_2)(q_5) = 0$	0

In C1, after the update with the first sentence, q_1 and q_2 do not change their EV, they fall under the OTHER-condition of Definition 1. I interpret q_3 as a strong PQ of the first sentence and, thus, 0.5 is added to its EV, while q_5 is a weak PQ of the first sentence and 0.3 is added. The question q_4 is the CQ of the first sentence and it is answered rather satisfactory, though not complete. Therefore, 0.8 is subtracted from the EV of q_4.

In C2, after the update with the second sentence, q_1 is interpreted as a weak PQ of the second sentence (probably triggered by the expression *the next evening*) and q_2 as a strong PQ. According to the Definition 1:PQ, 0.3 and 0.5, respectively, are added to their EVs. The EV of q_3 falls under the OTHER-condition and does not change. The EV of q_4 is further reduced since it addresses again the CQ of the second sentence. The value of q_5 is set to 0 since the second sentence fully answers q_5. After the application of f_e, the values are normalized for them to still sum up to 1.

In C3, both EVs of q_1 and q_2 increase, triggered by the discourse progressive element *then* (I treat them as PQs of the third sentence triggered by *then*). I take q_1 to be a stronger PQ than q_2. However, as long as they increase, it would not change the outcome with respect to the acceptability of the cleft if they were switched or if they both received +0.5. Question q_3, q_4 and q_5 fall under the OTHER-condition and keep their EV. Again, the values are normalized.

Now, we could assume $e_{def} = 0.25$ and $e_{cl} = 0.1$ and the model would predict the cleft in (20-a) to be acceptable (Definition 4), since the EV of the CQ cq_{cl} of the cleft falls below e_{def} and above e_{cl}:

$$e_{def} > f_e(C_3)(cq_{cl}) = f_e(C_3)(q_3) > e_{cl}$$

The canonical sentence in (20-b), on the other hand, has the same CQ as the cleft ($cq_{can} = cq_{cl}$), but its EV still does not exceed e_{def}:

$$f_e(C_3)(cq_{can}) = f_e(C_3)(q_3) < e_{def}$$

Hence, the canonical sentence, as a default case, is predicted to be unacceptable.

The most important message to take away from this example is that the EV of q_3, which is later identified with the CQ of the cleft, decreases from update to update just because the EVs of other questions (q_1 and q_2) increase. This is how the length of the inquiry is naturally incorporated as a predictor of acceptability of the cleft. It is, however, not the length per se, but also the relative expectedness of other questions after each update. If the new questions had a low EV or were already answered, the cleft would be predicted to be less acceptable and the canonical sentence might be preferred.

Contra Velleman et al. (2012), my model does not require the cleft to provide a maximal answer to the question it addresses. The model can, therefore, account for examples like (20), where the answer does not mean that Lena did not talk to anybody else. Nevertheless, it does allow the cleft to be exhaustive or provide a maximal answer. Maximality might be a side effect of the discourse function of clefts. My admittedly speculative explanation is that the reader/hearer might pragmatically infer the following: Given that the author/speaker bothered to pick up a question that was already settled or decreasing in expectedness, she probably has a complete/satisfactory answer to it, which would justify addressing it even though it is not expected.

I will discuss the example (3) and (5), repeated in (22) and (23), from the introduction in less detail, but will provide the gist of it.

(22) Lena hat gestern auf der Party mit einem Typen$_1$ gesprochen. Die beiden haben viel gelacht und sich direkt für den nächsten Abend verabredet. **Lena hat ihm sogar ein Geheimnis verraten.**
'Yesterday at the party, Lena talked to some guy$_1$. The two of them laughed a lot and they agreed to meet again the next evening. ***Lena even told him a secret.'***

a. ?Es war Peter$_1$, mit dem sie gesprochen hat.
it was Peter$_1$ with whom she talked has
'It was Peter$_1$ she talked to.'

b. ?Sie hat mit Peter$_1$ gesprochen.
She has with Peter$_1$ talked
'She talked to Peter$_1$.'

Example (22) is predicted by the model, given that the update *Lena even told him a secret* evokes the PQ *What was the secret?* with a very high EV. This would mean that the EV of the PQ *Which guy did Lena talk to?* would be pushed below the cleft threshold e_{cl} and the cleft would be predicted to be an unacceptable discourse move, as well as the canonical sentence of course. Finally, example (23) is easy to explain.

(23) **Lena ist gestern Abend bei der Party angekommen und hat erstmal einen leckeren Cocktail getrunken.** Danach hat sie mit ihrer Freundin Andrea getanzt und die beiden hatten sehr viel Spaß. Dann ist Lena glücklich nach Hause gefahren.
'Lena arrived at the party yesterday and first of all she had a tasty cocktail. *Thereafter, she danced with her friend Andrea and the two of them had a lot of fun. Then, Lena went home happily.'*

a. ?Es war ein Bloody Mary, den sie getrunken hat.
it was a Bloody Mary that she drunk has
'It was a Bloody Mary she drank.'

b. ?Sie hat einen Bloody Mary getrunken.
She has a Bloody Mary drunk
'She drank a Bloody Mary.'

The PQ *Which cocktail did Lena drink?*, which is evoked by the first sentence, can be assumed to be a rather weak PQ. Hence, it receives a low EV. Given the intervening material, that again raises new questions, this already low value will be pushed down further, most likely below the cleft threshold e_{cl}. Therefore, both the canonical sentence and the cleft are inappropriate discourse moves.

5 Conclusion

In this paper, I presented an approach to German *es*-clefts that analyzes them embedded in a broader discourse context and discusses how they structure the discourse. The analysis was based on four constructed examples, that were inspired by examples of clefts from novels, some of which constituted the rare case of the cleft being acceptable while the canonical sentence was not.

In order to describe the discourse function of the cleft, I presented a discourse model that is based on the QUD stack, as assumed by Roberts (2012). My model departed from the original by assuming a QUD set that does not only include current questions but also implicit or potential questions that were evoked by the preceding text. Furthermore, I added the concept of expectedness, which describes for each possible question in a given commitment space how strongly the reader/hearer expects that question to be addressed.

This model is capable of predicting why *es*-clefts in German can be used to refer back to questions that are not particularly pressing at that point in the conversation. And it explains why canonical sentences cannot be used if the question they address is not expected enough. And most importantly, it incorporates the progression of discourse updates and how that affects the acceptability of clefts. To my knowledge, previous approaches had not captured that. Moreover, the model provides an alternative explanation for existing puzzles in the literature, in particular, example (18) and (19) by Velleman et al. (2012).

Previous approaches struggle to account for the differences between the cleft and its canonical equivalent, presented in example (20), (22) and (23), since they cannot be explained by the need to mark exhaustivity, focus, or correction. Even though my approach does not focus on exhaustivity, it is compatible with the observation that clefts do quite frequently express exhaustivity. I analyze exhaustivity as a side effect of the discourse function of the cleft. I suggested that exhaustivity could be a pragmatic inference of the discourse function of the cleft, namely that an author/speaker would only address an unexpected question if she/he had a satisfying answer to it. The fact that the exhaustivity inference was shown to be cancelable (e.g., by Horn 1981) speaks in favor of it being a pragmatic inference anyway. This issue needs further investigations on the interaction of the expectedness of addressed questions and exhaustivity.

Another open issue is correction, which is frequently expressed by a cleft (Destruel and Velleman 2014; Destruel et al. 2019). This is problematic for my account because in corrections the cleft addresses a question that has already been answered and, thus, received the EV 0. My model would, therefore, incorrectly predict that a cleft cannot address that question. However, this seems to be a more general problem of incorporating revisions of a statement made by one of the discourse participants into the discourse model, which exceeds the scope of this paper. It is a promising extension to be investigated in future research, though.

Furthermore, example (4), repeated in (24), remains to be explained.

(24) Lena hat gestern auf der Party mit einem Typen$_1$ gesprochen. Die beiden haben viel gelacht und sich direkt für den nächsten Abend verabredet. Lena hat ihm sogar ein Geheimnis verraten. **Dann ist Lena glücklich nach Hause gefahren.**
'Yesterday at the party, Lena talked to some guy$_1$. The two of them laughed a lot and they agreed to meet again the next evening. Lena even told him a secret. ***Then, Lena went home happily.****'*

a. Es war Peter$_1$, mit dem sie gesprochen hat.
it was Peter$_1$ with whom she talked has
'It was Peter$_1$ she talked to.'

b. ?Sie hat mit Peter$_1$ gesprochen.
She has with Peter$_1$ talked
'She talked to Peter$_1$.'

In this example, there is a pressing question Q: *What was the secret?* intervening between the cleft and the question that the cleft addresses. My model would predict Q to have very high EV that pushes the value of the CQ of the cleft below the cleft threshold and would make the cleft unacceptable. The sentence in bold, however, seems to raise the EV of Q making the cleft acceptable again. As pointed out by a reviewer, the sentence in bold causes the feeling that the speaker is ignoring Q. Hence, the explanation of the cleft's acceptability is rather based on the relation between Q and the sentence in bold than on the relation between Q and the cleft. In order to capture this example, one would probably have to incorporate the effect of a topic shift into the model, following, e.g., van Kuppevelt (1995).

Furthermore, a reviewer pointed out the following example of an acceptable cleft in the context of a cleft question.

(25) A: Who is it that Lena talked to?
B: It is Peter that she talked to.

My model would incorrectly predict the cleft to be degraded given that its CQ (*Who did Lena talk to?*) is very expected to be addressed in the context of the cleft question. I assume that the acceptability of the cleft in this example arises from its interaction with the cleft question. Thus, we first need to understand the discourse effect of a cleft question before modeling the interaction of a cleft and a cleft question. I leave this for future research.

Finally, an important next step would be to determine the variables of the model by conducting a series of suitable experiments, that could be inspired by Westera and Rohde (2019). Once those values are approximated, the model makes testable predictions.

References

Beaver, D.I., Clark, B.Z.: Sense and Sensitivity: How Focus Determines Meaning. Explorations in Semantics, vol. 5. Wiley-Blackwell, Hoboken (2008)

Büring, D., Križ, M.: It's that, and that's it! Exhaustivity and homogeneity presuppositions in clefts (and definites). Semant. Pragmatics **6**(6), 1–29 (2013)

Cohen, A., Krifka, M.: Superlative quantifiers and meta-speech acts. Linguist. Philos. **37**(1), 41–90 (2014). https://doi.org/10.1007/s10988-014-9144-x

De Veaugh-Geiss, J.P., Zimmermann, M., Onea, E., Boell, A.-C.: Contradicting (not-) at-issueness in exclusives and clefts: an empirical study. In: Proceedings of SALT, vol. 25, pp. 373–393 (2015)

Destruel, E., Beaver, D.I., Coppock, E.: It's not what you expected! The surprising nature of cleft alternatives in French and English. Front. Psychol. **10**, 1–16 (2019)

Destruel, E., Velleman, L.: Refining contrast: empirical evidence from the English it-cleft. Empirical Issues Syntax Semant. **10**, 197–214 (2014)

Kiss, K.É.: Identificational focus versus information focus. Language **74**(2), 245–273 (1998)

Horn, L.: Exhaustiveness and the semantics of clefts. In: Proceedings of NELS, vol. 11, pp. 125–142 (1981)

Kamali, B., Krifka, M.: Focus and contrastive topic in questions and answers, with particular reference to Turkish. Theor. Linguist. **46**(1–2), 1–71 (2020)

Krifka, M.: Bias in commitment space semantics: declarative questions, negated quetions, and question tags. In: Proceedings of SALT, vol. 25, pp. 328–345 (2015)

van Kuppevelt, J.: Discourse structure, topicality and questioning. J. Linguist. **31**(1), 109–147 (1995)

Onea, E.: Potential Questions at the Semantics-Pragmatics Interface, Volume 33 of Current Research in the Semantics/Pragmatics Interface. Brill, Leiden, Boston (2016)

Onea, E.: Exhaustivity in it-clefts. In: Cummins, C., Katsos, N. (eds.) The Oxford Handbook of Experimental Semantics and Pragmatics, pp. 401–417. Oxford University Press, Oxford (2019)

Roberts, C.: Information structure in discourse: towards an integrated formal theory of pragmatics. Semant. Pragmatics **5**(6), 1–69 (2012)

Simons, M., Beaver, D.I., Roberts, C., Tonhauser, J.: The best question: explaining the projection behavior of factives. Discourse Process. **54**(3), 187–206 (2017)

Tonhauser, J., Beaver, D.I., Degen, J.: How projective is projective content? Gradience in projectivity and at-issueness. J. Semant. **35**(3), 495–542 (2018)

Tönnis, S.: German es-clefts in discourse. A question-based analysis involving expectedness. Ph.D. thesis, Graz University (2021)

Tönnis, S., Fricke, L.M., Schreiber, A.: Argument asymmetry in german cleft sentences. In Köllner, M., Ziai, R. (eds.) Proceedings of ESSLLI 2016 Student Session, pp. 208–218 (2016)

Velleman, D., Beaver, D.I., Destruel, E., Bumford, D., Onea, E., Coppock, L.: It-clefts are IT (inquiry terminating) constructions. In: Proceedings of SALT, vol. 22, pp. 441–460 (2012)

Wedgwood, D., Pethő, G., Cann, R.: Hungarian 'focus position' and English it-clefts: the semantic underspecification of 'focus' readings (2006)

Westera, M., Rohde, H.: Asking between the lines: elicitation of evoked questions in text. In: Proceedings of the 22nd Amsterdam Colloquium, pp. 397–406 (2019)

Zimmermann, M.: The grammatical expression of focus in West Chadic: variation and uniformity in and across languages. Linguistics **49**(5), 1163–1213 (2011)

Sources

Sendker, J.-P.: Herzenhören. Heyne, München (2012)

Extensions in Compositional Semantics

Thomas Ede Zimmermann(✉)

Goethe University Frankfurt, Frankfurt am Main, Germany
tezimmer@uni-frankfurt.de

Abstract. The paper scrutinizes the very notion of extension, which is central to many contemporary approaches to natural language semantics. The starting point is a puzzle about the connection between learnability and extensional compositionality, which is frequently made in semantics textbooks: given that extensions are not part of linguistic knowledge, how can their interaction serve as a basis for explaining it? Before the puzzle is resolved by recourse to the set-theoretic nature of intensions, a few clarifying observations on extensions are made, starting from their relation to (and the relation between) reference and truth. Extensions are then characterized as the result of applying a certain heuristic method for deriving contributions to referents and truth-values, which also gives rise to the familiar hierarchy of functional types. Moreover, two differences between extensions and their historic ancestors, Frege's *Bedeutungen*, are pointed out, both having repercussions on the architecture of compositional semantics: while the index-dependence of extensions invites a weak, unattested form of 'non-uniform' compositionality, *Bedeutungen* do not; and while the former are semantic values of expressions, the latter pertain to occurrences and, as a result, give rise to a universal principle of extensional compositionality. However, unlike extensions, they are of no help in resolving the initial mystery about learnability.

Keywords: Extensions · Intensions · Truth · Reference · Compositional semantics · Frege · Carnap · Montague

1 A Puzzle About Extensions

The following mystery has been part of the linguists' agenda ever since the beginnings of generative grammar (cf. [6]):

(Q1) How come speakers can identify indefinitely many linguistic expressions?

According to generative folklore, grammars come with (a characterization of) finitely many primitive elements, from which finitely many operations (or rules) derive ever more complex expressions in a stepwise fashion. The details of this process are moot, as is the nature of the elements involved – strings, trees, derivational histories, or what have you. However, there is general agreement that the syntax of a given language largely comes down to an inductive definition

© The Author(s), under exclusive license to Springer Nature Switzerland AG 2022
A. Özgün and Y. Zinova (Eds.): TbiLLC 2019, LNCS 13206, pp. 148–172, 2022.
https://doi.org/10.1007/978-3-030-98479-3_8

of its expressions.[1] Hence identifying the expressions of a language only requires speakers to know finitely many primitives and finitely many ways of combining them – which lets (Q1) appear far less mysterious.

Inductive definitions, then, help mitigating potential learnability worries in syntax. But semantics is plagued with similar problems. Indeed, ever since the early days of generative grammar, (Q1) has been extended from the individuation of linguistic forms to their interpretation (cf. [27]):

(Q2) How come speakers can grasp the meanings of indefinitely many linguistic expressions?

There appears to be wide agreement within the semantic community that at least one natural and plausible strategy of accounting for (Q2) is by way of *compositionality*:

(C) The meaning of a compound expression derives from combining the meanings of its immediate parts.

The details of (C) are open to debate. In particular, the pertinent part-whole relation may be determined by surface syntax or on a separate (syntactic) Logical Form level. However, as long as it is well-founded and the number of ultimate parts and ways of combining them is finite, then so is the knowledge speakers need to acquire in order to interpret the expressions of a language, thus resolving part of the mystery (Q2).

There are other ways of accounting for (Q2) than adopting (C), as there are other reasons for adopting (C) than accounting for (Q2).[2] However, introductory textbooks usually do make a connection between the two, motivating compositionality by appeal to learnability.[3] Yet when it comes to the specifics of semantic analysis, *meanings* quickly give way to *semantic values* that are especially tailored for the process of meaning composition and devoid of any distracting additional features. *Characters* and *intensions* are cases in point: they determine truth and reference (more about which in the next section), but they avowedly leave the expressive dimension of meaning out of account, given that it does not interfere with the compositional process.[4] According to the truth-conditional approach, their knowledge is part of language mastery. Indeed, inasmuch as the conditions under which a sentence S of a language L is true can be modeled by a set of points in Logical Space, knowing what S means

[1] I shun Chomsky's [6, p. 24] possibly more restrictive term 'recursive device' for reasons given by Tomalin [55, p. 307].

[2] For instance, those provided by Janssen [23, Sect. 4] and Szabó [53, Sect. 3], respectively.

[3] See, e.g., the textbooks by Dowty, Wall & Peters [10, pp. 4–10]; Heim & Kratzer [22, pp. 2f.]; Chierchia & McConnell-Ginet [5, pp. 6–8]; as well as Zimmermann & Sternefeld [63, pp. 58f.].

[4] I am assuming familiarity with the basic architecture of Kaplanian two-dimensional semantics along the lines of [26]. As to expressive meaning and its relation to compositional values, see Sect. 7 of Gutzmann's survey article [20].

amounts to knowing which proposition S expresses – i.e. its intension. In general, though, this requires additional contextual knowledge, on top of mastering L: the character of S, which is known to any L-speaker, only determines S's intension given a context of utterance. However, speakers cannot always be expected to be acquainted with all aspects of the utterance context that are needed to identify the intensions of deictic expressions. But they may be expected to know in what way the intensions of expressions depend on the utterance context, which is why a slightly more involved version of (C) is called for in the general case:

(C_χ) The character of a compound expression derives from combining the characters of its immediate parts.

The compositionality of characters (C_χ) suggests a straightforward route to a partial answer to (Q2): as far as those parts of meaning that determine truth and reference are concerned, speakers can acquire them by associating a character with each of the finitely many words in the lexicon and an operation on characters with each of the finitely many ways of forming compound expressions from their immediate parts. Moreover, (C_χ) has an interesting consequence for the large fragment of *eternal* expressions that do not contain any deictic (or, more generally, context-dependent) expressions and whose intension will thus be the same across all contexts. Hence speakers of L can identify the intension of a sentence S on the basis of their knowledge of S's character and thus solely in virtue of their linguistic competence. Since the same goes for S's sub-expressions, the following adaption of (C), with the superscript indicating its restricted range, may indeed help providing a partial answer to (Q2):

(C_ι^-) The intension of an eternal compound expression derives from combining the intensions of its immediate parts.

To begin with, the intensions of all non-deictic words are known to all speakers, as the values of the (constant) characters they have learned to associate with these words. Moreover, to combine the intensions of eternal expressions, speakers may apply the pertinent character combinations to the values of the (constant) characters of their parts and then obtain the value of the resulting (constant) character. Hence, as far as the intensions of eternal expressions are concerned, speakers can acquire them as a by-product of acquiring the compositional interpretation of their characters.

It should be noted that it is crucial for the above reasoning about (C_χ) and (C_ι^-) that characters and, as a consequence, certain pertinent intensions are known to speakers by virtue of their language mastery. Given this assumption, there is nothing *per se* wrong with trading meanings for semantic values when making a connection between compositionality and learnability. However, the semantic values that usually enter the equations and adorn the analysis trees in the semantic literature are neither characters nor intensions, but *extensions*. In particular, textbooks tend to feature a principle of *extensional compositionality* that holds throughout a large part of any language:

(C_ε^-) The extension of a compound extensional expression derives from combining the extensions of its immediate parts.

More often than not, though, the extension of an expression is unknown to the majority of speakers. As a case in point, given that the extensions of (declarative) sentences are their truth-values, knowing them all would come close to omniscience – hardly a prerequisite to linguistic proficiency. To be sure, (C_ε^-) is not supposed to apply to all expressions of a given language. As Frege [15] famously pointed out, the principle stops short at attitude reports and other (nowadays called) *intensional* environments. However, a substantial part of language does respect (C_ε^-), and it this *extensional fragment* that introductory semantics courses use as their didactic starting point. Still, restricting (C_ε^-) to extensional environments does not bring it closer to linguistic (as opposed to empirical) knowledge: even though all parts of a sentence like *No planet outside our solar system contains plants* conform to (C_ε^-), the extensions of some of them are unknown to speakers of English. Hence the connection between (C_ε^-) and (Q2) is not as straightforward as in the case of (C_χ) and (C_ι^-). This, then, is the puzzle announced in the section header:

(P) How does the compositionality of extensions bear on speakers' ability to grasp the meanings of indefinitely many linguistic expressions?

As it turns out (and the reader may already have noticed), (P) is not a substantial problem; a straightforward solution will be presented in due course. So why bother? Two reasons. For one thing, it is important to realize that there is a gap between motivating the compositionality of meaning in terms of learnability on the one hand, and observing the compositionality of extensions throughout a large portion of language on the other. Anyone who follows this line of reasoning should therefore be prepared to address this gap and close it; in particular, those of us who teach semantics at an introductory level ought to be aware of this commitment. For another thing, and perhaps more importantly, in order to address (P), one needs to have a clear understanding of what extensions are in the first place, arriving at which is the principal aim of this paper.

In the main body of this paper, the very concept of extension will be scrutinized and teased apart from various related yet importantly distinct constructions. More specifically, Sects. 2 and 3 address what I (and others) take to be the core of the theory of extension and intension. The discussion will be mostly in line with Montague's [36] and [37]. In particular, it will be based on the possible-worlds version of intensions (and thus certain extensions: see below), rather than Carnap's [4] original state descriptions. Moreover, extensions will not be sharply distinguished from Frege's [15] *Bedeutungen*, since much of what will be said applies to both concepts alike. Some less obvious differences will, however, be addressed in Sect. 4. The final section returns to the puzzle about (C_ε^-). As it turns out, unlike Carnapian intensions, Fregean *Bedeutungen* stand in the way of a straightforward answer to (Q2).

2 Truth and Reference

2.1 Extension and Reference

The term *Bedeutung* as used by Frege is commonly rendered by *reference*.[5] Indeed, in certain prototypical cases, an expression does refer to its extension. However, there are at least two reasons why the two notions of reference and extension are better kept apart:

- For one thing, an expression may have an extension that is distinct from its referent. Names and descriptions, if analysed as quantifiers, are cases in point. Thus, for instance, to say that the extension of the name *Batumi* is the set of all sets (of individuals) that contain the city of Batumi as a member does not mean that that name refers to a set, let alone that that city is a set. Rather, *Batumi* is the name of a city, and it is that city that the name refers to. However, there is a close relation between the referent and the extension of the name: the latter determines the former in a canonical way. In fact, part of the predictive power of semantic theory rests on the fact that at least the referents of some expressions can be gleaned from their extensions – even if they do not coincide with them.[6]
- For another thing, many linguistic expressions do not have a referent in the first place, and assigning them an extension does not make them refer to it (or anything else, for that matter). Determiners and (declarative) sentences are cases in point. It is hard to see what is won by proclaiming disjointness the referent of *no*. And even though it is a constitutive feature of the theory of extension and intension (as I understand it) that the extensions of declarative sentences are truth-values, saying that *two plus two equals four* refers to the same object as the description *the truth-value of all tautologies* is at best a puzzling or eccentric way of speaking but certainly not a result of semantic or conceptual analysis.

[5] – at least since the first English edition [18] of [15], due to Max Black. Later translations traded *reference* for the even more unfortunate term *meaning* (the closest English cognate of *Bedeutung* in ordinary German), thus making the English text as rough a read as the original; Beaney [2] tells the whole history. In his defense of Black's original translation, Bell [3, p. 193] speculates that 'employing everyday if somewhat misleading words, [Frege] believed that fewer readers would be repelled and his works would gain a wider audience'. I concur, but then Montague's [36], [37] and Putnam's [44] use of *extension* for *Bedeutung*, philologically questionable though it may be, fares even better in view of the fact that it lacks any non-technical meaning or connotation. Of course, Carnap [4, §29] had scrupulously laid out the differences between his distinction and Frege's.

[6] In an earlier paper [60], I have said more on the distinction between theory-internal values like extensions and their relation to theory-external objects like referents. To be sure, all this is about *semantic* reference – a concept whose coherence I do not wish to dispute, but that needs to be distinguished from the purely pragmatic concept of speaker's reference, as famoulsy pointed out by Kripke [32].

So rather than stretching the notion of reference beyond recognition, one should only apply it where it makes immediate sense. This certainly includes the *individual* reference of proper names and (non-empty) definite descriptions, but also the *divided* reference of common nouns and sentence predicates, which may be construed as distributing individual reference across their extensions.[7] However, it clearly does not extend to sentences and their truth-values. If a less misleading cover term for both is needed, *extension* does not appear to be the worst choice, even as a translation of *Bedeutung*. For the most part of this paper, excepting Sect. 4, I will follow this tradition and also use the same notation for both, writing '$[\![A]\!]^i$' for the extension of expression A relative to an index i (= world, time, world-time pair,...).

Why would anyone want to put truth-values and referents in the same category? To be sure, both concern and depend on the objects linguistic expressions are about. Moreover, the syntactic environments in which substitution of co-referential nominals preserve reference and truth-value seem to be the same as those in which substitution of materially equivalent clauses do. Arguably, this remarkable fact had been Frege's primary motive to treat them on a par, but it is hardly a clue as to why truth-values should be the sentential counterparts of nominal referents.[8] What, then, if anything, unifies the two? The answer to this question depends on what is taken to be the prototypical case of reference. More specifically, it is the difference between individual and divided reference that is at stake. I will take the two options in turn.

2.2 Truth and Individual Reference

If the starting point for the motivation of truth-values is individual rather than divided reference, the index-dependence of both the referents of definite descriptions and the truth-values of sentences may be exploited to assimilate the latter to the former. Given an index i, the phrase *the open door* refers to the sole, or perhaps the most prominent, object in i that is a door and standing open in i.[9] Identifying the extension of that description with its referent, one may characterize $[\![\textit{the open door}]\!]^i$ as that object in i that fits the description *[the] open door* in i. Extrapolating from this example (and glossing over a lot of messy details), the following recipe for determining the extension of a definite description D emerges, where the ι-operator is to be understood along Fregean lines.[10]

---

[7] Quine, who introduced the term [45, pp. 8ff.], took *divided reference* to be epistemologically prior to individual reference. Traditionally, the term *extensio[n]* has chiefly been used for divided reference; cf. Frisch's [19, pp. 183ff.] detailed account.

[8] The same point has been made by Tugendhat [56, pp. 178f.]. (In a recent paper, Richard Heck and Robert May [21] offer an interesting reconstruction of Frege's own motives, relating them to the analysis of (Boolean) connectives.)

[9] – or in any situation corresponding to i, given that in general indices are tuples of situational parameters, in the sense of [59, Sect. 8].

[10] ...as defined by Heim & Kratzer [22, p. 75, (5)]. A Russellian construal of the definite article ([47], [37, p. 393]) would have to distinguish between extension and referent,

(1) $[\![D]\!]^i = \iota x.\ x$ is in i and x matches D at i

A straightforward way of adapting (1) to sentences S and their truth-values is by taking them to refer to the entire index:[11]

(2) $[\![S]\!]^i = \iota j.\ j$ is in i and j matches S at i

(2) might be hard to parse, but it can still be made sense of. To begin with, the containment relation between indices may be construed as identity, assuming that no index can be part of another, distinct one (with a qualification to be made in the next paragraph). Next, the notion of *matching* is naturally construed by taking propositions expressed by sentences as sets of indices they are true of. And the potentially alleged index-dependence of this match may either be taken as redundant or as a result of diagonalizing a character.[12] I concentrate on the former option, if only to avoid two-dimensional complications, and reformulate (2) as:

(3) $[\![S]\!]^i = \iota j.\ j = i$ and S is true of j

According to (3), the extension of a sentence S is the index i at which it is evaluated – provided that S is true of i; otherwise the extension of S is undefined. Hence one may identify the truth-values accordingly: the evaluation index i corresponds to truth, whereas some default value # may mark absence of truth as undefinedness. That the former comes out as situation-dependent while the latter is not even a proper object ought to be regarded as cosmetic irritations that should not blind anyone to the achievement of (2) and (3): they bring out the parallelism between nominal and sentential extensions.

The reformulation (3) of (2) turns on the identification of index-inclusion and identity. This may be natural for some construals of indices – worlds or parameter-tuples, say – but it is a dubious move when it comes to a more naturalistic evaluation of expressions relative to situations with rich mereological structure. One way to go is to shift the starting point of the analogy between referents and truth-values from (1), which is restricted to descriptions headed by singular count-nouns, to (4), which also covers plurals and mass nouns and takes the part-whole relation $\sqsubseteq$ into account (following Sharvy [49, p. 612]):

(4) $[\![D]\!]^i = \iota x.\ x$ fits D in i and for all y: if y fits D in i, then $y \sqsubseteq x$

The sentential counterpart of (4) can be obtained by confining the situations j that fit a sentence S at a point i to parts of i:

but still provide the same two possibilities, depending on whether the extension of the head noun is a singleton.

[11] A sketch of this characterization of truth-values can be found in a handbook article [61, p. 191], which follows my earlier class notes [57, pp. 108f.]; Fabian [12, p. 86] and Leonardi [34] have expressed similar ideas.

[12] More precisely, in a two-dimensional framework. (2) could be the result of evaluating the character $[\lambda i.\lambda k.\iota j.\ j$ is in k and j matches S at $i]$ at context i.

(5) $[\![S]\!]^i = \iota j.\ S$ is true of $j \sqsubseteq i$ and for all k: if S is true of $k \sqsubseteq i$, then $k \sqsubseteq j$

As long as sentences S express 'persistent' propositions (as postulated by Kratzer [29, p. 616ff.]), (5) will pick out i if S is true and nothing otherwise, as desired.

2.3 Truth and Divided Reference

Both (3) and (5) adapt to divided reference if the ι-operator is replaced by set (or λ-) abstraction. However, a more compelling analogy between nominal referents and truth-values emerges if sentence predicates are construed as (distributively) referring to the (individual) referents of potential (singular) subjects: the predicate *snore* thus (distributively) refers to all (singular) referents u of (singular) subjects x of whom it can be truly said that x snores, which may be comprised in one single set $\{u$: u snores$\}$. Although this identification of the referents of verb phrases has come under attack since the days of Frege and Carnap, I will stick to it here, if only for simplicity.[13] In any case, the divided individual reference of sentence predicates does not directly extend to other kinds of verbal constituents: transitive verbs like *kiss* distributively refer to pairs of individuals and ditransitive verbs like *give* distribute their reference over triples, although they side with transitives, once partially saturated with an (indirect) object, as in *give a student*. Indeed, there is a familiar parallelism between the number n of (nominal) arguments $x_1, \ldots, x_n$ of an n-place predicate P and the length of the tuples $(u_1, \ldots, u_n)$ it distributively refers to: the predicate *kiss* thus refers to the pairs (u_1, u_2) that are the respective referents of subjects and objects such that x_1 *kisses* x_2 holds true. More generally, the referents of an n-place predicate P may be comprised in its *satisfaction set*, which may in turn be identified with its extension:[14]

(6) $[\![P]\!]^i = \{(u_1, \ldots, u_n) \mid P(x_1, \ldots, x_n)$ is satisfied by $u_1, \ldots, u_n$ at $i\}$

Since an n-place predicate may be construed as a sentence that lacks n (nominal) arguments, a sentence may be construed as a 0-place predicate. Adapting (6) to the limiting case $n = 0$ thus leads to the following characterization of the extension of a (declarative) sentence S:

(7) $[\![S]\!]^i = \{(\,) \mid S$ is satisfied at $i\}$

Here $(\,)$ is the (only) 0-tuple, which will be identified with the empty set $\emptyset$, and satisfaction boils down to (index-relative) truth *simpliciter*, given that there are

[13] There are good reasons for analyzing verb meanings in terms of events, as famously pointed out by Davidson [8] as well as his pre- and successors Ramsey [46] and Parsons [41]. Even so, much of what will later be said about extensions in general, carries over to event semantics.

[14] The parallelism also plays a role in Frege's [14] construction of verbal extensions to be addressed in the next section. The characterization (6) is reminiscent of Tarski's [54] account of predicate logic, as is the downward generalization (7) due to Carnap [4, §6-1]), but neither can be found in Frege's writings, where individual referents were taken as basic and predicate extensions were derived from them.

no satisfiers. Hence the extensions of all true and all false sentences come out as $\{\emptyset\}$ and $\emptyset$, respectively, and could thus be taken as the set-theoretic surrogates of the truth-values.[15]

3 Determining Extensions

Though the derivation of truth-values as extensions of 0-place predicates appears more cogent than their justification by analogy to the referents of definite descriptions, I will follow semantic tradition and make the connection between language and world in terms of truth and individual reference. The exponents of this connection are truth-valuable sentences and referential terms, whose parts systematically contribute to the truth-values and referents they have. According to Fregean tradition, they do so in a compositional way: the extensions of sentences and terms – i.e., their referents and truth-values – are obtained by suitably combining the extensions of their immediate parts, which in turn are obtained by suitably combining the extensions of their parts, etc.[16] For this to work, all expressions need to have extensions that they can contribute in the composition process. What, then, are the extensions of expressions that neither refer nor have truth-values? The answer to this question is usually given by construing contributions as functions:[17]

[15] This is so because the set in (6) contains all (and only) the objects of the form $(u_1, \ldots, u_n)$ such that the clause to the right of the abstraction operator '|' holds; hence the set in (7), where $n = 0$, contains all (and only) the objects of the form $\emptyset$ such that S is satisfied (at i). So if S is true (at i), then $[\![S]\!]^i$ contains precisely the objects of the form $\emptyset$, i.e. $[\![S]\!]^i = \{\emptyset\}$, because there is only one such object, viz. $\emptyset$; but if S is false (at i), then nothing (and a fortiori, nothing of the form $\emptyset$) satisfies the clause to the right of '|', and thus $[\![S]\!]^i = \emptyset$. It so happens that the two sets $\emptyset$ and $\{\emptyset\}$ are the set-theoretic representatives of the numbers 0 and 1, given the standard ordinal construction by von Neumann [38] – or, in fact, Zermelo (cf. [11, pp. 133f.]). Moreover, in view of this representation, the truth-functional connectives come out as the familiar set-theoretic Boolean operations on $\emptyset$ and $\{\emptyset\}$ – conjunction as intersection, disjunction as union, etc.

[16] This characterization of extensions generalizes the basis of Frege's [13, p. x] infamous context principle, according to which sentences should form the starting point of semantic analysis, from truth-evaluability to referentiality. Without this generalization, functional contributions risk becoming either indeterminate or non-well-founded, in view of the relativity of the function-argument relation (aka 'flip-flop' [42, p. 375]). Though it is unlikely that this was Frege's motive for generalizing truth (-values) to reference, it is helpful for the set-theoretic construal of (FH1). Critical discussions of the tension between compositionality and contextuality can be found in papers by Pelletier [43] and Janssen [24].

[17] Frege's idea of 'reifying' contributions to semantic values is actually independent of the choice of extensions and can also be used starting from truth-values and divided referents, or even from individuals and propositions; in the latter case, the so-called 'Russellian' hierarchy of denotations ensues, which turns out to be equivalent to (8) below, modulo some suitable coding like Kaplan's Russelling of Frege-Churches [25].

(FH1) *1st Fregean Heuristics* cf. [14]
Unless determined independently, the extension of an expression X is that function that assigns to every extension of a (possible) sister constituent Y of X the extension of the mother constituent $X + Y$.

Applications of (FH1) abound and ought to be familiar from semantics textbooks; see, e.g., [22, Chaps. 2 and 6]. In particular, the extensions of (i) verb phrases, (ii) transitives, (iii) ditransitives,... can be derived as (curried) (i) unary, (ii) binary, (iii) ternary,... relations. In all these cases, Y would be a referential nominal, whose extension coincides with its referent. In case (i), $X + Y$ is a sentence with subject Y and predicate X, whose extension coincides with its truth-value.[18] For (ii), Y would be the direct object of X and $X + Y$ would be a verb phrase, whose extension has been determined by an earlier application of (i); the result of (ii) could in turn be fed into (iii), where $X + Y$ is a complex transitive with indirect object Y; and so on. As a result, the extensions obtained for verbal constituents are curried versions of the divided referents assumed in Sect. 2.3. Still, the truth-values cannot be derived in the same way, since they have been taken for granted from the very first step (i) of applying (FH1).

Part of the power of (FH1) lies in the fact that it can be iterated *ad libitum*, as illustrated in steps (ii) and (iii) above: once suitable extensions of one kind of expression have been identified, the result can be fed into further applications of (FH1), thus proceeding from immediate constituents to indirect ones. This way a large number of ever more abstract extensions are obtained.

Some remarks on the status of (FH1) are in order. To begin with, it is not part of semantic theory but merely a *discovery procedure* (in the sense of [6, p. 51]) that helps identifying possible candidates for contributions to referents and truth-values. If everything goes well, it will output theoretical entities that may play the role of extensions in a compositional account of the language under scrutiny. In fact, as an immediate consequence of (FH1), the extensions of the mother constituents $X + Y$ locally satisfy ($\mathrm{C}_{\varepsilon}^{-}$) in that they can be obtained from those of their immediate parts: $[\![X+Y]\!]^i = [\![X]\!]^i([\![Y]\!]^i)$. Thus the success of (FH1) as a strategy of determining extensions may lead to the impression that functional application is a (cognitively or linguistically) privileged combination of extensions, when it is actually the result of a specific way of reconstructing (some) contributions to referents and truth-values in terms of (mathematical) functions.

The procedure (FH1) is far from being perfect:

- It is not deterministic: which extension (FH1) outputs for a given expression X depends on the syntactic environment in which it is applied. Quantificational nominals X like *everyone* are a case in point. Their extensions are

[18] Hence $X+Y$ need not be the concatenation of (the terminal strings of) X and Y; in particular, the order may be reversed, and additional morpho-syntactic interactions may take place. It should also be mentioned that binarity is not essential: (FH1) easily generalizes to n-ary constructions, where $n \geq 1$. The derivation of the extensions of coordinating conjunctions as binary connectives is a case in point.

usually determined by taking the sister Y to be a verb phrase and X as its subject so that the outcome is the familiar (characteristic function of a) set of predicate extensions. However, in principle the procedure could also apply where X is the object of a transitive verb, whence its extension comes out as a function from binary to unary relations.

- Even arbitrarily many iterations of (FH1) need not cover all expressions X of a given language, because no sister nodes Y and mothers $X+Y$ can be found both of whose extensions have been identified before. The immediate constituents of English determiner phrases like *every person* are a case in point. Neither seems to be part of other syntactic constellations that would allow determining its extension beforehand. Hence independent considerations are required to find appropriate extensions, like the assumed equivalence of nouns (*person*) and corresponding predicate nominals (*be a person*).
- (FH1) fails to produce results in the absence of *extensionality* (aka *extensional substitutivity*), when the extension of the mother node $X+Y$ does not depend on that of sister Y, as substitution of Y by some co-extensional Z brings out: $[\![X+Y]\!]^i \neq [\![X+Z]\!]^i$, though $[\![Y]\!]^i = [\![Z]\!]^i$. Attitude verbs X like *know* and their clausal complements are a case in point: two materially equivalent sentences Y and Z may lead to distinct sets of attitude holders when embedded under the same verb.

It is, of course, the third imperfection that gave rise to an amendment of (FH1) to save compositionality:

(FH2) *2nd Fregean Heuristics* cf. [15]
In the absence of extensionality, the extension of X is that function that assigns to every intension of a (possible) sister constituent Y of X the extension of the mother constituent $X+Y$.

According to (FH2), intensions come to the rescue when substitution failures challenge extensional compositionality. This repair strategy raises two problems. Firstly, how are intensions individuated? And in the second place, is there any guarantee that they do not pose additional substitution problems? As to the first question I will, until further notice, follow the popular, broadly Carnapian strategy of identifying intensions with (set-theoretic) functions from indices to extensions. Unfortunately, this decision has negative repercussions on the second question: most of the environments that give rise to extensional substitution problems are also sensitive to intensional substitution. I will again follow semantic tradition and ignore this complication in the hope it can be resolved by pragmatic considerations (see, e.g., Stalnaker's [50,51] suggestions); however, the topic will be briefly revisited in Sect. 5.

Even though (FH2) inherits the other two imperfections of (FH1), together the two discovery procedures go a long way towards identifying extensions (and thereby intensions) for all expressions. However, since an expression whose extensional contribution has been determined in one environment, may also occur in many other environments, this generalization is bound to lead to more compositional combinations than functional application (or its intensional variant resulting from (FH2)); see [58] for pertinent considerations. But to the extent that the functional contributions constructed according to these heuristics are sufficient, the following peculiar form of compositionality emerges, which Montague [36, pp. 75f.] dubbed *Frege's functional principle*:

(FFP) The extension of a compound expression derives from combining the extensions or intensions of its immediate parts, depending on whether they exhibit extensional substitutivity.

Given the Fregean heuristics of identifying them, extensions come in two kinds: the *basic* ones that connect language with the world; and the *derived* extensions that are obtained by applying (FH1) or (FH2). The former are the truth-values and the individual referents. The latter are functions assigning mother extensions to sister extensions or intensions; moreover, in the latter case one of the sisters also contributes its intension to the maternal extension. Hence the extensional contributions may be arranged in a hierarchy of intensional types:[19]

(8) *The hierarchy of extensional contributions* cf. [37]

- Individuals and truth-values are extensional contributions of type e and t, respectively;
- functions from extensional contributions of any type a to extensional contributions of any type b are extensional contributions of type (a, b);
- functions from the set of indices to extensional contributions of any type a are extensional contributions of type (s, a).

The (allegedly Fregean) principle (7) must not be confused with a weaker principle to the effect that intensions behave compositionally, which – applied to his 'Sinne' and restricted to eternal expressions – Frege apparently took to be obvious:

(C_ι) The intension of a compound expression derives from combining the intensions of its immediate parts.

(C_ι) is much stronger than the principle (C_ι^-) above, which was restricted to eternal expressions. Still, it is frequently taken to be valid, quite independently of any learnability considerations. Thus, in Kaplan's [26, p. 510] two-dimensional account of context-dependence, (C_ι) comes in the guise of a ban on so-called *monsters* – environments in which the extension of a mother node would depend on the whole character of one of its daughters. On the other hand, (C_ι) is slightly weaker than (FH2) as will be argued next.

---

[19] The term *extensional contribution* is meant to avoid the paradoxical ring which the more common term '(possible) extension' has in view of the third clause of (8). – As Klev [28, p. 75] pointed out, the split between e and t goes against the spirit of Frege's view of truth-values as referents; but it has become an important ingredient in type-logical reconstructions of Frege's theory of reference, starting with Church [7].

4 Extension and *Bedeutung*

In this section, I wish to address some potential terminological confusions surrounding the very term *extension* and its relation to Frege's *Bedeutung*. It is well-known that, despite certain commonalities, Carnap's [4] intensions, like its modern descendants [37], are much more coarse-grained than Frege's [15] *Sinne*. Given that the extensions of 'intensional operators' (like modal and attitude verbs) have intensions in their domain, this granularity gap also affects the difference between extensions and *Bedeutungen*. Moreover, since the derived extensions are set-theoretic functions from the hierarchy (8), they too are less fine-grained than their Fregean counterparts. We will get back to these differences in Sect. 5, arguing that this fine-grainedness blocks the road to a straightforward solution to the initial puzzle (P). In the current section two less obvious differences and their impact on semantic analysis will be scrutinized.

4.1 The Arguments of *extension*

The technical term *extension* denotes a binary function from parameters and expressions to extensional contributions,[20] where the parameters comprise anything that the extension of a given expression may depend on: model, assignment, context, index, etc. Suppressing all parameters but the index, the extension of *extension* (in semantics parlance – as opposed to the hairdresser's usage, which Daniel Hole reminded me of in correspondence), is that function f that satisfies $f(X)(i) = [\![X]\!]^i$ for any expression X of the object language (non-technical English, say) and any index i; likewise, assuming the rigidity of proper names [31], the extension of the complex nominal *the extension of 'Batumi'* is the city of Batumi. However, like all relational or functional nouns of English as well as its cognates in other languages, the technical term *extension* has a number of additional usages obtained by dropping arguments. As a rule, these missing arguments are interpreted either by reference to some contextually given object or by existential quantification over such objects.[21] Thus the functional noun *capital* can refer to the singleton of Tbilisi or Atlanta if the contextually salient state happens to be one of the Georgias; and in predicative position the same noun may denote the set of all capitals of a given region – the German federal states, say: {Berlin, Munich, Hamburg, Düsseldorf,...}. For expository purposes, the contextual and the existential construals may be distinguished by subscripts 'i' and '$\exists$', where the former denotes the pertinent

[20] Depending on how expressions are individuated, a further argument place specifying the object language may also be needed.

[21] The availability of these construals, which are presumably structural ambiguities or systematic polysemies, appears to be restricted, one potential factor being functionality. Thus, e.g., while the plural description *the fathers* may be used to refer to a group determined by existential quantification, an analogous reading of *the cousins* seems hard to hear. Although restrictions on the omission of verb arguments have been studied (notably by Sæbø [48]), I am not aware of any account of these asymmetries in the nominal domain.

index that the latter quantifies over. Since the noun *extension* has two independently elidable argument slots, disambiguation may proceed by a superscript (for the expression) and a subscript (for the index). Thus, e.g., *the* $extension_i$ *of 'It's raining'* may refer to the truth-value of *It's raining* at a contextually given index i; the plural *the* $extensions_\exists$ *of 'the number of planets'* can denote the set $\{n| \ (\exists i) \ [\![\textit{the number of planets}]\!]^i = n\}$; the statement *Truth-values are* $extensions^\exists_\exists$ says that $\{0,1\} \subseteq \{x|(\exists X)(\exists i) \ x = [\![X]\!]^i\}$ (where the existential quantifiers are likely to be contextually restricted); etc.

Turning to the principle (C_ε^-) now, it appears that *extension* is the relevant reading for both occurrences of the surface form *extension[s]* in it: the expression argument is overt (in both occurrences) and existential quantification over the index would obliterate its intended co-reference across the occurrences – mother and daughters are meant to be evaluated at the same index, of course. However, there is still a subtle ambiguity hidden in the formulation (C_ε^-): the evaluation index i shared by both occurrences of $extension_i$ may or may not coincide with the index at which the combination of the daughters' extension is performed. Of course, due to the generic character of (C_ε^-), i is not one contextually salient index but intended to be arbitrary, or universally quantified.

To see what is at stake here, one may consider an analogy involving a well-studied functional noun (cf. [35]):

(9) a. The temperature can be read off from a thermometer.
    b. The temperature can be read off from www.wunderground.com.

On its most prominent reading, (9-a) expresses that a thermometer read off at a given (local and temporal) position, will provide the temperature at that position. While (9-b) can also be construed along these lines, an equally obvious reading ensues if *temperature* relates to an aforementioned distant time and place – e.g., in response to a detective's question concerning the weather at the scene of a crime. Using the above disambiguation device, the difference between the two constellations in (9) are brought out in:

(10) a. The temperature_i can be read off_i from a thermometer.
     b. The temperature_i can be read off_j from www.wunderground.com.

In (10-a) and (10-b), the second subscript stands for the index of evaluation of the predicate *can be read off*, i.e., the time and place at which a potential reading takes place. Like the first index, it is likely to be construed as universally quantified. Hence the crucial difference between the two sentences consists in the coreference between that position and the argument of the functional noun *temperature*: to get the temperature at a position, you may use a thermometer *at that position* or consult a certain website *even when at some potentially different position.*

While the disambiguations in (10) correspond to the most plausible readings of the sentences in (9), the opposite distribution of subscripts is certainly also possible: (9-a) may be taken to express that a thermometer can be used at one position to determine the temperature at another one, though this is unlikely by what we know about how thermometers work; and similarly for (9-b). Semantically, both sentences are ambiguous as to how the index of evaluation of the predicate and the argument of the noun *temperature* are related. This, then, is the very kind of ambiguity that can also be observed in (C_{ε}^{-}):[22]

(11) a. The extension$_i$ of a compound expression derives from combining$_i$ the extensions$_i$ of its immediate parts.
b. The extension$_i$ of a compound expression derives from combining$_j$ the extensions$_i$ of its immediate parts.

Using semantic notation, the two readings (11-a) and (11-b) of (C_{ε}^{-}) can be expressed by the following two equations:

(12) a. $[\![X+Y]\!]^i = [\![X]\!]^i \oplus_i [\![Y]\!]^i$
b. $[\![X+Y]\!]^i = [\![X]\!]^i \oplus_j [\![Y]\!]^i$

In (12) the plus sign indicates an arbitrary syntactic construction in which the constituents X and Y stand, and the encircled plus sign denotes the corresponding combination of extensions (assuming there is one). The equations are to be understood as holding for any indices i and j. As a consequence, the index-dependence of the operation $\oplus$ in (12-b) is spurious: replacing j with any other index k, results in the same combination of the extensions $[\![X]\!]^i$ and $[\![Y]\!]^i$: $[\![X]\!]^i \oplus_j [\![Y]\!]^i = [\![X+Y]\!]^i = [\![X]\!]^i \oplus_k [\![Y]\!]^i$, given that both j and k are supposed to satisfy the equation (12-b).[23] The difference between the two versions of (C_{ε}^{-}), then, depends on whether this combination is itself index-dependent. While (12-b) implies (12-a), the two ways of reading the principle (C_{ε}^{-}) of extensional compositionality are certainly not equivalent: (12-a) would be satisfied if $\oplus$ sometimes combined the daughter extensions in one way, by conjunction say, and sometimes in another, perhaps by disjunction [62, p. 282]. A glance at the semantic literature reveals that such constructions appear to be unheard of, so that the stronger, *uniform* version of (extensional) compositionality (12-b) ought to be preferred over the weaker (12-a). However, there is nothing in the general set-up that would force this choice. So if (12-b) rather than (12-a) is to be understood as a defining principle for the extensional fragment of a language, it better be motivated somehow.

[22] It should be noted that the subscript on the noun *extension* does *not* relate to the point of evaluation of its extension but to an argument of the latter. The point of evaluation only becomes relevant when it comes to counterfactual extensions due to different underlying meanings, as in the following variation of a Kripkean [31, p. 289] theme: *People might have spoken a language in which the extension of 'two plus two equals four' was 0*, where *extension* needs to be interpreted at certain counterfactual points, though its index argument could also relate to the actual utterance situation.

[23] More precisely, $\oplus_j$ and $\oplus_k$ coincide on all daughter extensions $[\![X]\!]^i$ and $[\![Y]\!]^i$. If the operations are extrapolated (cf. [62, pp. 280f.]), they may still diverge on 'ineffable' contributions.

To see how Frege's *Bedeutungen* fare better in this respect, it is worth looking at the source of the (binary) functionality of the noun *extension*: by definition, x is the *extension of* an expression X *at* an index i iff x is the (unique) value that X's intension assigns to the argument i. As a consequence, *extension* and *intension* may be seen as having the very same extension.[24] To sees this, one may observe that *extension* can be interpreted as denoting a function that is successively applied to expressions X and indices i of the object-language, to yield the extension of X relative to i. Using double slashes for the interpretation of the meta-language and suppressing its extension-determining parameters ('...'), we thus have:

(13) $/\!/ extension /\!/^{\cdots} (X)(i) = [\![X]\!]^i$

At the same time, the meta-linguistic noun *intension* may be taken to denote a function that, when applied to any object-linguistic expression X, yields X's intension:

(14) $/\!/ intension /\!/^{\cdots} (X) = \lambda i.\ [\![X]\!]^i$

But then successive application of either of the functions specified in (13) and (14) to any X and i in their domains yields the same result $[\![X]\!]^i$, which means that:

(15) $/\!/ extension /\!/^{\cdots} = /\!/ intension /\!/^{\cdots}$

Their co-extensionality notwithstanding, the two nouns differ in their syntactic behaviour. In particular, if omitted, the second (index) argument of *extension* may be supplied contextually or existentially, as in the case of the first (expression) argument. However, the second argument of *intension* cannot be so construed; it is syntactically invisible. As a consequence the noun *intension* always refers to the function, whereas the noun *extension* never does; for there is no way of construing the omission of its index argument by abstraction.

There is nothing unusual, let alone paradoxical, about this relationship, which any functional noun bears to a corresponding 'abstract' count noun that denotes its extension: *square root* vs. *square root function*, *sum* vs. *addition*, etc. One might say that the members of such pairs are co-designative without being synonymous, because they resist substitution, if only for syntactic reasons.[25] However, Frege's *Bedeutung* and *Sinn* do not stand in this relationship, for the simple reason that the latter does not denote a function in the first place. For although the *Bedeutung* of a given expression generally depends on the circumstances, it

[24] Or so I am assuming, be it for dramatic effect; see [63, pp. 101f.] (with a correction in https://tinyurl.com/ybtw3oh9) for an interpretation of functional nouns along these lines, which easily (though not inevitably) generalizes to binary functionals as in (13).

[25] The situation is vaguely reminiscent of the reference to an unsaturated function by means of a saturated term, which Frege [17] seems to have excluded on principle, albeit for dubious reasons (as Parsons [40] argued). However, in the case at hand, both nouns would count as syntactically unsaturated in that they lack a determiner.

is not defined as the result of applying a certain function – sense, intension, or whatever – to (a representation of) those circumstances: *Bedeutungen* are *Bedeutungen of* something, but not *at* something – there just is no index argument. As a consequence, the index-dependent reformulations of (C_ε^-) with *Bedeutungen* in lieu of *extensions* should look like this:

(11) c. The *Bedeutung* of a compound expression derives from combining$_j$ the *Bedeutungen* of its immediate parts.

(12) c. $[\![X + Y]\!] = [\![X]\!] \oplus_j [\![Y]\!]$

Hence, like (12-b), (12-c). is a principle of uniform compositionality in that the *Bedeutungen* are combined independently of the index at which they are determined. But in this case there is no alternative non-uniform reading like (12-a). Of course, this does not mean that such a non-uniform combination could not be formulated in terms of Fregean *Bedeutungen*. In fact, it could, by explicitly mentioning (possible) circumstances or indices. But nothing like the non-uniform principle (11-a) suggests itself as a reading of (C_ε^-). In other words, uniform compositionality falls out of a Fregean construal of *Bedeutungen* without having to be motivated or postulated ([62, p. 284]). And it even goes beyond the extensional fragment, as will be argued now.

4.2 Occurrence and *Bedeutung*

By virtue of the last clause, the hierarchy defined in (8) not only contains all possible extensions of expressions but also their intensions. This is so because it is meant to cover whatever an expression may contribute to the referent of a definite description or the truth-value of a sentence in which it occurs; and according to (FFP), this contribution may consist in its extension or in its intension. In particular, the contribution an expression makes depends on the position in which it occurs: it is *occurrence-dependent.* For the sake of definiteness, an occurrence x of an expression X in a (host) expression Y may be identified with a pair $x = (p, Y)$, where p is the structural position of x in Y (cf. von Stechow's [52] account). Then x's *(extensional) contribution to* Y will be either (a) X's extension or (b) X's intension, as determined by the following induction on p: if p marks an immediate constituent of Y, then (a) applies just in case X exhibits extensional substitutivity in the construction of Y; and if p is the position of an indirect constituent of Y with mother $y = (q, Z)$, then (a) applies iff the contribution of y to Y is Z's extension and X exhibits extensional substitutivity in the construction of Z. Hence extensional contributions are extensions throughout the extensional fragment, whereas intensions take over throughout all intensional environments.

Somewhat confusingly, Frege [15] used the German noun *Bedeutung* to denote extensional contributions, distinguishing between direct [*gerade*] ones, which consist in *Bedeutungen*, and indirect [*ungerade*] ones, which denote *Sinne*. So *Bedeutung* not only displays the systematic ambiguity typical of functional nouns, it is also polysemous in an unsystematic way: apart from acting on *expressions*, the functions it denotes may also apply to their *occurrences*. More specifically, the relevant reading *Bedeutung*$_{loc}$ may be construed as denoting a function mapping occurrences to their extensional contributions.[26] Given this construal, the following compositionality principle holds without exception:

(C_{loc}) The *Bedeutung*$_{loc}$ of a compound expression derives from combining the *Bedeutungen*$_{loc}$ of its immediate parts.

On the basis of the above characterization of extensional contributions, (C_{loc}) immediately follows from (FFP) and (C_ι), if adapted to *Bedeutungen*. In their original form, these principles sum up the central compositionality properties of possible worlds semantics. However, though frequently attributed to Frege, (FFP) is not the only way of understanding the pertinent passages in [15]. In fact, a more popular reading has it that iterated intensional embeddings necessitate ever more indirect contributions (cf. Kripke's account [33, p. 183], but then again also Parsons' doubts [39]). Thus in a sentence like (16), the doubly underlined clause would not contribute its sense to the Bedeutung$_{loc}$ (the sense) of the singly underlined clause, but its 'indirect sense':

(16) Sue believes that Billy suspects that most plush toys are former pets.

Following this line of analysis, the full hierarchy of intensions or senses of types $a, (s, a), (s, (s, a)), \ldots$ would have to be invoked to account for the interpretation of all occurrences of a single expression, which seems to undermine any attempts to explain learnability in terms of compositionality (as Davidson [9, p. 136] pointed out). One may thus see Kripke's [30] approach to intensional operators as quantifying over the evaluation points of the embedded material while introducing a new index dependence, as an escape from this analytic impasse. Yet the fact that multiply embedded clauses as in (16) may be construed as simultaneously depending on more than one index (as observed by Bäuerle [1]), might be taken as a reason to re-evaluate the (allegedly) Fregean strategy of higher-order indirectness; see [62, pp. 291ff.] for more on this perspective.

5 Back to the Puzzle

Learnability had also been the focus of the initial puzzle, repeated here for the reader's convenience:

(P) How does the compositionality of extensions bear on speakers' ability to grasp the meanings of indefinitely many linguistic expressions?

[26] I continue to assume the interpretation of functional nouns mentioned in fn. 24 above. – Incidentally, Frege did not use the term *occurrence* (or its German cognate *Vorkommen*) but spoke of the way in which the expressions are used: 'die Wörter werden in der ungeraden Rede *ungerade* gebraucht' [in indirect speech, words are used *indirectly*] [15, p. 28].

The answer is embarrassingly simple. In a nutshell, the connection is made by the intensions of the expressions involved. By definition (and as mentioned above), extensions are the values that corresponding intensions assign to indices. So if the extensions behave compositionally (as they do throughout the extensional fragment), this means that at any index i, the extension of any mother constituent $[\![X + Y]\!]^i$ can be obtained by suitably combining the extensions $[\![X]\!]^i$ and $[\![Y]\!]^i$ of its daughters. It suffices to consider the more general, non-uniform case:

(12-a) $[\![X + Y]\!]^i = [\![X]\!]^i \oplus_i [\![Y]\!]^i$

The crucial observation is that, if (12-a) holds, the mother *intension* $[\![X + Y]\!]^\wedge$ can be obtained from the *intensions* $[\![X]\!]^\wedge$ and $[\![Y]\!]^\wedge$ of the daughters too: at any index i, the value that $[\![X + Y]\!]^\wedge$ assigns to i is determined by the extensions of the daughters, which are in turn determined by their intensions:

(17) $[\![X + Y]\!]^\wedge(i)$
$= [\![X + Y]\!]^i$
$= [\![X]\!]^i \oplus_i [\![Y]\!]^i$
$= [\![X]\!]^\wedge(i) \oplus_i [\![Y]\!]^\wedge(i)$

Since (17) holds for all indices i, the entire function $[\![X + Y]\!]^\wedge$ can now be collected by functional abstraction:

(18) $[\![X + Y]\!]^\wedge$
$= \lambda i.\ [\![X + Y]\!]^\wedge(i)$
$= \lambda i.\ [\![X]\!]^\wedge(i) \oplus_i [\![Y]\!]^\wedge(i)$

which is a 'pointwise' specification of the mother intension in terms of the daughter intensions. In particular, the pertinent combination $\hat{\oplus}$ of intensions comes out as:

(19) $\hat{\oplus} = \lambda f.\ \lambda g.\lambda i. f(i) \oplus_i g(i)$

The operation in (19) not only combines intensions so as to guarantee (12-a), it is also unique in this respect:[27] if there were an alternative distinct operation $\ominus$ to the same effect, then due to the extensionality of set-theoretic functions, it would have to differ from $\oplus$ when applied to at least some intensions $[\![X]\!]^\wedge$ and $[\![Y]\!]^\wedge$, and an index i:

(20) $[[\![X]\!]^\wedge \oplus [\![Y]\!]^\wedge](i) \neq [[\![X]\!]^\wedge \ominus [\![Y]\!]^\wedge](i)$

But this cannot be, given that, by intensional compositionality, both sides of the inequality in (20) come down to $[\![X + Y]\!]^\wedge(i)$. Hence the conclusion from (12-a) to (19) not only shows that extensional compositionality implies intensional compositionality, but also that extensional composition determines intensional

[27] The same reservations concerning extrapolation and effability as in fn. (23) apply, though.

composition: from the way the extensions combine (in extensional constructions), one can conclude how the corresponding intensions do. Hence to the extent that identifying the intensions of expressions is a matter of linguistic knowledge (as argued in Sect. 1), extensional compositionality helps explaining how this knowledge is acquired: even though speakers generally do not know what the extensions of the expressions they (or others) use are, as long as they behave compositionally, they also know how their intensions combine and can thus identify them compositionally. Of course, the reasoning is restricted to the extensional fragment. To go beyond it, further strategies such as the (alleged) Fregean strategy (FFP) of employing intensions as *ersatz* extensions are required. But the large domain of extensional constructions is a good starting point, especially for semantic rookies.

While the above reasoning is by no means original, it is well worth remembering whenever a connection is made between extensional compositionality and learnability. It is also worth pointing out that, for the 'pointwise' determination of intensional compositionality in (18) and (19) to work, it is essential that intensions are defined as set-theoretic functions. Fregean senses, despite their structural similarity to intensions, do not support anything like the inference from (12-a) to (19). In fact, their compositionality is quite independent of the compositional behavior of extensions and intensions. If say, *John or Mary* shares its sense with its converse *Mary or John* (as Frege may have taken for granted), replacing one with the other might well result in a subtle sense difference, even if the environment is extensional. Of course, there may be independent reasons for ruling out such non-compositional behavior, but neither extensional nor intensional compositionality does the job. And even if senses behave compositionally (as Frege seems to have assumed), extensional compositionality underdetermines their compositional behavior, as a simple permutation argument shows.[28] For concreteness, one may consider two expressions with the same intension but a sense difference that is inherited to a larger expression:

(21) $[\![\textit{Some oculists are occultists}]\!]^{\$}$
$= [\![\textit{some oculists}]\!]^{\$} \otimes [\![\textit{are occultists}]\!]^{\$}$
$\neq [\![\textit{Some eye-doctors are occultists}]\!]^{\$}$
$= [\![\textit{some eye-doctors}]\!]^{\$} \otimes [\![\textit{are occultists}]\!]^{\$}$

...where '$[\![X]\!]^{\$}$' denotes the sense of an expression X and $\otimes$ is the relevant sense composition, which corresponds to a fully extensional construction:

(22) $[\![\textit{Some oculists are occultists}]\!]^{i}$
$= [\![\textit{some oculists}]\!]^{i} \oplus [\![\textit{are occultists}]\!]^{i}$
$= [\![\textit{some eye-doctors}]\!]^{i} \oplus [\![\textit{are occultists}]\!]^{i}$
$= [\![\textit{Some eye-doctors are occultists}]\!]^{i}$

[28] As Kai Wehmeier pointd out to me, Frege used a similar argument in his Grundgesetze [16, pp. 10] to illustrate the underdetermination of functions by their courses of values, of which the relation between senses and intensions may be seen as a special case.

However, unlike the corresponding intensional combination that could be defined as in (19), the sense operation $\otimes$ is not uniquely determined by the (universally quantified) equations in (22). To see this, one could define an operation $\ominus$ that behaves like $\otimes$ except that:

$$x \ominus [\![\textit{are occultists}]\!]^{\$} = y \otimes [\![\textit{are occultists}]\!]^{\$},$$

whenever $x, y \in \{[\![\textit{some oculists}]\!]^{\$}, [\![\textit{some eye-doctors}]\!]^{\$}\}$. Hence $\ominus$ swaps the senses of the two determiner phrases in the above environment. But the move from $\otimes$ to $\ominus$ would not affect the observations in (22): the intensions determined by the permuted senses are the same anyway. Hence, as far as the extensional behavior in (22) is concerned, $\ominus$ is as good a hypothesis for the underlying sense composition as is $\otimes$. As a consequence, extensional compositionality, though perfectly sound as a basis for resolving the puzzle (P) in terms of intensions, should not be employed to motivate the compositionality of Fregean senses.

6 Conclusion

Extensions feature prominently in contemporary semantic theory, owing their pivotal position largely to their compositional behavior. Textbooks tend to present compositionality as the keystone to the learnability of meaning and illustrate it by combining extensions. As argued in Sect. 1, this way of proceeding leaves a puzzling explanatory gap (P): why would the compositional behavior of extensions be relevant to learnability, given that they are not part of linguistic knowledge? The answer to this question, finally given in Sect. 5, lies in the specific relation between extensions and intensions. In many cases, knowledge of the latter *is* part of linguistic knowledge and thus *their* compositionality may be seen as offering a route to semantic learnability; and due to their very definition as set-theoretic abstractions from extensions, the compositional behavior of intensions may piggyback on extensional compositionality. Hence to the extent that extensions do behave compositionally, their compositional behavior guarantees intensional compositionality and may thus form the basis of an account of semantic learnability. The textbook practice is thus fully legitimate, even though an explicit justification is not always provided. This is the first take-home message of the above, admittedly somewhat encyclopedic, exposition of the role of extensions in semantic theory.

Due to their set-theoretic nature, extensions differ from their ancestors, Frege's *Bedeutungen*, in a number of important respects. As a case in point, once we go beyond the extensional fragment, *Bedeutungen* diverge from the corresponding extensions, inheriting the fine-grainedness of their argument *Sinne*. More importantly, the latter are not derived by set-theoretic abstraction from corresponding (possible) *Bedeutungen*: whereas the intension of an expression is determined by the totality of its possible extensions, there is no way of deriving its *Sinn* from its potential *Bedeutungen*. As a consequence, *Bedeutungen* are susceptible to a natural and unambiguous concept of extensionality – a mild advantage of Frege's approach over the theory of extension and intension (as argued in Sect. 4.1). Yet the commonalities between extensions and *Bedeutungen* seem to outweigh their differences: both make the connection between word and world by generalizing the notions of reference and truth-value from definite descriptions and declarative sentences to arbitrary categories (as explained in Sect. 2); both are conceptually and formally simpler than full-fledged meanings or contents of expressions; both extensions and *Bedeutungen* display the same kind of compositional interaction with informational content as represented by, respectively, intensions and Fregean senses (cf. Sect. 3); and whereas knowledge of the latter values is often part of the mastery of a language, the extensions and *Bedeutungen* of most expressions are unknown to its speakers. In view of these parallels, it may seem that the compositionality of *Bedeutungen* is every bit as relevant to learnability as extensional compositionality. But it is not: whether or not *Bedeutungen* behave compositionally is independent of the compositionality of Fregean senses; rather than being derived from extensional compositionality, the compositionality of senses needs to be assumed on independent grounds. The second take-home message, then, is a warning to those who prefer to have their semantic values more finely grained than extensions and intensions: the compositional behavior of extensions (or *Bedeutungen*) provides no evidence for the compositionality of linguistic meaning, so conceived.

Acknowledgment. This paper is a slightly revised and expanded version of my talk at the Tbilisi Symposium. Predecessors had been presented at the Universities of Massachusetts (Amherst), California (San Diego, Los Angeles, Santa Cruz), Cologne, and Göttingen in 2018 and 2019. I am indebted to my audiences for many remarks and objections that helped improving the quality of the argumentation. Special thanks go to Dolf Rami and Kai Wehmeier for enlightening discussions on Frege's doctrine of *Sinn* and *Bedeutung*; to Ramona Hiller and Jan Köpping for help with the TeX file; to Hans-Martin Gärtner for spotting a number of errors in the pre-final version of the manuscript; to two anonymous reviewers for various suggestions concerning the presentation of the material; and to Ayline Heller for proof-reading the final manuscript. Any remaining errors and shortcomings are home-grown.

References

1. Bäuerle, R.: Pragmatisch-semantische Aspekte der NP-Interpretation. In: Faust, M., Harweg, R., Lehfeldt, W., Wienold, G. (eds.) Allgemeine Sprachwissenschaft, Sprachtypologie und Textlinguistik, pp. 121–131. Narr, Tübingen (1983)
2. Beaney, M.: Translating "Bedeutung" in Frege's Writings: A Case Study and Cautionary Tale in the History and Philosophy of Translation. In: Ebert, P.A., Rossberg, M. (eds.) Essays on Frege's Basic Laws of Arithmetic, pp. 567–636. Oxford University Press (2019)
3. Bell, D.: On the Translation of Frege's Bedeutung. Analysis **40**(4), 191–195 (1980)
4. Carnap, R.: Meaning and Necessity. University of Chicago Press, Chicago & London (1947)
5. Chierchia, G., McConnell-Ginet, S.: Meaning and Grammar, 2nd edn. MIT Press, Cambridge, Mass (1990)
6. Chomsky, N.: Syntactic Structures. Mouton & Co, The Hague (1957)

7. Church, A.: A Formulation of the Logic of Sense and Denotation. In: Henle, P. (ed.) Structure, Method, and Meaning, pp. 3–24. Liberal Arts, New York (1951)
8. Davidson, D.: The logical form of action sentences. In: Rescher, N. (ed.) The Logic of Decision and Action, pp. 81–95. University of Pittsburgh Press, Pittsburgh (1967)
9. Davidson, D.: On saying that. Synthese **19**, 130–146 (1968)
10. Dowty, D., Wall, R., Peters, S.: Introduction to Montague Semantics. Reidel, Dordrecht (1981)
11. Ebbinghaus, H.D.: Ernst Zermelo. An Approach to His Life and Work. In cooperation with Volker Peckhaus. Springer, Berlin & Heidelberg (2007)
12. Fabian, R.: Sinn und Bedeutung von Namen und Sätzen: eine Untersuchung zur Semantik Gottlob Freges. Verband der Wiss. Ges. Österreichs, Vienna (1975)
13. Frege, G.: Die Grundlagen der Arithmetik. Koebner, Breslau (1884)
14. Frege, G.: Function und Begriff. Pohle, Jena (1891)
15. Frege, G.: Über Sinn und Bedeutung. Zeitschrift für Philosophie und philosophische Kritik NF **100**(1), 25–50 (1892)
16. Frege, G.: Grundgesetze der Arithmetik. vol. 1. Pohle, Jena (1893)
17. Frege, G.: Über Begriff und Gegenstand. Vierteljahresschrift für wissenschaftliche Philosophie **16**, 192–205 (1892)
18. Frege, G.: Sense and Reference. Philosophical Review **57**(3), 209–230 (1948)
19. Frisch, J.C.: Extension and Comprehension in Logic. Philosophical Library, New York (1969)
20. Gutzmann, D.: Dimensions of Meaning. In: Gutzmann, D., Matthewson, L., Meier, C., Rullmann, H., Zimmermann, T.E. (eds.) The Wiley Blackwell Companion to Semantics vol. 1, pp. 589–617. Wiley & Sons Ltd. (2021)
21. Heck, R.K., May, R.C.: The Birth of Semantics. Journal for the History of Analytical Philosophy **8**(6), 1–31 (2020)
22. Heim, I., Kratzer, A.: Semantics in Generative Grammar. Blackwell Publishers Ltd., Oxford (1998)
23. Janssen, T.M.: Compositionality (with an appendix by B. H. Partee). In: van Benthem, J., ter Meulen, A.G.B. (eds.) Handbook of Logic and Language, pp. 417–473. Elsevier, Amsterdam (1997)
24. Janssen, T.M.: Frege, Contextuality and Compositionality. Journal of Logic, Language and Information **10**, 115–136 (2001)
25. Kaplan, D.: How to Russell a Frege-Church. Journal of Philosophy **72**, 716–729 (1975)
26. Kaplan, D.: Demonstratives. An Essay on the Semantics, Logic, Metaphysics and Epistemology of Demonstratives and Other Indexicals. In: Almog, J., Perry, J., Wettstein, H. (eds.) Themes from Kaplan, pp. 481–563. Oxford University Press, Oxford (1989)
27. Katz, J.J., Fodor., J.A.: The structure of a semantic theory. Language 39(2), 170–210 (1963)
28. Klev, A.M.: Categories and Logical Syntax. Ph.D. thesis, Leiden University (2014)
29. Kratzer, A.: An investigation of the lumps of thought. Linguistics and Philosophy **12**, 607–653 (1989)
30. Kripke, S.A.: Semantical considerations on modal logic. Acta Philosophica Fennica **16**, 83–94 (1963)
31. Kripke, S.A.: Naming and Necessity. In: Davidson, D., Harman, G. (eds.) Semantics of Natural Language, pp. 253–355. Reidel, Dordrecht (1972)

32. Kripke, S.A.: Speaker's Reference and Semantic Reference. In: French, P., Uehling, T., Wettstein, H. (eds.) Midwest Studies in Philosophy, Vol II: Studies in the Philosophy of Language, pp. 255–276. University of Minnesota Press, Minneapolis (1977)
33. Kripke, S.A.: Frege's theory of sense and reference: Some exegetical notes. Theoria **74**(3), 181–218 (2008)
34. Leonardi, P.: The Names of the True. In: Coliva, A., Leonardi, P., Moruzzi, S. (eds.) Eva Picardi on Language, Analysis and History, pp. 67–85. Springer, Cham (2018). https://doi.org/10.1007/978-3-319-95777-7_4
35. Löbner, S.: The Partee Paradox. Rising Temperatures and Numbers. In: Gutzmann, D., Matthewson, L., Meier, C., Rullmann, H., Zimmermann, T.E. (eds.) The Blackwell Companion to Semantics, vol. 4, pp. 2239–2264. Wiley & Sons, Oxford (2021)
36. Montague, R.: Pragmatics and Intensional Logic. Synthèse **22**, 68–94 (1970)
37. Montague, R.: Universal Grammar. Theoria **36**(3), 373–398 (1970)
38. von Neumann, J.: Zur Einführung der transfiniten Zahlen. Acta Literarum ac scientarium Regiae Universitatis Hungaricae Francisco-Josephinae, Sectio scientiarium mathematicarum **1**, 199–208 (1923)
39. Parsons, T.: Frege's Hierarchies of Indirect Senses and the Paradox of Analysis. In: P. French, T., Uehling, T., Wettstein, H. (eds.) Midwest Studies in Philosophy VI: The Foundations of Analytic Philosophy, pp. 37–57. University of Minnesota Press, Minneapolis (1981)
40. Parsons, T.: Why Frege should not have said "The Concept Horse is not a Concept." History of Philosophy Quarterly **3**, 449–465 (1986)
41. Parsons, T.: Events in the semantics of English. A study in subatomic semantics. MIT Press, Cambridge, Mass. (1990)
42. Partee, B., Rooth, M.: Generalized conjunction and type ambiguity. In: Bäuerle, R., Schwarze, C., Stechow, A. von. (eds.) Meaning, Use, and Interpretation of Language, pp. 361–383. De Gruyter, Berlin (1983)
43. Pelletier, F.J.: Did Frege Believe Frege's Principle? Journal of Logic, Language and Information **10**(1), 87–114 (2001)
44. Putnam, H.: The meaning of "meaning" .In: Gunderson, K. (ed.) Language, Mind, and Knowledge, pp. 131–193. University of Minnesota Press, Minneapolis (1975)
45. Quine, W.V.O.: Word and Object. MIT Press, Cambridge, Mass (1960)
46. Ramsey, F.P.: Facts and propositions. Proceedings of the Aristotelian Society, Suppl. **7**, 153–170 (1927)
47. Russell, B.: On denoting. Mind **14**, 479–493 (1905)
48. Saebø, K.J.: Anaphoric presuppositions and zero anaphora. Linguistics and Philosophy **19**, 187–209 (1996)
49. Sharvy, R.: A more general theory of definite descriptions. The Philosophical Review **89**, 607–627 (1980)
50. Stalnaker, R.: The Problem of Logical Omniscience. I. Synthese **89**, 425–440 (1991)
51. Stalnaker, R.: The Problem of Logical Omniscience, II. In: Context and Content, pp. 255–273. Oxford University Press, Oxford (1999)
52. Stechow, A. von: Occurrence-interpretation and Context-Theory. In: Gambarara, D., Piparo, F.l., Ruggiero, G. (eds.) Linguaggi e formalizzazioni, pp. 307–347. Bulzoni, Rome (1979)
53. Szabó, Z.G.: Compositionality. Stanford Encyclopedia of Philosophy (2017), https://plato.stanford.edu/archives/sum2017/entries/compositionality/
54. Tarski, A.: Der Wahrheitsbegriff in den formalisierten Sprachen. Studia Philosophica I, 236–405 (1936)

55. Tomalin, M.: Syntactic Structures and Recursive Devices: A Legacy of Imprecision. Journal of Logic, Language and Information **20**(3), 297–315 (2011)
56. Tugendhat, E.: The Meaning of 'Bedeutung' in Frege. Analysis **30**(6), 177–189 (1970)
57. Zimmermann, T.E.: Grundzüge der Semantik (2002), https://user.uni-frankfurt.de/~tezimmer/Zimmermann/Skript.02.pdf
58. Zimmermann, T.E.: Compositionality Problems and how to Solve Them. In: Werning, M., Hinzen, W., Machery, E. (eds.) The Oxford Handbook of Compositionality, pp. 81–106. Oxford University Press, Oxford (2012)
59. Zimmermann, T.E.: Context dependence. In: Maienborn, C., Heusinger, K.v., Portner, P. (eds.) Handbook of Semantics. Volume 3, pp. 2360–2407. De Gruyter, Berlin/New York (2012)
60. Zimmermann, T.E.: Equivalence of Semantic Theories. In: Schantz, R. (ed.) Prospects for Meaning, pp. 629–649. De Gruyter, Berlin (2012)
61. Zimmermann, T.E.: Intensionale Semantik. In: Kompa, N. (ed.) Handbuch Sprachphilosophie, pp. 187–197. J.B. Metzler, Stuttgart (2015)
62. Zimmermann, T.E.: Fregean Compositionality. In: Rabern, B., Ball, D. (eds.) The Science of Meaning, pp. 276–305. Oxford University Press, Oxford (2018)
63. Zimmermann, T.E., Sternefeld, W.: Introduction to Semantics. An Essential Guide to the Composition of Meaning. De Gruyter, Berlin (2013)

Embedded Questions are Exhaustive Alright, but...

Malte Zimmermann[1](✉), Lea Fricke[2], and Edgar Onea[2]

[1] Universität Potsdam, 14476 Potsdam, Germany
mazimmer@uni-potsdam.de
[2] Karl-Franzens-Universität, 8010 Graz, Austria

Abstract. We present two novel diagnostics for gauging the exhaustivity level of German *wh*-interrogatives embedded under the predicates *wissen* 'know' and *überraschen* 'surprise'. The readings available in combination with the concessive particle combination *SCHON...aber* 'alright...but' and the Q-adverb *teilweise* 'partially' provide evidence that embedded *wh*-interrogatives under veridical and distributive *wissen* 'know' have a weakly exhaustive (WE) reading as their basic semantic interpretation [19]. The logically stronger strongly exhaustive (SE) reading is a pragmatic enrichment that can be cancelled by *SCHON...aber*. In our event-based analysis, *know* + *wh* expresses the maximal plurality of sub-events of knowing the individual answers to the question. Under the cognitive-emotive attitude verb *überraschen* 'surprise', which is not obligatorily distributive, *wh*-interrogatives allow for two types of WE-interpretations, distributive and non-distributive. The *SCHON...aber*-diagnostic shows the logically stronger distributive WE-reading to be a pragmatic enrichment. In view of (novel) experimental evidence that *surprise* + *wh* allows for SE-interpretations, we follow [12] and tentatively analyze *surprise* + *wh* as expressing a psychological state caused by a complex situation, or subparts or missing parts thereof.

Keywords: Embedded questions · Exhaustivity · Q-adverbs · Discourse particles · Pragmatic enrichment

This work was funded by the Deutsche Forschungsgemeinschaft (DFG, German Research Foundation), Priority Program SPP 1727 XPRAG.de, Project 'Exhaustiveness in embedded questions across languages' (Onea, Zimmermann). We would like to thank the audience at TbiLLC13 as well as two anonymous reviewers for valuable feedback. All remaining errors are our own.

© The Author(s), under exclusive license to Springer Nature Switzerland AG 2022
A. Özgün and Y. Zinova (Eds.): TbiLLC 2019, LNCS 13206, pp. 173–194, 2022.
https://doi.org/10.1007/978-3-030-98479-3_9

1 Introduction

This paper takes a fresh look at the different exhaustivity levels of *wh*-interrogatives embedded under the veridical and distributive predicate *wissen* 'know', and under the cognitive-emotive and non-distributive *überraschen* 'surprise', cf. (1), (2).[1]

(1) Nino **weiß**, [wer getanzt hat].
'Nino **knows** who danced.'

(2) Es **überraschte** Nino, [wer getanzt hat].
'It **surprised** Nino who danced.'

The discussion will be based on two novel empirical diagnostics regarding the interaction of embedded *wh*-interrogatives with the concessive particle combination *SCHON...aber* 'alright...but' and the Q-adverb *teilweise* 'partially', as shown in (3).

(3) Nino weiß **SCHON/teilweise**, [wer getanzt hat].
'Nino knows who danced alright, Nino knows who danced alright.
/Nino knows in part who danced.'

A highly debated issue in question semantics is which of the observable surface readings of varying exhaustivity (strongly exhaustive [SE], intermediate exhaustive [IE], weakly exhaustive [WE]) are underlying semantic interpretations, and which ones are mere pragmatic inferences, if any. To this end, we will investigate the interpretive effect of particle combinations and Q-adverbs on the interpretation of interrogatives under *know* and *surprise*. We will show that insertion of the particle combination *SCHON...aber* blocks the generation of some pragmatic implicatures. From this, we conclude that exhaustivity inferences of *wh*-interrogatives that are blocked by the presence of the particle combination are pragmatic inferences. The Q-adverb *teilweise* 'partially', by contrast, operates on truth-conditional semantic content proper. We conclude that exhaustivity inferences targeted by *teilweise* must be part of the truth-conditional semantic content of embedded *wh*-interrogatives. Applying the two diagnostics to *wh*-interrogatives embedded under *wissen* 'know' and *überraschen* 'surprise', we find the following: First, SE-readings under *wissen* 'know' are pragmatic inferences that are derived from a weaker semantic interpretation [19, 40] under an internal subject perspective [13, 39]. This internal perspective follows from the novel general pragmatic *Principle of Attitude Report Verification (PARV)*. Second, the observable distributive readings with *überraschen* 'surprise' result from pragmatic strengthening of a relatively

[1] The distributivity of *wissen* 'know' is evidenced by the fact that knowledge of who danced in *s* will entail knowledge of every individual that danced in *s*: In a situation *s* with three individuals, Berit, Daniel and Malte, that danced, the truth of (1) entails that Nino knows that Berit danced and that Daniel danced and that Malte danced. By contrast, [24] was the first to show that *überraschen* 'surprise' is non-distributive, as one can be surprised by the composition of a group (e.g., that B and D and M and all danced together) without being surprised at the individual dancers; see §2 for more discussion of the semantics of *know* and *surprise*.

weak underlying semantic interpretation, which can be cast in terms of an existential WE-semantics [14, 34, 35], or by analyzing cognitive-emotive attitude verbs like *surprise* as predicates operating on facts/situations rather than propositions/questions [12].

The article is structured as follows: Sect. 2 provides background information on the exhaustivity of *wh*-interrogatives and the interpretive effects of *SCHON...aber* and *teilweise.* Section 3 presents the novel empirical findings for *wh*-interrogatives embedded under *wissen* 'know' (henceforth: *know* + *wh*), and it sketches an event-based analysis of *know* as operating over the plural sum of knowledge sub-events, effectively giving rise to a semantic WE-interpretation. Section 4 presents the novel empirical findings and a preliminary analysis of *wh*-interrogatives embedded under *überraschen* 'surprise'. Section 5 concludes.

2 Background: Exhaustive Force, Particles, and Q-Adverbs

This section provides background information on the variable interpretation of embedded *wh*-interrogatives as weakly or strongly exhaustive (Sect. 2.1), on the interpretive effects of the discourse particle *SCHON* 'alright' in combination with concessive *aber* 'but' (Sect. 2.2), and on the semantic import of the Q-adverb *teilweise* 'partially' (Sect. 2.3).

2.1 Different EXH-Force Under *Know* and *Surprise*: SE vs. WE

The surface interpretation of sentences with embedded *wh*-interrogatives can vary in the exhaustive force of the embedded interrogative, depending in part on the meaning of the embedding predicate. Consider (1) with *wissen* 'know' in a scenario with four individuals, Mary, Alex, Paul and Anna. Of these four, Mary and Alex danced, and Paul and Anna did not. The two readings of (1) of interest differ in how much information Nino must have regarding who did and who did not dance. On the strongly exhaustive reading [13], she must have complete information regarding the entire answer space, namely that Mary and Alex danced, and Paul and Anna did not. On the weakly exhaustive reading [19], it suffices for (1) to be true that Nino's information state is complete with respect to the positive answer space: She would only need to know that Mary and Alex danced. Moreover, non-exhaustive readings [41] with *know* are blocked by the inherent distributivity or homogeneity of this predicate [4, 24].[2]

[2] We focus on WE- and SE-readings in the discussion to come, in which we derive the SE-reading from the WE-reading, which we take to be the semantic basis of any semantic theory of embedded questions. The additional intermediate exhaustive reading (IE) is a strengthened WE-reading with the additional requirement that the subject have no false beliefs about individuals that are not in the extension of the embedded predicate. For (1), this would require that Nino does not (falsely) believe of Paul or Anna that they danced. We have nothing of substance to say about the IE-reading in this paper and will therefore remain silent on how it derives from the WE-reading. [40] derives IE-readings by applying an exhaustivity operator. Alternatively, there may be a *no-false belief* constraint as part of the semantics of the embedding verb *know*, which is veridical, i.e. truth-bound, so that the WE-reading with *know* is indistinguishable from the so-called IE-reading, as proposed by [36, 37] and [22] for other embedding predicates, such as *predict*. Throughout, we will continue to use the traditional label WE-reading in connection

The cognitive-emotive attitude predicate *überraschen* 'surprise' differs semantically from *wissen* 'know' in several ways. This has repercussions for the interpretation of embedded *wh*-interrogatives. For one, *surprise* is not obligatorily distributive [24]. So, (2) could still be true if Nino did not expect both Mary and Alex to dance at the same party (because they are rivals and never dance if the other does) even though she is not surprised by Mary's dancing per se, nor by Alex's. Given non-distributivity, *surprise* + *wh* may give rise to different readings than *know* + *wh* in the above scenario. The different readings will crucially depend on Nino's prior expectations. On the distributive WE-reading (WE_dist), Nino didn't expect Mary nor Alex to dance, so that her surprise is complete with respect to the positive answer space of *Who danced?*. A non-distributive WE-reading (WE_nondist; cf. [14, 34, 35]) obtains if it's just Alex that Nino didn't expect to dance. Now her surprise is directed at the positive answer space in a non-distributive manner. In addition, there may be two SE-readings with *surprise*, which make reference to the full logical answer space including the non-dancers: the non-distributive SE-reading (SE_nondist) obtains if Nino is not surprised by the actual dancers, but she did expect Anna to dance as well, contrary to fact. Finally, the distributive SE-reading (SE_dist) would require Nino to be surprised by everybody who danced and by everybody who didn't (= complete counter-expectation).

Notice that *know* and *surprise* also exhibit different entailment patterns [32, 40], i.a. *Know* is upward entailing so that SE entails WE: If Nino knows who was and who was not at the party (SE), it follows that she knows who was at the party (WE). The same entailment does not hold for *surprise*: If it surprises Nino who did and who did not dance (SE), it does not follow that she is surprised by who actually danced (WE). The surprise may be directed exclusively at the non-dancers.

The literature offers different views on the available interpretations of *wh*-interrogatives under *know*. In [13], all embedded *wh*-interrogatives denote propositions inducing a full partition of the entire logical space. In this partitioning question semantics, all embedded *wh*-interrogatives are predicted to be strongly exhaustive. For [17], the SE-reading with interrogatives under *know* follows from the lexical partitioning semantics of the matrix predicate, such that (1) will be true iff Mary knows the complete answer to who was at the party, and that this is the complete answer. Differences aside, both accounts only predict SE-readings for *know* + *wh*. This strong position is problematic on at least two counts: Firstly, whereas SE-readings are indeed prominent with *know*, other embedding verbs such as *predict*, *tell*, or *announce* allow for weaker interpretations, which cannot be modelled in a partition semantics [4, 17]. Secondly, recent experimental work has found the weaker IE/WE-readings (i.e. to know the complete positive answer and nothing more) to be readily available with an acceptability rate of >90% even with English *know* and French *savoir* 'know' [6, 8]. In addition, [7] provide experimental evidence for both WE_nondist and SE_nondist-readings with interrogatives under *surprise*. The experiments in [6] and [7] involved picture matching and acceptability judgments with an external, participant-centered perspective.

---

with *know* and *surprise*, where it should be understood as (empirically) equivalent to the label IE-reading in the case of *know*, as in [40] modulo our non-commitment regarding the derivation of IE.

Novel Experimental Evidence. In two experiments with novel setups, we were able to replicate to some extent these findings for German *wissen* 'know' and *überraschen* 'surprise'. In a contradiction experiment [10], we tested for the obligatoriness of SE-readings with *wissen* 'know' predominantly from the internal perspective of the attitude holder. Participants had to judge the contradictoriness of discourse sequences, such as (4) (in italics), in which the SE-reading is explicitly negated in the final clause.

(4) Context: [Anna, Beth, Chloe, Doro, Emma and Franzi share a flat in Berlin.] On the long weekend, they organized a games night. [Their former flatmate] Jannick was there as well. During games night, they mixed drinks.
Jannick knows who out of the flatmates mixed a cocktail, but he doesn't know that Emma and Franzi did not mix a cocktail.

If only SE-readings were available under *wissen*, such sequences should be systematically judged as contradictory. Conversely, if participants judge them as non-contradictory, this constitutes evidence for the WE/IE-reading. The results show that more than 25% of all cases were judged as non-contradictory, indicating that WE/IE-readings are available to some extent.

The second experiment was carried out for a range of matrix predicates in German, including *wissen* 'know' and *überraschen* 'surprise' [11]. Target sentences were objects of bets, and compensation was performance based, so that participants were actively engaged through a financial incentive. Again, the linguistic items and contexts were designed such that target sentences had to be judged from the internal perspective of the attitude holder, while the external perspective of the addressee had to be taken into account as well. This design targets the optimal reading from a communication-oriented perspective; see [11] for details on the experimental setup. The descriptive results for the two predicates of interest are as follows: For *wissen*, there was evidence for a WE/IE-reading in 46% of all cases, as opposed to a ceiling 100% for SE. For *überraschen*, there was evidence for the two WE-readings (WE_dist: 100%, WE_nondist: 96%), but, interestingly, also for the SE_nondist-reading at a robust level of 58%. The availability of SE_nondist will play a crucial role in the analysis of *überraschen* in Sect. 4.

Previous Analyses of Flexible SE/WE-Interpretations. There is ample evidence from introspection and experiments that the interpretation of *wh*-interrogatives is flexible between SE and WE under *wissen* 'know', and variable between three surface interpretations under *überraschen* 'surprise'. The literature offers different ways to account for this flexibility, with different sources for the observed variability in exhaustive force. [3] derive the variability from two covert answer operators *ANS1* (giving rise to WE) and *ANS2* (deriving SE), which both operate on an unconstrained interpretation of the interrogative in terms of Hamblin-alternatives [16]. [22] postulate covert EXH-operators either in the embedded interrogative (deriving SE) or in the matrix clause (deriving WE/IE). [40] derives an IE-interpretation as the only available semantic reading by placing covert EXH in the matrix clause. SE-readings are derived as a pragmatic enrichment via a hearer-based (excluded middle) competence assumption. Finally, [39] posit a lexical ambiguity in the attitude verb *know* as expressing an internal perspective (SE-reading) or external perspective (WE/IE-reading), respectively. Our analysis of *know* + *wh* in Sect. 3 will incorporate core ingredients from the last two accounts.

2.2 The Interpretive Effect of *SCHON...Aber*: Implicature Blocking

According to [42], the German discourse particle *schon* 'alright' is a modal comparative degree operator that commits the speaker to the truth of the prejacent proposition *p*, after weighing the circumstantial evidence in favor of *p* against the evidence for its polar counterpart $\neg p$. In general, the presence of *schon* indicates that there may be some reason to doubt the validity of *p*. Because it expresses polar comparison, *schon,* and accented *SCHON* in particular, are commonly found in verum focus contexts [18]. In combination with the (implicit) concessive particle *aber* 'but' in a subsequent clause, accented *SCHON* has an additional effect on interpretation: It consists in the blocking of pragmatic implicatures based on prototypicality or relevance.[3] Consider (5A), which gives rise to the relevance-based implicature that *Levan is not hungry* in the absence of *SCHON*. With *SCHON*, this implicature is blocked. Likewise, B's implicit question in (6) is whether she can get petrol, so that A's response without *SCHON* would give rise to the relevance-based conversational implicature that the petrol station is open and sells petrol. This implicature is blocked in the presence of *SCHON*, thereby indicating that the implicit question is answered in the negative: no petrol available.

(5) Q: Is Levan hungry? Has he had breakfast?
A: Er hat (**SCHON**) gefrühstückt (, **aber**...)
'He's had breakfast (**alright, but** ...)'

(6) Context: B tells A that she needs petrol and asks about a petrol station nearby.
A: Es gibt hier (**SCHON**) eine Tankstelle (, **aber** ...)
'There is a petrol station (**alright, but** ...)'

Crucially, *SCHON...aber* does not block scalar implicatures. In (7), its presence does not rescue the impending contradiction between the implicature (*not all*) and its contradiction in the subsequent clause (*all*).

(7) #Cleo hat **SCHON** *einige* Kekse gegessen, **aber** eigentlich hat sie *alle* gegessen.
'Cleo has eaten *some* cookies **alright, but** actually she's eaten *all* of them.'

[3] How exactly this blocking of implicatures should be modelled is an open question. It seems to us that the presence of *SCHON* in a sentence is understood by the hearer as a cue suggesting that (a certain type of) implicatures should not be derived in the first place. However, for the purposes of this paper a somewhat weaker formulation would also suffice: *SCHON* is licit in contexts in which certain types of implicatures are cancelled with an upcoming *aber* ('but') construction. We will use the stronger claim in this paper for explicitness.

We speculate that the insensitivity to scalar implicatures follows from the polar comparative nature of modal *SCHON* [42], and from the fact that the scalar alternative (*C ate **all** the cakes*) logically entails the literal meaning *p* (*C ate **some** cakes*): Adding implicature-blocking *SCHON* to a proposition *p* normally constitutes a reason for doubting *p*, but in the case of the scalar *not-all* implicature in (7) the validity of the unblocked alternative (*all the cakes*) casts no doubt on the entailed *p* (*some cakes*). For this reason, the presence of *SCHON* is unmotivated as there is no contradiction.

Moreover, modal *SCHON* does not resolve lexical ambiguities, as shown for the German homonym *Bank* ('bench' or 'bank') in (8). (8) can only be understood in jest (☺) as a play of words, i.e., at a meta-linguistic level.

(8) Ich kenne **SCHON** eine *Bank* hier in der Nähe, die Deutsche *Bank*, **aber** auf der kannst du nicht bequem sitzen. ☺
'I know a bank nearby **alright**, Deutsche *Bank*, **but** it's not comfy to sit on.'

Finally, *SCHON...aber* does not affect truth-conditional semantic content. Its presence in (9) does not lead to a rejection of the claim that at least five beers were drunk:

(9) #Ich habe **SCHON** fünf Bier getrunken, **aber** eigentlich nur drei.
'I drank five beers alright, but actually only three.'

The insensitivity of *SCHON...aber* to semantic content will play an important role in our semantic analysis of *wh*-interrogatives under *know* and *surprise*. In particular, we can conclude that any inferences blocked by the presence of *SCHON...aber* are not semantic entailments, but mere pragmatic implicatures triggered by considerations of prototypicality or relevance. Pragmatically, the presence of *SCHON...aber* indicates that a prototypical default does not obtain, which in turn casts doubt on the truth of the prejacent *p* by the semantic meaning of *SCHON* as a modal degree operator.

2.3 The Meaning of *Teilweise* 'Partially': Quantifying Over Pluralities

In contrast to *SCHON...aber*, the Q-adverb *teilweise* 'partially, in parts' is a quantificational modifier operating on truth-conditional semantic content. For the purposes of this paper, there are three important aspects to the meaning of *teilweise*:

Firstly, *teilweise* affects the truth-conditions. Whereas Nino must have eaten all of the (contextually salient) Khachapuris for (10) to be true, (11) will already be true if Nino ate only a subset of them. More generally, sentences with *teilweise* are true if a subpart of the theme-related eventualities in question are instantiated.

(10) Nino hat die Khachapuris gegessen.
'Nino ate the Khachapuris.'

(11) Nino hat die Khachapuris **teilweise** gegessen.
'Nino ate the Khachapuris **partially**.'

Secondly, we assume that *teilweise* only excludes maximal eventualities in the pragmatics, as (11) will also be true in situations in which Nino ate all of the Khachapuris. Having said this, we concede that it is quite misleading to use *teilweise* in a situation in which Nino ate all of the Khachapuris, but there is good evidence for the assumption that the partiality associated with *teilweise* is a pragmatic effect. (12) will be true if Ninos granny is pleased with Nino eating some or (even better!) all of the Khachapuris. Moreover, there is a clear contrast between *teilweise* vs. *nur teilweise*, as shown in (13).

(12) Wenn Nino die Khachapuris **teilweise** gegessen hat, ist ihre Oma zufrieden.
'If Nino ate the Khachapuris **partially**, her granny will be pleased.'

(13) Nino hat die Khachapuris **nur teilweise** gegessen.
'Nino ate the Khachapuris **only partially**.'

Thirdly, *teilweise* only operates on pluralities of discrete eventualities, which must be tied to atomic entities in the individual domain, as expressed by plural count NPs. As a result, *teilweise* in (14) cannot be used to express that Nino ate only part of the soup, unlike the part-whole modifier *zum Teil* 'in part'. The only felicitous reading of (14) is one in which Nino ate the soup in discrete portions (possibly together with others).

(14) #Nino hat die Suppe **teilweise** gegessen.
'#Nino **partially** ate the soup.'

(11) and (14) show that *teilweise* is not lexically connected to question embedding, but to plural eventualities. Most importantly, all these requirements can only be fulfilled if the atomic pluralities are targeted by *teilweise* in the process of semantic composition.

There are several conceivable ways to implement the semantics of *teilweise*. An obvious possibility would be to follow [4] or [24] in assuming that *teilweise* is a run-of-the-mill adverbial quantifier that takes individuals, propositions, or eventualities as its arguments. An alternative would be to implement *teilweise* as a quantifier that takes a plurality as argument and returns a part of that plurality for the further compositional procedure, [2]. Here, we opt for an event-semantic analysis, though, in which *teilweise* operates on mereological part-whole structures. Providing arguments in support of our analysis goes beyond the scope of this paper. For its core arguments, not much hinges on the particular choice of analysis for *teilweise*, as long as it accounts for the three main empirical observations. For this reason, the present analysis should be considered a mere handy tool for formally implementing the essential points viz. the semantics of embedded questions. For the same reason, we refrain from a detailed compositional analysis. Most of what follows could be restated in any analysis that takes questions to denote a Hamblin/Karttunen style set of alternatives.

For explicitness, we analyze the Q-adverb *teilweise* as a quantificational part-whole modifier of a verbal projection that operates over plural mereological sub-event structures. We assume that a clause with *teilweise* will be true iff there exists some sub-event *e* of a complex plural event *e'*. In (11), this plural event is the maximal eating event of a contextually given maximal set of khachapuris, which is formally derived by the sum-formation operator ⊕. The Neo-Davidsonian event-semantic representation of (11) is shown in (15), with *TH* = *Theme*; see also [2, 26], i.a.

(15) $\exists e. e \sqsubseteq e' \wedge e' = \oplus\{e | eat(e) \wedge TH(e) \in \oplus\{x | khachapuri(x)\}\} \wedge AG(e, N)$

To conclude, the meaning parts targeted by *teilweise* constitute semantic content proper. Any inferences that are not affected by *teilweise* must be considered pragmatic implicatures. In Sect. 3, we will employ this diagnostic to show that SE-readings with *know* + *wh* must be pragmatic implicatures, and not semantic entailments!

3 *Wissen* 'Know' + *Wh*: Data and Analysis

This section presents novel empirical data on the interpretation of *wh*-interrogatives embedded under *wissen* 'know'. In Sect. 3.1, we present evidence from the interpretation of such interrogatives in combination with *SCHON...aber* and with *teilweise* that shows that their basic semantic interpretation is the WE-reading. We will put forward an event-based semantic analysis of *know* + *wh* in Sect. 3.2. The SE-reading, in turn, is not an independent semantic reading, but derived from the WE-reading by way of pragmatic enrichment. Our analysis in Sect. 3.3 will take up ideas by [40] and [39], but we will put the ingredients together in a different manner.

3.1 Novel Evidence on *Know* + *Wh*: IE is Semantic, but SE is Pragmatic!

Looking first at the interpretive effect of *teilweise*, we find that this Q-adverb only ranges over the positive alternatives in the question, i.e. the complete set of true answers constituting the WE-reading [19]. The semantic effect of *teilweise* is to turn this WE-interpretation into a non-exhaustive question interpretation. Consider (16) and recall from Sect. 2.3 that *teilweise* only operates on truth-functional semantic content. (16) will be true if Nino knows for only part of the dancers that they danced, i.e., her knowledge is non-exhaustive regarding the WE-interpretation. As a result, the follow-up in (16a) is licit. Crucially, the alternative follow-up in (16b), in which Nino's knowledge is shown to be incomplete regarding the entire answer space including negative answers (= SE), is NOT felicitous. But it should be if SE-readings were bona fide semantic entailments, thus making (16) semantically ambiguous. The infelicity of (16b) thereby constitutes negative evidence against the analysis of SE as a semantic entailment.

(16) Nino weiß nur **teilweise**, wer getanzt hat, weil sie nicht weiß, ...
'Nino knows only **partially** who danced because she doesn't know...'
a. ..., dass Levan getanzt hat.
'... that Levan danced.
b. # ..., dass David **nicht** getanzt hat.
'... that David **didn't** dance.'

Next, consider the effect of *SCHON...aber* in (17). Here, the particle combination indicates that the SE-inference blocked. This is compatible with the felicitous follow-up in (17b), which is directed at the negative answer space (= part of the SE-denotation), and which improves significantly in the presence of *SCHON...aber* as opposed to its counterpart without. In contrast, as *SCHON...aber* cannot operate on the semantic content of the clause, cf. (9), it cannot be used to turn the underlying WE-reading into a non-exhaustive reading, viz. the infelicity of (17a), which marks Nino's knowledge as incomplete regarding the WE-denotation.

(17) Nino weiß **SCHON**, wer getanzt hat, **aber** sie weiß nicht, ...
'Nino knows who danced **alright**, **but** she doesn't know...'
a. #..., dass Levan getanzt hat. b. ..., dass David **nicht** getanzt hat.
'... that Levan danced.' '... that David didn't dance.'

In sum, the infelicity of (17a) constitutes negative evidence that the WE-reading is the underlying semantic interpretation of *know* + *wh*, whereas the felicity of (17b) constitutes positive evidence that SE is a mere pragmatic implicature. The data in (18) and (19) illustrate the same point (follow-ups in English for reasons of space):

(18) Nino weiß **teilweise** wer getanzt hat,
'Nino knows partially who danced,
a. # ... but she doesn't know that this is all. (SE violation #)
b. ... but she doesn't know of all dancers that they danced. (WE violation OK)

(19) Nino weiß **SCHON** wer getanzt hat, **aber**
'Nino knows who danced alright, but
a. ... she doesn't know that this is all. (SE violation OK)
b. # ... she doesn't know of all dancers that they danced. (WE violation #)

3.2 An Event-Semantic Analysis of WE-Readings with *Wissen* 'Know'

In our event-semantic account of the basic semantic WE-reading of *know* + *wh*, completeness of the answer is aspectually derived via event summation. We suggest the lexical entry in (20) for *wissen* 'know', using event composition with knowledge events and content arguments, as suggested by [29] and [26]. According to (20), for x to know (the answer to) Q means that x is in an attitudinal state e that is composed of the maximal sum of K(nowledge) substates e' that have the individual positive answers p to Q as their content.[4]

(20) $[\![wissen]\!]^{w} = \lambda Q.\lambda x.\lambda e.e = \oplus\{e' | p \in Q \wedge K_w(e') \wedge Content_w(e',p)\} \wedge AttitudeHolder_w(e,x)$

We also assume that the denotation Q of *wh*-interrogative clauses is the set of Hamblin-alternatives [16]. Given the veridicality and factivity of the knowledge attitude, we moreover assume that only true propositions can be known, i.e., that only true propositions in w

[4] We assume that K is a primitive *knowledge* predicate over eventualities.

can form the content of a knowledge eventuality; in other words $K_w(e') \wedge Content_w(e', p)$ can only be true iff $w \in p$. Finally, we assume that the $\oplus$-operator is part of the lexical aspect of *wissen*, making *e* the maximal possible knowledge eventuality concerning the question *Q*. This derives weak exhaustiveness for *Q* as an aspectual phenomenon, thereby eliminating the need for a covert ANS-operator [3, 17]: $\oplus$ sums in *e* the sub-states of knowledge of all true propositions in *Q*. For (1), this results in an event predication over the stative eventuality of *x* knowing the complete list of dancers, or rather the complete list of true propositions of the form *y danced*, as shown in (21) for the world of evaluation *w*. Further application of (21) to the denotation of *Nino* and subsequent existential closure over events will yield the complete meaning of (1).

(21)

$$\lambda x.\lambda e.e = \\ \oplus\{e'|p \in \{\lambda v.dance(y,v)|y \in Human_w\} \wedge K_w(e') \wedge Content_w(e',p)\} \wedge \\ AttitudeHolder_w(e,x)$$

The analysis in (20) and (21) directly extends to *know* + *that* when *that*-CPs are modelled as singleton sets of sets of worlds ⟨⟨⟨s, t⟩, t⟩⟩ [5]. Notice, too, that the event maximality imposed by $\oplus$ makes the eventuality bounded, which explains the old puzzle of why stative verbs of knowledge are crosslinguistically marked as perfective/telic, such as e.g. in Finnish [21] or in Hausa [27].

Applying the Q-adverb *teilweise* 'partially' to (21), and following the logic from Sect. 2.3, we derive the meaning of (18) in (22). (22) specifies a sub-event *e* of the maximal knowledge eventuality *e'* regarding the question *Who danced?*, and *x* is the attitude holder of this knowledge sub-eventuality *e*.

(22)

$$\lambda x.\lambda e.\mathrm{e} \sqsubseteq \mathrm{e}' \wedge e' = \\ \oplus\{e'|p \in \{\lambda v.dance(x,v)|x \in Human\} \wedge K_w(e') \wedge Content_w(e',p)\} \wedge \\ AttitudeHolder_w(e,x)$$

Feeding in the subject meaning and existential closure over events yields the correct meaning for (18). In sum, combining *teilweise* and *know* + *wh* results in a non-exhaustive semantic interpretation. We turn to pragmatic strengthening from WE to SE next.

3.3 Pragmatics: Strengthening to SE

As mentioned in Sect. 2.1, the SE-reading of *know* + *wh* does not only entail knowledge of the complete answer to the question, but also the knowledge that this is the complete answer [17]. In other words, *to know*-SE entails not only that the attitude holder knows the complete answer, but also that she knows that this is the complete answer, cf. [17]. In the event-semantic reformulation of *know* + *wh* in (23), this is represented in terms of two conjoined knowledge eventualities, where the second eventuality *e*" captures the missing component that turns the formula into a valid representation of SE-knowledge.

(23)

$$\exists e.e = \oplus\{e'|p \in \{\lambda v.dance(y,v)|y \in Human\} \wedge K_w(e') \wedge Content_w(e',p)\} \wedge AttitudeHolder_w(e,Nino) \wedge \exists e''[K_w(e'') \wedge Content_w(e'',\lambda v.e = \oplus\{e'|p \in \{\lambda w.dance(y,w)|y \in Human\} \wedge K_v(e') \wedge Content_v(e',p)\}) \wedge AttitudeHolder_w(e'',Nino)]$$

In view of the evidence for the blocking of pragmatic implicatures and SE-readings with *SCHON...aber* presented in Sect. 2.3 and Sect. 3.1, we propose to analyze the strengthened SE-reading in (23) as a pragmatic enrichment of (22). This enrichment follows from a hearer-based pragmatic preference for interpreting 3rd person attitude reports from the *internal* 1st person-perspective of the attitude holder. To capture this preference, we propose the novel general pragmatic principle PARV in (24).

(24) PRINCIPLE OF ATTITUDE REPORT VERIFICATION (PARV): In lack of further evidence, assume that if the utterance "S has the attitude X" is true, S is in a state of mind that allows her to truthfully utter: "I have the attitude X".

With PARV, the SE-reading of (1) (*Nino weiß, wer getanzt hat* 'Nino knows who danced') is derived from its underlying semantic WE-interpretation in (21) by the defeasible assumption that Nino is able to confirm (1) by uttering (25), i.e. that she knows that she knows the WE-reading, and not just part of it. Crucially, such 1st person knowledge reports are always SE, as is evidenced by the infelicity of the subsequent follow-up, which contradicts the 1st person SE-knowledge. The obligatory SE-construal with 1st person attitude reports follows from the fact that the reporting 1st person attitude holder must know that the summed (WE) knowledge eventuality is the complete knowledge state regarding *Q*, for else she cannot rule out that her knowledge is incomplete. In the formula in (24), this is captured in the occurrence of the second event *e*".

(25) Nino: Ich weiß, wer getanzt hat...
'Nino: I know who danced...'
...but I don't know everybody who danced.

Notice that (24) is mute on negative embedders, such as *keine Ahnung haben* 'be unaware' in (26), in which case the speaker cannot commit to the embedded content. Such predicates trigger logical scale reversal, such that the SE-interpretation is no longer an independent and logically stronger entailment, but rather entailed by semantic WE. If Nino is already unaware of the complete list of dancers in *w* (WE), it follows that she is also unaware of the complete list of dancers and non-dancers (SE).

(26) Nino hat keine Ahnung, wer getanzt hat.
'Nino is unaware who danced.'

Although not semantic in nature, the PARV-driven SE-reading is the default surface interpretation of *know* + *wh* in the absence of further evidence, which makes it difficult to cancel in the absence of context information or explicit discourse marking. This is evidenced, for instance, by the fact that SE-violations with *know* + *wh* were rated as contradictory in almost 75% of all cases in the contradiction experiment in [10] reported in Sect. 2.1. However, same as other prototypicality-based implicatures (Sect. 2.2), the default pragmatic SE-enrichment can be blocked by the particle combination *SCHON...aber*, as illustrated in (17) in Sect. 3.1.

More generally, PARV captures the implicit hearer-based assumption that attitude holders will normally be reported to have an attitude X if they are *de se* aware of having X. In such cases, they could explicitly commit to X in the form of a 1st person report. Presumably, the PARV-driven preference for evaluating attitude reports from the internal perspective of the attitude holder is due to the fact that attitudes are mental objects located in the holder's mind, for which the best or most reliable kind of evidence is a commitment by the attitude holder in the form of a 1st person report. If so, PARV would be connected to more general cognitive mechanisms associated with Theory of Mind [30]. Importantly, PARV in (24) is best considered a general interpretive principle that is not tied to questions per se, but which is also active, for instance, in the resolution of *de re/dicto*-ambiguities: In full parallel to SE-readings with *know* + *wh*, DPs contained in 3rd person attitude reports receive *de dicto* readings by default, and they must be *de dicto* in 1st person reports, cf. (27). In particular, the *de re* reading of (27a) is verified by a situation in which Rico owns a ruby which he falsely believes to be a worthless glass stone. The speaker may use the term *a ruby* to refer to that ruby, and correctly report that Rico knows that he owns that object. Crucially, Rico cannot report of himself that he owns a ruby, as long as he is not aware of the fact that this stone is in fact a ruby, cf. (27b). The contrast can be replicated with definite descriptions, too.

(27) a. Rico knows that he owns a ruby, but he is not aware it's a ruby.
3rd person: cancelled default reading = *de dicto*

b. Rico: #I know that I own a ruby, but I am not aware it's a ruby.
1st person: obligatory *de dicto*

Likewise, *de se*-pronouns as commonly found with logophoric construals [20, 28] are also tied to the internal perspective of the attitude holder. Given these observations, the pragmatic SE-enrichment with *wh*-interrogatives under *know* appears to be just another instance of perspective-dependent interpretation in natural language.

Our proposal to derive SE-readings by way of pragmatic enrichment is similar in spirit to the account in [40], but it differs in how the enrichment is triggered. [40] derives SE-readings from IE-readings via a hearer-based (excluded middle) competence assumption (CA). However, this is problematic, as the exact content of CA is unclear. On the formulation in (28a), CA is already equivalent to the SE-reading of (1), resulting in circularity. The formulation in (28b) does not generalize to other SE-compatible verbs, such as the verbs of saying *predict* or *tell*, as predictions or statements do not follow from beliefs. Another issue with the analysis of [40] is that it assumes a NEG-raising like property of predicates like *know*,[5] even though no evidence exists to this assumption – in fact, it seems that exactly the contrary is supported by facts.

(28) *Competence assumption (CA) by addressee of (1):*
a. Nino knows for everybody whether they danced or not.
b. Nino has some belief about everybody whether they danced or not.

This being said, there are some valid concerns as to whether PARV can also handle speech act verbs correctly. After all, PARV is limited to verbs of propositional attitude. For speech act verbs, it no longer holds true that the main evidence for their truth is in fact in the mind of the subject, as speech act verbs have public effects. But then again, (i.) speech act verbs tend to have less of a bias towards SE-readings; (ii.) even with speech act verbs it is essential what the subject meant when making her utterance, cf. the *de re/de dicto* ambiguities in (29); and (iii.) there are no sufficient empirical data for teasing apart the attitude component and the quotational aspects of speech act verbs [37] as would be necessary for an in-depth evaluation of PARV.

(29) Nino predicted that the winner will be the spy.
a. Nino: "The winner will be the spy." *De dicto*
Nino: "I predicted that the winner will be a spy."
b. Nino: "The winner will be Rico." (incidentally, Rico is a spy!) *De re*
Nino: "??I predicted that the winner will be a spy but I was not aware of it."

In deriving SE-readings as an effect of assuming an internal perspective, we adopt a core idea of [39], first traces of which are already found in [13]. [39] also link the weaker WE- (for them: IE) and the SE-reading to the external and internal perspectives of speaker and attitude holder, respectively. They do so, however, by treating the attitude predicate *know* as semantically ambiguous between [± internal perspective]. Their account in terms of a lexical ambiguity clashes with the above argument against semantic SE, though, and in particular with the observation that *SCHON...aber* cannot be exploited for disambiguating semantic ambiguities, cf. (8). Moreover, the availability of WE- and SE-readings with other question-embedding verbs (*predict*, *tell*...) [6, 11, 22]) would necessitate the assumption of a systematic WE/SE-ambiguity in the lexicon of such verbs, an undesirable consequence. In view of these findings, and given the observable parallels to other perspective-dependent phenomena in 1st and 3rd person reports, we consider our pragmatic account superior.

[5] According to [40], NEG-raising is the crucial step for deriving the SE-reading from underlying IE. The IE-reading guarantees that for any false alternative p, the subject does not believe p. By NEG-raising, now we move from the proposition that the subject does not believe p to the proposition that the subject does in fact believe *not* p. In other words, NEG-raising transforms the non-belief of false alternatives into a positive belief that false alternatives are false.

3.4 Conclusion on *Wissen* 'know' + *Wh*

In this section, we presented two novel diagnostics shedding light on the underlying semantic interpretation of *wh*-interrogatives under the veridical and homogeneous/distributive attitude verb *wissen* 'know'. The combination of such interrogatives with the Q-adverb *teilweise* 'partially' and the particle combination *SCHON...aber* 'alright...but' shows that their underlying semantic interpretation is WE, whereas the SE-reading is a pragmatic enrichment. In Sect. 3.3, we argued that this pragmatic enrichment is triggered by a default tendency to interpret 3rd person attitude reports from the attitude holder's 1st person internal perspective. We also suggested that the same enrichment process is at work in the derivation of *de dicto* readings and logophoricity effects.

4 *Überraschen* 'Surprise' + *Wh*: Data and Analysis

This section presents novel empirical data on the interpretation of *wh*-interrogatives embedded under the cognitive-emotive attitude verb *überraschen* 'surprise'. In Sect. 4.1, we consider the interpretation of *surprise* + *wh* in combination with *SCHON...aber* and *teilweise*. Our findings provide novel evidence for the claim in [34, 35], and [14] that they come with a fairly weak non-distributive, or non-homogeneous semantic WE_nondist interpretation, which can be pragmatically strengthened to WE_dist. Again, such pragmatic strengthening is blocked in the presence of *SCHON...aber*. Sect. 4.2 discusses the interpretation of *surprise* + *wh* from a theoretical perspective. We discuss a shortcoming of the existential WE-interpretation à la [14], and we end by sketching a tentative analysis of *überraschen* 'surprise' and other cognitive-emotive attitude verbs as denoting a cognitive-emotive attitude towards a fact, or a proposition-dependent or proposition-exemplifying situation à la [12, 25], and [1].

4.1 Novel Evidence: WE_nondist is Semantic, but WE_dist is Pragmatic!

Recall from Sect. 2.1 that *wh*-interrogatives under *überraschen* 'surprise' allow for two WE-construals of different logical strength. In (2), the attitude holder Nino may be surprised by each and every individual in the positive answer space of dancers (= WE_dist). Alternatively, she may be surprised by just some of the dancers (WE_nondist), cf. [14, 34]. WE_dist logically entails WE_nondist.

If we add the concessive particle combination *SCHON...aber*, we find that it blocks the logically stronger WE_dist interpretation, which involves surprise at each individual answer. This is evidenced by the felicitous follow-up in (30a) vs. (30b), in which the presence of *SCHON...aber* does not serve to cancel a semantic entailment.

(30) Es hat Nino **SCHON** überrascht, wer getanzt hat, ...
'It surprised Nino alright who danced ...'
a. ...**aber** es hat sie nicht bei jedem Tänzer überrascht
'... but she wasn't surprised at every dancer.'
b. #...**aber** sie war gar nicht überrascht.
'... but she wasn't surprised at all.'

Secondly, the Q-adverb *teilweise* 'partially' is difficult to interpret with *surprise* + *wh*, if not outright degraded, in the absence of other suitable plural expressions, cf. (31). This combination is also not readily attested in corpora:

(31) ??Es überrascht Nino **teilweise**, wer getanzt hat.
??'It partially surprises Nino who danced.'

As *teilweise* operates over plural events only, cf. (14), it is conceivable that the deviant status of (31) is due to the absence of such event pluralities with *surprise* + *wh*.

4.2 Towards a Non-propositional Analysis of *Surprise* + *Wh*

A classic way of deriving WE_nondist-readings for *surprise* + *wh* would consist in adopting an existential analysis with weak exhaustive force à la [34, 35], and [14]. *Überraschen* 'surprise' would take the WE-set of minimal (believed to be) true answers *Q* as its complement and map these to true iff there is at least one proposition *p* in this set such that the attitude holder did not expect this proposition to be true in *w*, cf. [14]:

(32) $[\![surprise]\!]^{w}(Q)(z)$ = True iff for all worlds w' compatible with z's past expectations in w, there is at least one $p \in \{q: q \in Q \wedge w \in p\}$ such that $w' \notin p$; defined if for all $p \in \{q: q \in Q \wedge w \in p\}$, z believes p in w.

Pragmatic strengthening to WE_dist would formally amount to replacing the existential quantifier in (32) with the universal quantifier. Informally, such pragmatic strengthening is licit as the strengthened readings still entail the truth of the underlying semantic entailment. They just depict particular ways of making (32) true. This is entirely parallel to what we find in the domain of adnominal quantifier scope in (33), in which the surface ∀∃-reading (*all the students watched a movie*) can be pragmatically strengthened to an inverse ∃∀-pseudoscope reading (*there is a movie that all the students watched*), which is again just a specific way of making the semantic ∀∃-reading true [31]:

(33) All the students have watched a/some movie.

Finally, the deviant status of (31) with *teilweise* may simply follow from semantic redundancy, as the underlying WE_nondist semantics already captures the incompleteness or subpart requirement of *teilweise*.

Alternatively, the deviant status of (31) may also follow from the inability of *teilweise* to access the subparts of individual situations with complex non-atomic substructure [23]. And indeed, there is some reason to believe so, as *surprise* can also give rise to SE_nondist-readings, which are not accounted for at all on the WE-analysis in (32) [7, 9, 11]. There is indeed some experimental evidence that the target of the surprise in cases of SE_nondist is not from the set of positive true answers that are accessed in (32). For illustration, consider the following example from the betting experiment [9, 11]. In the betting experiment, participants could decide to cash in a betting slip, or not, depending on how they interpreted the meaning of a sentence with a *wh*-interrogative embedded under *surprise*, cf. (34a). The truth-value judgment underlying participants' choices is made on the basis of a 1st person report of the attitude holder (here: Tiffany), cf. (34b), and of information about the circumstantial facts, cf. (34c).

(34) a. BET: Tiffany war überrascht, wer von den Teilnehmerinnen und Teilnehmern in der Sendung eine Heuschrecke gegessen hat.
'It surprised Tiffany who of the participants ate a grasshopper on the show.'

b. Tiffany: "I often think about the show, in which Freddy and Alessa bravely ate a grasshopper and the other three refused to do it. I expected that Carlo and Sophie would also eat a grasshopper on the show. After all, the two of them are generally quite flexible when it comes to food."

c. Facts:

Alessa	Carlo	Freddy	Mara	Sophie ate a grasshopper.
YES	NO	YES	NO	NO

In the setting in (34bc), the surprise of Mary is directed at the negative answer space: What is unexpected is that Carlo and Sophie did NOT eat the grasshopper. Crucially, the WE-based lexical entry for *surprise* in (32) predicts the bet to be false in this SE_nondist-setting, so that participants should not cash it in. This prediction stands in stark contrast to participants' behavior, who opted for cashing in in 58% of all cases, where cashing in is equivalent to judging (34a) true in the SE_nondist setting (34bc).

The availability of SE_nondist readings for *surprise* + *wh* casts some serious doubt on the adequacy of the WE-meaning representation in (32). For this reason, we would like to raise the possibility that *überraschen* 'surprise', and other cognitive-emotive attitude verbs, such as *be glad*, *be happy*, *be worried* etc., differ from *know* (and other epistemic attitude verbs) in a more fundamental way. Following [12], we would like to propose that such predicates do not select for a set of propositions (a question meaning), but rather for – what [12] call – a fact, or an exemplified or situated proposition [1, 25]. On this line of thought, the attitude of surprise may be conceptualized as a psychological state that is caused by potentially complex situations and their overall constitution or make-up, including missing subparts.[6] Put differently, we think of the meaning of *surprise* and of other emotive-cognitive factives as lexically decomposable into a causing eventuality and a primitive emotional state (here: surprisal) caused by the eventuality.

It is important to see that this means that the actual states of *surprisal* or happiness or worry etc. are primitive neuropsychological or emotional states, as typically assumed in language processing [15]. They are not phenomenologically intentional in that they do not have a propositional attitude argument. The impression of intentionality, i.e., the directedness towards a proposition or situation, is the result of associating the causing propositional attitude or cognitive attitude towards a situation with the resulting state.

[6] This is reminiscent of [38]'s notion of surprise as being directed at the overall size and constitution of the answer, except that the propositional notion of answer is replaced with a directly observable situation with unexpected subparts or unexpectedly missing subparts.

The general pattern for the meaning of cognitive-emotive factives is formally captured in (35), where the causing stimulus *s* could stand for a situation or a fact; see above.

(35) *s* surprises *X* = *X*'s acquaintance with *s* causes *X* to experience surprisal

The surprisal is then not caused by a belief in the truth of a proposition, but more directly by becoming acquainted with some situation or fact. In this vein, surprise can also be triggered non-verbally, e.g., by the content of pictures and photographs, or by the absence of content on such pictures, which are visual representations of complex situations. The famous picture of Lenin giving a speech in front of a revolutionary crowd in Sverdlov Square, Moscow, which was later purged of Trotzki's presence, constitutes a striking example of surprise by the absence of content. As a result, there are different ways of making (36) true:

(36) The Communist Party members were surprised by [what the picture showed].
  i. by what it showed (WE: surprise at visible content, e.g. Lenin)
  ii. by what it didn't show (SE_nondist: surprise at missing content: Trotzki)

There are other kinds of evidence pointing towards a different semantic status of epistemic and cognitive-emotive attitude verbs. *Surprise* can take situation-referring DPs or depictive DPs as arguments (37a), whereas *wissen* 'know' cannot (37b).

(37) a. *Der Krach/Das Bild* überraschte Nino.
  '*The noise/the picture* surprised Nino.'

  b. *Nino weiß *den Krach/das Bild.*
  'Nino knows *#the noise/the picture.*'

Secondly, the situation argument is directly expressed with the mandatory pronoun *es* 'it' with *überraschen* in (38a), whereas such a pronominal reference is at best optional with *wissen* 'know' in (38b).

(38) a. *(**Es**$_7$) überrascht Nino, wer getanzt hat.
  'It surprised Nino who danced.'

  b. Nino weiß ($^{??}$**es**$_7$), wer getanzt hat.
  'Nino knows it who danced.'

The empirical differences in (37) and (38) motivate a different semantic analysis for *überraschen* and other cognitive-emotive verbs in which they do not operate directly on the propositional content of the *wh*-interrogative. Following ideas in [12], and in particular [1] on the cognitive-emotive attitude predicate *interesting*, *überraschen* 'surprise' can be analyzed as directly selecting for a situation *s* such that *s* is a stimulus situation or fact that is part of a larger situation *s'* that (fully) resolves the *wh*-interrogative meaning

Q, and s causes a surprisal e of x in w, as tentatively shown in (39). For a situation to resolve a *wh*-question meaning, the situation must contain sufficient information for allowing at least for a partial answer to the *wh*-question.

(39) $[\![surprise]\!]^w = \lambda Q.\lambda x.\lambda s.\exists e \exists s'.s \sqsubseteq s' \wedge resolves(s', Q) \wedge cause(s, \mathrm{e}) \wedge surprisal(\mathrm{e}) \wedge holder(\mathrm{e}, x)$

Importantly, our theory of *surprise* naturally predicts that *surprise* has both a stative and an achievement reading, as shown in (40). For the stative reading (40a), the aspectual modification targets the resulting surprisal state whereas the achievement reading (40b) focuses on the causation event.

(40) a. I am surprised that... b. It surprises me that...

Given that a situation can cause surprisal by its size or by its general make-up or constitution [38], the denotation in (39) is general enough to be compatible with WE_dist, WE_nondist and SE_nondist readings alike. In the default case, this underspecified interpretation will be pragmatically enriched to the strongest logical reading, namely WE_dist, which expresses surprisal at all relevant subparts of the situation. Same as with *wissen* 'know', such pragmatic enrichment is blocked in the presence of *SCHON...aber*. Finally, the Q-adverb *teilweise* can only operate on semantically plural sums of eventualities, but not on the internal subparts (or lumps, [23]) of a complex situation, cf. the soup-eating situation by Nino in (14) above. This accounts for the observed infelicity of *teilweise* in combination with *surprise* + *wh*, where the surprise is directed at a complex situation. In order to express partial surprise, i.e., surprise at the subparts of a complex situation, we require the part-whole modifier *zum Teil* 'in part', which CAN operate on the material subparts of individual situations:

(41) Nino ist **zum Teil** überrascht, wer getanzt hat.
'It surprises Nino in part who danced.'

We postpone a more detailed situation-based analysis of *überraschen* 'surprise' to another occasion, and we conclude by pointing the interested reader to a recent analysis in [25] of depictive verbs like *imagine* as taking proposition-dependent situations as complements. As *imagine* can select for *wh*-interrogatives, too, it is tempting to aim at a unified analysis of different situation-selecting attitude verbs.

5 Conclusions and Theoretical Implications

In this paper, we investigated the interpretation of *wh*-interrogative clauses embedded under the attitude predicates *wissen* 'know' and *überraschen* 'surprise' in interaction with the particle combination *SCHON...aber* 'alright...but' and the Q-adverb *teilweise* 'partially'. We have shown that *SCHON...aber* does not operate on semantic content, but rather blocks the emergence of pragmatic implicatures based on considerations of relevance or prototypicality. The Q-adverb *teilweise*, by contrast, operates on semantic content by presenting an event as a mereological subpart of some plural sum event. Applying these novel empirical diagnostics to *know* + *wh*, we found that SE-inferences with *know* + *wh* are pragmatic in nature, whereas the logical weaker WE-inferences are semantic in nature. Applying the same diagnostics to *surprise* + *wh*, we found that the WE_dist reading under *surprise* is pragmatic and the result of default pragmatic strengthening. We also saw that the existence of both WE_dist and WE_nondist readings with *surprise* is accounted for on an existential WE-analysis à la [19] and [14], but the unexpected emergence of SE_nondist-readings is not! This led us to tentatively propose a fact- or situation-based reanalysis of cognitive-emotive attitude verbs like *überraschen* 'surprise' à la [12], on which the denotation of *surprise* does not operate on a set of propositions, i.e. the set of true answers in *w*, but on a fact that is situated or exemplified by the Karttunen-meaning of the *wh*-interrogative.

The general theoretical repercussions of our endeavor are as follows. We have presented novel empirical evidence that the meaning of embedded *wh*-interrogatives is indeed underspecified in the form of a set of Hamblin-alternatives, cf. [3]. Moreover, the observation that there is no inherent distributivity or homogeneity component built into the meaning of such *wh*-clauses argues against the obligatory presence of a max-operator in *wh*-clauses, pace [33]. Likewise, we have argued that the exhaustivity effects frequently observed with embedded questions are not located in the denotation of the *wh*-interrogatives themselves, for instance in the form of covert ANS(wer)- or EXH-operators. Instead, they follow from the aspectual semantics of the embedding attitude predicates. As a result, some attitude verbs such as cognitive-emotive *surprise* only come with very weak exhaustivity requirements, whereas the complete WE-interpretation with epistemic *know* is the result of sum formation over knowledge sub-events. The corresponding SE-inferences are not semantically derived. Finally, we tentatively suggested that cognitive-emotive attitude verbs may express a relation not to sets of propositions, but to proposition-dependent situations or facts, which may also be expressed in the form of plain nominal DPs.

References

1. Abenina-Adar, M.: Interesting interrogatives. In: Franke, M., et al. (eds.) Proceedings of Sinn und Bedeutung 24, pp. 1–16. Open Journal Systems, Konstanz (2020)
2. Beck, S., Sharvit, Y.: Pluralities of questions. J. Semant. **19**, 105–157 (2002)
3. Beck, S., Rullmann, H.: A flexible approach to exhaustivity in questions. Nat. Lang. Semant. **7**, 249–298 (1999)
4. Berman, S.: On the semantics and logical form of wh-clauses. Ph.D. thesis, University of Massachusetts at Amherst (1991)
5. Ciardelli, I., Groenendijk, J., Roelofsen, F.: Inquisitive Semantics. Oxford University Press, Oxford (2018)
6. Cremers, A., Chemla, E.: A psycholinguistic study of the exhaustive readings of embedded questions. J. Semant. **33**, 49–85 (2016)
7. Cremers, A., Chemla, E.: Experiments on the acceptability and possible readings of questions embedded under emotive-factives. Nat. Lang. Semant. **25**, 223–261 (2017)

8. Cremers, A., Tieu, L., Chemla, E.: Children's exhaustive readings of questions. Lang. Acquis. **24**(4), 343–360 (2017)
9. Fricke, L., Blok, D.: Exhaustiveness in embedded questions. An experimental comparison of four predicates of embedding. Talk Presented at XPrag.de, Berlin (2019)
10. Fricke, L., Bombi, C., Blok, D., Zimmermann, M.: The pragmatic status of strong exhaustive readings of embedded questions. Poster presented at DGfS 42, Hamburg (2020)
11. Fricke, L., Blok, D., Zimmermann, M., Onea, E.: The pragmatics of embedded questions. An experimental comparison of four predicates of embedding. Manuscript (2020)
12. Ginzburg, J., Sag, I.: Interrogative investigations: the form, meaning and use of English interrogatives. CSLI, Stanford (2000)
13. Groenendijk, J., Stokhof, M.: Studies on the semantics of questions and the pragmatics of answers. Ph.D. thesis, University of Amsterdam (1984)
14. Guerzoni, E., Sharvit, Y.: A question of strength: on NPIs in interrogative clauses. Linguist. Philos. **30**(3), 361–391 (2007)
15. Hale, J.: A probabilistic early parser as a psycholinguistic model. In: Proceedings of the 2nd Meeting of the North American chapter of the Association for Computational Linguistics on Language Technologies, pp. 1–8 (2001)
16. Hamblin, C.: Questions in Montague English. Found. Lang. **10**, 41–53 (1973)
17. Heim, I.: Interrogative semantics and Karttunen's semantics for know. In: Buchalla, R., Mittwoch, A. (eds.) Proceedings of the Ninth Annual Conference and the Workshop on Discourse of IATL, pp. 128–144. Academon, Jerusalem (1994)
18. Höhle, T.: Über Verum-Fokus im Deutschen. In: Jacobs, J. (ed.) Informationsstruktur und Grammatik, pp. 112–141. Springer, Wiesbaden (1992). https://doi.org/10.1007/978-3-663-12176-3_5
19. Karttunen, L.: Syntax and semantics of questions. Linguist. Philos. **1**, 3–44 (1977)
20. Kiemtoré, A.: Issues in Jula complementation: structure(s), relation(s), and matter(s) of interpretation. Ph.D. thesis, Universität Stuttgart (2022)
21. Kiparsky, P.: Partitive case and aspect. In: Butt, M., Geuder, W. (eds.) The Projection of Arguments. Lexical and Compositional Features, pp. 265–317. CSLI, Stanford (1998)
22. Klinedinst, N., Rothschild, D.: Exhaustivity in questions with non-factives. Semant. Pragmatics **4**(2), 1–23 (2011)
23. Kratzer, A.: An investigation of the lumps of thought. Linguist. Philos. **12**(5), 607–653 (1989)
24. Lahiri, U.: Questions and Answers in Embedded Contexts. Oxford University Press, Oxford (2002)
25. Liefke, K.: A propositionalist semantics for imagination and depiction reports. In: Franke, M., et al. (eds.) Proceedings of Sinn und Bedeutung 24, pp. 515–532. Open Journal Systems, Konstanz (2020)
26. Moulton, K.: Natural selection and the syntax of clausal complementation. Ph.D. thesis, UMass Amherst (2009)
27. Newman, P.: The Hausa Language. Yale University Press, New Haven (2000)
28. Pearson, H.: The sense of self: topics in the semantics of de se expressions. Ph.D. thesis, Harvard University (2013)
29. Pietroski, P.M.: On explaining that. J. Philos. **97**, 655–662 (2000)
30. Premack, D., Woodruff, G.: Does the chimpanzee have a theory of mind? Behav. Brain Sci. **1**(4), 515–526 (1978)
31. Reinhart, T.: Quantifier scope. How labor is divided between QR and choice functions. Linguist. Philos. **20**, 335–397 (1997)
32. Romero, M.: Surprise-predicates, strong exhaustivity and alternative questions. In: Proceedings of SALT, vol. 25, pp. 225–245 (2015)
33. Rullmann, H.: Maximality in the semantics of wh-constructions. Ph.D. thesis, UMass Amherst (1995)

34. Sharvit, Y.: Embedded questions and 'de dicto' readings. Nat. Lang. Seman. **10**, 97–123 (2002)
35. Sharvit, Y., Guerzoni, E.: Reconstruction and its problems. In: Proceedings of the 14th Amsterdam Colloquium, pp. 205–210. Universiteit van Amsterdam, Amsterdam (2003)
36. Spector, B.: Exhaustive interpretations: what to say and what not to say. Unpublished paper, presented at the LSA Workshop on Context and Content, Cambridge, 15 July 2005 (2005)
37. Spector, B.: Aspects de la pragmatique des opérateurs logiques. Université Paris Diderot dissertation, Paris (2006)
38. Theiler, N.: A multitude of answers. Embedded questions in typed inquisitive semantics. MSc thesis, University of Amsterdam (2014)
39. Theiler, N., Roelofsen, F., Aloni, M.: A uniform semantics for declarative and interrogative complements. J. Semant. **35**(3), 409–466 (2018)
40. Uegaki, W.: Interpreting questions under attitudes. Ph.D. thesis, MIT (2015)
41. Xiang, Y.: Interpreting questions with non-exhaustive answers. Ph.D. dissertation, Harvard University (2006)
42. Zimmermann, M.: Wird schon stimmen! A degree operator analysis of schon. J. Semant. **35**(4), 687–739 (2018)

Logic and Computation

Lyndon Interpolation for Modal μ-Calculus

Bahareh Afshari[1,2](✉) and Graham E. Leigh[2]

[1] Institute for Logic, Language and Computation, University of Amsterdam, Amsterdam, The Netherlands
bahareh.afshari@gu.se
[2] Department of Philosophy, Linguistics and Theory of Science, University of Gothenburg, Gothenburg, Sweden
graham.leigh@gu.se

Abstract. It is known that the modal μ-calculus has the Craig interpolation property, indeed uniform interpolation. We prove Lyndon interpolation for the calculus, a strengthening of Craig interpolation which is not implied by uniform interpolation. The proof utilises 'cyclic' sequent calculus and provides an algorithmic construction of interpolants from valid implications. This direct approach enables us to derive a correspondence between the shape of interpolants and existence of sequent calculus proofs.

Keywords: Lyndon interpolation · Modal μ-calculus · Cyclic proofs · Sequent calculus

1 Introduction

The modal μ-calculus is an extension of modal logic by two quantifiers, μ and ν, that bind propositional variables. The formulæ $\mu x A$ and $\nu x A$ are interpreted over Kripke frames (labelled transition systems) respectively as the least and greatest fixed points of the function $x \mapsto A(x)$. Modal μ-calculus can thus be thought of as a logic that allows for restricted second-order quantification while still maintaining computationally attractive properties such as decidability of validity and the finite model property. Moreover, many program logics (LTL, PDL, CTL, etc.) can be embedded into μ-calculus making it an important metatheory [3,4].

Modal logics are known to widely enjoy interpolation (see e.g. [7,23]) and μ-calculus does so in a very strong sense: given a formula A and a finite set of propositions and modality operators L, there exists a formula I (the interpolant) in the language L such that $A \to I$ is valid and for *every* formula B whose common language with A lies within L, $A \to B$ is valid if and only if $I \to B$ is valid. This property, called *uniform interpolation*, easily implies *Craig interpolation*:

Supported by the Knut and Alice Wallenberg Foundation [2015.0179] and the Swedish Research Council [2016-03502 & 2017-05111]. The authors wish to express their gratitude to the anonymous referees for their comments and suggestions.

© The Author(s), under exclusive license to Springer Nature Switzerland AG 2022
A. Özgün and Y. Zinova (Eds.): TbiLLC 2019, LNCS 13206, pp. 197–213, 2022.
https://doi.org/10.1007/978-3-030-98479-3_10

if $A \to B$ is valid then there is a formula I in the language common to A and B such that both $A \to I$ and $I \to B$ are valid. *Lyndon interpolation* [14] is a strengthening of Craig interpolation where a proposition is considered to be in the common language only if it occurs with the same polarity in both A and B. Note that uniform interpolation does not entail Lyndon interpolation.

Uniform interpolation for modal μ-calculus was established by D'Agostino and Hollenberg in [6]. Their proof involves both semantic and syntactic arguments. The authors utilise (disjunctive) modal automata from [9] to show that a form of propositional quantifier, known as bisimulation quantifiers [16,21], is representable in modal μ-calculus and can be used to define interpolants.

Aside from the method of propositional quantification, there are a number of other ways to approach interpolation in non-classical and modal logics (see e.g. [5,12]) among which is the syntactical approach via sequent calculus. If one has to hand a (complete) sequent calculus that admits elimination of cuts then (Craig) interpolants can often be constructed by recursion over the cut-free derivations. Indeed, there is an intimate connection between interpolation and the existence of various forms of sequent calculi [8,13].

Following the proof-theoretic approach to interpolation, in this paper we show how to directly extract interpolants for modal μ-calculus from sequent calculus. The first proof system to consider is Kozen's axiomatisation [11] which expands the standard axioms of the modal system K by regeneration and induction rules for the least (μ) and greatest (ν) fixed point quantifiers:

$$\begin{aligned} \nu\text{-regeneration:}\quad & \nu \mathsf{x} A(\mathsf{x}) \to A(\nu \mathsf{x} A(\mathsf{x})) \\ \mu\text{-regeneration:}\quad & A(\mu \mathsf{x} A(\mathsf{x})) \to \mu \mathsf{x} A(\mathsf{x}) \\ \nu\text{-induction:}\quad & B \to A(B) \vdash B \to \nu \mathsf{x} A(\mathsf{x}) \\ \mu\text{-induction:}\quad & A(B) \to B \vdash \mu \mathsf{x} A(\mathsf{x}) \to B \end{aligned}$$

Completeness of Kozen's axiomatisation was established by Walukiewicz [22]. The proof, imitated for the natural sequent formulation of the system, makes essential use of cut and it remains a significant open problem whether Kozen's formulation without cut is also complete. But aside from cut, the induction rules themselves do not preserve interpolants, so an alternate proof calculus is still needed.

We will instead utilise a finitary and complete 'circular' proof system introduced in [1] and utilised in [2]. This system discards Kozen's induction rules in favour of an inference that focuses on repeating sequents

$$\dfrac{\begin{matrix}[\vdash \Gamma] \\ \vdots \\ \vdash \Gamma\end{matrix}}{\vdash \Gamma}\,\mathsf{dis}$$

The sequent at the top in brackets is understood as an assumption of the proof which is discharged. Applications of such a rule are subject to a global soundness condition: after identifying leaves with their point of discharge, every infinite path through the resulting graph must feature an infinite formula trace along which a greatest fixed point formula of 'outermost' importance is regenerated. In other words, the discharging rules provide a finitary description of a refutation tableaux in the sense of [15]. The question of designing such a calculus rests on encoding the global soundness condition within the syntactic confines of (finitary) sequent calculi, so that correctness of a cyclic proof becomes a local requirement and not based on properties of the infinite tree described by the cyclic proof. This can be achieved by adopting an annotated sequent calculus. In the cyclic calculi of Jungteerapanich [10] and Stirling [20] each formula in the proof (and also each sequent) is annotated by a word from a finite set of *names* for ν-quantified variables in such a way that correctness of the proof can be inferred by checking a syntactic property of names in the finite proof. Refining the calculus of [20], the authors, in [1], present a cyclic proof system in which correctness of the discharge inference can be inferred directly from the syntactic shape of the discharged sequent. It is this latter calculus, named Circ in [1], that we utilise in this paper.

Unbeknown to the authors a proof of Lyndon interpolation for the modal μ-calculus, using a similar approach, was announced by Shamkanov in [17]. The unpublished result utilises ill-founded proofs and their automata induced regularisations [18].

2 Modal μ-Calculus

Formulæ of the modal μ-calculus are specified by the grammar:

$$A = \top \mid \bot \mid p \mid \bar{p} \mid \mathsf{x} \mid A \wedge A \mid A \vee A \mid \langle \mathfrak{a} \rangle A \mid [\mathfrak{a}]A \mid \mu \mathsf{x} A \mid \nu \mathsf{x} A$$

where p ranges over a set Prop of **propositions**, x over a set Var of **variable** symbols and $\mathfrak{a}$ over a set Act of **action** symbols. Formulæ $\top$ and $\bot$ are called **constants**, p and $\bar{p}$ are **literals**, and $\mu\mathsf{x}A$ and $\nu\mathsf{x}A$ are **quantified** formulæ. Roman letters A, B, C, etc. range over formulæ. The set of **free** and **bound** variables of a formula, $\mathrm{FV}(A)$ and $\mathrm{BV}(A)$ respectively, are defined as usual. A formula with no free variables is **closed**.

2.1 Syntactic Considerations

By definition, variables cannot occur in negated contexts. This is important as it excuses the need for a syntactic restriction on forming quantified formulæ. In the present framework, negation is simulated by duality of the connectives, modalities and quantifiers. We write $\overline{A}$ for the **dual** of A, given by

$$\begin{array}{lllll} \overline{A \wedge B} = \overline{A} \vee \overline{B} & \overline{[\mathfrak{a}]A} = \langle \mathfrak{a} \rangle \overline{A} & \overline{\mu \mathsf{x} A} = \nu \mathsf{x} \overline{A} & \overline{\mathsf{x}} = \mathsf{x} & \overline{\top} = \bot \\ \overline{A \vee B} = \overline{A} \wedge \overline{B} & \overline{\langle \mathfrak{a} \rangle A} = [\mathfrak{a}]\overline{A} & \overline{\nu \mathsf{x} A} = \mu \mathsf{x} \overline{A} & \overline{\overline{p}} = p & \overline{\bot} = \top \end{array}$$

Implication is introduced as a defined connective, $A \to B = \overline{A} \vee B$. Given A, B and x, $A[\mathsf{x}/B]$ denotes the formula obtained by replacing every free occurrence of x in A by B and renaming bound variables of A to avoid variable capture. This will also be written as $A(B)$ if the choice of x is clear from the context. The notion of sub-formula most useful for μ-calculus is given by the Fischer–Ladner closure, $\mathrm{FL}(\cdot)$, defined as follows

$$\begin{aligned}
\mathrm{FL}(l) &= \{l\} \quad l \text{ a constant, literal or variable} \\
\mathrm{FL}(A \circ B) &= \{A \circ B\} \cup \mathrm{FL}(A) \cup \mathrm{FL}(B) \quad \circ \in \{\wedge, \vee\} \\
\mathrm{FL}((\mathfrak{a})A) &= \{(\mathfrak{a})A\} \cup \mathrm{FL}(A) \quad (\mathfrak{a}) \in \{\langle \mathfrak{a} \rangle, [\mathfrak{a}]\} \\
\mathrm{FL}(\sigma \mathsf{x} A) &= \{\sigma \mathsf{x} A\} \cup \mathrm{FL}(A[\mathsf{x}/\sigma \mathsf{x} A]) \quad \sigma \in \{\mu, \nu\}
\end{aligned}$$

For the purposes of this article, a **sub-formula** of A is any formula $B \in \mathrm{FL}(A)$.

Fix a formula A. A literal l is said to occur in A if $l \in \mathrm{FL}(A)$; an action symbol $\mathfrak{a}$ occurs in A if there is a modal sub-formula $\langle \mathfrak{a} \rangle B$ or $[\mathfrak{a}]B$ of A. The **language** of A, denoted $L(A)$, is the set of literals and action symbols that occur in A. For example, the language of $A = \mu \mathsf{x}((p \vee [\mathfrak{a}](\mathsf{x} \wedge \overline{q})) \vee (\overline{p} \wedge \langle \mathfrak{b} \rangle \mathsf{x}))$ is $L(A) = \{p, \overline{p}, \overline{q}, \mathfrak{a}, \mathfrak{b}\}$. In particular, $q \notin L(A)$. If $L(A) \subseteq L$ we say A is in the language L.

The relative position of free and bound variables in a formula generates a relation on variable symbols called the **subsumption** relation. Given A, this is the binary pre-order $<_A$ on variables generated by setting $\mathsf{x} <_A \mathsf{y}$ if $\mathsf{x} \neq \mathsf{y}$ and x occurs free in the scope of the quantifier $\sigma \mathsf{y}$ in A. In general, $<_A$ need not be well-founded, but every formula is α-convertible to a formula with well-founded subsumption relation. For convenience, we assume a single, global, well-founded pre-order $<$ on Var and restrict attention to formulæ A with $<_A \subseteq <$. We say x **subsumes** y if $\mathsf{x} \leq \mathsf{y}$.

2.2 Semantics

Formulæ are interpreted over labelled transition systems (LTS), tuples $\mathcal{S} = \langle S, \{\rightarrow_{\mathfrak{a}}\}_{\mathfrak{a}}, \lambda \rangle$, where S is a non-empty set of **states**, $\rightarrow_{\mathfrak{a}} \subseteq S \times S$ is the **accessibility** relation on S, and $\lambda \colon \mathsf{Prop} \rightarrow \mathit{Pow}(S)$ is a **labelling** of states by propositions, given as a function from the set of propositions to the power set of S. A **valuation** over the LTS $\mathcal{S}$ is a function $v \colon \mathsf{Var} \rightarrow 2^S$ assigning a set of states to each variable symbol. Given a valuation v and $T \subseteq S$, we write $v[\mathsf{x} \mapsto T]$ for the valuation v' such that $v'(\mathsf{x}) = T$ and $v'(\mathsf{y}) = v(\mathsf{y})$ for $\mathsf{y} \in \mathsf{Var} \setminus \{\mathsf{x}\}$. The **denotation** of A over $\mathcal{S}$ and v, denoted $\|A\|_v^{\mathcal{S}}$, is a subset of S defined inductively on A:

- For literals, constants and variables:

$$\|\mathsf{x}\|_v^{\mathcal{S}} = v(\mathsf{x}) \quad \|\top\|_v^{\mathcal{S}} = S \quad \|\bot\|_v^{\mathcal{S}} = \emptyset \quad \|p\|_v^{\mathcal{S}} = \lambda(p) \quad \|\overline{p}\|_v^{\mathcal{S}} = S \setminus \lambda(p)$$

- For logical connectives and modal formulæ:

$$\begin{aligned}
\|A \wedge B\|_v^{\mathcal{S}} &= \|A\|_v^{\mathcal{S}} \cap \|B\|_v^{\mathcal{S}} & \|[\mathfrak{a}]A\|_v^{\mathcal{S}} &= \{s \in S : \forall t \in S((s \rightarrow_{\mathfrak{a}} t) \Rightarrow t \in \|A\|_v^{\mathcal{S}})\} \\
\|A \vee B\|_v^{\mathcal{S}} &= \|A\|_v^{\mathcal{S}} \cup \|B\|_v^{\mathcal{S}} & \|\langle \mathfrak{a} \rangle A\|_v^{\mathcal{S}} &= \{s \in S : \exists t \in S((s \rightarrow_{\mathfrak{a}} t) \wedge t \in \|A\|_v^{\mathcal{S}})\}
\end{aligned}$$

- For quantified formulæ:

$$\|\mu \mathsf{x} A\|_v^{\mathcal{S}} = \bigcap\{T \subseteq S : \|A\|_{v[\mathsf{x}\mapsto T]}^{\mathcal{S}} \subseteq T\}$$
$$\|\nu \mathsf{x} A\|_v^{\mathcal{S}} = \bigcup\{T \subseteq S : T \subseteq \|A\|_{v[\mathsf{x}\mapsto T]}^{\mathcal{S}}\}$$

Observe that $\|\overline{A}\|_v^{\mathcal{S}} = S \backslash \|A\|_{\overline{v}}^{\mathcal{S}}$ where $\overline{v}$ is the valuation defined by $\overline{v}(\mathsf{x}) = S \backslash v(\mathsf{x})$. In particular, for A closed, $\overline{A}$ denotes the negation of A in the usual sense. We write $\mathcal{S}, s \models A$ if $s \in \|A\|_v^{\mathcal{S}}$ for every valuation v. A formula A is **valid** if $\mathcal{S}, s \models A$ for every LTS $\mathcal{S}$ and state s.

2.3 Annotations and Sequents

In the next section we introduce a labelled sequent calculus for the modal μ-calculus. These labels, henceforth called **annotations**, are finite words built from a fixed set of names for fixed point variables. To each variable symbol x we associate an infinite set N_x of **names (for** x**)** which is assumed to be partitioned into two infinite sets: N_x^V, the **variable names**, and N_x^A, the **assumption names** for x. We assume $N_\mathsf{x} \cap N_\mathsf{y} = \emptyset$ if $\mathsf{x} \neq \mathsf{y}$. Symbols x, y, etc. (also with indices) range over variable names, and $\hat{x}$, $\hat{y}$, etc. over assumption names. The set of all assumption names is denoted $N_\mathsf{A} = \bigcup_{\mathsf{x}\in\mathsf{Var}} N_\mathsf{x}^\mathsf{A}$, and the set of all names is $N = \bigcup_{\mathsf{x}\in\mathsf{Var}} N_\mathsf{x}$. Given $\hat{x} \in N$ we let $N_{\hat{x}} = N_\mathsf{x}$ where x is such that $\hat{x} \in N_\mathsf{x}$.

For a set $M \subseteq N$ of names, M^* is the set of finite words in M and includes the empty word ϵ. The (reflexive) sub-word relation on N^* is denoted $\sqsubseteq$ and defined by $x_1 \cdots x_m \sqsubseteq a$ iff there exists $a_0, \ldots, a_m \in N^*$ such that $a = a_0 x_1 a_1 \cdots x_m a_m$. For $M \subseteq N$ and $a \in N$, $a \restriction M$ is the maximal word in M^* which is a sub-word of a. The global subsumption ordering $<$ extends to names in the natural way: for $x, y \in N$, $x \leq y$ ($x < y$) if $x \in N_\mathsf{x}$, $y \in N_\mathsf{y}$ and $\mathsf{x} \leq \mathsf{y}$ (resp. $\mathsf{x} < \mathsf{y}$). If $a \in N^*$ we write $a \leq \mathsf{x}$ if $a \in M^*$ where $M = \bigcup_{\mathsf{y}\leq\mathsf{x}} N_\mathsf{y}$.

We now introduce the notions of annotated formula and sequent required for the present work.

Definition 1. *An **annotation** is a non-repeating word $a \in N^*$, i.e., $a = x_0 \cdots x_m$ where each x_i is a name and $x_i = x_j$ iff $i = j$. An **annotated formula** is a pair (a, A), henceforth written A^a, where A is a formula and $a \in N^*$ is an annotation. A **sequent** is an expression $a : \Gamma$ where $a \in N_\mathsf{A}^*$ is an annotation in assumption names (only) and $\Gamma = \{A_1^{a_1}, \ldots, A_n^{a_n}\}$ is a finite set of annotated formulæ such that $a_i \restriction N_\mathsf{A} \sqsubseteq a$ for each i. The word a is called the **control** of the sequent.*

Unannotated formulæ are identified with their annotation by the empty word and finite sets of unannotated formulæ with sequents with empty control, in which case the control is omitted. Annotations play a purely syntactic role and are ignored for all semantic considerations. When there is no cause for confusion we refer to annotated formulæ simply as 'formulæ'. Finite sets of annotated formulæ are denoted Γ, Δ, etc. As usual for sequent calculi, formulæ are identified with singleton sets and comma is used in place of union of sets of formulæ. Substitution extends to sequents in the obvious way, setting $(a : \Gamma)[\mathsf{x}/A]$ to be the sequent $a : \{B[\mathsf{x}/A]^b : B^b \in \Gamma\}$.

$\mathsf{Ax1}: p, \overline{p}$ $\qquad$ $\mathsf{Ax2}: \top$ $\qquad$ $\mathsf{Ax3}: \mu \mathsf{x} A, \nu \mathsf{x} \overline{A}$

$$\frac{\Gamma, B, C}{\Gamma, B \vee C} \vee \qquad \frac{\Gamma, B \quad \Gamma, C}{\Gamma, B \wedge C} \wedge \qquad \frac{\Gamma, A}{\langle \mathsf{a} \rangle \Gamma, [\mathsf{a}] A} \mathsf{mod} \qquad \frac{\Gamma}{\Gamma, A} \mathsf{weak}$$

$$\frac{\Gamma, A[\mathsf{x}/\sigma \mathsf{x} A]}{\Gamma, \sigma \mathsf{x} A} \sigma \qquad \frac{\Gamma, A[\mathsf{x}/\overline{\Gamma}]}{\Gamma, \nu \mathsf{x} A} \mathsf{ind} \qquad \frac{\Gamma, \overline{A} \quad \Gamma, A}{\Gamma} \mathsf{cut}$$

Fig. 1. Axioms and rules of Koz, where $\sigma \in \{\mu, \nu\}$ and $\overline{\Gamma}$ denotes $\bigwedge_{A \in \Gamma} \overline{A}$.

3 Sequent Calculi

Kozen [11] introduced an axiomatisation for the modal μ-calculus which was proved complete by Walukiewicz [22]. Presented as a (one-sided) sequent calculus, Kozen's system, denoted Koz, extends the natural formulation of the modal logic K by fixed point and induction inferences for the two quantifiers (see Fig. 1), with the restriction that all formulæ are closed. The induction rule ind formalises the semantic argument leading to $B \to \nu \mathsf{x} A$ being valid, namely that for every transition system $\mathcal{S}$, if $\|B\|^{\mathcal{S}} \subseteq \|A(B)\|^{\mathcal{S}}$ then $\|B\|^{\mathcal{S}} \subseteq \|\nu \mathsf{x} A(\mathsf{x})\|^{\mathcal{S}}$.

To obtain a syntactic proof of interpolation one typically argues by induction on the proof witnessing the valid implication, showing that for each rule of inference an interpolant for the conclusion can be constructed from interpolant(s) for the premise(s). Both cut and induction rules, however, violate the sub-formula property needed to constrain the language of the interpolant, thus it is important to work instead with an analytic sequent calculus.

3.1 A Circular Proof System

We will utilise the cut-free sequent calculus introduced in [1] by the name of Circ. As discussed in the introduction, Circ is a cyclic proof system in the sense that in addition to axioms, leaves of a (closed) proof may be non-axiomatic sequents provided the sequents are repeated on the path descending to the root. This additional rule allows to omit the problematic induction and cut rules at the cost of finding a sound notion of repeat. For this we utilise the annotated sequents defined in Sect. 2.3. The inference rules of Circ are presented in Fig. 2. Note that axioms do not depend on annotations. The logical rules $\vee$, $\wedge$, mod and weak have the same form as in Koz but incorporate annotations which are carried directly from premise(s) to conclusion; an instance of weak introduces an arbitrary annotated formula to the conclusion provided the conclusion remains an annotated sequent.

Notable restrictions on annotations occur in the quantifier rules μ and ν_x, the expansion rule exp, and the discharging inference $\mathsf{dis}_{\hat{x}}$. The first of these allows to deduce $a : \Gamma, \mu \mathsf{x} A^b$ from $a : \Gamma, A(\mu \mathsf{x} A)^b$ provided the annotation b names only variables subsuming x. The dual quantifier inference ν_x specifies the same fixed point property for ν-quantified formulæ, but in addition, allows dropping a single

$$\mathsf{Ax1}\quad a : p^b, \overline{p}^c \qquad\qquad \mathsf{Ax2}\quad a : \top^b$$

$$\frac{a : \Gamma, B^b, C^b}{a : \Gamma, (B \vee C)^b}\vee \qquad \frac{a : \Gamma, B^b \quad a : \Gamma, C^b}{a : \Gamma, (B \wedge C)^b}\wedge$$

$$(b \leq \mathsf{x})\frac{a : \Gamma, A[\mathsf{x}/\mu\mathsf{x}A]^b}{a : \Gamma, \mu\mathsf{x}A^b}\mu \qquad \frac{a : \Gamma, A^b}{a : \langle\mathsf{a}\rangle\Gamma, [\mathsf{a}]A^b}\mathsf{mod}$$

$$(b \leq x \in N_{\mathsf{x}}^{\mathsf{V}})\frac{a : \Gamma, A[\mathsf{x}/\nu\mathsf{x}A]^{bx}}{a : \Gamma, \nu\mathsf{x}A^b}\nu_x \qquad \frac{a : \Gamma}{a : \Gamma, A^b}\mathsf{weak}$$

$$(\forall i \leq k.\, b_i \sqsubseteq a_i)\frac{b_0 : A_1^{b_1}, \ldots, A_k^{b_k}}{a_0 : A_1^{a_1}, \ldots, A_k^{a_k}}\mathsf{exp}$$

$$\begin{array}{c}\left[a\hat{x} : \Gamma, A_0^{a_0\hat{x}x_0}, \ldots, A_k^{a_k\hat{x}x_k}\right]^{\hat{x}} \\ \vdots \\ (x_0, \ldots, x_k \in N_{\hat{x}}^{\mathsf{V}})\dfrac{a\hat{x} : \Gamma, A_0^{a_0\hat{x}}, \ldots, A_k^{a_k\hat{x}}}{a : \Gamma, A_0^{a_0}, \ldots, A_k^{a_k}}\mathsf{dis}_{\hat{x}}\end{array}$$

Fig. 2. Axioms and rules of Circ.

name for the quantified variable from the annotation. In the presence of the rule exp, removing a name from the premise is optional so a ν inference analogous to μ is also available. The rule exp, called the **expansion rule**, allows annotations to be arbitrarily expanded, subject to the result being an (annotated) sequent. The final rule, $\mathsf{dis}_{\hat{x}}$, referred to as the **discharge rule**, deserves further explanation. Ignoring annotations, the inference marks proof cycles:

$$\begin{array}{c}[\Gamma, A_0, \ldots, A_k] \\ \vdots \\ \dfrac{\Gamma, A_0, \ldots, A_k}{\Gamma, A_0, \ldots, A_k}\mathsf{dis}\end{array}$$

The control and annotations restricts which simple cycles are admissible. Reading from root to leaf, the inference expresses that, given a sequent $a : \Delta$, it is possible to append the control and (some) annotations by a fresh assumption name $\hat{x}$ (for the variable x, say), and find a repeat of the sequent further up in the proof at which the same control and annotations are witnessed but where each occurrence of the name $\hat{x}$ has been extended by some other name for x. Since the names $(x_i)_i$ in the rule $\mathsf{dis}_{\hat{x}}$ are required to be variable names (rather than assumption names), it follows there has occurred instances of the rules $(\nu_{x_i})_i$ along this path. These properties are essential to deducing soundness of applications of the inference, a matter which we expand upon below.

Definition 2 (Proofs). *A proof (in the calculus* Circ*) is a finite tree π of sequents locally correct with respect to the inference rules of Fig. 2 subject to the* *restriction that for every assumption name $\hat{x}$ there is no more than one instance of* $\mathsf{dis}_{\hat{x}}$ *in* π. *The sequent labelled at a vertex* α *of* π *is denoted* $\pi[\alpha]$. *The* ***sub-proof*** *of* π *rooted at* α, *written* π_α, *is the maximal sub-tree of* π *with root* α. *The* ***conclusion*** *of* π *is the sequent* $\pi[\rho_\pi]$ *annotating the root* ρ_π *of* π.

A leaf λ *of a proof* π *is* ***axiomatic*** *if the sequent* $\pi[\lambda]$ *is an instance of* Ax1 *or* Ax2; *otherwise,* λ *is an* ***assumption (of*** π***)***. *The assumptions are further classified into two groups:*

- *A* ***discharged***, *or* ***closed***, ***assumption*** *of* π *is an assumption* λ *such that* $\pi[\lambda]$ *matches the pattern of a discharged assumption for some occurrence of an inference* $\mathsf{dis}_{\hat{x}}$ *on the path from* λ *to* ρ_π.
- *An* ***open assumption*** *of* π *is any assumption which is not discharged in the sense above.*

A proof is ***closed*** *if all assumptions are discharged. Given a finite set* $\mathcal{A} \cup \{a : \Gamma\}$ *of annotated sequents and proof* π, *we write* $\mathcal{A} \vdash_\pi a : \Gamma$ *to express that the conclusion of* π *is the sequent* $a : \Gamma$ *and all open assumptions in* π *are sequents in* $\mathcal{A}$. *Explicit mention of the proof* π *will be dropped in cases it is not relevant. In particular,* $\vdash a : \Gamma$ *denotes the existence of a closed proof of* $a : \Gamma$.

As an example, we present a cyclic proof of the axiom Ax3 from the system Koz. For the result it is necessary to prove a more general property which we now formulate. Let $A = A(\mathsf{x}_0, \ldots, \mathsf{x}_n)$ be any formula such that $\mathrm{FV}(A) \subseteq \{\mathsf{x}_0, \ldots, \mathsf{x}_n\}$ and assume that for every $i \leq n$ and every $\mathsf{y} \in \mathrm{BV}(A)$, $\mathsf{x}_i < \mathsf{y}$. We claim that for all closed formulæ $(C_i)_{i \leq n}$ and any three annotations $a, c, d \in (\bigcup_{i \leq n} N_{\mathsf{x}_i})^*$ such that $a : A^c, \overline{A}^d$ is a sequent, we have

$$\{a : C_i^c, \overline{C}_i^d \mid i \leq n\} \vdash a : A(C_0, \ldots, C_n)^c, \overline{A(C_0, \ldots, C_n)}^d \tag{1}$$

The claim is established by induction on the formula A. For A quantifier-free, the argument is straightforward. We consider the case $A = \nu\mathsf{y}B(\mathsf{y}, \mathsf{x}_0, \ldots, \mathsf{x}_n)$ and make the simplifying assumption that $n = 0$. Let $\mathcal{A} = \{a : C^c, \overline{C}^d\}$ be the single assumption of the proof in (1) and let $\hat{y} \in N_{\mathsf{y}}^{\wedge}$ and $y \in N_{\mathsf{y}}^{\vee}$ be fresh names for y. By the induction hypothesis we have

$$\mathcal{A}, \{a\hat{y} : A(C)^{c\hat{y}y}, \overline{A(C)}^d\} \vdash a\hat{y} : B(A(C), C)^{c\hat{y}y}, \overline{B(A(C), C)}^d$$

Note, an application of exp has been inserted at uses of the assumption $\mathcal{A}$ as compared to the induction hypothesis. To the conclusion of this proof we now apply the μ-rule to the formula $\overline{B(A(C), C)}^d$, yielding $\overline{A(C)}^d$, followed by ν_y to the formula $B(A(C), C)^{c\hat{y}y}$ eliminating the variable name y. Then the conclusion and the assumption $a\hat{y} : A(C)^{c\hat{y}y}, \overline{A(C)}^d$ are of the required form to apply the discharge inference:

$$\dfrac{\dfrac{\dfrac{\begin{matrix}[a\hat{y} : A(C)^{c\hat{y}y}, \overline{A(C)}^d]^{\hat{y}} \\ \vdots \\ a\hat{y} : B(A(C), C)^{c\hat{y}y}, \overline{B(A(C), C)}^d\end{matrix}}{a\hat{y} : B(A(C), C)^{c\hat{y}y}, \overline{A(C)}^d}\,\mu}{a\hat{y} : A(C)^{c\hat{y}}, \overline{A(C)}^d}\,\nu_y}{a : A(C)^c, \overline{A(C)}^d}\,\mathsf{dis}_{\hat{y}}$$

Hence, $\mathcal{A} \vdash a : A(C)^c, \overline{A(C)}^d$ is deduced.

Theorem 1. Circ *is sound and complete for the modal* μ*-calculus. I.e., a formula* A *is valid iff there exists a closed proof with conclusion* A*. In addition, there is an effective procedure for constructing proofs of valid formulæ.*

The proof of Theorem 1 presented in [1] provides a procedure for obtaining proofs from valid sequents: Given an arbitrary valid sequent, Stirling's goal-orientated tableaux rules from [20] can be applied to yield a finite proof in his sequent calculus; from there, a process of proof-regularisation generates a closed proof in Circ [1, Theorem 5.4].

Soundness of Circ-proofs can be deduced via a reduction to Niwinski–Walukiewicz refutation tableaux [15]. A proof can be considered as finite representations of ill-founded proof-trees obtained by identifying discharged assumptions with the sequents at which they are discharged. Formally, given a proof π which discharges an assumption

$$\dfrac{\begin{matrix}[a\hat{x} : \Gamma, A_0^{a_0\hat{x}x_0}, \ldots, A_m^{a_m\hat{x}x_m}]^{\hat{x}} \\ \vdots \\ a\hat{x} : \Gamma, A_0^{a_0\hat{x}}, \ldots, A_m^{a_m\hat{x}}\end{matrix}}{a : \Gamma, A_0^{a_0}, \ldots, A_m^{a_m}}\,\mathsf{dis}_{\hat{x}}$$

we may consider the infinite unravelling of this assumption:

$$\dfrac{\begin{matrix}\vdots \\ \dfrac{a\hat{x} : \Gamma, A_0^{a_0\hat{x}}, \ldots, A_m^{a_m\hat{x}}}{a\hat{x} : \Gamma, A_0^{a_0\hat{x}x_0}, \ldots, A_m^{a_m\hat{x}x_m}}\,\mathsf{exp} \\ \vdots \\ \dfrac{a\hat{x} : \Gamma, A_0^{a_0\hat{x}}, \ldots, A_m^{a_m\hat{x}}}{a\hat{x} : \Gamma, A_0^{a_0\hat{x}x_0}, \ldots, A_m^{a_m\hat{x}x_m}}\,\mathsf{exp} \\ \vdots \\ a\hat{x} : \Gamma, A_0^{a_0\hat{x}}, \ldots, A_m^{a_m\hat{x}}\end{matrix}}{a : \Gamma, A_0^{a_0}, \ldots, A_m^{a_m}}\,\mathsf{dis}_{\hat{x}}$$

The only role of the discharge inference in this ill-founded 'proof' is to introduce (reading from conclusion to premise) the assumption name $\hat{x}$ which now persists throughout this infinite path along with all other assumption names occurring in a. Indeed, every infinite path through this 'unravelled' proof can be expressed in this form for some choice of control a and discharge rule. By weak König's Lemma there exists an infinite ancestor trace $\vec{B} = (B_i^{b_i})_{i\in\omega}$ through each such path which starts at a formula $B_0^{b_0} \in \{A_0^{a_0}, \ldots, A_m^{a_m}\}$ and such that every b_i for $i \geq 1$ contains $\hat{x}$, from which it follows that the most subsuming variable of B_0 that is unravelled infinitely often in $\vec{B}$ is a ν-variable.

The following lemma will prove useful.

Lemma 1. *If* $\vdash a : \Gamma$ *then there exists a closed proof of* $a : \Gamma$ *such that for every sequent* $b : \Delta$ *in the proof which is not the conclusion of an application of* exp*:*

1. *b comprises precisely the assumption names present in Δ;*
2. *for each formula $C^c \in \Delta$, the annotation c*
   (a) *contains only names of variables bound in C, and*
   (b) *is weakly increasing in the subsumption ordering, i.e., if $c = c_0xc_1yc_2$ then $x \leq y$.*

Proof. Suppose $\vdash_\pi a : \Gamma$. We argue that the desired proof can be obtained by restricting every annotation occurring in π to the form required of the lemma by inserting, where appropriate, instances of exp. In order to fulfil conditions 1 and 2(a) it will suffice to replace all annotations and controls in π by the maximal subword satisfying these criteria. To fulfil 2(b) these annotations are further refined by replacing $c = x_0 \ldots x_k$ by the maximal subword $c^- = x_{i_0} \cdots x_{i_l} \sqsubseteq c$ such that for every $j \leq l$, $x_{i_j} \leq x_{i_j+1} \cdots x_k$.

The described transformation preserves axioms, logical, modal and structural rules, though in the case of binary inferences, instances of exp may need to be inserted at premises. The quantifier rules ν_x and μ are also unproblematic. It therefore suffices to consider an instance of $\mathsf{dis}_{\hat{x}}$ in π:

$$\dfrac{\begin{array}{c}\left[a\hat{x} : \Gamma, A_0^{a_0\hat{x}x_0}, \ldots, A_k^{a_k\hat{x}x_k}\right]^{\hat{x}} \\ \vdots\ (\pi_\alpha) \\ a\hat{x} : \Gamma, A_0^{a_0\hat{x}}, \ldots, A_k^{a_k\hat{x}}\end{array}}{a : \Gamma, A_0^{a_0}, \ldots, A_k^{a_k}}\ \mathsf{dis}_{\hat{x}}$$

We assume the annotations $a, a_0, \ldots, a_k$ have been reduced as per the lemma. Observe that the specified restriction of the annotation $a_i\hat{x}$ contains the name $\hat{x}$ iff $\hat{x}$ names a variable x bound in A_i. In particular, we require to check that if x is not bound in some A_i then removing $\hat{x}$ from the annotation in the premise one can still make an application of $\mathsf{dis}_{\hat{x}}$ to discharge the open assumption. Let α mark the premise of the final inference in π and let $X = \nu\mathsf{x}X_0$ be the quantified formula which $\hat{x}$ names. Wlog, we may assume at least one leaf is discharged by this inference or else $\hat{x}$ can be removed from the whole sub-proof

π_α along with this application of $\mathsf{dis}_{\hat{x}}$ without consequence. Thus, we have $X \in \mathrm{FL}(A_0 \vee \ldots \vee A_k)$, so let us assume $X \notin \mathrm{FL}(A_i)$ for each $i < k_X$ and $X \in \mathrm{FL}(A_i)$ for $k_X \leq i \leq k$. Then this instance of dis can be replaced by an instance in which $\hat{x}$ annotates the formulæ $A_{k_X}, \ldots, A_k$ only: the proof given by the induction hypothesis ensures that assumptions discharged by this rule in π_α have the appropriate form to allow discharging by the new inference.

4 Extracting the Interpolant

Our main result is effective Lyndon interpolation for our sequent calculus:

Theorem 2 (Lyndon interpolation). *Let A, B be closed formulæ and suppose $A \vee B$ is valid. Then there exists a closed formula I in the language $L(\overline{A}) \cap L(B)$, effectively computable from A and B, and proofs $\vdash A, I$ and $\vdash \overline{I}, B$.*

By completeness, we may assume a closed proof π of A, B from which we will construct the interpolant I and witnessing proofs

$$\vdash_{\pi_A} A, I \qquad \vdash_{\pi_B} \overline{I}, B$$

It is convenient to place some simplifying restrictions on π before proceeding with the proof. The first of these will be the assumption $\mathrm{BV}(A) \cap \mathrm{BV}(B) = \emptyset$ which guarantees that $\mathrm{FL}(A) \cap \mathrm{FL}(B)$ comprises quantifier-free formulæ only.[1] However, we will work with the stronger assumption that all annotated subformulæ of A and B are marked such that we can consider $\mathrm{FL}(A)$ and $\mathrm{FL}(B)$ as having no formulæ in common. The simplest way to manage this is to first transform the proof $\vdash_\pi A, B$ into a (closed) proof of $\epsilon : A^a, B^b$ where $a, b \in N^\vee$ are two (fresh) names which subsume all bound variables in the two formulæ and prefix all annotations in π. In such a proof, at every sequent the A-ancestors and B-ancestors will be explicitly isolated by their annotation. Such a proof can be constructed either via the completeness argument sketched above or directly from π. We will briefly sketch the latter approach. Starting at the root, which is annotated $\epsilon : A^a, B^b$, the construction attempts to prefix the annotation of all formulæ in π by either a or b. Let us call a sequent annotated in this way a **marking**. Given annotated formulæ Γ we let $\Gamma^a = \{G^{ag} \mid G^g \in \Gamma\}$ and similarly Γ^b. Given a proof π with conclusion Δ and a partition $\Delta = \Delta_0 \cup \Delta_1$, the intention is to construct a proof π_{Δ_0} with conclusion Δ_0^a, Δ_1^b whose open assumptions are markings of assumptions in π. This proceeds by induction on π. For most inferences, the desired marking of the premise(s) is immediate given a marking of the conclusion. In cases in which the natural process would annotate a single formula by both names, the marking by a is preferred and the b-marking is eliminated from the premise by inserting an instance of weak. Note however, our assumption on variables means that such a formula is quantifier-free, whereby the assumption of Lemma 1 implies the formula has otherwise empty annotation.

[1] We leave to the interested reader how to handle the situation that A and B have bound variables in common.

The only non-trivial inference is the discharge rule where selecting a marking of the conclusion will induce markings of the premise and (discharged) assumptions which need not coincide. Suppose we are considering the following instance of discharge in π.

$$\dfrac{\begin{array}{c}\left[c\hat{x} : \Gamma, C_0^{c_0\hat{x}x_0}, \dots, C_k^{c_k\hat{x}x_k}\right]^{\hat{x}} \\ \vdots \\ c\hat{x} : \Gamma, C_0^{c_0\hat{x}}, \dots, C_k^{c_k\hat{x}}\end{array}}{c : \Gamma, C_0^{c_0}, \dots, C_k^{c_k}}\,\mathsf{dis}_{\hat{x}}$$

Consider a marking of the conclusion $\Gamma, C_0^{c_0}, \dots, C_k^{c_k}$ given by Δ_0^a, Δ_1^b. The C_i-formulæ all contain a common quantified sub-formula (the variable which $\hat{x}$ names), whence one of Δ_0 or Δ_1 contains all the formulæ $C_0^{c_0}, \dots, C_k^{c_k}$. Let us suppose this is Δ_1, and let $\Gamma_1 = \Gamma \cap \Delta_1$. By the induction hypothesis, this marking lifted to the premise induces a marking of each associated closed assumption: suppose the sequent at a discharged leaf λ is assigned the marking $\Delta_{\lambda,0}^a, \Delta_{\lambda,1}^b$. Note, as before $\{C_0^{c_0\hat{x}x_0}, \dots, C_k^{c_k\hat{x}x_k}\} \subseteq \Delta_{\lambda,1}$. In order to apply a single instance of $\mathsf{dis}_{\hat{x}}$ that discharges all these leaves, we require that $\Delta_{\lambda,0} = \Delta_0$ for each λ. If this is not satisfied of an assumption λ then in place of closing the assumption it is necessary to insert a copy of the marked proof $\pi_{\Delta_{\lambda,0}}$ with an additional instance of discharge (where $\Gamma_{\lambda,1} = \Gamma \cap \Delta_{\lambda,1}$):

$$\dfrac{\begin{array}{c}\dfrac{\begin{array}{c}\vdots\ \pi_{\Delta_{\lambda,0}} \\ c\hat{x}\hat{x}' : \Delta_{\lambda,0}^a, \Gamma_{\lambda,1}^b, C_0^{bc_0\hat{x}\hat{x}'}, \dots, C_k^{bc_k\hat{x}\hat{x}'}\end{array}}{c\hat{x} : \Delta_{\lambda,0}^a, \Gamma_{\lambda,1}^b, C_0^{bc_0\hat{x}x_0}, \dots, C_k^{bc_k\hat{x}x_k}}\,\mathsf{dis}_{\hat{x}'} + \mathsf{exp} \\ \vdots\ \pi_{\Delta_0} \\ c\hat{x} : \Delta_0^a, \Gamma_1^b, C_0^{bc_0\hat{x}}, \dots, C_k^{bc_k\hat{x}}\end{array}}{c : \Delta_0^a, \Gamma_1^b, C_0^{bc_0}, \dots, C_m^{bc_m}}\,\mathsf{dis}_{\hat{x}}$$

Assumptions in $\pi_{\Delta_{\lambda,0}}$ corresponding to the original discharge will be assigned markings. Any that match either the marking at λ or the root can be discharged, otherwise further unravellings are required. In general, an application of dis may need to be unfolded 2^n times where $n = |\mathrm{FL}(A) \cap \mathrm{FL}(B)|$ (these are the only formulæ which may appear on either side of a marking).

Thus, in the following we assume a closed proof π of the sequent A^a, B^b and that every formula in π features either a or b in its annotation (and never both) and each assumption name occurs only in annotations prefixed by a or b (but never both). Let $S = g : \Gamma$ be a sequent in π. The **splitting** of S is the pair of annotated sequents $S_A = g_A : \Gamma_A$ and $S_B = g_B : \Gamma_B$ where $\Gamma = \{C^{ac} : C^c \in \Gamma_A\} \cup \{C^{bc} : C^c \in \Gamma_B\}$ and g_A (g_B) is the restriction of g to assumption names in Γ_A (resp. Γ_B). Thus, $(\epsilon : \{A\}, \epsilon : \{B\})$ is the splitting of $\epsilon : A^a, B^b$.

Let D be the set of vertices of π which are conclusions to instances of discharge. Let $\{\mathsf{x}_\alpha \mid \alpha \in D\}$ be a set of fresh variable symbols. For each assumption leaf λ in π we let $\lambda^* \in D$ be the corresponding vertex at which λ is discharged. To each sequent $\pi[\alpha] = g : \Gamma$ in π we associate:

1. a formula I_α over variables indexed in D where x_β is free in I_α only if β is on the path from α to ρ_π.
2. proofs $\mathcal{A}^A_\alpha \vdash_{\pi^A_\alpha} g_A : \Gamma_A, I_\alpha$ and $\mathcal{A}^B_\alpha \vdash_{\pi^B_\alpha} g_B : \overline{I_\alpha}, \Gamma_B$ where for each open assumption λ of π_α there exists (unique) open assumptions $(d_A : \Delta_A, \mathsf{x}_{\lambda^*}) \in \mathcal{A}^A_\alpha$ and $(d_B : \mathsf{x}_{\lambda^*}, \Delta_B) \in \mathcal{A}^B_\alpha$ such that $d_A : \Delta_A$ and $d_B : \Delta_B$ is the splitting of the sequent $\pi[\lambda]$.

Given the above, the desired interpolant will be the (closed) formula $I = I_{\rho_\pi}$ associated to the root of π. The construction is determined by a case distinction on the final inference rule of π_α:

Ax1 There are three cases depending on the splitting: $\{p, \overline{p}\} \subseteq \Gamma_A$, $\{p, \overline{p}\} \subseteq \Gamma_B$ or $(l, \overline{l}) \in \Gamma_A \times \Gamma_B$ where $l \in \{p, \overline{p}\}$. Define $I_\alpha = \bot$, $\top$ or $\overline{l}$ respectively: the desired proofs π^A_α and π^B_α are then instances of axioms.

Ax2 Analogous to the case above: I_α is either $\top$ or $\bot$.

Assumption In this case α is an assumption leaf. Define $I_\alpha = \mathsf{x}_{\alpha^*}$. The two proofs π^A_α and π^B_α are the (open) assumptions specified by the splitting of $\pi[\alpha]$ and requirement 2 above.

Disjunction Let β be the premise of α and define $I_\alpha = I_\beta$. The desired proofs π^A_α and π^B_α immediately follow from π^A_β and π^B_β.

Conjunction Let β and γ be the two premises of α. If the principal formula of this inference falls in the sequent Γ_A, set $I_\alpha = I_\beta \vee I_\gamma$; otherwise, $I_\alpha = I_\beta \wedge I_\gamma$. Again, the proofs π^A_α and π^B_α are easily constructed from the induction hypothesis.

Modality Let β be the unique premise. Suppose $\pi[\alpha] = g : \langle\mathfrak{a}\rangle\Gamma, [\mathfrak{a}]C^{bc}$. If $\Gamma_A = \emptyset$ set $I_\alpha = \top$, otherwise $I_\alpha = [\mathfrak{a}]I_\beta$. Notice that if $\mathfrak{a} \notin L(\overline{A}) \cap L(B)$ then indeed $\Gamma_A = \emptyset$. In the case $\Gamma_A = \emptyset$ (whence $g_A = \epsilon$) the two proofs π^A_α and π^B_α can be constructed directly:

$$(\mathsf{Ax2})\ \epsilon : I_\alpha \qquad \dfrac{\begin{matrix}\pi_\alpha \\ g : \langle\mathfrak{a}\rangle\Gamma, [\mathfrak{a}]C^c\end{matrix}}{g : \overline{I_\alpha}, \langle\mathfrak{a}\rangle\Gamma, [\mathfrak{a}]C^c}\,\mathsf{weak}$$

Otherwise, π^A_α and π^B_α are constructed from premises π^A_β and π^B_β respectively:

$$\dfrac{\begin{matrix}\pi^A_\beta \\ g_A : \Gamma_A, I_\beta\end{matrix}}{g_A : \langle\mathfrak{a}\rangle\Gamma_A, I_\alpha}\,\mathsf{mod} \qquad \dfrac{\begin{matrix}\pi^B_\beta \\ g_B : \overline{I_\beta}, \Gamma_B, C^c\end{matrix}}{g_B : \overline{I_\alpha}, \langle\mathfrak{a}\rangle\Gamma_B, [\mathfrak{a}]C^c}\,\mathsf{mod}$$

The case C^c is marked as C^{ac} is symmetric and we choose $I_\alpha = \langle\mathfrak{a}\rangle I_\beta$.

Discharge Let β be the premise of this rule and suppose $\pi[\beta]$ has the form $g\hat{x} : \Gamma, C_0^{bc_0\hat{x}}, \dots, C_k^{bc_k\hat{x}}$ (the case each C_i is marked with a is symmetric). Let π^A_β and π^B_β be given with assumptions $\mathcal{A}^A_\beta$ and $\mathcal{A}^B_\beta$ as in condition 2 above. Define

$$\mathcal{A}^A_\alpha = \{(d : \Delta, \mathsf{x}_\gamma) \in \mathcal{A}^A_\beta : \gamma \neq \alpha\} \qquad \mathcal{A}^B_\alpha = \{(d : \mathsf{x}_\gamma, \Delta) \in \mathcal{A}^B_\beta : \gamma \neq \alpha\}$$

Set $I_\alpha = \nu \mathsf{x}_\alpha I_\beta$. We claim there exist proofs

$$\mathcal{A}^A_\alpha \vdash_{\pi^A_\alpha} g_A : \Gamma_A, I_\alpha \qquad \mathcal{A}^B_\alpha \vdash_{\pi^B_\alpha} g_B : \Gamma_B, C_0^{c_0}, \ldots, C_k^{c_k}, \overline{I_\alpha}.$$

We begin with the proof π^B_α. Let $\mathcal{B} = \mathcal{A}^B_\beta \setminus \mathcal{A}^B_\alpha$. We have, by assumption,

$$\mathcal{A}^B_\alpha, \mathcal{B} \vdash_{\pi^B_\beta} g_B\hat{x} : \Gamma_B, C_0^{c_0\hat{x}}, \ldots, C_k^{c_k\hat{x}}, \overline{I_\beta}.$$

Each sequent $S \in \mathcal{B}$ has the form $S = g_B\hat{x} : \Gamma_B, C_0^{c_0\hat{x}x_0}, \ldots, C_k^{c_k\hat{x}x_k}, \mathsf{x}_\alpha$ for some $x_0, \ldots, x_k \in N^{\mathsf{V}}_{\hat{x}}$. Uniformly substituting in π^B_β every occurrence of x_α by $\overline{I_\alpha}$ and applying the inference μ at the root yields a proof

$$\mathcal{A}^B_\alpha, \mathcal{B}[\mathsf{x}_\alpha/\overline{I_\alpha}] \vdash_{\pi^B_\beta} g_B\hat{x} : \Gamma_B, C_0^{c_0\hat{x}}, \ldots, C_k^{c_k\hat{x}}, \overline{I_\alpha}.$$

An application of $\mathsf{dis}_{\hat{x}}$ will discharge the assumptions $\mathcal{B}[\mathsf{x}_\alpha/\overline{I_\alpha}]$ and provide the desired proof.
The case of building π^A_α is similar. By assumption we have

$$\mathcal{A}^A_\alpha, \{g_A : \Gamma_A, \mathsf{x}_\alpha\} \vdash_{\pi^A_\beta} g_A : \Gamma_A, I_\beta.$$

Notice that the discharged name $\hat{x}$ does not occur in this proof since it names a variable in B. We now introduce annotations to the formula I_β. Pick $(\hat{x}, x) \in N^{\mathsf{A}}_{\mathsf{x}_\alpha} \times N^{\mathsf{V}}_{\mathsf{x}_\alpha}$. Expanding the control of every sequent in π^A_β by $\hat{x}$ and annotating all I_β sub-formulæ in the proof by $\hat{x}x$, we obtain a proof

$$\mathcal{A}^A_\alpha, \{g_A\hat{x} : \Gamma_A, \mathsf{x}_\alpha^{\hat{x}x}\} \vdash_{\pi'} g_A\hat{x} : \Gamma_A, I_\beta^{\hat{x}x}.$$

The sequents in $\mathcal{A}^A_\alpha$ do not require further annotations as instances of exp can be inserted at leaves. In π', replace every occurrence of x_α by I_α, yielding

$$\mathcal{A}^A_\alpha, \{g_A\hat{x} : \Gamma_A, I_\alpha^{\hat{x}x}\} \vdash_{\pi''} g_A\hat{x} : \Gamma_A, I_\beta^{\hat{x}x}[\mathsf{x}/I_\alpha].$$

Appending the rule ν_x to the root and discharging the assumptions by an application of $\mathsf{dis}_{\hat{x}}$ gives the proof π^A_α.

Remaining rules The quantifier and structural rules are unary and, like the case of $\vee$, choosing $I_\alpha = I_\beta$ where β is the immediate premise suffices.

This completes the proof of Theorem 2. Analysing the construction, bounds on the sizes of I, π_A and π_B are readily obtained:

Corollary 1. *Suppose* $\vdash_\pi A^a, B^b$ *and every formula in* π *contains* a *or* b *in its annotation. There exists an interpolant* I *and associated proofs* π_A *and* π_B *whose sizes are linearly bounded in the number of vertices in* π.

5 On the Form of the Interpolant

The interpolant constructed in the previous section is structurally identical to the proof witnessing the interpolated implication and from this observation one can immediately infer results on the logical form of the interpolant. For example, given a proof π of A, B, a conjunction-free interpolant is constructed if π avoids the conjunction rule on any ancestor of B and the disjunction rule on any ancestor of A. An action label occurs in the interpolant only if a modality rule for this action is used in the π. Note that although we generate the interpolant from an unravelling of π (and not directly from π), these two properties of proofs are preserved through unravellings.

As it is the quantifiers that are the source of expressibility and complexity in the μ-calculus, it is of interest to examine the quantifier structure of interpolants and their dependence on the structure of proofs. A simple example is given by the next lemma.

Lemma 2. *If the sequent $\epsilon : A, B$ has a closed proof which does not feature assumption names for variables in A (B) then there is a Π_1 (resp. Σ_1) formula I such that $A \vee I$ and $\overline{I} \vee B$ are valid.*

A proof is n**-open** if the control of every sequent in π is a word of length no greater than n. Note, there exists an n-open proof for a sequent iff there exists a proof π such that for every $\alpha \in \pi$ the set of controls from open assumptions in π_α has cardinality at most n.

Lemma 3. *If A^a, B^b admits an n-open proof there is an interpolant with at most n distinct variables.*

Proof. The interpolant constructed in the previous section utilised a (fresh) variable symbol x_α for each application of discharge in the proof, and a variable x_α occurs free in the generated interpolant I_β just if α is the conclusion of an instance of discharge on the path strictly between β and the root. Thus, given an n-open proof the construction can be carried out assuming at most n many distinct variable symbols.

Let us call a proof **separated** if every control contains names for variables from at most one formula in the conclusion. A separated proof may still utilise applications of the discharge rule for variables from different formulæ but if an assumption is open in the use of a discharge rule then the two variables associated to these discharges are from the same formula in the conclusion.

Lemma 4. *If A, B admits a separated proof and $BV(A) \cap BV(B) = \emptyset$ then there exists an alternation-free interpolant.*

We recall the alternation-free fragment is the class of μ-formulæ for which the subsumption ordering does not relate variables bound by different quantifiers.

Proof. Since separated proofs are closed under the unravelling of closed assumptions (described in the previous section) we may assume a separated closed proof of A^a, B^b with the usual restrictions. We remark that in the construction the interpolant I acquires a quantifier ν only at instances of the discharge rule associated to variables from B and a μ-quantifier at instances of the discharge rule associate to A-variables. Since the proof is assumed to be separated, at no point will a sub-formula I_α of the interpolant be generated that contains free two variables x and y such that x is ν-quantified in I and y is μ-quantified.

An immediate consequence of Lemma 4 is the following:

Lemma 5. *If every valid implication in alternation-free μ-calculus admits a separated proof then alternation-free μ-calculus has the Lyndon interpolation property.*

Although we leave open the question of whether alternation-free μ-calculus has interpolation, we remark that valid implications between Σ_1/Π_1 formulæ readily admit separated proofs, yielding an interpolation theorem for the first level of the alternation-free hierarchy [24].

Theorem 3. *Every valid implication between formulæ in the Boolean closure of $\Pi_1 \cup \Sigma_1$ admits an alternation-free interpolant.*

6 Discussion

We have shown how to extract interpolants for modal μ-formulæ via sequent calculus. The proof rests on the cyclic proof calculus called Circ introduced in [1] which is both cut-free and analytic. Our approach is similar to the method used to obtain interpolation for Gödel–Löb provability logic from a cyclic proof system over S4 [19].

A natural continuation of this work consists of adapting the above methods to other modal and temporal logics with fixed points. Our result depends on two important properties of the modal μ-calculus: the existence of an analytic cyclic proof system and the expressive capabilities to define interpolants from (nested) back-edges. For fragments of the modal μ-calculus in which only a subset of fixed points are available the latter will be more constraining whereas for richer systems such as μ-calculus with converse modalities, adequate cyclic calculi are problematic. We believe both pursuits can offer fruitful contributions to interpolation techniques and an understanding of these logics. An outstanding question is whether uniform interpolation can likewise be deduced proof-theoretically.

References

1. Afshari, B., Leigh, G.E.: Finitary proof systems for Kozen's μ. Oberwolfach Preprint Series 2016-26, Mathematisches Forschungsinstitut Oberwolfach (2016)
2. Afshari, B., Leigh, G.E.: Cut-free completeness for modal μ-calculus. In: Proceeding of Thirty-Second Annual ACM/IEEE Symposium on Logic in Computer Science (LICS). Lecture Notes in Computer Science. Springer, Cham (2017)
3. Bradfield, J., Stirling, C.: Modal mu-calculi. In: Blackburn, P., van Benthem, J., Wolter, F. (eds.) Handbook of Modal Logic, pp. 721–756. Elsevier, Amsterdam (2007)
4. Bradfield, J., Walukiewicz, I.: The μ-calculus and model checking. In: Clarke, E., Henzinger, T., Veith, H., Bloem, R. (eds.) Handbook of Model Checking, pp. 871–919. Springer, Cham (2018). https://doi.org/10.1007/978-3-319-10575-8_26
5. D'Agostino, G.: Interpolation in non-classical logics. Synthese **164**(3), 421–435 (2008)

6. D'Agostino, G., Hollenberg, M.: Logical questions concerning the μ-calculus. J. Symb. Log. **65**(1), 310–332 (2000)
7. Gabbay, D.M., Maksimova, L.: Interpolation and Definability. Modal and Intuitionistic Logic. Oxford University Press, Oxford (2005)
8. Iemhoff, R.: Uniform interpolation and the existence of sequent calculi. Ann. Pure Appl. Log. **170**(11), 102711 (2019)
9. Janin, D., Walukiewicz, I.: Automata for the modal μ-calculus and related results. In: Wiedermann, J., Hájek, P. (eds.) MFCS 1995. LNCS, vol. 969, pp. 552–562. Springer, Heidelberg (1995). https://doi.org/10.1007/3-540-60246-1_160
10. Jungteerapanich, N.: A tableau system for the modal μ-calculus. In: Giese, M., Waaler, A. (eds.) TABLEAUX 2009. LNCS (LNAI), vol. 5607, pp. 220–234. Springer, Heidelberg (2009). https://doi.org/10.1007/978-3-642-02716-1_17
11. Kozen, D.: Results on the propositional μ-calculus. Theoret. Comput. Sci. **27**, 333–354 (1983)
12. Kracht, M.: Modal consequence relations. In: Blackburn, P., van Benthem, J., Wolter, F. (eds.) Handbook of Modal Logic, vol. 3, pp. 491–545. Elsevier, Amsterdam (2007)
13. Kuznets, R.: Proving Craig and Lyndon interpolation using labelled sequent calculi. In: Michael, L., Kakas, A. (eds.) JELIA 2016. LNCS (LNAI), vol. 10021, pp. 320–335. Springer, Cham (2016). https://doi.org/10.1007/978-3-319-48758-8_21
14. Lyndon, R.C.: Ian interpolation theorem in the predicate calculus. Pac. J. Math. **9**, 129–142 (1959)
15. Niwinski, D., Walukiewicz, I.: Games for the μ-calculus. Theor. Comput. Sci. **163**(1&2), 99–116 (1996)
16. Pitts, A.M.: On an interpretation of second order quantification in first order intuitionistic propositional logic. J. Symb. Logic **57**(1), 33–52 (1992)
17. Shamkanov, D.S.: Lyndon interpolation for the modal μ-calculus. In: Third Workshop Gentzen Systems and Beyond (abstract) (2014)
18. Shamkanov, D.S.: Private communication (2019)
19. Shamkanov, D.S.: Circular proofs for the Gödel-Löb provability logic. Math. Notes **96**(3), 575–585 (2014). https://doi.org/10.1134/S0001434614090326
20. Stirling, C.: A tableau proof system with names for modal μ-calculus. In: Voronkov, A., Korovina, M.V. (eds.) HOWARD-60, pp. 306–318 (2014)
21. Visser, A.: Bisimulations, model descriptions and propositional quantifiers. Logic Group Preprint Series No. 161, Utrecht (1996)
22. Walukiewicz, I.: Completeness of Kozen's axiomatisation of the propositional μ-calculus. Inf. Comput. **157**, 142–182 (2000)
23. Zakharyaschev, M., Wolter, F., Chagrov, A.V.: Advanced modal logic. In: Gabbay, D.M., Guenthner, F. (eds.) Handbook of Philosophical Logic, pp. 83–266. Springer, Dordrecht (2001). https://doi.org/10.1007/978-94-017-0454-0_2
24. Zenger, L.: Proof theory for fragments of the modal μ-calculus. Master's thesis, University of Amsterdam, The Netherlands (2021)

Decidable and Undecidable Problems for First-Order Definability and Modal Definability

Philippe Balbiani[1(✉)] and Tinko Tinchev[2]

[1] Toulouse Institute of Computer Science Research, CNRS and Toulouse University, Toulouse, France
philippe.balbiani@irit.fr

[2] Faculty of Mathematics and Informatics, Sofia University St. Kliment Ohridski, Sofia, Bulgaria

Abstract. The core of this paper is Chagrova's Theorems about first-order definability of given modal formulas and modal definability of given elementary conditions. We consider classes of frames for which modal definability is decidable and classes of frames for which first-order definability is trivial. We give a new proof of Chagrova's Theorem about modal definability and sketches of proofs of new variants of Chagrova's Theorem about modal definability.

Keywords: First-order definability · Modal definability · Chagrova's theorems

1 Introduction

The question of the correspondence between elementary conditions and modal formulas is concomitant with the creation of the relational semantics of modal logic, frames serving as interpretation structures both for first-order formulas in the signature with one binary predicate and equality and for propositional modal formulas in the language with one box. Kripke [22] already observed that some elementary conditions possess a modal correspondent: transitivity vs $\Box p \rightarrow \Box\Box p$, symmetry vs $p \rightarrow \Box\Diamond p$, etc. Less than 20 years have elapsed between Kripke's observation and the development of Correspondence Theory culminating in the publication of the book "Modal Logic and Classical Logic" [5]: in 1975, Sahlqvist [26] isolated a large set of modal formulas which guarantee completeness with respect to first-order definable classes of frames whereas van Benthem [4] and Goldblatt [17] independently noticed that McKinsey formula $\Box\Diamond p \rightarrow \Diamond\Box p$ has no first-order correspondent.

Since the first-order conditions corresponding to Sahlqvist formulas are effectively computable [6, Section 3.6], it is natural to ask whether Sahlqvist fragment contains all modal formulas possessing first-order correspondents. This question has received a negative answer, the conjunction $(\Box\Diamond p \rightarrow \Diamond\Box p) \wedge (\Box p \rightarrow \Box\Box p)$ possessing a first-order correspondent while not being equivalent to a Sahlqvist formula. See [6, Example 3.57 and Exercise 3.7.1] for details. See also [18] for an extension of the Sahlqvist set of modal formulas. Hence, owing to the significance of Correspondence Theory, it is natural to ask whether the following problems are decidable:

© The Author(s), under exclusive license to Springer Nature Switzerland AG 2022
A. Özgün and Y. Zinova (Eds.): TbiLLC 2019, LNCS 13206, pp. 214–236, 2022.
https://doi.org/10.1007/978-3-030-98479-3_11

First-order definability: determine whether a given modal formula possesses a first-order correspondent,

Modal definability: determine whether a given first-order sentence possesses a modal correspondent.

This question has received a negative answer, the limitative results in this topic having been firstly obtained by Chagrova in her doctoral thesis [11] and then further developed in [7–9,12]. Chagrova's results (henceforth called Chagrova's Theorems) have been obtained by reductions from accessibility problems in Minsky machines and by the use of the frames presented in [8, Figures 1 and 2].

In Chagrova's Theorems, when we are talking about first-order sentences corresponding to modal formulas, we mean that they correspond with respect to the class of all frames. Thus, immediately, there is the question whether Chagrova's Theorems still hold if one consider restricted classes of frames. Giving rise to the modal logic **S5**, the class of all partitions is perhaps the most simple class of frames that one may conceive of. The simple character of the class of all partitions also appears within the context of first-order definability: every modal formula being equivalent in this class to a modal formula of degree at most 1, it follows from a remark of van Benthem [5, Lemma 9.7] that the class of all partitions gives rise to a trivial first-order definability problem. As for the modal definability problem, Balbiani and Tinchev [2] have proved that it is **PSPACE**-complete with respect to the class of all partitions when the modal language is extended by the universal modality.

Other classes of frames of simple character are the classes giving rise to the modal logics **KD45** (the class of all serial, transitive and Euclidean frames) and **K45** (the class of all transitive and Euclidean frames). As for the class of all partitions and for the same reason, Georgiev [15,16] has proved that the first-order definability problem is trivial with respect to these classes whereas the modal definability problem is **PSPACE**-complete. The most important computational property shared by the modal logics **S5**, **KD45** and **K45** is the **NP**-completeness of the satisfiability problem. The satisfiability problem of **K5** is **NP**-complete too and this modal logic shares many computational properties with the modal logics **S5**, **KD45** and **K45** as well, for instance the polysize model property. Nevertheless, with respect to the class of all **K5**-frames (the class of all Euclidean frames), although the first-order definability problem is still trivial, the modal definability problem becomes undecidable [1].

The core of this paper will be Chagrova's Theorems about first-order definability and modal definability. In Sect. 3, we will consider classes of frames for which modal definability is decidable. In particular, we will demonstrate a new result—namely, Theorem 1—saying that the problem of deciding modal definability of first-order sentences with respect to the class of all partitions is **PSPACE**-complete. In Sect. 4, we will consider classes of frames for which first-order definability is trivial. In particular, we will demonstrate a new result—namely, Theorem 2—saying that the problem of deciding first-order definability of modal formulas with respect to the class of all reflexive, transitive and connected frames with finitely many clusters is trivial. In Sect. 5, using standard methods in model theory such as relativization of first-order formulas and reduct of frames, we will give a new proof of Chagrova's Theorem about modal definability and we will give sketches of proofs of new variants of Chagrova's Theorem about modal definability. We assume the reader is at home with the basic tools and techniques in model theory and modal logics. For more on them, see [14, 19] and [6, 10, 21].

2 Preliminaries

We introduce a handful of definitions that will be useful throughout the paper.

2.1 Frames

For all sets E, $\|E\|$ will denote the cardinality of E. A *frame* is a structure $\mathcal{F} = (W, R)$ where W is a nonempty set of *states* and R is a binary relation on W. For all frames $\mathcal{F} = (W, R)$, for all s in $\mathcal{F}$ and for all subsets S of $\mathcal{F}$, let $R(s) = \{t \in W : sRt\}$ and $R(S) = \bigcup\{R(s) : s \in S\}$. For all frames $\mathcal{F} = (W, R)$ and for all s in $\mathcal{F}$, let $R^\star(s) = \bigcup\{R^n(s) : n \in \mathbb{N}\}$ where $R^0(s) = \{s\}$ and for all $n \geq 1$, $R^n(s) = R(R^{n-1}(s))$. For all frames $\mathcal{F} = (W, R)$, we say $\mathcal{F}$ is *rooted* if there exists s in $\mathcal{F}$ such that $R^\star(s) = W$. In that case, we say s is a *root* of $\mathcal{F}$. For all frames $\mathcal{F} = (W, R)$ and for all s in $\mathcal{F}$, the *subframe of $\mathcal{F}$ generated from s* is the frame $\mathcal{F}_s = (W_s, R_s)$ where $W_s = R^\star(s)$ and R_s is the restriction of R to W_s. Obviously, s is a root of $\mathcal{F}_s$. In a frame $\mathcal{F} = (W, R)$, we will say that

- R is *reflexive* if for all s in $\mathcal{F}$, sRs,
- R is *serial* if for all s in $\mathcal{F}$, there exists t in $\mathcal{F}$ such that sRt,
- R is *symmetric* if for all s, t in $\mathcal{F}$, if sRt then tRs,
- R is *transitive* if for all s, t, u in $\mathcal{F}$, if sRt and tRu then sRu,
- R is *Euclidean* if for all s, t, u in $\mathcal{F}$, if sRt and sRu then tRu and uRt,
- R is *connected* if for all s, t, u in $\mathcal{F}$, if sRt and sRu then either tRu, or uRt.

The frame $\mathcal{F} = (W, R)$ is *reflexive* (respectively *serial, symmetric, transitive, Euclidean, connected)* if R is reflexive (respectively serial, symmetric, transitive, Euclidean, connected). The frame $\mathcal{F} = (W, R)$ is a *partition* if R is reflexive, symmetric and transitive. The partition $\mathcal{F} = (W, R)$ is *bounded* if there exists a positive integer n such that for all s in $\mathcal{F}$, $\|R(s)\| \leq n$. For all bounded partitions $\mathcal{F} = (W, R)$, let $n_\mathcal{F}$ be the least positive integer n such that for all s in $\mathcal{F}$, $\|R(s)\| \leq n$. The partition $\mathcal{F} = (W, R)$ is *small* if there exists a positive integer π such that for all s in $\mathcal{F}$, $\|\{R(t) : t \in W$ and $\|R(s)\| = \|R(t)\|\}\| \leq \pi$. For all small partitions $\mathcal{F} = (W, R)$, let $\pi_\mathcal{F}$ be the least positive integer π such that for all s in $\mathcal{F}$, $\|\{R(t) : t \in W$ and $\|R(s)\| = \|R(t)\|\}\| \leq \pi$. In this paper, we will consider the following classes of frames: the class $\mathcal{C}_{all}$ of all frames, the class $\mathcal{C}_E$ of all Euclidean frames, the class $\mathcal{C}_{sE}$ of all serial and Euclidean frames, the class $\mathcal{C}_{tE}$ of all transitive and Euclidean frames, the class $\mathcal{C}_{stE}$ of all serial, transitive and Euclidean frames and the class $\mathcal{C}_{par}$ of all partitions. We will also consider other classes of frames: the class $\mathcal{C}_{rtc}$ of all reflexive, transitive and connected frames and the class $\mathcal{C}_{rtc}^{\omega}$ of all reflexive, transitive and connected frames $\mathcal{F}$ such that for all s in $\mathcal{F}$, $\mathcal{F}_s$ contains finitely many clusters. Remind that for all reflexive, transitive and connected frames $\mathcal{F} = (W, R)$, a *cluster* is an equivalence class modulo the equivalence relation $\simeq_\mathcal{F}$ on $\mathcal{F}$ such that for all s, t in $\mathcal{F}$, $s \simeq_\mathcal{F} t$ iff sRt and tRs.

2.2 Modal Language and Truth

Modal Language. Let us consider a countable set **PVAR** of *propositional variables* (denoted $p, q, \ldots$). The set $\mathcal{L}_{\mathbf{MF}}$ of all *modal formulas* (denoted $\varphi, \psi, \ldots$) is inductively defined as follows:

- $\varphi, \psi ::= p \mid \bot \mid \neg\varphi \mid (\varphi \vee \psi) \mid \Box\varphi,$

where p ranges over **PVAR**. We define the other Boolean constructs as usual. The modal formula $\Diamond\phi$ is obtained as the well-known abbreviation: $\Diamond\phi ::= \neg\Box\neg\phi$. We adopt the standard rules for omission of the parentheses. The *degree* of the modal formula φ (in symbols $\mathtt{deg}(\varphi)$) is the nonnegative integer inductively defined as usual [6, Definition 2.28]. The set of all *subformulas* of the modal formula φ (in symbols $\mathtt{sf}(\varphi)$) is the set of modal formulas inductively defined as follows:

- $\mathtt{sf}(p) = \{p\}$,
- $\mathtt{sf}(\bot) = \{\bot\}$,
- $\mathtt{sf}(\neg\varphi) = \{\neg\varphi\} \cup \mathtt{sf}(\varphi)$,
- $\mathtt{sf}(\varphi \vee \psi) = \{\varphi \vee \psi\} \cup \mathtt{sf}(\varphi) \cup \mathtt{sf}(\psi)$,
- $\mathtt{sf}(\Box\varphi) = \{\Box\varphi\} \cup \mathtt{sf}(\varphi)$.

The set of all *boxed subformulas* of the modal formula φ (in symbols $\mathtt{sf}^{\Box}(\varphi)$) is the set of modal formulas inductively defined as follows:

- $\mathtt{sf}^{\Box}(p) = \emptyset$,
- $\mathtt{sf}^{\Box}(\bot) = \emptyset$,
- $\mathtt{sf}^{\Box}(\neg\varphi) = \mathtt{sf}^{\Box}(\varphi)$,
- $\mathtt{sf}^{\Box}(\varphi \vee \psi) = \mathtt{sf}^{\Box}(\varphi) \cup \mathtt{sf}^{\Box}(\psi)$,
- $\mathtt{sf}^{\Box}(\Box\varphi) = \{\Box\varphi\} \cup \mathtt{sf}^{\Box}(\varphi)$.

As is well-known, for all modal formulas φ, $\|\mathtt{sf}^{\Box}(\varphi)\| + 1 \leq \|\mathtt{sf}(\varphi)\|$.

Truth. A *valuation* on a frame $\mathcal{F} = (W, R)$ is a function V assigning to each propositional variable p a subset $V(p)$ of W. The *satisfiability* of a modal formula φ at a state s with respect to a valuation V in a frame $\mathcal{F} = (W, R)$ (in symbols $\mathcal{F}, V, s \models \varphi$) is inductively defined as follows:

- $\mathcal{F}, V, s \models p$ iff $s \in V(p)$,
- $\mathcal{F}, V, s \not\models \bot$,
- $\mathcal{F}, V, s \models \neg\varphi$ iff $\mathcal{F}, V, s \not\models \varphi$,
- $\mathcal{F}, V, s \models \varphi \vee \psi$ iff either $\mathcal{F}, V, s \models \varphi$, or $\mathcal{F}, V, s \models \psi$,
- $\mathcal{F}, V, s \models \Box\varphi$ iff for all states t in $\mathcal{F}$, if sRt then $\mathcal{F}, V, t \models \varphi$.

As a result, $\mathcal{F}, V, s \models \Diamond\varphi$ iff there exists a state t in $\mathcal{F}$ such that sRt and $\mathcal{F}, V, t \models \varphi$. A modal formula φ is *true* with respect to a valuation V in a frame $\mathcal{F}$ (in symbols $\mathcal{F}, V \models \varphi$) if φ is satisfied at all states with respect to V in $\mathcal{F}$. A modal formula φ is *valid* in a frame $\mathcal{F}$ (in symbols $\mathcal{F} \models \varphi$) if φ is true with respect to all valuations on $\mathcal{F}$. A modal formula φ is *valid* in a class $\mathcal{C}$ of frames (in symbols $\mathcal{C} \models \varphi$) if φ is valid in all frames in $\mathcal{C}$. A frame $\mathcal{F}$ is *weaker* than a frame $\mathcal{F}'$ (in symbols $\mathcal{F} \preceq \mathcal{F}'$) if for all modal formulas φ, if $\mathcal{F} \models \varphi$ then $\mathcal{F}' \models \varphi$. For all positive integers n, let

- $\psi_n ::= \bigwedge\{\Diamond p_i : 0 \leq i \leq n\} \to \bigvee\{\Diamond(p_i \wedge p_j) : 0 \leq i < j \leq n\}$.

It is a well-known fact that for all positive integers n and for all partitions $\mathcal{F}$, $\mathcal{F} \models \psi_n$ iff $\mathcal{F}$ is bounded and $n_{\mathcal{F}} \leq n$.

Generated Subframes. A frame $\mathcal{F}' = (W', R')$ is a *generated subframe* of a frame $\mathcal{F} = (W, R)$ (in symbols $\mathcal{F} \rightarrowtail \mathcal{F}'$) if $W' \subseteq W$ and

- for all s', t' in $\mathcal{F}'$, if $s'R't'$ then $s'Rt'$,
- for all s' in $\mathcal{F}'$ and for all t in $\mathcal{F}$, if $s'Rt$ then t is in $\mathcal{F}'$ and $s'R't$.

The least generated subframe of a frame $\mathcal{F} = (W, R)$ generated by a state s in $\mathcal{F}$ is the frame $\mathcal{F}_s = (W_s, R_s)$ where $W_s = R^\star(s)$ and R_s is the restriction of R to W_s. Generated subframes give rise to the following results:

Proposition 1 (Generated subframes Theorem). *If the frame $\mathcal{F}'$ is a generated subframe of the frame $\mathcal{F}$ then $\mathcal{F} \preceq \mathcal{F}'$.*

Proof. See [6, Theorem 3.14 (ii)].

Proposition 2. *Let $\mathcal{F} = (W, R)$ be a frame, s be a state in $\mathcal{F}$, V be a valuation on $\mathcal{F}$ and V_s be the restriction of V to W_s. For all modal formulas φ and for all t in W_s, $\mathcal{F}, V, t \models \varphi$ iff $\mathcal{F}_s, V_s, t \models \varphi$.*

Proof. By induction on φ.

Disjoint Unions. The frame $\mathcal{F}' = (W', R')$ is the *disjoint union* of a family of frames $\mathcal{F}_i = (W_i, R_i)$ where i ranges over a nonempty set I if for all $i, j \in I$, if $i \neq j$ then $W_i \cap W_j = \emptyset$, $W' = \bigcup\{W_i : i \in I\}$ and $R' = \bigcup\{R_i : i \in I\}$. Disjoint unions give rise to the following result:

Proposition 3 (Disjoint unions Theorem). *If the frame $\mathcal{F}'$ is the disjoint union of a family of frames $\mathcal{F}_i$ where i ranges over a nonempty set I then for all $i \in I$, $\mathcal{F}' \preceq \mathcal{F}_i$.*

Proof. Suppose the frame $\mathcal{F}'$ is the disjoint union of a family of frames $\mathcal{F}_i$ where i ranges over a nonempty set I. Let $i \in I$. Obviously, $\mathcal{F}_i$ is a generated subframe of $\mathcal{F}'$. Hence, by Proposition 1, $\mathcal{F}' \preceq \mathcal{F}_i$

Bounded Morphic Images. A frame $\mathcal{F}' = (W', R')$ is a *bounded morphic image* of a frame $\mathcal{F} = (W, R)$ (in symbols $\mathcal{F} \twoheadrightarrow \mathcal{F}'$) if there exists a function f assigning to each state s in $\mathcal{F}$ a state $f(s)$ in $\mathcal{F}'$ such that

- f is surjective,
- for all s, t in $\mathcal{F}$, if sRt then $f(s)R'f(t)$,
- for all s in $\mathcal{F}$ and for all t' in $\mathcal{F}'$, if $f(s)R't'$ then there exists t in $\mathcal{F}$ such that sRt and $f(t) = t'$.

In that case, the function f is a *surjective bounded morphism.* Bounded morphic images give rise to the following result:

Proposition 4 (Bounded morphic images Theorem). *If the frame $\mathcal{F}'$ is a bounded morphic image of the frame $\mathcal{F}$ then $\mathcal{F} \preceq \mathcal{F}'$.*

Proof. See [6, Theorem 3.14 (iii)].

2.3 First-Order Language and Truth

First-Order Language. Let us consider a countable set **IVAR** of *individual variables* (denoted x, y, ...). The set $\mathcal{L}_{\mathbf{FOF}}$ of all *first-order formulas* (denoted A, B, ...) is inductively defined as follows:

- $A, B ::= \mathbf{R}(x, y) \mid x = y \mid \neg A \mid (A \vee B) \mid \forall x A$,

where x and y range over **IVAR**. We define the other Boolean constructs as usual. The first-order formula $\exists x A$ is obtained as the well-known abbreviation: $\exists x A ::= \neg\forall x \neg A$. We adopt the standard rules for omission of the parentheses. For all first-order formulas A, let $\mathtt{fiv}(A)$ be the set of all free individual variables occurring in A. A first-order formula A is a *sentence* if $\mathtt{fiv}(A) = \emptyset$. The *quantifier rank* of the first-order formula A (in symbols $\mathtt{qr}(A)$) is the nonnegative integer inductively defined as usual [14, Chapter 1]. The *relativization* of a first-order formula C with respect to a first-order formula A and an individual variable x (in symbols $(C)^A_x$) is inductively defined as follows:

- $(\mathbf{R}(y, z))^A_x$ is $\mathbf{R}(y, z)$,
- $(y = z)^A_x$ is $y = z$,
- $(\neg C)^A_x$ is $\neg(C)^A_x$,
- $(C \vee D)^A_x$ is $(C)^A_x \vee (D)^A_x$,
- $(\forall y C)^A_x$ is $\forall y(A[x/y] \rightarrow (C)^A_x)$.

In the above definition, $A[x/y]$ denotes the first-order formula obtained from the first-order formula A by replacing every free occurrence of the individual variable x in A by the individual variable y. From now on, when we write $(C)^A_x$, we will always assume that the sets of individual variables occurring in A and C are disjoint. The reader may easily verify by induction on the first-order formula C that $\mathtt{fiv}((C)^A_x) \subseteq (\mathtt{fiv}(A) \setminus \{x\}) \cup \mathtt{fiv}(C)$. Hence, if C is a sentence then $\mathtt{fiv}((C)^A_x) \subseteq \mathtt{fiv}(A) \setminus \{x\}$.

Truth. An *assignment* on a frame $\mathcal{F}$ is a function g assigning to each individual variable x a state $g(x)$ in $\mathcal{F}$. The *update* of an assignment g on a frame $\mathcal{F}$ with respect to a state s in $\mathcal{F}$ and an individual variable x (in symbols g^x_s) is the assignment g^x_s on $\mathcal{F}$ such that $g^x_s(x) = s$ and for all individual variables $y \neq x$, $g^x_s(y) = g(y)$. Given a frame $\mathcal{F}$, for all nonnegative integers n, for all states $s_1, \ldots, s_n$ in $\mathcal{F}$ and for all individual variables $x_1, \ldots, x_n$, $g^{x_1 \ldots x_n}_{s_1 \ldots s_n}$ is the assignment g' on $\mathcal{F}$ inductively defined as follows

- if $n = 0$ then $g' = g$,
- if $n \geq 1$ then $g' = (g^{x_1 \ldots x_{n-1}}_{s_1 \ldots s_{n-1}})^{x_n}_{s_n}$.

The *satisfiability* of a first-order formula A with respect to an assignment g in a frame $\mathcal{F} = (W, R)$ (in symbols $\mathcal{F}, g \models A$) is inductively defined as follows:

- $\mathcal{F}, g \models \mathbf{R}(x, y)$ iff $g(x) R g(y)$,
- $\mathcal{F}, g \models x = y$ iff $g(x) = g(y)$,
- $\mathcal{F}, g \models \neg A$ iff $\mathcal{F}, g \not\models A$,
- $\mathcal{F}, g \models A \vee B$ iff either $\mathcal{F}, g \models A$, or $\mathcal{F}, g \models B$,
- $\mathcal{F}, g \models \forall x A$ iff for all states s in $\mathcal{F}$, $\mathcal{F}, g^x_s \models A$.

As a result, $\mathcal{F}, g \models \exists x A$ iff there exists a state s in $\mathcal{F}$ such that $\mathcal{F}, g_s^x \models A$. A first-order formula A is *valid* in a frame $\mathcal{F}$ (in symbols $\mathcal{F} \models A$) if A is satisfied with respect to all assignments in $\mathcal{F}$. A first-order formula A is *valid* in a class $\mathcal{C}$ of frames (in symbols $\mathcal{C} \models A$) if A is valid in all frames in $\mathcal{C}$. For all positive integers n, let

– $B_n ::= \forall x_0 \ldots \forall x_n (\bigwedge\{\mathbf{R}(x_i, x_j) : 0 \leq i < j \leq n\} \rightarrow \bigvee\{x_i = x_j : 0 \leq i < j \leq n\})$.

It is a well-known fact that for all positive integers n and for all partitions $\mathcal{F}$, $\mathcal{F} \models B_n$ iff $\mathcal{F}$ is bounded and $n_{\mathcal{F}} \leq n$.

Lemma 1. *Let A be a sentence. The following conditions are equivalent:*

1. $\mathcal{C}_{par} \models A$,
2. *for all small and bounded partitions $\mathcal{F}$, if $n_{\mathcal{F}}, \pi_{\mathcal{F}} \leq \mathtt{qr}(A)$ then $\mathcal{F} \models A$,*

Proof. (1 ⇒ 2) Obvious.
(2 ⇒ 1) Suppose $\mathcal{C}_{par} \not\models A$. Hence, there exists a partition $\mathcal{F}$ such that $\mathcal{F} \not\models A$. Let $\mathcal{F}'$ be the bounded partition obtained from $\mathcal{F}$ by eliminating in all equivalence classes, as many states as it is needed so that the size of each equivalence class becomes at most equal to $\mathtt{qr}(A)$. Obviously, $n_{\mathcal{F}'} \leq \mathtt{qr}(A)$. Moreover, Duplicator wins the Ehrenfeucht-Fraïssé game $G_{\mathtt{qr}(A)}(\mathcal{F}, \mathcal{F}')$[1]. Thus, for all sentences B, if $\mathtt{qr}(B) \leq \mathtt{qr}(A)$ then $\mathcal{F} \models B$ iff $\mathcal{F}' \models B$. Since $\mathcal{F} \not\models A$, $\mathcal{F}' \not\models A$. Let $\mathcal{F}''$ be the small and bounded partition obtained from $\mathcal{F}'$ by eliminating for all positive integers π, as many equivalence classes as it is needed so that the number of equivalence classes of size π becomes at most equal to $\mathtt{qr}(A)$. Obviously, $n_{\mathcal{F}''}, \pi_{\mathcal{F}''} \leq \mathtt{qr}(A)$. Moreover, Duplicator wins the Ehrenfeucht-Fraïssé game $G_{\mathtt{qr}(A)}(\mathcal{F}', \mathcal{F}'')$. Consequently, for all sentences B, if $\mathtt{qr}(B) \leq \mathtt{qr}(A)$ then $\mathcal{F}' \models B$ iff $\mathcal{F}'' \models B$. Since $\mathcal{F}' \not\models A$, $\mathcal{F}'' \not\models A$.

Lemma 2. *The problem of deciding the $\mathcal{C}_{par}$-validity of $\mathcal{L}_{\mathbf{FOF}}$-formulas is* **PSPACE**-*complete.*

Proof. By Lemma 1, a sentence A is not $\mathcal{C}_{par}$-valid iff there exists a small and bounded partition $\mathcal{F}$ such that $n_{\mathcal{F}}, \pi_{\mathcal{F}} \leq \mathtt{qr}(A)$ and $\mathcal{F} \not\models A$. Hence, in order to determine whether a given sentence A is $\mathcal{C}_{par}$-valid, it suffices to execute the following procedure:

```
procedure val(A)
begin
for all small and bounded partitions F such that n_F, π_F ≤ qr(A), call
MC(F, A);
if all these calls are accepting then accept;
otherwise, reject;
end
```

where the call $\mathtt{MC}(\mathcal{F}, A)$ is accepting iff $\mathcal{F} \models A$. Obviously, the call $\mathtt{val}(A)$ is accepting iff A is $\mathcal{C}_{par}$-valid. Since the procedure MC can be implemented in polynomial space [27,30], the procedure val can be implemented in polynomial space. Thus, the

[1] Ehrenfeucht-Fraïssé games constitute a useful tool for characterizing frames modulo elementary equivalence. See [14, Chapter 2] for a general introduction.

problem of deciding the $\mathcal{C}_{par}$-validity of $\mathcal{L}_{\mathbf{FOF}}$-formulas is in **PSPACE**. As for the **PSPACE**-hardness of the problem of deciding the $\mathcal{C}_{par}$-validity of $\mathcal{L}_{\mathbf{FOF}}$-formulas, it immediately follows from the **PSPACE**-hardness of the membership problem in the first-order theory of pure equality [28].

Relativization. Let $\mathcal{F}$, $\mathcal{F}'$ be frames. $\mathcal{F}'$ is a *relativized reduct* of $\mathcal{F}$ if there exists a first-order formula A, there exists an individual variable x and there exists an assignment g on $\mathcal{F}$ such that $\mathcal{F}'$ is the restriction of $\mathcal{F}$ to the set of all states s in $\mathcal{F}$ such that $\mathcal{F}, g_s^x \models A$. In that case, we say $\mathcal{F}'$ is the *relativized reduct* of $\mathcal{F}$ with respect to A, x and g. Relativized reducts give rise to the following result:

Proposition 5 (Relativization Theorem). *Let $\mathcal{F}$, $\mathcal{F}'$ be frames, A be a first-order formula, x be an individual variable and g be an assignment on $\mathcal{F}$. If $\mathcal{F}'$ is the relativized reduct of $\mathcal{F}$ with respect to A, x and g then for all first-order formulas $C(y_1, \ldots, y_n)$ and for all assignments g' on $\mathcal{F}'$, $\mathcal{F}, g^{y_1 \ldots y_n}_{g'(y_1) \ldots g'(y_n)} \models (C(y_1, \ldots, y_n))^A_x$ iff $\mathcal{F}', g' \models C(y_1, \ldots, y_n)$.*

Proof. See [19, Theorem 5.1.1].

2.4 Modal Definability and First-Order Definability

Let $\mathcal{C}$ be a class of frames. A sentence A is *modally definable* with respect to $\mathcal{C}$ if there exists a modal formula φ such that for all frames $\mathcal{F}$ in $\mathcal{C}$, $\mathcal{F} \models A$ iff $\mathcal{F} \models \varphi$. In that case, we say φ is a *modal definition* of A with respect to $\mathcal{C}$. A modal formula φ is *first-order definable* with respect to $\mathcal{C}$ if there exists a first-order sentence A such that for all frames $\mathcal{F}$ in $\mathcal{C}$, $\mathcal{F} \models \varphi$ iff $\mathcal{F} \models A$. In that case, we say A is a *first-order definition* of φ with respect to $\mathcal{C}$. Table 1 contain examples of the correspondence between modal formulas and sentences.

Table 1. Examples of the correspondence between modal formulas and sentences.

φ	A
$p \to \Diamond p$	"**R** is reflexive"
$\Diamond\Diamond p \to \Diamond p$	"**R** is transitive"
$p \to \Box\Diamond p$	"**R** is symmetric"
$\Diamond\top$	"**R** is serial"
$\Diamond p \to \Box\Diamond p$	"**R** is Euclidean"

Proposition 6. *Let $\mathcal{C}$ be a class of frames. For all modal formulas φ, if there exists a modal formula ψ such that $\mathcal{C} \models \varphi \leftrightarrow \psi$ and $\mathtt{deg}(\psi) \leq 1$ then φ is first-order definable with respect to $\mathcal{C}$.*

Proof. See [5, Lemma 9.7].

3 Modal Definability: Decidable Cases

We consider classes of frames for which modal definability is decidable: $\mathcal{C}_{tE}$, $\mathcal{C}_{stE}$ and $\mathcal{C}_{par}$. For the purpose of proving the decidability of modal definability with respect to $\mathcal{C}_{par}$, we need to consider the following lemmas.

Lemma 3. *Let $\mathcal{F} = (W, R), \mathcal{F}' = (W', R')$ be bounded partitions. If $n_{\mathcal{F}} \geq n_{\mathcal{F}'}$ then for all modal formulas φ, if $\mathcal{F} \models \varphi$ then $\mathcal{F}' \models \varphi$.*

Proof. Suppose $n_{\mathcal{F}} \geq n_{\mathcal{F}'}$. Let φ be a modal formula. Suppose $\mathcal{F} \models \varphi$ and $\mathcal{F}' \not\models \varphi$. Hence, there exists a valuation V' on $\mathcal{F}'$ and there exists a state s' in $\mathcal{F}'$ such that $\mathcal{F}', V', s' \not\models \varphi$. Thus, by Proposition 2, $\mathcal{F}'_{s'}, V'_{s'}, s' \not\models \varphi$ where $V'_{s'}$ is the restriction of V' to $W_{s'}$. Consequently, $\mathcal{F}'_{s'} \not\models \varphi$. Obviously, $n_{\mathcal{F}'} \geq n_{\mathcal{F}'_{s'}}$. Since $n_{\mathcal{F}} \geq n_{\mathcal{F}'}$, $n_{\mathcal{F}} \geq n_{\mathcal{F}'_{s'}}$. Hence, let s be a state in $\mathcal{F}$ such that $\|R(s)\| \geq \|R'_{s'}(s')\|$. Since $\mathcal{F} \models \varphi$, by Proposition 1, $\mathcal{F}_s \models \varphi$. Moreover, $\mathcal{F}_s \twoheadrightarrow \mathcal{F}'_{s'}$. Thus, by Proposition 4, $\mathcal{F}'_{s'} \models \varphi$: a contradiction.

Lemma 4. *Let $\mathcal{F}, \mathcal{F}'$ be bounded partitions such that $n_{\mathcal{F}} \geq n_{\mathcal{F}'}$. For all sentences A, if A is modally definable with respect to $\mathcal{C}_{par}$ and $\mathcal{F} \models A$ then $\mathcal{F}' \models A$.*

Proof. Let A be a sentence. Suppose A is modally definable with respect to $\mathcal{C}_{par}$ and $\mathcal{F} \models A$. Hence, there exists a modal formula φ such that for all partitions $\mathcal{F}''$, $\mathcal{F}'' \models A$ iff $\mathcal{F}'' \models \varphi$. Since $\mathcal{F} \models A$, $\mathcal{F} \models \varphi$. Since $n_{\mathcal{F}} \geq n_{\mathcal{F}'}$, by Lemma 3, $\mathcal{F}' \models \varphi$. Since for all partitions $\mathcal{F}''$, $\mathcal{F}'' \models A$ iff $\mathcal{F}'' \models \varphi$, $\mathcal{F}' \models A$.

Lemma 5. *Let A be a sentence. If $\mathcal{C}_{par} \not\models A$ and $\mathcal{C}_{par} \not\models \neg A$ then A is modally definable with respect to $\mathcal{C}_{par}$ iff there exists a positive integer n such that $n < \mathtt{qr}(A)$ and for all bounded partitions $\mathcal{F}$, $\mathcal{F} \models A$ iff $n \geq n_{\mathcal{F}}$.*

Proof. Suppose $\mathcal{C}_{par} \not\models A$ and $\mathcal{C}_{par} \not\models \neg A$.
($\Rightarrow$) Suppose A is modally definable with respect to $\mathcal{C}_{par}$. Let $N = \{n_{\mathcal{F}} : \mathcal{F}$ is a bounded partition such that $\mathcal{F} \models A\}$. Since $\mathcal{C}_{par} \not\models A$ and $\mathcal{C}_{par} \not\models \neg A$, by Lemma 1, there exists bounded partitions $\mathcal{G}'$ and $\mathcal{G}''$ such that $n_{\mathcal{G}'} \leq \mathtt{qr}(A)$, $\mathcal{G}' \not\models A$ and $\mathcal{G}'' \not\models \neg A$. Hence, $\mathcal{G}'' \models A$. Since A is modally definable with respect to $\mathcal{C}_{par}$, by Lemma 4, $n_{\mathcal{G}'}$ is strictly greater than all positive integers in N. Moreover, $n_{\mathcal{G}''} \in N$. Thus, $N \neq \emptyset$. Since $n_{\mathcal{G}'}$ is strictly greater than all positive integers in N, N possesses a maximal element. Let $n = \max N$. Since $n_{\mathcal{G}'} \leq \mathtt{qr}(A)$ and $n_{\mathcal{G}'}$ is strictly greater than all positive integers in N, $n < \mathtt{qr}(A)$. For the sake of the contradiction, suppose there exists a bounded partition $\mathcal{H}$ such that either $\mathcal{H} \models A$ and $n < n_{\mathcal{H}}$, or $\mathcal{H} \not\models A$ and $n \geq n_{\mathcal{H}}$. In the former case, $n_{\mathcal{H}}$ is in N. Consequently, $n \geq n_{\mathcal{H}}$: a contradiction. In the latter case, by Lemma 4, $n_{\mathcal{H}}$ is strictly greater than all positive integers in N. Hence, $n < n_{\mathcal{H}}$: a contradiction.
($\Leftarrow$) Suppose there exists a positive integer n such that $n < \mathtt{qr}(A)$ and for all bounded partitions $\mathcal{F}$, $\mathcal{F} \models A$ iff $n \geq n_{\mathcal{F}}$. For the sake of the contradiction, suppose there exists a partition $\mathcal{G} = (W, R)$ such that either $\mathcal{G} \models A$ and $\mathcal{G} \not\models \psi_n$, or $\mathcal{G} \not\models A$ and $\mathcal{G} \models \psi_n$, ψ_n being the modal formula defined in Sect. 2.2. In the former case, since for all partitions $\mathcal{F}$, $\mathcal{F} \models \psi_n$ iff $\mathcal{F}$ is bounded and $n_{\mathcal{F}} \leq n$, if $\mathcal{G}$ is bounded then

$n_{\mathcal{G}} > n$. Thus, there exists a state s in $\mathcal{G}$ such that $\|R(s)\| \geq n+1$. Since $n < \mathtt{qr}(A)$, $n+1 \leq \mathtt{qr}(A)$. Let $\mathcal{G}'$ be the bounded partition obtained from $\mathcal{G}$ by eliminating in all equivalence classes, as many states as it is needed so that the size of each equivalence class becomes at most equal to $\mathtt{qr}(A)$. As the reader can check, Duplicator wins the Ehrenfeucht-Fraïssé game $G_{\mathtt{qr}(A)}(\mathcal{G}, \mathcal{G}')$. Consequently, for all sentences B, if $\mathtt{qr}(B) \leq \mathtt{qr}(A)$ then $\mathcal{G} \models B$ iff $\mathcal{G}' \models B$. Since $\mathcal{G} \models A$, $\mathcal{G}' \models A$. Since there exists a state s in $\mathcal{G}$ such that $\|R(s)\| \geq n+1$ and $n+1 \leq \mathtt{qr}(A)$, $n_{\mathcal{G}'} \geq n+1$. Hence, $n_{\mathcal{G}'} > n$. Since for all bounded partitions $\mathcal{F}$, $\mathcal{F} \models A$ iff $n \geq n_{\mathcal{F}}$, $\mathcal{G}' \not\models A$: a contradiction. In the latter case, since for all partitions $\mathcal{F}$, $\mathcal{F} \models \psi_n$ iff $\mathcal{F}$ is bounded and $n_{\mathcal{F}} \leq n$, $\mathcal{G}$ is bounded and $n_{\mathcal{G}} \leq n$. Since for all bounded partitions $\mathcal{F}$, $\mathcal{F} \models A$ iff $n \geq n_{\mathcal{F}}$, $\mathcal{G} \models A$: a contradiction. As a result, we obtain that for all partitions $\mathcal{G}$, $\mathcal{G} \models A$ iff $\mathcal{G} \models \psi_n$. Thus, A is modally definable with respect to $\mathcal{C}_{par}$.

Lemma 6. *Let A be a sentence. The following conditions are equivalent:*

- *A is modally definable with respect to $\mathcal{C}_{par}$,*
- *one of the following conditions holds:*
  - $\mathcal{C}_{par} \models A$,
  - $\mathcal{C}_{par} \models \neg A$,
  - *there exists a positive integer n such that $n < \mathtt{qr}(A)$ and for all bounded partitions $\mathcal{F}$, $\mathcal{F} \models A$ iff $n \geq n_{\mathcal{F}}$.*

Proof. By Lemma 5, using the fact that if $\mathcal{C}_{par} \models A$ then A corresponds to the modal formula $\top$ with respect to $\mathcal{C}_{par}$ and if $\mathcal{C}_{par} \models \neg A$ then A corresponds to the modal formula $\bot$ with respect to $\mathcal{C}_{par}$.

Lemma 7. *Let A be a sentence. If $\mathcal{C}_{par} \not\models A$ and $\mathcal{C}_{par} \not\models \neg A$ then A is modally definable with respect to $\mathcal{C}_{par}$ iff there exists a positive integer n such that $n < \mathtt{qr}(A)$ and $\mathcal{C}_{par} \models A \leftrightarrow B_n$, B_n being the sentence defined in Sect. 2.3.*

Proof. Suppose $\mathcal{C}_{par} \not\models A$ and $\mathcal{C}_{par} \not\models \neg A$.
($\Rightarrow$) Suppose A is modally definable with respect to $\mathcal{C}_{par}$. Since $\mathcal{C}_{par} \not\models A$ and $\mathcal{C}_{par} \not\models \neg A$, by Lemma 6, there exists a positive integer n such that $n < \mathtt{qr}(A)$ and for all bounded partitions $\mathcal{F}$, $\mathcal{F} \models A$ iff $n \geq n_{\mathcal{F}}$. Hence, for all bounded partitions $\mathcal{F}$, $\mathcal{F} \models A$ iff $\mathcal{F} \models B_n$. Thus, for all bounded partitions $\mathcal{F}$, $\mathcal{F} \models A \leftrightarrow B_n$. Consequently, by Lemma 1, $\mathcal{C}_{par} \models A \leftrightarrow B_n$.
($\Leftarrow$) Suppose there exists a positive integer n such that $n < \mathtt{qr}(A)$ and $\mathcal{C}_{par} \models A \leftrightarrow B_n$. Hence, for all bounded partitions $\mathcal{F}$, $\mathcal{F} \models A \leftrightarrow B_n$. Thus, for all bounded partitions $\mathcal{F}$, $\mathcal{F} \models A$ iff $\mathcal{F} \models B_n$. Consequently, for all bounded partitions $\mathcal{F}$, $\mathcal{F} \models A$ iff $n \geq n_{\mathcal{F}}$. Since $n < \mathtt{qr}(A)$, by Lemma 6, A is modally definable with respect to $\mathcal{C}_{par}$.

Lemma 8. *Let A be a sentence. The following conditions are equivalent:*

- *A is modally definable with respect to $\mathcal{C}_{par}$,*
- *one of the following conditions holds:*
  - $\mathcal{C}_{par} \models A$,
  - $\mathcal{C}_{par} \models \neg A$,
  - *there exists a positive integer n such that $n < \mathtt{qr}(A)$ and $\mathcal{C}_{par} \models A \leftrightarrow B_n$.*

Proof. By Lemma 7, using the fact that if $\mathcal{C}_{par} \models A$ then A corresponds to the modal formula $\top$ with respect to $\mathcal{C}_{par}$ and if $\mathcal{C}_{par} \models \neg A$ then A corresponds to the modal formula $\bot$ with respect to $\mathcal{C}_{par}$.

Lemma 9. *Let A be a sentence. The following conditions are equivalent:*

- $\mathcal{C}_{par} \models A$,
- $B_{\mathtt{qr}(A)} \to A$ *is modally definable with respect to* $\mathcal{C}_{par}$.

Proof. ($\Rightarrow$) Suppose $\mathcal{C}_{par} \models A$. Hence, $\mathcal{C}_{par} \models B_{\mathtt{qr}(A)} \to A$. Thus, $B_{\mathtt{qr}(A)} \to A$ corresponds to the modal formula $\top$ with respect to $\mathcal{C}_{par}$. Consequently, $B_{\mathtt{qr}(A)} \to A$ is modally definable with respect to $\mathcal{C}_{par}$.
($\Leftarrow$) Suppose $B_{\mathtt{qr}(A)} \to A$ is modally definable with respect to $\mathcal{C}_{par}$. For the sake of the contradiction, suppose $\mathcal{C}_{par} \not\models A$. Hence, by Lemma 1, there exists a bounded partition $\mathcal{F}$ such that $n_{\mathcal{F}} \leq \mathtt{qr}(A)$ and $\mathcal{F} \not\models A$. Thus, $\mathcal{F} \models B_{\mathtt{qr}(A)}$. Since $\mathcal{F} \not\models A$, $\mathcal{F} \not\models B_{\mathtt{qr}(A)} \to A$. Consequently, $\mathcal{C}_{par} \not\models B_{\mathtt{qr}(A)} \to A$. Moreover, obviously, $\mathcal{C}_{par} \not\models \neg(B_{\mathtt{qr}(A)} \to A)$. Since $B_{\mathtt{qr}(A)} \to A$ is modally definable with respect to $\mathcal{C}_{par}$, by Lemma 6, there exists a positive integer n such that $n < \mathtt{qr}(B_{\mathtt{qr}(A)} \to A)$ and for all bounded partitions $\mathcal{G}$, $\mathcal{G} \models B_{\mathtt{qr}(A)} \to A$ iff $n \geq n_{\mathcal{G}}$. Hence, $n \leq \mathtt{qr}(A)$. Let $\mathcal{F}'$ be a bounded partition such that $n_{\mathcal{F}'} > \mathtt{qr}(A)$. Thus, $\mathcal{F}' \not\models B_{\mathtt{qr}(A)}$. Consequently, $\mathcal{F}' \models B_{\mathtt{qr}(A)} \to A$. Since for all bounded partitions $\mathcal{G}$, $\mathcal{G} \models B_{\mathtt{qr}(A)} \to A$ iff $n \geq n_{\mathcal{G}}$, $n \geq n_{\mathcal{F}'}$. Since $n_{\mathcal{F}'} > \mathtt{qr}(A)$, $n > \mathtt{qr}(A)$: a contradiction.

As a result,

Theorem 1. *The problem of deciding the modal definability with respect to* $\mathcal{C}_{par}$ *of* $\mathcal{L}_{\mathbf{FOF}}$*-formulas is* **PSPACE***-complete.*

Proof. By Lemma 8, a sentence A is modally definable with respect to $\mathcal{C}_{par}$ iff either $\mathcal{C}_{par} \models A$, or $\mathcal{C}_{par} \models \neg A$, or there exists a positive integer n such that $n < \mathtt{qr}(A)$ and $\mathcal{C}_{par} \models A \leftrightarrow B_n$. Hence, in order to determine whether a given sentence A is modally definable with respect to $\mathcal{C}_{par}$, it suffices to execute the following procedure:

```
procedure MD(A)
begin
call val(A);
if this call is accepting then accept;
otherwise, call val(¬A);
if this call is accepting then accept;
otherwise, for all positive integers n such that n<qr(A), call val(A ↔
Bn);
if one of these calls is accepting then accept;
otherwise, reject;
end
```

Obviously, the call MD(A) is accepting iff A is modally definable with respect to $\mathcal{C}_{par}$. Since the procedure `val` can be implemented in polynomial space, the procedure `MD` can be implemented in polynomial space. Thus, the problem of deciding the modal definability with respect to $\mathcal{C}_{par}$ of $\mathcal{L}_{\mathbf{FOF}}$-formulas is in **PSPACE**. As for the **PSPACE**-hardness of the problem of deciding the modal definability with respect to $\mathcal{C}_{par}$ of $\mathcal{L}_{\mathbf{FOF}}$-formulas, it immediately follows from Lemmas 2 and 9.

An interesting question is the following: when the ordinary language of modal logic is extended either with the universal modality, or with the difference modality, is the problem of deciding the modal definability with respect to $\mathcal{C}_{par}$, $\mathcal{C}_{tE}$ and $\mathcal{C}_{stE}$ of $\mathcal{L}_{\mathbf{FOF}}$-formulas still decidable? If the answer is "yes", is this problem still **PSPACE**-complete? The answers to these questions have been given in [2, 15, 16].

Proposition 7. *When the ordinary language of modal logic is extended with the universal modality, the problem of deciding the modal definability with respect to $\mathcal{C}_{par}$, $\mathcal{C}_{tE}$ and $\mathcal{C}_{stE}$ of $\mathcal{L}_{\mathbf{FOF}}$-formulas is* **PSPACE***-complete.*

4 First-Order Definability: Trivial Cases

In this section, we consider classes of frames for which first-order definability is trivial: $\mathcal{C}_{tE}$, $\mathcal{C}_{stE}$ and $\mathcal{C}_{par}$. We take as well a special interest in $\mathcal{C}_{E}$, $\mathcal{C}_{sE}$ and $\mathcal{C}^{\omega}_{rtc}$ and we prove that they give rise to a trivial first-order definability problem too. It is a well-known fact that with respect to $\mathcal{C}_{tE}$, $\mathcal{C}_{stE}$ and $\mathcal{C}_{par}$, every modal formula is equivalent to a modal formula of degree less than or equal to 1. As a result,

Proposition 8. *The problem of deciding first-order definability with respect to $\mathcal{C}_{tE}$, $\mathcal{C}_{stE}$ and $\mathcal{C}_{par}$ is trivial: every modal formula is first-order definable with respect to $\mathcal{C}_{tE}$, $\mathcal{C}_{stE}$ and $\mathcal{C}_{par}$.*

Proof. By Proposition 6.

The reader may ask whether there exists classes of frames with respect to which the problem of deciding first-order definability is trivial and there exists modal formulas equivalent to no modal formula of degree less than or equal to 1. It is a well-known fact that with respect to $\mathcal{C}_{E}$ and $\mathcal{C}_{sE}$, every modal formula is equivalent to a modal formula of degree less than or equal to 2 but some modal formula is equivalent to no modal formula of degree less than or equal to 1. Nevertheless,

Proposition 9. *The problem of deciding first-order definability with respect to $\mathcal{C}_{E}$ and $\mathcal{C}_{sE}$ is trivial: every modal formula is first-order definable with respect to $\mathcal{C}_{E}$ and $\mathcal{C}_{sE}$.*

Proof. Since $\mathcal{C}_{E}$ contains $\mathcal{C}_{sE}$, it suffices to prove that (Π) every modal formula is first-order definable with respect to $\mathcal{C}_{E}$. The proof of (Π) has been presented by Balbiani *et al.* [1]. It is based on the following line of reasoning. For all frames $\mathcal{F} = (W, R)$ in $\mathcal{C}_{E}$ and for all states s in $\mathcal{F}$, exactly one of the following conditions holds:

- $R_s = \emptyset$,
- $R_s = W_s \times W_s$,
- $R_s = (\{s\} \times S) \cup (T \times T)$ for some nonempty subsets S and T of $W_s \backslash \{s\}$ such that $S \subseteq T$.

When $\mathcal{F}$ is finite, for all states s in $\mathcal{F}$, $\mathcal{F}_s$ can be exactly characterized by a triple $\sigma = (\sigma_1, \sigma_2, \sigma_3)$ in $\{0, 1\} \times \mathbb{N}^2$: σ_1 will be the number of irreflexive states in $\mathcal{F}_s$; σ_2 will be the number of states accessible from s in 1 step; σ_3 will be the number of states accessible from s either in 1 step, or in 2 steps. When $R_s = \emptyset$, this triple will be such that $\sigma_1 = 1$, $\sigma_2 = 0$ and $\sigma_3 = 0$. When $R_s = W_s \times W_s$, this triple will be such that $\sigma_1 = 0$, $\sigma_2 \geq 1$ and $\sigma_3 = \sigma_2$. When $R_s = (\{s\} \times S) \cup (T \times T)$ for some nonempty subsets S and T of $W_s \backslash \{s\}$ such that $S \subseteq T$, this triple will be such that $\sigma_1 = 1$, $\sigma_2 \geq 1$ and $\sigma_3 \geq \sigma_2$. A *type* is a triple $\sigma = (\sigma_1, \sigma_2, \sigma_3)$ in $\{0, 1\} \times \mathbb{N}^2$ such that one of the following conditions holds:

- $\sigma_1 = 1$, $\sigma_2 = 0$ and $\sigma_3 = 0$,
- $\sigma_1 = 0$, $\sigma_2 \geq 1$ and $\sigma_3 = \sigma_2$,
- $\sigma_1 = 1$, $\sigma_2 \geq 1$ and $\sigma_3 \geq \sigma_2$.

Obviously, for all types $\sigma = (\sigma_1, \sigma_2, \sigma_3)$, one can construct a finite rooted frame $\mathcal{F}_\sigma = (W_\sigma, R_\sigma)$ in $\mathcal{C}_E$ which is characterized by σ. Moreover, for all types $\sigma = (\sigma_1, \sigma_2, \sigma_3)$, one can write a first-order formula $A_\sigma(x)$ such that for all assignments g on F_σ, if $g(x)$ is equal to the root of $\mathcal{F}_\sigma$ then $F_\sigma, g \models A_\sigma(x)$. For all types $\sigma = (\sigma_1, \sigma_2, \sigma_3)$, x is the only individual variable freely occurring in the first-order formula $A_\sigma(x)$ associated to it. Given a type $\sigma = (\sigma_1, \sigma_2, \sigma_3)$, how is constructed the finite rooted frame $\mathcal{F}_\sigma = (W_\sigma, R_\sigma)$ and how is written the first-order formula $A_\sigma(x)$? We will answer later in this section to a similar question within the context of the first-order definability problem with respect to $\mathcal{C}^{\omega}_{rtc}$. Now, for all modal formulas φ, let $\Delta(\varphi) = \{\sigma : \sigma = (\sigma_1, \sigma_2, \sigma_3)$ is a type such that $\mathcal{F}_\sigma \not\models \varphi$ and $\sigma_3 \leq \|\mathtt{sf}(\varphi)\|\}$. Obviously, for all modal formulas φ, $\Delta(\varphi)$ is finite. The finite rooted frame $\mathcal{F}_\sigma = (W_\sigma, R_\sigma)$ and the first-order formula $A_\sigma(x)$ associated to a given type $\sigma = (\sigma_1, \sigma_2, \sigma_3)$ possess interesting properties. For example[2],

Lemma 10. *For all types $\sigma = (\sigma_1, \sigma_2, \sigma_3)$ and for all assignments g on $\mathcal{F}_\sigma$, if $g(x)$ is the root of $\mathcal{F}_\sigma$ then $\mathcal{F}_\sigma, g \models A_\sigma(x)$.*

Lemma 11. *Let $\mathcal{F}$ be a frame in $\mathcal{C}_E$ and g be an assignment on $\mathcal{F}$. For all types $\sigma = (\sigma_1, \sigma_2, \sigma_3)$, if $\mathcal{F}, g \models A_\sigma(x)$ then there exists a surjective bounded morphism $f : \mathcal{F}_{g(x)} \twoheadrightarrow \mathcal{F}_\sigma$ such that $f(g(x))$ is the root of $\mathcal{F}_\sigma$.*

Lemma 12. *Let φ be a modal formula. For all frames $\mathcal{F}$ in $\mathcal{C}_E$, if $\mathcal{F} \not\models \varphi$ then there exists a type $\sigma = (\sigma_1, \sigma_2, \sigma_3)$ such that $\mathcal{F}_\sigma \not\models \varphi$, $\sigma_3 \leq \|\mathtt{sf}(\varphi)\|$ and $\mathcal{F} \models \exists x\, A_\sigma(x)$.*

In Lemmas 10, 11 and 12, $\mathcal{F}_\sigma$ denotes the finite rooted frame in $\mathcal{C}_E$ associated to σ and $\mathcal{F}_{g(x)}$ denotes the subframe of $\mathcal{F}$ generated from $g(x)$. For all modal formulas φ, let A_φ be the first-order formula $\neg\exists x \bigvee\{A_\sigma(x) : \sigma \in \Delta(\varphi)\}$. Notice that for all modal formulas φ, A_φ is a sentence. Given a modal formula φ, the reason for our interest in the sentence A_φ is the following result:

Lemma 13. *Let φ be a modal formula. For all frames $\mathcal{F}$ in $\mathcal{C}_E$, the following conditions are equivalent:*

- $\mathcal{F} \models \varphi$,
- $\mathcal{F} \models A_\varphi$.

[2] Lemmas 10, 11 and 12 assert the properties that are needed for proving Proposition 9. Their proofs have been given with full details in [1]. Similar properties needed for proving Theorem 2 below are asserted in Lemmas 14, 15 and 16. Their proofs are given with full details below.

This ends the proof of Proposition 9.

The reader may ask whether there exists classes of frames with respect to which every modal formula is first-order definable and for all $n \in \mathbb{N}$, there exists modal formulas equivalent to no modal formula of degree less than or equal to n. It is a well-known fact that with respect to $\mathcal{C}^{\omega}_{rtc}$, for all $n \in \mathbb{N}$, some modal formula is equivalent to no modal formula of degree less than or equal to n. Nevertheless,

Theorem 2. *The problem of deciding first-order definability with respect to $\mathcal{C}^{\omega}_{rtc}$ is trivial: every modal formula is first-order definable with respect to $\mathcal{C}^{\omega}_{rtc}$.*

Proof. We will follow a line of reasoning similar to the line of reasoning sketched in the proof of Proposition 9. For all frames $\mathcal{F}$ in $\mathcal{C}^{\omega}_{rtc}$ and for all states s in $\mathcal{F}$, $\mathcal{F}_s$ contains finitely many clusters. When $\mathcal{F}$ is finite, for all states s in $\mathcal{F}$, $\mathcal{F}_s$ can be exactly characterized by a finite nonempty sequence $\sigma = (\sigma_1, \ldots, \sigma_a)$ of positive integers. In this proof, a *type* is a finite nonempty sequence $\sigma = (\sigma_1, \ldots, \sigma_a)$ of positive integers. For all types $\sigma = (\sigma_1, \ldots, \sigma_a)$, let $\|\sigma\| = \sigma_1 + \ldots + \sigma_a$. For all types $\sigma = (\sigma_1, \ldots, \sigma_a)$, let $\mathcal{F}_\sigma = (W_\sigma, R_\sigma)$ be the $\mathcal{C}^{\omega}_{rtc}$-frame such that $W_\sigma = \{(i,k) : 1 \leq i \leq a$ and $1 \leq k \leq \sigma_i\}$ and R_σ is the binary relation on W_σ such that for all $(i,k), (j,l)$ in W_σ, $(i,k) R_\sigma (j,l)$ iff $i \leq j$. For all types $\sigma = (\sigma_1, \ldots, \sigma_a)$, let $A_\sigma(x)$ be the first-order formula $\exists x_{1,1} \ldots \exists x_{1,\sigma_1} \ldots \exists x_{a,1} \ldots \exists x_{a,\sigma_a} B_\sigma$ where B_σ is the conjunction of the following formulas:

- $x = x_{1,1} \vee \ldots \vee x = x_{1,\sigma_1}$,
- $x_{i,k} \neq x_{j,l}$ for all $(i,k), (j,l)$ in W_σ such that either $i \neq j$, or $k \neq l$,
- $\mathbf{R}(x_{i,k}, x_{j,l})$ for all $(i,k), (j,l)$ in W_σ such that $i \leq j$,
- $\neg\mathbf{R}(x_{j,l}, x_{i,k})$ for all $(i,k), (j,l)$ in W_σ such that $i < j$,
- $\forall y(\mathbf{R}(x,y) \rightarrow \bigvee\{\mathbf{R}(y, x_{i,k}) : (i,k)$ is in $W_\sigma\})$.

Notice that for all types $\sigma = (\sigma_1, \ldots, \sigma_a)$, x is the only individual variable freely occurring in $A_\sigma(x)$. Now, for all modal formulas φ, let $\Delta(\varphi) = \{\sigma : \sigma = (\sigma_1, \ldots, \sigma_a)$ is a type such that $\mathcal{F}_\sigma \not\models \varphi$ and $\|\sigma\| \leq 3.\|\mathtt{sf}(\varphi)\|\}$. Obviously, for all modal formulas φ, $\Delta(\varphi)$ is finite. The finite rooted frame $\mathcal{F}_\sigma = (W_\sigma, R_\sigma)$ and the first-order formula $A_\sigma(x)$ associated to a given type $\sigma = (\sigma_1, \ldots, \sigma_a)$ possess interesting properties. The following result will play in this proof the role played by Lemma 10 in the proof of Proposition 9.

Lemma 14. *For all types $\sigma = (\sigma_1, \ldots, \sigma_a)$ and for all assignments g on $\mathcal{F}_\sigma$, if $g(x)$ is in $\{(1,1), \ldots, (1,\sigma_1)\}$ then $\mathcal{F}_\sigma, g \models A_\sigma(x)$.*

Proof. Let $\sigma = (\sigma_1, \ldots, \sigma_a)$ be a type and g be an assignment on $\mathcal{F}_\sigma$. Suppose $g(x)$ is in $\{(1,1), \ldots, (1,\sigma_1)\}$. Let g' be the assignment on $\mathcal{F}_\sigma$ such that

- $g'(x) = g(x)$,
- $g'(x_{i,k}) = (i,k)$ for all (i,k) in W_σ,
- for all individual variables $z \neq x$, if $z \neq x_{i,k}$ for all (i,k) in W_k then $g'(z) = g(z)$.

Since $g(x)$ is in $\{(1,1),\ldots,(1,\sigma_1)\}$,

- either $g'(x) = g'(x_{1,1})$, ..., or $g'(x) = g'(x_{1,\sigma_1})$,
- $g'(x_{i,k}) \neq g'(x_{j,l})$ for all $(i,k),(j,l)$ in W_σ such that either $i \neq j$, or $k \neq l$,
- $R_\sigma(g'(x_{i,k}), g'(x_{j,l}))$ for all $(i,k),(j,l)$ in W_σ such that $i \leq j$,
- not $R_\sigma(g'(x_{j,l}), g'(x_{i,k}))$ for all $(i,k),(j,l)$ in W_σ such that $i < j$,
- for all (j,l) in W_σ, if $R_\sigma(g'(x),(j,l))$ then there exists (i,k) in W_σ such that $R_\sigma((j,l), g'(x_{i,k}))$.

Hence, $\mathcal{F}_\sigma, g' \models B_\sigma$. Since g' is an assignment on $\mathcal{F}_\sigma$ such that $g'(x) = g(x)$ and for all individual variables $z \neq x$, if $z \neq x_{i,k}$ for all (i,k) in W_k then $g'(z) = g(z)$, $\mathcal{F}_\sigma, g \models A_\sigma(x)$.

The following result will play in this proof the role played by Lemma 11 in the proof of Proposition 9.

Lemma 15. *Let $\mathcal{F} = (W,R)$ be a frame in $\mathcal{C}^\omega_{rtc}$ and g be an assignment on $\mathcal{F}$. For all types $\sigma = (\sigma_1,\ldots,\sigma_a)$, if $\mathcal{F}, g \models A_\sigma(x)$ then there exists a surjective bounded morphism $f : \mathcal{F}_{g(x)} \twoheadrightarrow \mathcal{F}_\sigma$ such that $f(g(x))$ is in $\{(1,1),\ldots,(1,\sigma_1)\}$.*

Proof. Let $\sigma = (\sigma_1,\ldots,\sigma_a)$ be a type. Suppose $\mathcal{F}, g \models A_\sigma(x)$. Let g' be an assignment on $\mathcal{F}$ such that

- $g'(x) = g(x)$,
- for all individual variables $z \neq x$, if $z \neq x_{i,k}$ for all (i,k) in W_k then $g'(z) = g(z)$,
- $\mathcal{F}, g' \models B_\sigma$.

Hence,

- either $g'(x) = g'(x_{1,1})$, ..., or $g'(x) = g'(x_{1,\sigma_1})$,
- $g'(x_{i,k}) \neq g'(x_{j,l})$ for all $(i,k),(j,l)$ in W_σ such that either $i \neq j$, or $k \neq l$,
- $R(g'(x_{i,k}), g'(x_{j,l}))$ for all $(i,k),(j,l)$ in W_σ such that $i \leq j$,
- not $R(g'(x_{j,l}), g'(x_{i,k}))$ for all $(i,k),(j,l)$ in W_σ such that $i < j$,
- for all states t in $\mathcal{F}$, if $R(g'(x),t)$ then there exists (i,k) in W_σ such that $R(t, g'(x_{i,k}))$.

Let C_1 be the cluster of $g'(x_{1,1}),\ldots,g'(x_{1,\sigma_1})$ in $\mathcal{F}_{g'(x)}$, ..., C_a be the cluster of $g'(x_{a,1}),\ldots,g'(x_{a,\sigma_a})$ in $\mathcal{F}_{g'(x)}$. By the above 5 itemized conditions,

- $g'(x)$ is in C_1,
- $\|C_i\| \geq \sigma_i$ for all i in $\{1,\ldots,a\}$,
- $C_i \ll C_j$ for all i,j in $\{1,\ldots,a\}$ such that $i \leq j$,
- not $C_j \ll C_i$ for all i,j in $\{1,\ldots,a\}$ such that $i < j$,
- for all states t in $W_{g'(x)}$, there exists a least element i in $\{1,\ldots,a\}$ such that $tR_{g'(x)}g'(x_{i,1})$, ..., $tR_{g'(x)}g'(x_{i,\sigma_i})$,

where $\ll$ is the reflexive, antisymmetric, transitive and connected relation between $\mathcal{F}$'s clusters such that for all $\mathcal{F}$'s clusters C, D, $C \ll D$ iff there exists states t, u in $\mathcal{F}$ such that $t \in C$, $u \in D$ and tRu. Let $f : W_{g'(x)} \longrightarrow W_\sigma$ be such that

- either $f(g'(x)) = (1,1)$, ..., or $f(g'(x)) = (1,\sigma_1)$,
- for all i in $\{1,\ldots,a\}$, $f_{|C_i}$ is a surjective function from C_i to $\{(i,k) : 1 \leq k \leq \sigma_i\}$,
- for all states t in $W_{g'(x)} \backslash (C_1 \cup \ldots \cup C_a)$, $f(t)$ is in $\{(i,k) : 1 \leq k \leq \sigma_i\}$ where i is the least element in $\{1,\ldots,a\}$ such that $tR_{g'(x)}g'(x_{i,1})$, ..., $tR_{g'(x)}g'(x_{i,\sigma_i})$.

Obviously, $f : \mathcal{F}_{g(x)} \twoheadrightarrow \mathcal{F}_\sigma$ is a surjective bounded morphism. Moreover, since $g'(x) = g(x)$, $f(g(x))$ is in $\{(1,1),\ldots,(1,\sigma_1)\}$.

The following result will play in this proof the role played by Lemma 12 in the proof of Proposition 9.

Lemma 16. *Let φ be a modal formula. For all frames $\mathcal{F}$ in $\mathcal{C}^{\omega}_{rtc}$, if $\mathcal{F} \not\models \varphi$ then there exists a type $\sigma = (\sigma_1,\ldots,\sigma_a)$ such that $\mathcal{F}_\sigma \not\models \varphi$, $\|\sigma\| \leq 3.\|\mathtt{sf}(\varphi)\|$ and $\mathcal{F} \models \exists x A_\sigma(x)$.*

Proof. Let $\mathcal{F} = (W,R)$ be a frame in $\mathcal{C}^{\omega}_{rtc}$. Suppose $\mathcal{F} \not\models \varphi$. Hence, there exists a valuation V on $\mathcal{F}$ and there exists a state s in $\mathcal{F}$ such that $\mathcal{F},V,s \not\models \varphi$. Since $\mathcal{F}$ is a $\mathcal{C}^{\omega}_{rtc}$-frame, $\mathcal{F}_s$ contains finitely many clusters. Moreover, s belongs to the first cluster of $\mathcal{F}_s$. For all states t in $\mathcal{F}_s$, let $B(t) = \{\Box\psi \in \mathtt{sf}^{\Box}(\varphi) : \mathcal{F}_s,V_s,t \models \Box\psi\}$ where V_s is the restriction of V to W_s. Notice that for all states t,u in $\mathcal{F}_s$, if tR_su then $B(t) \subseteq B(u)$. Let $n \geq 1$ and $t_1,\ldots,t_n$ be states in $\mathcal{F}_s$ such that

- for all states t in $\mathcal{F}_s$, there exists i in $\{1,\ldots,n\}$ such that $B(t) = B(t_i)$,
- for all i,j in $\{1,\ldots,n\}$, if $i < j$ then $B(t_i)$ is strictly contained in $B(t_j)$.

Notice that $n \leq \|\mathtt{sf}^{\Box}(\varphi)\|+1$. Thus, $n \leq \|\mathtt{sf}(\varphi)\|$. Moreover, for all i,j in $\{1,\ldots,n\}$, if $i < j$ then $t_iR_st_j$ and not $t_jR_st_i$. For all i in $\{1,\ldots,n\}$, let $CB(t_i) = \{C(u) : u$ is a state in $\mathcal{F}_s$ such that $B(u) = B(t_i)\}$. Obviously, for all i in $\{1,\ldots,n\}$, $C(t_i) \in CB(t_i)$. For all i in $\{1,\ldots,n\}$, let u_i be a state in the last cluster of $CB(t_i)$. For all i in $\{1,\ldots,n\}$, let $\alpha_i \geq 0$ and $\Box\psi_{i,1},\ldots,\Box\psi_{i,\alpha_i}$ be a list of $\mathtt{sf}^{\Box}(\varphi) \backslash B(t_i)$ when $i = n$ and a list of $B(t_{i+1}) \backslash B(t_i)$ otherwise. Obviously, $\alpha_1 + \ldots + \alpha_n \leq \|\mathtt{sf}^{\Box}(\varphi)\|$. Consequently, $\alpha_1 + \ldots + \alpha_n + 1 \leq \|\mathtt{sf}(\varphi)\|$. For all i in $\{1,\ldots,n\}$ and for all j in $\{1,\ldots,\alpha_i\}$, let $v_{i,j}$ in $C(u_i)$ be such that $\mathcal{F}_s,V_s,v_{i,j} \not\models \psi_{i,j}$. For all i in $\{1,\ldots,n\}$, let τ_i be the cardinality of $\{s,u_i\} \cup \{v_{i,1},\ldots,v_{i,\alpha_i}\}$ when s is in $C(u_i)$ and the cardinality of $\{u_i\} \cup \{v_{i,1},\ldots,v_{i,\alpha_i}\}$ otherwise. Obviously, for all i in $\{1,\ldots,n\}$, $\tau_i \leq \alpha_i + 2$. Let σ be $(\tau_1,\ldots,\tau_n)$ when s is in $C(u_1)$ and $(1,\tau_1,\ldots,\tau_n)$ otherwise. Obviously, $\|\sigma\| \leq \tau_1 + \ldots + \tau_n + 1$. Since for all i in $\{1,\ldots,n\}$, $\tau_i \leq \alpha_i + 2$, $\|\sigma\| \leq \alpha_1 + \ldots + \alpha_n + 2.n + 1$. Since $n \leq \|\mathtt{sf}(\varphi)\|$ and $\alpha_1 + \ldots + \alpha_n + 1 \leq \|\mathtt{sf}(\varphi)\|$, $\|\sigma\| \leq 3.\|\mathtt{sf}(\varphi)\|$. Moreover, by construction of σ, $\mathcal{F}$ obviously satisfies the sentence $\exists x A_\sigma(x)$. In the end, let us notice that $\mathcal{F}_\sigma$ is isomorphic to $\mathcal{F}' = (W',R')$ where $W' = \{s,u_1,\ldots,u_n\} \cup \{v_{1,1},\ldots,v_{1,\alpha_1},\ldots,v_{n,1},\ldots,v_{n,\alpha_n}\}$ and R' is the restriction of R to W'. More important is that, as the reader can prove it by induction on ψ, for all $\psi \in \mathtt{sf}(\varphi)$ and for all $w' \in W'$, $\mathcal{F}',V',w' \models \psi$ iff $\mathcal{F},V,w \models \psi$ where V' is the restriction of V to W'. Since $\mathcal{F},V,s \not\models \varphi$, $\mathcal{F}',V',s \not\models \varphi$. Hence, $\mathcal{F}' \not\models \varphi$. Since $\mathcal{F}_\sigma$ is isomorphic to $\mathcal{F}'$, $\mathcal{F}_\sigma \not\models \varphi$.

For all modal formulas φ, let A_φ be the first-order formula $\neg\exists x \bigvee\{A_\sigma(x) : \sigma \in \Delta(\varphi)\}$. Notice that for all modal formulas φ, A_φ is a sentence. Given a modal formula φ, the reason for our interest in the sentence A_φ is the following result:

Lemma 17. *Let φ be a modal formula. For all frames $\mathcal{F}$ in $\mathcal{C}^{\omega}_{rtc}$, the following conditions are equivalent:*

- $\mathcal{F} \models \varphi$,
- $\mathcal{F} \models A_{\varphi}$.

Proof. Let $\mathcal{F} = (W, R)$ be a frame in $\mathcal{C}^{\omega}_{rtc}$.
($\Rightarrow$) Suppose $\mathcal{F} \models \varphi$ and $\mathcal{F} \not\models A_{\varphi}$. Hence, there exists an assignment g on $\mathcal{F}$ such that $\mathcal{F}, g \models \exists x \bigvee\{A_{\sigma}(x) : \sigma \in \Delta(\varphi)\}$. Thus, there exists a state s in $\mathcal{F}$ such that $\mathcal{F}, g^x_s \models \bigvee\{A_{\sigma}(x) : \sigma \in \Delta(\varphi)\}$. Consequently, there exists $\sigma \in \Delta(\varphi)$ such that $\mathcal{F}, g^x_s \models A_{\sigma}(x)$. Hence, $\mathcal{F}_{\sigma} \not\models \varphi$. Moreover, by Lemma 15, $\mathcal{F}_s \twoheadrightarrow \mathcal{F}_{\sigma}$. Since $\mathcal{F} \models \varphi$, by Proposition 1, $\mathcal{F}_s \models \varphi$. Since $\mathcal{F}_s \twoheadrightarrow \mathcal{F}_{\sigma}$, by Proposition 4, $\mathcal{F}_{\sigma} \models \varphi$: a contradiction.
($\Leftarrow$) Suppose $\mathcal{F} \models A_{\varphi}$ and $\mathcal{F} \not\models \varphi$. Thus, by Lemma 16, there exists a type τ such that $\mathcal{F}_{\tau} \not\models \varphi$, $\|\tau\| \leq \|3.\mathtt{sf}(\varphi)\|$ and $\mathcal{F} \models \exists x\ A_{\tau}(x)$. Consequently, τ is in $\Delta(\varphi)$. Let g be an assignment on $\mathcal{F}$. Since $\mathcal{F} \models \exists x\ A_{\tau}(x)$, $\mathcal{F}, g \models \exists x\ A_{\tau}(x)$. Hence, there exists $s \in W$ such that $\mathcal{F}, g^x_s \models A_{\tau}(x)$. Since τ is in $\Delta(\varphi)$, $\mathcal{F}, g^x_s \models \bigvee\{A_{\sigma}(x) : \sigma \in \Delta(\varphi)\}$. Thus, $\mathcal{F}, g \models \exists x \bigvee\{A_{\sigma}(x) : \sigma \in \Delta(\varphi)\}$. Consequently, $\mathcal{F}, g \not\models A_{\varphi}$. Hence, $\mathcal{F} \not\models A_{\varphi}$: a contradiction.

This ends the proof of Theorem 2.

5 Chagrova's Theorem About Modal Definability

In this section, we give a new proof of Chagrova's Theorem about modal definability and we give sketches of proofs of new variants of Chagrova's Theorem about modal definability.

5.1 A New Proof of Chagrova's Theorem About Modal Definability

Firstly, we give a new proof of Chagrova's Theorem about modal definability. Our strategy will be as follows:

- remind the reduction of Kalmár [20] of the problem of deciding the validity in $\mathcal{C}_{all}$ of sentences from an arbitrary first-order language to the problem of deciding the validity in $\mathcal{C}_{all}$ of sentences from the first-order language $\mathcal{L}_{\mathbf{FOF}}$,
- prove that the problem of deciding the validity in $\mathcal{C}_{all}$ of sentences from the first-order language $\mathcal{L}_{\mathbf{FOF}}$ is reducible to the problem of deciding the modal definability with respect to $\mathcal{C}_{all}$.

Proposition 10. *The problem of deciding the validity in $\mathcal{C}_{all}$ of sentences from an arbitrary first-order language is reducible to the problem of deciding the validity in $\mathcal{C}_{all}$ of sentences from the first-order language $\mathcal{L}_{\mathbf{FOF}}$.*

Proof. See [20].

Proposition 11. *The problem of deciding the validity in $\mathcal{C}_{all}$ of sentences from the first-order language $\mathcal{L}_{\mathbf{FOF}}$ is reducible to the problem of deciding modal definability with respect to $\mathcal{C}_{all}$.*

Proof. Let C be a sentence from the first-order language $\mathcal{L}_{\mathbf{FOF}}$. Let D be the sentence $\exists y\,(\exists x\; y \neq x \wedge \neg(C)_x^{y\neq x})$. We demonstrate $\mathcal{C}_{all} \models C$ iff D is modally definable with respect to $\mathcal{C}_{all}$.
($\Rightarrow$) Suppose $\mathcal{C}_{all} \models C$. For the sake of the contradiction, suppose D is not modally definable with respect to $\mathcal{C}_{all}$. We have to consider 2 cases.

1st case: $\mathcal{C}_{all} \models \neg D$. Hence, D corresponds to the modal formula $\bot$ with respect to $\mathcal{C}_{all}$. Thus, D is modally definable with respect to $\mathcal{C}_{all}$: a contradiction.

2nd case: $\mathcal{C}_{all} \not\models \neg D$. Consequently, there exists a frame $\mathcal{F}$ such that $\mathcal{F} \not\models \neg D$. Hence, $\mathcal{F} \models D$. Let g be an assignment on $\mathcal{F}$. Since $\mathcal{F} \models D$, $\mathcal{F}, g \models D$. Thus, there exists a state s in $\mathcal{F}$ such that $\mathcal{F}, g_s^y \models \exists x\; y \neq x$ and $\mathcal{F}, g_s^y \not\models (C)_x^{y\neq x}$. Consequently, $\mathcal{F}$ possesses a relativized reduct $\mathcal{F}'$ with respect to $y \neq x$, x and g_s^y. Hence, by Proposition 5, $\mathcal{F}, g_s^y \models (C)_x^{y\neq x}$ iff $\mathcal{F}', g \models C$. Since $\mathcal{F}, g_s^y \not\models (C)_x^{y\neq x}$, $\mathcal{F}', g \not\models C$. Thus, $\mathcal{F}' \not\models C$. Consequently, $\mathcal{C}_{all} \not\models C$: a contradiction.
($\Leftarrow$) Suppose D is modally definable with respect to $\mathcal{C}_{all}$. Hence, there exists a modal formula φ such that for all frames $\mathcal{G}$, $\mathcal{G} \models D$ iff $\mathcal{G} \models \varphi$. For the sake of the contradiction, suppose $\mathcal{C}_{all} \not\models C$. Thus, there exists a frame $\mathcal{F}_0$ such that $\mathcal{F}_0 \not\models C$. Let g be an assignment on $\mathcal{F}_0$. Since $\mathcal{F}_0 \not\models C$, $\mathcal{F}_0, g \not\models C$. Let $\mathcal{F} = (W, R)$ be the frame defined by $W = \{s\}$ and $R = \emptyset$ where s is a new state. Let $\mathcal{F}'$ be the disjoint union of $\mathcal{F}_0$ and $\mathcal{F}$. Obviously, $\mathcal{F}_0$ is the relativized reduct of $\mathcal{F}'$ with respect to $y \neq x$, x and g_s^y. Consequently, by Proposition 5, $\mathcal{F}', g_s^y \models (C)_x^{y\neq x}$ iff $\mathcal{F}_0, g \models C$. Since $\mathcal{F}_0, g \not\models C$, $\mathcal{F}', g_s^y \not\models (C)_x^{y\neq x}$. Since $\mathcal{F}$ consists of a single state, $\mathcal{F} \not\models D$. Since $\mathcal{F}'$ is the disjoint union of $\mathcal{F}_0$ and $\mathcal{F}$, $\mathcal{F}', g_s^y \models \exists x\; y \neq x$. Since $\mathcal{F}', g_s^y \not\models (C)_x^{y\neq x}$, $\mathcal{F}', g \models D$. Hence, $\mathcal{F}' \models D$. Since for all frames $\mathcal{G}$, $\mathcal{G} \models D$ iff $\mathcal{G} \models \varphi$, $\mathcal{F}' \models \varphi$. Since $\mathcal{F}'$ is the disjoint union of $\mathcal{F}_0$ and $\mathcal{F}$, by Proposition 3, $\mathcal{F} \models \varphi$. Since φ is a modal definition of D with respect to $\mathcal{C}_{all}$, $\mathcal{F} \models D$: a contradiction.

This tight relationship between the problem of deciding the validity in $\mathcal{C}_{all}$ of sentences from the first-order language $\mathcal{L}_{\mathbf{FOF}}$ and the problem of deciding modal definability with respect to $\mathcal{C}_{all}$ constitutes the key result of our method. Notice that there are 2 modal-related constraints in the proof of Proposition 11. The 1st constraint is that the modal language contains a formula like $\bot$ which is valid in no frame. We have used this constraint at the beginning of the ($\Rightarrow$) part of the proof. The 2nd constraint is that the modal language does not contain modalities like the universal modality and the difference modality which prevent from using the Disjoint unions Theorem. We have used this constraint at the end of the ($\Leftarrow$) part of the proof. Now, we infer the following result:

Corollary 1 (Chagrova's Theorem about modal definability). *The problem of deciding modal definability with respect to $\mathcal{C}_{all}$ is undecidable.*

Proof. By Propositions 10 and 11.

5.2 Proofs of New Variants of Chagrova's Theorem About Modal Definability

Secondly, we give sketches of proofs of new variants of Chagrova's Theorem about modal definability. In the proof of Proposition 11, the unique occurrences of the subformulas $\exists x\ y \neq x$ and $\neg(C)_x^{y \neq x}$ in the sentence D associated to the given sentence C play specific roles. More precisely, in the ($\Rightarrow$) direction of the proof of Proposition 11, $\exists x\ y \neq x$ is used to show the existence of some relativized reduct $\mathcal{F}'$ of $\mathcal{F}$ whereas $\neg(C)_x^{y \neq x}$ is used to infer that C does not hold in $\mathcal{F}'$ by means of the Relativization Theorem between $\mathcal{F}$ and $\mathcal{F}'$. The truth is that in this direction of the proof of Proposition 11, the Relativization Theorem is used to infer some information about $\mathcal{F}'$, namely $\mathcal{F}', g \not\models C$, from some other information about $\mathcal{F}$, namely $\mathcal{F}, g_s^y \not\models (C)_x^{y \neq x}$. As for the ($\Leftarrow$) direction of the proof of Proposition 11, the Relativization Theorem is used to infer some information about $\mathcal{F}'$, namely $\mathcal{F}', g_s^y \not\models (C)_x^{y \neq x}$, from some other information about $\mathcal{F}_0$, namely $\mathcal{F}_0, g \not\models C$. This use of the Relativization Theorem is possible and leads to a contradiction with the assumption that D is modally definable with respect to $\mathcal{C}_{all}$ because $\mathcal{F}'$ has been constructed from $\mathcal{F}_0$ in such a way that

- $\mathcal{F}_0$ is the relativized reduct of $\mathcal{F}'$ with respect to appropriate syntactic and semantics elements,
- $\mathcal{F}'$ is the disjoint union of $\mathcal{F}_0$ and some other frame.

In [3], the above line of reasoning has been generalized to restricted classes of frames such as the class of all reflexive frames, the class of all symmetric frames, etc. The common property of these classes of frames is their *stability* where a class $\mathcal{C}$ of frames is *stable* if there exists a first-order formula A, there exists an individual variable x and there exists a sentence B such that

(a) for all frames $\mathcal{F}$ in $\mathcal{C}$, for all assignments g on $\mathcal{F}$ and for all frames $\mathcal{F}'$, if $\mathcal{F}'$ is the relativized reduct of $\mathcal{F}$ with respect to A, x and g then $\mathcal{F}'$ is in $\mathcal{C}$,
(b) for all frames $\mathcal{F}_0$ in $\mathcal{C}$, there exists frames $\mathcal{F}, \mathcal{F}'$ in $\mathcal{C}$ and there exists an assignment g on $\mathcal{F}$ such that $\mathcal{F}_0$ is the relativized reduct of $\mathcal{F}$ with respect to A, x and g, $\mathcal{F} \models B$, $\mathcal{F}' \not\models B$ and $\mathcal{F} \preceq \mathcal{F}'$.

In this case, (A, x, B) is a *witness of the stability of* $\mathcal{C}$. The following result proved in [3] states that if $\mathcal{C}$ is stable then the problem of deciding the modal definability of sentences with respect to $\mathcal{C}$ is at least as difficult as the problem of deciding the validity of sentences in $\mathcal{C}$.

Proposition 12. *If $\mathcal{C}$ is stable then the problem of deciding the validity of sentences from the first-order language $\mathcal{L}_{\mathbf{FOF}}$ in $\mathcal{C}$ is reducible to the problem of deciding the modal definability of sentences with respect to $\mathcal{C}$.*

As a result, if one wants to show that the problem of deciding the modal definability of sentences with respect to a class $\mathcal{C}$ of frames is undecidable, a possible strategy is the following:

- prove that the problem of deciding the validity of sentences from the first-order language $\mathcal{L}_{\mathbf{FOF}}$ in $\mathcal{C}$ is undecidable,

– find a first-order formula A, an individual variable x and a sentence B such that (A, x, B) is a witness of the stability of $\mathcal{C}$.

Obviously, if $\mathcal{C}$ is the class of all frames satisfying a universal elementary condition then $\mathcal{C}$ satisfies the condition (a) defining stability with respect to any first-order formula A, any individual variable x and any sentence B. It is quite easy to see why. Suppose $\mathcal{C}$ is the class of all frames satisfying a universal elementary condition. Let $\mathcal{F}$ be a frame in $\mathcal{C}$, g be an assignment on $\mathcal{F}$ and $\mathcal{F}'$ be a frame. Suppose $\mathcal{F}'$ is the relativized reduct of $\mathcal{F}$ with respect to A, x and g. This means that $\mathcal{F}'$ is the restriction of $\mathcal{F}$ to the set of all states s in $\mathcal{F}$ such that $\mathcal{F}, g^x_s \models A$. Since $\mathcal{C}$ is the class of all frames satisfying a universal elementary condition and $\mathcal{F}$ is in $\mathcal{C}$, $\mathcal{F}'$ is in $\mathcal{C}$. In other respect, if $\mathcal{C}$ is closed under taking disjoint unions, generated subframes and bounded morphic images then $\mathcal{C}$ satisfies the condition (b) defining stability with respect to the first-order formula $A := x_1 \neq x$, the individual variable x and the sentence $B := \exists y \exists z y \neq z$. It is quite easy to see why. Suppose $\mathcal{C}$ is closed under taking disjoint unions, generated subframes and bounded morphic images. Let $\mathcal{F}_0$ be a frame in $\mathcal{C}$. We have to consider 2 cases.

1st case: $\mathcal{F}_0$ is serial. Let $\mathcal{F}' = (W', R')$ be the frame defined by $W' = \{s'\}$ and $R' = \{(s', s')\}$ where s' is a new state. Since $\mathcal{F}_0$ is serial, obviously, $\mathcal{F}'$ is a bounded morphic image of $\mathcal{F}_0$. Since $\mathcal{C}$ is closed under taking bounded morphic images and $\mathcal{F}_0$ is in $\mathcal{C}$, $\mathcal{F}'$ is in $\mathcal{C}$. Let $\mathcal{F}$ be the disjoint union of $\mathcal{F}_0$ and $\mathcal{F}'$. Since $\mathcal{C}$ is closed under taking disjoint unions, $\mathcal{F}_0$ is in $\mathcal{C}$ and $\mathcal{F}'$ is in $\mathcal{C}$, $\mathcal{F}$ is in $\mathcal{C}$. Since $\mathcal{F}'$ consists of a single state, $\mathcal{F}' \not\models B$. Since $\mathcal{F}$ is the disjoint union of $\mathcal{F}_0$ and $\mathcal{F}'$, $\mathcal{F} \models B$. Let g be an assignment on $\mathcal{F}$ such that $g(x_1) = s'$. Obviously, $\mathcal{F}_0$ is the relativized reduct of $\mathcal{F}$ with respect to A, x and g. Finally, since $\mathcal{F}$ is the disjoint union of $\mathcal{F}_0$ and $\mathcal{F}'$, $\mathcal{F} \preceq \mathcal{F}'$.

2nd case: $\mathcal{F}_0$ is not serial. Let $\mathcal{F}' = (W', R')$ be the frame defined by $W' = \{s'\}$ and $R' = \emptyset$ where s' is a new state. Since $\mathcal{F}_0$ is not serial, obviously, $\mathcal{F}'$ is isomorphic to a generated subframe of $\mathcal{F}_0$. Since $\mathcal{C}$ is closed under taking generated subframes and $\mathcal{F}_0$ is in $\mathcal{C}$, $\mathcal{F}'$ is in $\mathcal{C}$. Let $\mathcal{F}$ be the disjoint union of $\mathcal{F}_0$ and $\mathcal{F}'$. Since $\mathcal{C}$ is closed under taking disjoint unions, $\mathcal{F}_0$ is in $\mathcal{C}$ and $\mathcal{F}'$ is in $\mathcal{C}$, $\mathcal{F}$ is in $\mathcal{C}$. Since $\mathcal{F}'$ consists of a single state, $\mathcal{F}' \not\models B$. Since $\mathcal{F}$ is the disjoint union of $\mathcal{F}_0$ and $\mathcal{F}'$, $\mathcal{F} \models B$. Let g be an assignment on $\mathcal{F}$ such that $g(x_1) = s'$. Obviously, $\mathcal{F}_0$ is the relativized reduct of $\mathcal{F}$ with respect to A, x and g. Finally, since $\mathcal{F}$ is the disjoint union of $\mathcal{F}_0$ and $\mathcal{F}'$, $\mathcal{F} \preceq \mathcal{F}'$.

The above remarks immediately show that $\mathcal{C}_{all}$ is stable. The truth is that

Proposition 13. *The following classes of frames are stable as well:* $\mathcal{C}_{\mathrm{E}}$, *the class of all reflexive frames, the class of all transitive frames, the class of all reflexive transitive frames, the class of all strict partial orders, the class of all partial orders, the class of all lattices, the class of all symmetric frames and the class of all reflexive symmetric frames.*

Proof. See [1,3] for details.

Gathering results from [13,23–25,29], one can prove that

Proposition 14. *The validity of sentences from the first-order language* $\mathcal{L}_{\mathrm{FOF}}$ *is undecidable in each of the following classes of frames:* $\mathcal{C}_{\mathrm{E}}$, *the class of all reflexive frames, the class of all transitive frames, the class of all reflexive transitive frames, the class of* *all strict partial orders, the class of all partial orders, the class of all lattices, the class of all symmetric frames and the class of all reflexive symmetric frames.*

Proof. See [1,3] for details.

As a corollary, one obtain the following variants of Chagrova's Theorem about modal definability.

Corollary 2 (Variants of Chagrova's Theorem about modal definability). *The problem of deciding modal definability with respect to the following classes of frames is undecidable: $\mathcal{C}_E$, the class of all reflexive frames, the class of all transitive frames, the class of all reflexive transitive frames, the class of all strict partial orders, the class of all partial orders, the class of all lattices, the class of all symmetric frames and the class of all reflexive symmetric frames.*

Proof. By Propositions 13 and 14.

6 Conclusion

The core of this paper has been Chagrova's Theorems about first-order definability of given modal formulas and modal definability of given elementary conditions. We have analyzed Chagrova's Theorems and we have tried to understand why their proofs cannot be easily repeated for proving the undecidability of first-order definability and modal definability with respect to restricted classes of frames. We have considered classes of frames for which modal definability is decidable, for instance $\mathcal{C}_{par}$, $\mathcal{C}_{tE}$ and $\mathcal{C}_{stE}$. We have considered classes of frames for which first-order definability is trivial, for instance $\mathcal{C}_{par}$, $\mathcal{C}_{tE}$ and $\mathcal{C}_{stE}$, but also $\mathcal{C}^{\omega}_{rtc}$. Using standard methods in model theory such as relativization of first-order formulas and reduct of frames, we have given a new proof of Chagrova's Theorem about modal definability and we have given sketches of proofs of new variants of Chagrova's Theorem about modal definability. Much remains to be done.

An obvious question is whether there exists other classes of frames for which modal definability is decidable. Is modal definability with respect to $\mathcal{C}_{rtc}$ decidable? What about first-order definability with respect to $\mathcal{C}_{rtc}$? Another question is whether there exists other classes of frames for which first-order definability is trivial. It is also of interest to consider restrictions or extensions of the ordinary language of modal logic. For example, one can consider the implication restriction of $\mathcal{L}_{\mathbf{MF}}$ based on the connectives $\rightarrow$ and $\Box$ or the tense extension of $\mathcal{L}_{\mathbf{MF}}$ based on the Boolean connectives and the modal connectives $\Box$ and $\Box^{-1}$. For such restrictions or extensions of $\mathcal{L}_{\mathbf{MF}}$, what is the computability of first-order definability and modal definability? And in the end, there is the question whether there exists classes of frames for which modal definability is decidable and first-order definability is undecidable.

Acknowledgement. We are indebted to the participants of TbiLLC 2019 for their helpful comments. We also make a point of thanking the referees for their feedback: their useful suggestions have been essential for improving the correctness and the readability of a preliminary version of this paper.

Funding. The preparation of this paper has been supported by *Bulgarian Science Fund* (Project *DN02/15/19.12.2016*).

References

1. Balbiani, P., Georgiev, D., Tinchev, T.: Modal correspondence theory in the class of all Euclidean frames. J. Log. Comput. **28**, 119–131 (2018)
2. Balbiani, P., Tinchev, T.: Definability over the class of all partitions. J. Log. Comput. **16**, 541–557 (2006)
3. Balbiani, P., Tinchev, T.: Undecidable problems for modal definability. J. Log. Comput. **27**, 901–920 (2017)
4. Van Benthem, J.: A note on modal formulas and relational properties. J. Symb. Log. **40**, 85–88 (1975)
5. Van Benthem, J.: Modal Logic and Classical Logic. Bibliopolis (1983)
6. Blackburn, P., de Rijke, M., Venema, Y.: Modal Logic. Cambridge University Press, Cambridge (2001)
7. Chagrov, A., Chagrova, L.: Algorithmic problems concerning first-order definability of modal formulas on the class of all finite frames. Stud. Logica. **55**, 421–448 (1995). https://doi.org/10.1007/BF01057806
8. Chagrov, A., Chagrova, L.: The truth about algorithmic problems in correspondence theory. In: Advances in Modal Logic, vol. 6, pp. 121–138. College Publications (2006)
9. Chagrov, A., Chagrova, L.: Demise of the algorithmic agenda in the correspondence theory? Log. Invest. **13**, 224–248 (2007)
10. Chagrov, A., Zakharyaschev, M.: Modal Logic. Oxford University Press, Oxford (1997)
11. Chagrova, L.: On the problem of definability of propositional formulas of intuitionistic logic by formulas of classical first-order logic. Doctoral thesis of the University of Kalinin (1989)
12. Chagrova, L.: An undecidable problem in correspondence theory. J. Symb. Log. **56**, 1261–1272 (1991)
13. Church, A., Quine, W.: Some theorems on definability and decidability. J. Symb. Log. **17**, 179–187 (1952)
14. Ebbinghaus, H.-D., Flum, J.: Finite Model Theory. Springer, Heidelberg (1995). https://doi.org/10.1007/3-540-28788-4
15. Georgiev, D.: Definability in the class of all **KD45**-frames–computability and complexity. J. Appl. Non-Class. Log. **27**, 1–26 (2017)
16. Georgiev, D.: Algorithmic methods for non-classical logics. Doctoral thesis of Sofia University St. Kliment Ohridski (2017)
17. Goldblatt, R.: First-order definability in modal logic. J. Symb. Log. **40**, 35–40 (1975)
18. Goranko, V., Vakarelov, D.: Elementary canonical formulæ: extending Sahlqvist's theorem. Ann. Pure Appl. Log. **141**, 180–217 (2006)
19. Hodges, W.: Model Theory. Cambridge University Press, Cambridge (1993)
20. Kalmár, L.: Zurückführung des Entscheidungsproblems auf den Fall von Formeln mit einer einzigen, binären, Funktionsvariablen. Compositio Math. **4**, 137–144 (1937)
21. Kracht, M.: Tools and Techniques in Modal Logic. Elsevier, Amsterdam (1999)
22. Kripke, S.: Semantical analysis of modal logic. I. Normal modal propositional calculi. Zeitschrift für mathematische Logik und Grundlagen der Mathematik **9**, 67–96 (1963)
23. Rogers, H.: Some results on definability and decidability in elementary theories. Doctoral thesis of the University of Princeton (1952)
24. Rogers, H.: Certain logical reduction and decision problems. Ann. Math. **64**, 264–284 (1956)

25. Rogers, H.: Theory of Recursive Functions and Effective Computability. McGraw-Hill (1967)
26. Sahlqvist, H.: Completeness and correspondence in the first and second order semantics for modal logic. In: Proceedings of the Third Scandinavian Logic Symposium, pp. 110–143. North-Holland (1975)
27. Stockmeyer, L.: The complexity of decision problems in automata theory. Doctoral thesis of the Massachusetts Institute of Technology (1974)
28. Stockmeyer, L.: The polynomial-time hierarchy. Theor. Comput. Sci. **3**, 1–22 (1977)
29. Tarski, A.: Undecidability of the theories of lattices and projective geometries. J. Symb. Log. **14**, 77–78 (1949)
30. Vardi, M.: The complexity of relational query languages. In: Proceedings of the 14th ACM Symposium on Theory of Computing, pp. 137–146. ACM (1982)

Topological Evidence Logics: Multi-agent Setting

Alexandru Baltag[1], Nick Bezhanishvili[1], and Saúl Fernández González[2](✉)

[1] ILLC, Universiteit van Amsterdam, Amsterdam, The Netherlands
[2] IRIT, Université de Toulouse, Toulouse, France
saul.fdez.glez@gmail.com

Abstract. We introduce a multi-agent topological semantics for evidence-based belief and knowledge, which extends the *dense interior semantics* developed in [2]. We provide the complete logic of this multi-agent framework together with *generic models* for a fragment of the language. We also define a new notion of group knowledge which differs conceptually from previous approaches.

1 Introduction

A semantic study of epistemic logics, the family of modal logics concerned with what an epistemic agent *believes* or *knows*, has been mostly conducted in the framework of relational structures (Kripke frames) [13]. These are sets of possible worlds connected by (epistemic or doxastic) accessibility relations. Knowledge (K) and belief (B) are thus modal operators which are interpreted via standard possible worlds semantics.

It is claimed in [13] that the accessibility relation for knowledge must be (minimally) reflexive and transitive. On the syntactic level, this demand translates into the fact that any logic for knowledge based on these frames must contain the axioms of S4. This, paired with the fact, famously proven by McKinsey and Tarski [14], that S4 is the logic of topological spaces under the *interior semantics* (see [4]), lays the ground for a topological treatment of knowledge. Moreover, McKinsey and Tarski [14] proved that certain *generic* spaces, such as the real line, spaces which intuitively lend themselves to be models for certain situations of knowledge, have S4 as their logic.

The semantics outlined in [14] treats the "knowledge" modality as the interior operator, which, if one thinks of the open sets as "pieces of evidence", adds an evidential dimension to the notion of knowledge that one could not get within Kripke frames (see [16] for lengthy discussion on this topic).

Under this interpretation, knowing a proposition amounts to having evidence for it. This can be an undesirable property, for it constitutes, arguably, an overly simplistic account of what knowledge is. As Gettier [11] argues, there is more to knowledge than 'true and justified belief'. Depending on the properties one ascribes to knowledge, belief and the relation thereof, one can get different epistemic logics, each with their axioms and rules. For certain applications, one

© The Author(s), under exclusive license to Springer Nature Switzerland AG 2022
A. Özgün and Y. Zinova (Eds.): TbiLLC 2019, LNCS 13206, pp. 237–257, 2022.
https://doi.org/10.1007/978-3-030-98479-3_12

would want, for instance, to operate within a framework in which misleading true evidence can lead to false beliefs (see [16,18] for more in-depth discussion).

Inspired by [18] and [6] a new topological semantics was introduced in [2] and explored in depth in [16]. This semantics allows one to talk about knowledge and belief, evidence (both "basic" and "combined") and a notion of *justification* via the dense-interior operator. Also an epistemic logic complete with respect to the proposed semantics has been given in [2] and [16]. The models for this logic based on the dense-interior semantics in topological spaces are called *topo-e-models.*

In [1] an analogue of the McKinsey-Tarski theorem was proved for the dense-interior semantics: the logic of topological evidence models is sound and complete with respect to any individual topological space (X, τ) which is *dense-in-itself, metrizable*, and homeomorphic to the disjoint union $(X, \tau) \sqcup (X, \tau)$.

The framework defined in [2] is single-agent. In this paper, we introduce a multi-agent topological evidence semantics which generalises the single-agent case and differs substantially from prior approaches. In this sense, we provide several logics of multi-agent models and give some conceptual and theoretical contributions for a notion of group knowledge in this framework.

Outline. In Sect. 2 we present the (one-agent) notion of topological evidence models introduced in [2] together with some relevant results. In Sect. 3 we introduce and justify our multi-agent setting, we show how it generalises the single-agent case and we provide the logic for several fragments of the language. In Sect. 4, we obtain "generic models", i.e., unique topological spaces whose logic under the semantics previously introduced is exactly the logic of all topological spaces. Section 5 discusses a notion of *group knowledge* in this setting, and gives a sound a complete logic of distributed knowledge. We conclude in Sect. 6.[1]

2 Single-Agent Topological Evidence Models

The relation between belief and knowledge has historically been one of the main focuses of epistemology. One would want to have a formal system that accounts for knowledge and belief together, which requires careful consideration regarding the way in which they interact. Canonically, knowledge has been thought of as "true, justified belief". However, Gettier's counterexamples of cases of true, justified belief which do not amount to knowledge shattered this paradigm [11].

Stalnaker [18] argues that a relational semantics is insufficient to capture Gettier's considerations in [11] and, trying to stay close to most of the intuitions of Hintikka in [13], provides an axiomatisation for a system of knowledge and belief in which knowledge is an S4.2 modality, belief is a KD45 modality and the following formulas can be proven: $B\phi \leftrightarrow \neg K \neg K\phi$ and $B\phi \leftrightarrow BK\phi$. "Believing p" is the same as "not knowing you don't know p" and belief becomes "subjective certainty", in the sense that the agent cannot distinguish whether she believes or knows p, and believing amounts to believing that one knows.

[1] This paper is based on Saúl Fernández González's Master's thesis [10].

A topological semantics in which knowledge is simply the interior modality (i.e., evaluating formulas on a topological space and setting $\|K\phi\| = \mathrm{Int}\,\|\phi\|$) proves insufficient to capture these nuances. In [2] a new semantics is introduced, building on the idea of *evidence models* of [6] which exploits the notion of evidence-based knowledge allowing to account for notions as diverse as *basic evidence* versus *combined evidence*, *factual*, *misleading* and *nonmisleading evidence*, etc. It is a semantics whose logic maintains a Stalnakerian spirit with regards to the relation between knowledge and belief, which behaves well dynamically and which does not confine us to work with "strange" classes of spaces.

This is the *dense-interior semantics*, defined on *topological evidence models.*

2.1 The Logic of Topological Evidence Models

We briefly present here the framework introduced in [2], see also [16]. Our language is now $\mathcal{L}_{\forall KB\Box\Box_0}$, which includes the modalities K (knowledge), B (belief), $[\forall]$ (infallible knowledge), $\Box_0$ (basic evidence), $\Box$ (combined evidence).

Definition 2.1 (The dense interior semantics). *We interpret sentences on* topological evidence models *(i.e. tuples (X, τ, E_0, V) where (X, τ, V) is a topological model and E_0 is a subbasis of τ) as follows: $x \in [\![K\phi]\!]$ iff $x \in \mathrm{Int}[\![\phi]\!]$ and* $\mathrm{Int}[\![\phi]\!]$ *is* dense[2]*; $x \in [\![B\phi]\!]$ iff $\mathrm{Int}[\![\phi]\!]$ is dense; $x \in [\![[\forall]\phi]\!]$ iff $[\![\phi]\!] = X$; $x \in [\![\Box_0\phi]\!]$ iff there is $e \in E_0$ with $x \in e \subseteq [\![\phi]\!]$; $x \in [\![\Box\phi]\!]$ iff $x \in \mathrm{Int}[\![\phi]\!]$. Validity is defined in the standard way.*

We see that "knowing" does not equate "having evidence" in this framework, but it is rather something stronger: in order for the agent to know P, she needs to have a piece of evidence for P which is *dense*, i.e., which has nonempty intersection with (and thus cannot be contradicted by) any other piece of evidence.

Fragments of the Logic. The following logics are obtained by considering certain fragments of the language (i.e. certain subsets of the modalities above)[3].

"K-only", $\mathcal{L}_K$	S4.2.
"Knowledge", $\mathcal{L}_{\forall K}$	S5 axioms and rules for $[\forall]$, plus S4.2 for K, plus $[\forall]\phi \to K\phi$ and $\neg[\forall]\neg K\phi \to [\forall]\neg K\neg\phi$.
"Combined evidence", $\mathcal{L}_{\forall\Box}$	S5 for $[\forall]$, S4 for $\Box$, plus $[\forall]\phi \to \Box\phi$.
"Evidence", $\mathcal{L}_{\forall\Box\Box_0}$	S5 for $[\forall]$, S4 for $\Box$, plus the axioms $\Box_0\phi \to \Box_0\Box_0\phi$, $[\forall]\phi \to \Box_0\phi$, $\Box_0\phi \to \Box\phi$, $(\Box_0\phi \wedge [\forall]\psi) \to \Box_0(\phi \wedge [\forall]\psi)$.

[2] A set $U \subseteq X$ is dense whenever $\mathrm{Cl}\,U = X$ or equivalently whenever $U \cap V \neq \varnothing$ for all nonempty open sets V.

[3] We recall that S4 is the least normal modal logic containing the axioms (T) $\Box\phi \to \phi$ and (4) $\Box\phi \to \Box\Box\phi$; that S5 is S4 plus the axiom (5) $\neg\Box\phi \to \Box\neg\Box\phi$, and that S4.2 is S4 plus the axiom (.2) $\Diamond\Box\phi \to \Box\Diamond\phi$.

We will refer to these logics respectively as $\mathsf{S4.2}_K$, $\mathsf{Logic}_{\forall K}$, $\mathsf{Logic}_{\forall\Box}$ and $\mathsf{Logic}_{\forall\Box\Box_0}$. K and B are definable in the evidence fragments[4], thus we can think of the logic of $\mathcal{L}_{\forall\Box\Box_0}$ as the "full logic".

3 Going Multi-agent

There have been different approaches to a multi-agent logic derived from the framework introduced in [2]. In [17], a two-agent logic with distributed knowledge was defined. However, the semantics of this approach seems to come with some conceptual problems which were discussed in [10]. Another approach, present in [16], generalises the one-agent case and is devoid of the aforementioned conceptual issues, yet it uses the semantics of subset space logic: sentences are evaluated at a pair (x, U) where x is a world and U is some neighbourhood of x.

The system introduced in the present section and expanded upon in the subsequent ones generalises the one-agent models while maintaining the underlying ideas to the single-agent case, where sentences are evaluated at worlds. We will limit ourselves to two agents for simplicity in the exposition. Extending these results to any finite number of agents is straightforward.

The Problem of Density. A first idea when attempting to incorporate a second epistemic agent would be to simply add a second topology to the single-agent framework and read things in the same way. That is, we could interpret sentences on bitopological spaces (X, τ_1, τ_2) where τ_1 and τ_2 are topologies defined on X, and we say, for $i = 1, 2$, that $x \in K_i\phi$ if and only there is a set $U \in \tau_i$ which is dense in τ_i such that $x \in U \subseteq \|\phi\|$. However, this approach is highly problematic because it requires the extra assumptions that the same set of worlds is epistemically accessible for both agents, and thus conflates infallible knowledge. This is discussed in more depth in [10]. Our proposal to eliminate these complications involves making explicit which worlds are compatible with an agent's information at world x. This is done via the use of partitions.

3.1 Topological-Partitional Models

In order to specify which worlds an agent considers possible, we can define the topologies which encode the evidence of the agents on a common space X, but we restrict, for each agent and at each world $x \in X$, the set of worlds epistemically accessible to the agent at x. We can still speak about density, but *locally.* A straightforward way to this is through the use of partitions.

Definition 3.1. *A* topological-partitional model *is a tuple*

$$\mathfrak{M} = (X, \tau_1, \tau_2, \Pi_1, \Pi_2, V)$$

where V is a valuation, τ_i is a topology defined on X and Π_i is a partition of X with the property that $\Pi_i \subseteq \tau_i$.

[4] $K\phi \equiv \Box\phi \wedge [\forall]\Box\Diamond\phi$ and $B\phi \equiv \neg K\neg K\phi$.

The worlds which are compatible with agent i's information at $x \in X$ are now precisely the worlds in the unique cell of the partition Π_i which includes x. The concept of justification comes now in the form of a local notion of density:

Definition 3.2. *For $x \in X$, let $\Pi_i(x)$ be the unique $\pi \in \Pi_i$ with $x \in \pi$. For $U \subseteq X$, let $\Pi_i[U] = \{\pi \in \Pi_i : \pi \cap U \neq \varnothing\} = \{\Pi_i(x) : x \in U\}$.*

A set $U \subseteq X$ is locally dense in $\pi \in \Pi_i$ *whenever $\pi \subseteq \mathrm{Cl}_{\tau_i} U$ or equivalently when every nonempty open set contained in π has nonempty intersection with U. We will say that a nonempty set U is* locally dense in Π_i *(or simply* locally dense *if there is no ambiguity) if $\mathrm{Cl}_{\tau_i} U = \bigcup \Pi_i[U]$. Equivalently, U is locally dense if it is locally dense in π for every $\pi \in \Pi_i[U]$.*

With this we can define a semantics for two-agent knowledge:

Definition 3.3 (Two-agent locally-dense-interior semantics). *Let*

$$\mathfrak{M} = (X, \tau_1, \tau_2, \Pi_1, \Pi_2, V)$$

be a topological-partitional model and let $x \in X$. As usual, we have $\|p\| = V(p)$, $\|\phi \wedge \psi\| = \|\phi\| \cap \|\psi\|$ and $\|\neg\phi\| = X \backslash \|\phi\|$. For $i = 1, 2$ set:

$$\begin{aligned} \mathfrak{M}, x \vDash K_i\phi \textit{ iff } & x \in \mathrm{Int}_{\tau_i} \|\phi\| \\ & \& \ \mathrm{Int}_{\tau_i} \|\phi\| \textit{is locally dense in} \, \Pi_i(x). \end{aligned}$$

Consider a topological-partitional model $(X, \tau_1, \tau_2, \Pi_1, \Pi_2, V)$ and set

$$\tau_i^* := \{U \in \tau_i : U \text{ is } \Pi_i - \text{locally dense}\} \cup \{\varnothing\}.$$

It is straightforward to check that the following holds:

Lemma 3.4. *(X, τ_1^*, τ_2^*) is an extremally disconnected bitopological space and the locally-dense-interior semantics on $(X, \tau_1, \tau_2, \Pi_1, \Pi_2, V)$ coincides with the interior semantics on $(X, \tau_1^*, \tau_2^*, V)$.*

In particular, given a topological-partitional model $(X, \tau_{1,2}, \Pi_{1,2}, V)$ in which every τ_i-open set is Π_i-locally dense, the locally-dense-interior semantics and the interior semantics coincide.

One last remark before proceeding with the main results: at first glance demanding each element $\pi \in \Pi_i$ to be open may seem as a very strong condition. For example, a connected space such as $\mathbb{R}$ does not admit any such partition other than the trivial one $\Pi_i = \{\mathbb{R}\}$. We could instead do the following:

i. Define topological-partitional models to have arbitrary partitions;
ii. Define $U \subseteq X$ to be locally dense at $\pi \in \Pi_i$ whenever $U \cap \pi$ is dense in the subspace topology $\tau_i|_\pi$;
iii. Set $x \in \|K_i\phi\|$ if and only if there exists $U \in \tau_i$ locally dense in $\Pi_i(x)$ with $x \in U \cap \Pi_i(x) \subseteq \|\phi\|$.

As it turns out, these models can be turned in a truth-preserving manner into topological-partitional models of the kind defined above. Indeed, let $\bar{\tau}_i$ be the topology generated by $\{U \cap \pi : U \in \tau_i, \pi \in \Pi_i\}$. Then clearly $\Pi_i \subseteq \bar{\tau}_i$ and it is a straightforward check that $(X, \tau_i, \Pi_i), x \vDash \phi$ under this semantics if and only if $(X, \bar{\tau}_i, \Pi_i), x \vDash \phi$ under the semantics in Definition 3.3.

For this reason, we will limit ourselves to the study of models with open partitions. Let us now look at an example:

Example 3.5. We have four possible worlds, $X = \{x_{11}, x_{01}, x_{10}, x_{00}\}$ and two agents, Alice and Bob, represented by a and b. Let us consider two propositions, p and q. Let $V(p) = P = \{x_{11}, x_{10}\}$ and $V(q) = \{x_{11}, x_{01}\}$. The actual world is x_{11}, in which p and q hold.

At q-worlds Alice only considers q-worlds possible, and at $\neg q$-worlds, she only considers $\neg q$-worlds possible. In addition to this, at p-worlds she has fallible evidence that p. At $\neg p$-worlds she does not receive this evidence.

The only worlds consistent with Bob's information are those in which $q \rightarrow p$ holds. Moreover, in p-worlds he has fallible evidence for p and in $\neg p$-worlds he has it for $\neg p$.

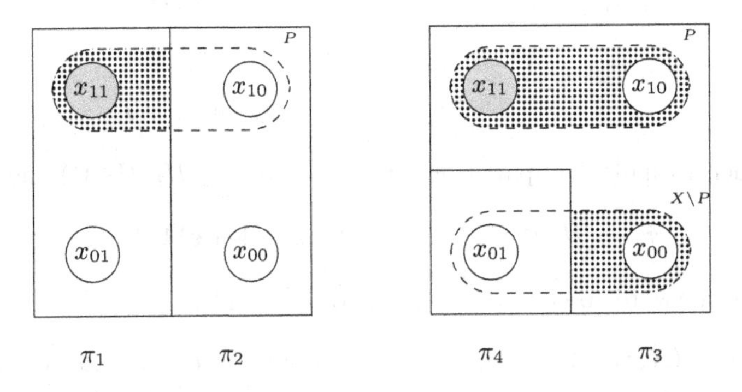

Fig. 1. The topology and partition of Alice (left) and Bob (right). The dotted areas are the proper open subsets of the cell of each partition which includes the actual world. We can see that x_{11} is in a π_1-locally dense open set contained in P but not in a π_3-locally dense one.

Let $\pi_1 = \{x_{11}, x_{01}\}$, $\pi_2 = \{x_{01}, x_{00}\}$, $\pi_3 = \{x_{11}, x_{10}, x_{00}\}$, $\pi_4 = \{x_{01}\}$. Alice's and Bob's partitions are respectively $\Pi_a = \{\pi_1, \pi_2\}$ and $\Pi_b = \{\pi_3, \pi_4\}$. Their topologies τ_a and τ_b are generated respectively by $\{\pi_1, \pi_2, P\}$ and $\{\pi_3, \pi_4, P, X \backslash P\}$ (see Fig. 1).

At the actual world x_{11}, Alice knows p yet Bob does not: indeed, $\{x_{11}\}$ is a τ_a-open set, locally dense in π_1 and contained in P, thus $K_a p$ holds. And any τ_b-open set contained in P is not locally dense, because it has empty intersection with the open set $\{x_{00}\}$, thus $\neg K_b p$ holds at x_{11}.

Certain topological spaces come equipped with open partitions, in the form of their *connected components*.

Definition 3.6. *Let* (X, τ) *be a topological space. A set* $U \subseteq X$ *is said to be* connected *if it does not contain a proper clopen subset.*

A connected component *of* (X, τ) *is a maximal connected subset of* X.

The following result can be found in any topology textbook (see e.g. [15]):

Lemma 3.7. *The connected components of* (X, τ) *coincide with the equivalence classes of the relation:* $x \sim y$ *if and only if there is a connected subset of* X *containing* x *and* y.

The following lemma, whose proof is straightforward, shows that the connected components of an Alexandroff space are always open:

Lemma 3.8. *Let* $(W, \leq)$ *be a preordered set. Then:*

i. *The connected components on* $(W, \mathrm{Up}(W))$ *are open and they coincide with the equivalence classes under the reflexive, transitive and symmetric closure of* $\leq$, *i.e. the following equivalence relation:* $x \sim y$ *if and only if there exist* $x_0, ..., x_n \in X$ *with* $x_0 = x, x_n = y$ *and* $x_k \leq x_{k+1}$ *or* $x_k \geq x_{k+1}$ *for* $0 \leq k \leq n-1$.
ii. *If* $(W, \leq)$ *is an* S4.2 *frame (i.e. if* $\leq$ *is a weakly directed preorder) we have:* $x \sim y$ *if and only if there exists some* $z \in W$ *such that* $x \leq z \geq y$.
iii. *If* $(W, \leq)$ *is a forest (i.e. if* $\leq$ *is the reflexive and transitive closure of some relation* $\prec$ *such that every element has at most one* $\prec$*-predecessor), then* $x \sim y$ *if and only if there exists some* $z \in W$ *such that* $x \geq z \leq y$.

Proof. (i). Let us see that $[x]_\sim$ is clopen and connected. Clearly it is both upward and downward closed. Moreover, if $\varnothing \neq U \subseteq [x]_\sim$ is a clopen set, take $y \in U$ and $z \in [x]_\sim$. Since there is a path of $\leq$ and $\geq$ from y to z and U is both an upset and a downset, we have that $z \in U$, thus $[x]_\sim$ is connected.

(ii). Take a path $(x_0 = x, x_1, ..., x_n = y)$ such that $x_k \leq x_{k+1}$ or $x_{k+1} \leq x_k$ for all $0 \leq k \leq n-1$, and note that $x_{k-1} \geq x_k \leq x_{k+1}$ implies that there exists a certain x'_k such that $x_{k-1} \leq x'_k \geq x_{k+1}$. Applying this successively we reach a chain $x = x'_0 \leq ... \leq x'_k \geq ... \geq x'_n = y$.

(iii). Similar to (ii.), noting that $x_{k-1} \prec x_k \succ x_{k+1}$ implies $x_{k-1} = x_{k+1}$.

For $x \in W$, we shall denote $\uparrow x := \{z \in W : x \leq z\}$. Note that item (ii) entails that each upset in a directed preorder is $\sim$-locally dense. Indeed, take x and y in the same equivalence class. Item (ii) gives us that $\uparrow x \cap \uparrow y \neq \varnothing$, thus every pair of nonempty upsets contained in the same connected component has nonempty intersection.

This fact plus the last item in Lemma 3.4 have an immediate consequence:

Corollary 3.9. *Let* $(X, \leq_1, \leq_2, \sim_1, \sim_2, V)$ *be a model in which each* $\leq_i$ *is a weakly directed preorder and* $\sim_i$ *is the equivalence relation given by:* $x \sim_i y$ *if and only if there exists* $z \in X$ *such that* $x \leq_i z \geq_i y$. *Then the locally-dense-interior semantics on this model coincide with the Kripke semantics on* $(X, \leq_1, \leq_2, V)$.

As an immediate consequence of this, plus the fact that $\mathsf{S4.2}_{K_1} + \mathsf{S4.2}_{K_2}$ is the logic of frames $(W, \leq_1, \leq_2)$ where each $\leq_i$ is a weakly directed preorder, we have:

Theorem 3.10. *$\mathsf{S4.2}_{K_1} + \mathsf{S4.2}_{K_2}$ is the logic of topological-partitional models for two agents.*

3.2 Other Fragments

Let us now consider other fragments of the logic. For this we add to our language the *infallible knowledge modalities* $[\forall]_i$, the *evidence modalities* $\Box_i$, and the *belief modalities* B_i, for $i = 1, 2$, and their respective duals $[\exists]_i$, $\Diamond_i$ and $\hat{B}_i$. We interpret these on topological-paritional models $(X, \tau_{1,2}, \Pi_{1,2}, V)$ as follows:

$$x \in \|[\forall]_i\phi\| \text{ iff } \Pi_i(x) \subseteq \|\phi\|;$$
$$x \in \|\Box_i\phi\| \text{ iff } x \in \mathrm{Int}_{\tau_i} \|\phi\|;$$
$$x \in \|B_i\phi\| \text{ iff } \mathrm{Int}_{\tau_i} \|\phi\| \text{ is locally dense in } \Pi_i(x).$$

Analogously to the one-agent case, we can check that the following equalities hold: $\|K_i\phi\| = \|\Box_i\phi \wedge [\forall]_i\Diamond_i\Box_i\phi\|$; $\|B_i\phi\| = \|\hat{K}_iK_i\phi\|$.

Much like in the one-agent framework, we are interested in looking at fragments of this logic. We will focus on the *knowledge fragment* $\mathcal{L}_{K_i\forall_i}$, the *knowledge-belief fragment* $\mathcal{L}_{K_iB_i}$, and the *factive evidence fragment* $\mathcal{L}_{\Box_i\forall_i}$.

The factive evidence fragment $\mathcal{L}_{\Box_i\forall_i}$. The logic for this fragment is $\mathsf{Logic}_{\Box_i\forall_i}$, which is the least normal modal logic which includes

- the axioms and rules of $\mathsf{S4}$ for $\Box_i$;
- the axioms and rules of $\mathsf{S5}$ for $[\forall_i]$;
- the axiom $[\forall_i]\phi \to \Box_i\phi$ for $i = 1, 2$.

Soundness for topological-partitional models is a rather simple check: the $\mathsf{S4}$ rules for the topological interior hold, for $\mathrm{Int}\, P \subseteq P \cap \mathrm{Int}\,\mathrm{Int}\, P$ and so do the $\mathsf{S5}$ rules for $[\forall]_i$, which are defined via equivalence relations. The fact that each equivalence class is open takes care of the axiom $[\forall]_i\phi \to \Box_i\phi$.

For completeness, we can use the Sahlqvist completeness theorem (see [9]) and note that the axioms of $\mathsf{Logic}_{\Box_i\forall_i}$ are Sahlqvist formulas and thus canonical and the canonical Kripke model for this logic is of the shape $(X, \leq_1, \leq_2, \sim_1, \sim_2)$, where each $\leq_i$ is a preorder (due to the $\mathsf{S4}$ axioms) and each $\sim_i$ constitutes an equivalence relation (due to the $\mathsf{S5}$ axioms). Moreover, the axiom $[\forall_i]\phi \to \Box_i\phi$ grants us that $x \leq_i y$ implies $x \sim_i y$ and thus that the $\sim_i$-equivalence classes are $\leq_i$-open sets. In other words, this canonical model is a topological-partitional model.

Therefore if $\phi \notin \mathsf{Logic}_{\Box_i\forall_i}$, then ϕ will be refuted in the canonical model, whence we have a topological-partitional model refuting it. And thus, we have completeness. $\Box$

The Knowledge Fragment $\mathcal{L}_{K_i\forall_i}$. The logic of the fragment with all the knowledge modalities, $K_1, K_2, [\forall]_1$ and $[\forall]_2$ is $\mathsf{Logic}_{K_i\forall_i}$, the least logic including the axioms and rules of S4 for each K_i, S5 for each $[\forall]_i$ plus the following axioms for $i = 1, 2$:

$$\text{(A) } [\forall]_i\phi \to K_i\phi;$$
$$\text{(B) } [\exists]_i K_i\phi \to [\forall]_i\hat{K}_i\phi.$$

Note that the .2 axiom for K_i is derivable from (A) and (B).

Soundness is a routine check, whereas for completeness we can again resort to the Sahlqvist theorem. The canonical model is of the shape $(X, \leq_1, \leq_2, \sim_1, \sim_2)$ where each $\leq_i$ is a weakly directed preorder and each $\sim_i$ is an equivalence relation. Moreover the Sahlqvist first order correspondent of axiom (A) gives us that $x \leq_i y$ implies $x \sim_i y$ and axiom (B) tells us that, if $x \sim_i y$, then there exists some z such that $x \leq_i z \geq_i y$. These two facts, together with item (ii) of Lemma 3.8, imply that the $\sim_i$-equivalence classes are exactly the $\leq_i$-connected components. And thus the Kripke semantics on this model coincide with the locally-dense-interior semantics on the topological-partitional model $(X, \tau_1, \tau_2, \Pi_1, \Pi_2)$ where $\tau_i = \mathrm{Up}\leq_i(X)$ and Π_i are the $\leq_i$-connected components. Completeness follows. □

The Knowledge-Belief Fragment $\mathcal{L}_{K_iB_i}$. The logic of the knowledge-belief fragment is $\mathsf{Stal}_1 + \mathsf{Stal}_2$ the least normal modal logic including the S4 axioms and rules for K_i plus the following axioms, for $i = 1, 2$:

(PI_i) $B_i\phi \to K_iB_i\phi$;	(NI_i) $\neg B_i\phi \to K_i\neg B_i\phi$;
(KB_i) $K_i\phi \to B_i\phi$;	(CB_i) $B_i\phi \to \neg B_i\neg\phi$;
(FB_i) $B_i\phi \to B_iK_i\phi$.	

We have that $\mathsf{S4.2}_{K_1} + \mathsf{S4.2}_{K_2} \cup \{B_i\phi \leftrightarrow \hat{K}_iK_i\phi : \phi \in \mathcal{L}_{K_iB_i}\} \subseteq \mathsf{Stal}_1 + \mathsf{Stal}_2$ and thus, if a formula ϕ in the language $\mathcal{L}_{K_iB_i}$ is not provable in $\mathsf{Stal}_1 + \mathsf{Stal}_2$, we can rewrite it as per into a formula in the language $\mathcal{L}_{K_i}$ which is not provable in $\mathsf{S4.2}_{K_1} + \mathsf{S4.2}_{K_2}$. By completeness of the latter, there is a topological-partitional countermodel for ϕ, and completeness of $\mathsf{Stal}_1 + \mathsf{Stal}_2$ follows. □

4 Generic Models for Two Agents

In their famous paper [14], McKinsey and Tarski prove that S4 is not only the logic of topological spaces when one considers the interior semantics (i.e. when one reads $\|K\phi\| = \mathrm{Int}\,\|\phi\|$), but that there are single topological spaces, such as the real line $\mathbb{R}$ or the rationals $\mathbb{Q}$, whose logic is precisely S4. In [1], the authors of this paper have been concerned with finding *generic models* such as these for the logic of single-agent topo-e-models. In this section we provide two examples of generic models for the multi-agent logic, i.e., two topological-partitional spaces whose logic is precisely $\mathsf{S4.2}_{K_1} + \mathsf{S4.2}_{K_2}$.

The Quaternary Tree $\mathcal{T}_{2,2}$. The quaternary tree $\mathcal{T}_{2,2}$ is a full infinite tree with two relations R_1 and R_2 such that each node of the tree has exactly four successors, two of them being R_1-successors and the other two being R_2-successors, as it appears in Fig. 2.

By setting T to be the set of points of $\mathcal{T}_{2,2}$ and $\leq_i$ to be the reflexive and transitive closure of R_i for $i = 1, 2$, we can see $\mathcal{T}_{2,2} = (T, \leq_1, \leq_2)$ as a birelational preordered frame.

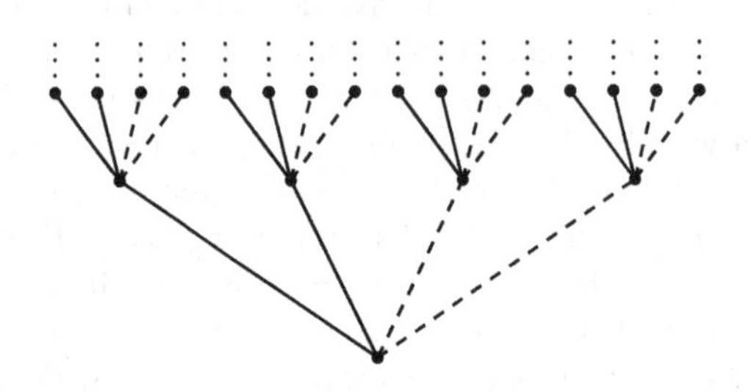

Fig. 2. The quaternary tree $\mathcal{T}_{2,2}$. R_1 and R_2 are represented respectively by the continuous and the dashed lines.

It is proven in [5] that the logic of this frame under the usual Kripke semantics is $\mathsf{S4} + \mathsf{S4}$. This result is a corollary of the following proposition which we will use in our proof:

Proposition 4.1 ([5]). *Given a finite frame $\mathfrak{F} = (W, \preceq_1, \preceq_2)$, where $\preceq_1$ and $\preceq_2$ are both preorders, there exists a p-morphism from $\mathcal{T}_{2,2}$ onto $\mathfrak{F}$, i.e., a surjective map $\mathsf{p} : \mathcal{T}_{2,2} \twoheadrightarrow \mathfrak{F}$ such that, for $i = 1, 2$, (i) $x \leq_i y$ implies $(\mathsf{p}\,x) \preceq_i (\mathsf{p}\,y)$, and (ii) $(\mathsf{p}\,x) \preceq_i v$ implies there exists $y \in T$ such that $x \leq_i y$ and $\mathsf{p}\,y = v$.*

Completeness of $\mathcal{T}_{2,2}$ with respect to $\mathsf{S4.2}_{K_1} + \mathsf{S4.2}_{K_2}$. Let us now bring this to our realm. We want to think of $\mathcal{T}_{2,2}$ as a topological-partitional model. For this, we turn to its connected components.

As per item (iii) of Lemma 3.8, we know that the connected components are given by the equivalence relation: $x \sim_i y$ if and only if there exists a z such that $x \geq_i z \leq_i y$. Note that for each $x \in \mathcal{T}_{2,2}$ and $i = 1, 2$, the set of $\leq_i$-predecessors of x forms a finite chain (and in particular, there is a least predecessor x_0 of x, which does not have any $\leq_i$ predecessors other than itself). These two facts give us the following characterisation:

Lemma 4.2. *The $\leq_i$-connected components of $\mathcal{T}_{2,2}$ are exactly the upsets of the form $\uparrow_i x_0$, where x_0 does not have any $\leq_i$-predecessors other than itself.*

Now, let $(W, \leq_1, \leq_2, V)$ be a finite model whose underlying frame is a rooted birelational weakly directed preorder. We can define a map $\mathsf{p} : \mathcal{T}_{2,2} \twoheadrightarrow W$ and a valuation $V^{\mathcal{T}_{2,2}}$ as above. Let σ_i be the topology of $\leq_i$-upsets of W and $\equiv_i$ be the equivalence relation determining the connected components. Recall that $\mathfrak{W} = (W, \sigma_{1,2}, \equiv_{1,2}, V)$ is a topological-partitional model in which every σ_i-open set is $\equiv_i$-locally dense. Moreover, we have:

Lemma 4.3. *For $x \in \mathcal{T}_{2,2}$, $w \in W$ and $i = 1, 2$, let $[x]_{\sim_i}$ and $[w]_{\equiv_i}$ be the respective equivalence classes (i.e., the respective connected components containing x and w). Then the following holds:*

i. *For any $x \in \mathcal{T}_{2,2}$, $\mathsf{p}[x]_{\sim_i} \subseteq [\mathsf{p}\, x]_{\equiv_i}$.*
ii. *Let $x_0 \in \mathcal{T}_{2,2}$ and let U be a (locally dense) σ_i-open set such that $\mathsf{p}\, x_0 \in U \subseteq [\mathsf{p}\, x_0]_{\equiv_i}$. Then $U' := \bigcup\{\uparrow_i x : x \sim_i x_0 \,\&\, \mathsf{p}\, x \in U\}$ is a locally dense upset such that $x_0 \in U' \subseteq [x_0]_{\sim_i}$.*

Proof. (i). Set $y \sim_i x$. Then there is some z such that $y \geq_i z \leq_i x$ and thus, since the map p preserves order, we have that $\mathsf{p}\, y \geq_i \mathsf{p}\, z \leq_i \mathsf{p}\, x$ and thus $\mathsf{p}\, y \equiv_i \mathsf{p}\, x$.

(ii). U' is an upset because it is a union of upsets and $x_0 \in U' \subseteq [x_0]_{\sim_i}$ by construction. Let us see that it is locally dense. Take some $z \in \mathcal{T}_{2,2}$ such that $\uparrow_i z \subseteq [x_0]_{\sim_i}$. Now, $\mathsf{p}(\uparrow_i z)$ is an open set (by opennes of p) and $\mathsf{p}(\uparrow_i z) \subseteq \mathsf{p}[x_0]_{\sim_i} \subseteq [\mathsf{p}\, x_0]_{\equiv_i}$. By local density of U there exists some $a \in U \cap \mathsf{p}(\uparrow_i z)$. That is, for some $z' \geq_i z$ we have $\mathsf{p}\, z' = a$ and $\mathsf{p}\, z' \in U$, thus by construction $z' \in \uparrow_i z \cap U'$ and thus $\uparrow_i z \cap U' \neq \varnothing$.

As a consequence:

Proposition 4.4. *For any $x \in \mathcal{T}_{2,2}$ and any formula ϕ in the language, $\mathcal{T}_{2,2}, x \vDash \phi$ if and only if $\mathfrak{W}, \mathsf{p}\, x \vDash \phi$.*

Proof. This is once again an induction on the structure of formulas in which the only involved case is the induction step corresponding to the K_i modalities.

Suppose $x \vDash K_i\phi$. Then there exists some locally dense open set U with $x \in U \subseteq [x]_{\sim_i}$ such that $y \vDash \phi$ for all $y \in U$. But then

$$\mathsf{p}\, x \in \mathsf{p}\, U \subseteq \mathsf{p}[x]_{\sim_i} \subseteq [\mathsf{p}\, x]_{\equiv_i},$$

this last inclusion given by (i) of the previous lemma, and $\mathsf{p}\, U$ is a locally dense open set in W: it is open because p is an open map and it is locally dense because every open set in W is locally dense. Moreover, for every $\mathsf{p}\, y \in \mathsf{p}\, U$ we have by induction hypothesis that $\mathsf{p}\, y \vDash \phi$. Thus $\mathsf{p}\, x \vDash K_i\phi$.

Conversely, suppose $\mathsf{p}\, x \vDash K_i\phi$. Then there exists a (locally dense) σ_i-open set U with $\mathsf{p}\, x \in U \subseteq [\mathsf{p}\, x]_{\equiv_i}$ such that $w \vDash \phi$ for all $w \in U$. But then by part (ii) of the previous lemma $U' := \bigcup\{\uparrow_i z : z \sim_i x \,\&\, \mathsf{p}\, z \in U\}$ is a locally dense upset such that $x \in U' \subseteq [x]_{\sim_i}$. Now take $y \in U'$. We have that $y \geq_i z$ for some $z \in [x]_{\sim_i}$ with $\mathsf{p}\, z \in U$. But since p is order preserving we have that $\mathsf{p}\, y \geq_i \mathsf{p}\, z$ and thus $\mathsf{p}\, y \in U$, which means that $\mathsf{p}\, y \vDash \phi$ and thus, by induction hypothesis, $y \vDash \phi$. This means that $U' \subseteq \|\phi\|^{\mathcal{T}_{2,2}}$ and thus $x \vDash K_i\phi$.

Completeness is now an immediate consequence.

Corollary 4.5. $\mathsf{S4.2}_{K_1} + \mathsf{S4.2}_{K_2}$ *is sound and complete with respect to the quaternary tree $(\mathcal{T}_{2,2}, \leq_1, \leq_2, \sim_1, \sim_2)$.*

The Product $\mathbb{Q}\times\mathbb{Q}$. Let us now show that it is possible to define two topologies and two equivalence relations on the product space $\mathbb{Q}\times\mathbb{Q}$ which make it into a generic topological-partitional space for $\mathsf{S4.2}_{K_1}+\mathsf{S4.2}_{K_2}$.

These topologies will be the vertical and horizontal topologies, which can be defined on a product $X\times Y$ and, in a way, "lift" the topologies of the components.

Definition 4.6. *Let* (X,τ) *and* (Y,σ) *be two topological spaces. The* horizontal *and* vertical *topologies,* τ_H *and* τ_V*, are the topologies on* $X\times Y$ *generated, respectively, by the bases*

$$\mathcal{B}_H=\{U\times\{y\}:U\in\tau,y\in Y\}\ \text{and}\ \mathcal{B}_V=\{\{x\}\times V:x\in X,V\in\sigma\}.$$

In particular, if we take both components to be $\mathbb{Q}$ with the natural topology, we obtain our bitopological space $(\mathbb{Q}\times\mathbb{Q},\tau_H,\tau_V)$. An important result about this space is the following:

Theorem 4.7 ([5]). $\mathsf{S4}+\mathsf{S4}$ *is the logic of* $(\mathbb{Q}\times\mathbb{Q},\tau_H,\tau_V)$ *under the interior semantics.*

Now we shall show there exists a partition on $\mathbb{Q}\times\mathbb{Q}$ which will give us the desired completeness result. Note that we cannot shelter ourselves in the connected components this time, for the connected components in $(\mathbb{Q}\times\mathbb{Q},\tau_H,\tau_V)$ are the singletons, which are not even open sets.

Let (X,τ_1,τ_2) be a bitopological space and $\mathfrak{Y}=(Y,\sigma_1,\sigma_2,\sim_1,\sim_2,V)$ be a topological partitional model. Moreover, let

$$f:(X,\tau_1,\tau_2)\twoheadrightarrow(Y,\sigma_1,\sigma_2)$$

be a surjective map which is open and continuous in both topologies. We shall call this an *onto interior map.* Define two equivalence relations $\equiv_1$ and $\equiv_2$ on X by:

$$x\equiv_i y \text{ if and only if } fx\sim_i fy.$$

Define a valuation on X by $V^f(p)=\{x\in X: fx\in V(p)\}$. The following holds:

Proposition 4.8. $\mathfrak{X}=(X,\tau_1,\tau_2,\equiv_1,\equiv_2,V^f)$ *is a topological evidence model and, for every formula* ϕ *in the language and every* $x\in X$ *we have that* $\mathfrak{X},x\vDash\phi$ *if and only if* $\mathfrak{Y},fx\vDash\phi$.

Proof. Checking that $\mathfrak{X}$ is a topological partitional model amounts to checking that each equivalence class is an open set. Let $[x]_{\equiv_i}$ be the equivalence class under $\equiv_i$ of some $x\in X$. Note that the image of this class coincides with the equivalence class of fx, i.e. $f[x]_{\equiv_i}=[fx]_{\sim_i}$. Indeed, $fy\in[fx]_{\sim_i}$ implies $y\in[x]_{\equiv_i}$, and thus $fy\in f[x]_{\equiv_i}$; conversely $y\in f[x]_{\equiv_i}$ implies $y=fx'$ for some $x'\equiv_i x$ and thus $y=fx'\sim_i fx$. Now, $[fx]_{\sim_i}$ is an equivalence class and thus an open set and, since f is continuous, $f^{-1}f[x]_{\equiv_i}$ is also an open set. So it suffices to show that $f^{-1}f[x]_{\equiv_i}=[x]_{\equiv_i}$. And indeed, if $z\in f^{-1}f[x]_{\equiv_i}$ then $fz\in f[x]_{\equiv_i}=[fx]_{\sim_i}$ which means that $fz\sim_i fx$ and thus $z\equiv_i x$.

The second result is an induction on formulas. For the propositional variables and the induction steps corresponding to the Boolean connectives the result is straightforward. Now suppose that for some ϕ it is the case that, for all x, $\mathfrak{X}, x \vDash \phi$ if and only if $\mathfrak{Y}, fx \vDash \phi$, and let $\mathfrak{X}, x \vDash K_i\phi$. This means that there exists some open set $U \in \tau_i$ such that $x \in U \subseteq \|\phi\|^{\mathfrak{X}}$ and U is locally dense in $[x]_{\equiv_i}$, i.e., for every nonempty open set $V \subseteq [x]_{\equiv_i}$, it is the case that $U \cap V \neq \varnothing$. But then we have that $fx \in f[U]$, the set $f[U]$ is open (by openness of f) which is contained in $f\|\phi\|^{\mathfrak{X}}$ (and thus, by induction hypothesis, in $\|\phi\|^{\mathfrak{Y}}$) and $f[U]$ is locally dense in $[fx]_{\sim_i}$. Indeed, suppose V is an open set contained in $[fx]_{\sim_i}$. then $f^{-1}[V]$ is an open set contained in $f^{-1}[fx]_{\sim_i} = [x]_{\equiv_i}$ which implies that there exists some $z \in f^{-1}[V] \cap U$ and thus some $fz \in V \cap f[U]$. Conversely, suppose $\mathfrak{Y}, fx \vDash K_i\phi$. There is an open set $U \subseteq \|\phi\|^{\mathfrak{Y}}$ which includes fx and which is locally dense on $[fx]_{\sim_i}$. Then $f^{-1}[U]$ is an open set including x which is contained in $f^{-1}\|\phi\|^{\mathfrak{Y}} = \|\phi\|^{\mathfrak{X}}$ and moreover it is locally dense on $[x]_{\equiv_i}$: indeed, if V is an open set contained in $[x]_{\equiv_i}$, then $f[V]$ is an open set contained in $[fx]_{\sim_i}$ and thus there exists some $y \in f[V] \cap [fx]_{\sim_i}$. But then $y = fz$ for some $z \in V$ and $z \in V \cap f^{-1}[fx]_{\sim_i} = V \cap [x]_{\equiv_i}$, whence $\mathfrak{X}, x \vDash K_i\phi$.

It is proven in [5] that there exists an onto map $f : \mathbb{Q} \times \mathbb{Q} \to \mathcal{T}_{2,2}$, open and continuous in both τ_H and τ_V. The previous proposition plus this fact grants us the existence of a partition which makes $\mathbb{Q} \times \mathbb{Q}$ a generic model for $\mathsf{S4.2}_{K_1} + \mathsf{S4.2}_{K_2}$.

Corollary 4.9. *Let $f : \mathbb{Q} \times \mathbb{Q} \to \mathcal{T}_{2,2}$ be some onto interior map. Define $(x, y) \equiv_i^f (x', y')$ iff $f(x, y)$ and $f(x', y')$ belong to the same $\leq_i$-connected component for $i = 1, 2$. Then $\mathsf{S4.2}_{K_1} + \mathsf{S4.2}_{K_2}$ is sound and complete with respect to*

$$(\mathbb{Q} \times \mathbb{Q}, \tau_H, \tau_V, \equiv_1^f, \equiv_2^f).$$

This existence result is not in itself very satisfactory, and it leads to the immediate question: what do these partitions look like? One could easily think of suitable candidates, (such as

$$\equiv_1 = \{\mathbb{Q} \times (\pi + k, \pi + k + 1) : k \in \mathbb{Z}\} \ \& \ \equiv_2 = \{(\pi + k, \pi + k + 1) \times \mathbb{Q} : k \in \mathbb{Z}\},$$

for instance) but none of the 'obvious' candidates seem to give us completeness (in the example above, $[\forall]_1\phi \to \Box_2\phi$ holds everywhere, while not being a theorem of the logic). This problem is open for future work.

5 Distributed and Common Knowledge

We have so far a multi-agent framework whose logic simply combines the axioms of the single-agent logic for each of the agents.

In the present section we consider the notions of *distributed* and *common* knowledge applied to this framework.

5.1 Distributed Knowledge

We can think of *distributed knowledge* as whatever the group knows implicitly, or whatever would become known if all the agents were to share their information. Not only does the group know ϕ if one agent in the group knows it, but the group also knows things that no individual agent knows yet can be derived from the information of several agents. For example, if agent 1 knows p to be the case, and agent 2 knows $p \rightarrow q$ to be the case, then together they know q, even if individually no one does.

In relational semantics, if $\mathcal{A}$ is a finite group of agents and, for each $a \in \mathcal{A}$, K_a is the Kripke modality corresponding to some relation R_a, then we can think of D as the Kripke modality corresponding to the relation $\bigcap_{a \in \mathcal{A}} R_a$.

Let us remark something here: given two preorders $\leq_1$ and $\leq_2$ defined on a set X, let τ_i be the topology of $\leq_i$-upwards closed sets for $i = 1, 2$. The Kripke semantics on $(X, \leq_1, \leq_2)$ correspond with the interior semantics on (X, τ_1, τ_2), and the collection of upwards-closed sets of the relation $\leq_1 \cap \leq_2$ is precisely the *join topology* $\tau_1 \vee \tau_2$, i.e., the least topology containing $\tau_1 \cup \tau_2$, or, equivalently, the topology generated by $\{U_1 \cap U_2 : U_i \in \tau_i\}$. We will be using join topologies in our approach.

A Problematic Approach. What exactly amounts to distributed knowledge in our framework? A very direct way to translate the ideas presented so far would be this: we say that $D\phi$ holds at w whenever agent 1 and agent 2 have each a piece of evidence which, when put together, constitute a justification for ϕ (i.e., a *locally dense* piece of evidence).

This approach, while intuitive, has two issues. On the one hand, it might be the case that an agent has a piece of evidence for ϕ which is dense in her topology (i.e., she knows ϕ) yet, when the evidence of both agents is put together, the corresponding evidence is no longer locally dense in the partition of the join topology (i.e., the group does not know ϕ).[5] Obviously, this is undesirable.

On the other hand, this notion reflects what the group could come to know if they put their evidence together and acted, in a way, as a collective agent. This is more an account of *implicit evidence* of the group rather than its *implicit knowledge*. According to [12],

> It is also often desirable to be able to reason about the knowledge that is distributed in the group, i.e., what someone who could combine the knowledge of all of the agents in the group would know. Thus, for example, if Alice knows ϕ and Bob knows $\phi \Rightarrow \psi$, then the knowledge of ψ is distributed among them, even though it might be the case that neither of them individually knows ψ. (...) [D]istributed knowledge corresponds to what a (fictitious) 'wise man' (one that knows exactly what each individual agent knows) would know.

[5] In [10] this is discussed in more depth and an example is provided.

The desired interpretation of 'distributed knowledge' here is that of a 'wise man' who has the information of what each agent knows, as opposed to what evidence they have. Thus, from this lens, one might want to keep misleading evidence out of the equation, and consider that this hypothetical 'wise man' forms his knowledge based not on what the agents have evidence for, but rather on what the agents *actually know*.

On this account, instead of each agent having a piece of *evidence* that, when combined together, constitute a *justification* for ϕ, we would want for each to have a justification which combine into a piece of evidence for ϕ. (For a more in-depth argument, see [10]).

There seem to be good reasons to stick to a notion of distributed knowledge which disregards the idea of 'putting evidence together' and which is based solely on the knowledge of the agents, whose logic would contain axioms like $K_1\phi \to D\phi$. In the following we present a way to have such a notion.

Our Proposal: The Semantics. We again have a language with two modal operators K_1 and K_2 for the knowledge of each agent plus an operator D for distributed knowledge.

Definition 5.1 (Semantics for D). *Let $\mathfrak{X} = (X, \tau_{1,2}, \Pi_{1,2}, V)$ be a topological-partitional model. We read $||p||$, $||\phi \wedge \psi||$ $||\neg\phi||$ and $||K_i\phi||$ as in Definition. 3.3, and:*

$$x \in ||D\phi|| \textit{ iff there exist } U_1 \in \tau_1, U_2 \in \tau_2 \textit{ such that}$$
$$U_i \textit{ is } \Pi_i - \textit{locally dense and } x \in U_1 \cap U_2 \subseteq ||\phi||.$$

While the problematic semantics outlined above amounted to reading distributed knowledge as the interior in the topology $(\tau_1 \vee \tau_2)^*$, what we are doing here is reading it as interior in $\tau_1^* \vee \tau_2^*$.

The Logic of Distributed Knowledge. Let Logic_{K_iD} be the least set of formulas containing:

- The S4.2 axioms and rules for K_1 and for K_2;
- The S4 axioms and rules for D;
- The axioms $K_i\phi \to D\phi$ for $i = 1, 2$.

Theorem 5.2. Logic_{K_iD} *is sound and complete with respect to topological - partitional models.*

We will dedicate the rest of this subsection to showing this fact.

Soundness. That every topological-partitional model satisfies the S4.2 axioms for K_i can be proven exactly as in Sect. 3.1. That D satisfies the S4 axioms is a consequence of D being read as $\mathrm{Int}_{\tau_1^* \vee \tau_2^*}$. And for the two extra axioms, if $x \models K_i\phi$, then there exists $U_i \in \tau_i^*$ with $x \in U_i \subseteq ||\phi||$. Let $j \neq i$ and, by taking $U_j = X$, which is a Π_j-locally dense τ_j-open set, we get $x \in U_i \cap U_j \subseteq ||\phi||$ and thus $x \models D\phi$.

Completeness. Let X be the set of maximal consistent sets over the language. We define R_i and R_D on X as follows: given $T, S \in X$,

$$TR_iS \text{ iff } K_i\phi \in T \text{ implies } \phi \in S \text{ for all } \phi \text{ in the language;}$$
$$TR_DS \text{ iff } D\phi \in T \text{ implies } \phi \in S \text{ for all } \phi \text{ in the language.}$$

Note that $R_D \subseteq R_i$ for $i = 1, 2$. Indeed, if TR_DS and $K_i\phi \in T$, then $D\phi \in T$ as per the axiom $K_i\phi \to D\phi$ and thus $\phi \in S$.

A *labelled path over* X is a path

$$\alpha = T_0 \xrightarrow{i_1} T_1 \xrightarrow{i_2} \ldots \xrightarrow{i_n} T_n,$$

where $T_0, \ldots, T_n \in X$ and $i_1, \ldots, i_n \in \{R_1, R_2, R_D\}$. Given $S \in X$ and a path $\alpha = T_0 \xrightarrow{i_1} T_1 \xrightarrow{i_2} \ldots \xrightarrow{i_n} T_n$, we define

$$\mathsf{last}\,\alpha := T_n \text{ and } \alpha \xrightarrow{i} S := T_0 \xrightarrow{i_1} T_1 \xrightarrow{i_2} \ldots \xrightarrow{i_n} T_n \xrightarrow{i} S.$$

Now, let $\mathcal{T}$ be the smallest set of labelled paths over X such that: (i.) The path T_0 (of length 0) belongs to $\mathcal{T}$; (ii.) For $i = 1, 2$, if $\alpha \in \mathcal{T}$ and $(\mathsf{last}\,\alpha)R_iT$, then $\alpha \xrightarrow{R_i} T \in \mathcal{T}$; (iii.) If $\alpha \in \mathcal{T}$ and $(\mathsf{last}\,\alpha)R_DT$, then $\alpha \xrightarrow{R_D} T \in \mathcal{T}$.

For $i = 1, 2, D$ we define: $\alpha \prec_i \beta$ if and only if $\alpha = \beta \xrightarrow{R_i} S$ for some $S \in X$.

We have thus given $\mathcal{T}$ the structure of a forest. Indeed, every $\alpha \in \mathcal{T}$ has at most one predecessor under $\prec_1 \cup \prec_2 \cup \prec_D$. Now let us define three preorders on $\mathcal{T}$: for $i = 1, 2$, let $\leq_i$ be the reflexive and transitive closure of $\prec_i \cup \prec_D$ and $\leq_D$ to be the reflexive and transitive closure of $\prec_D$. Note that by construction $\leq_D = \leq_1 \cap \leq_2$.

Now let us see what the $\leq_1$- and $\leq_2$-connected components look like. By part (iii) of Lemma 3.8, we know that the connected components of the topology of upsets of $\leq_i$ ($i = 1, 2$) are given by the equivalence relation: $\alpha \sim_i \beta$ iff there exists γ such that $\alpha \geq_i \gamma \leq_i \beta$. The definition of $\leq_i$ plus the fact that $R_D \subseteq R_i$ entail that $(\mathsf{last}\,\gamma)R_i(\mathsf{last}\,\alpha)$ and $(\mathsf{last}\,\gamma)R_i(\mathsf{last}\,\beta)$. Therefore we have the following result:

Lemma 5.3. *If α and β belong to the same $\leq_i$-connected component on $\mathcal{T}$, then $\mathsf{last}\,\alpha$ and $\mathsf{last}\,\beta$ belong to the same R_i-connected component in X.*

Moreover, there is an alternative characterisation of the connected components, similar to that in Lemma 4.2, which we will find useful:

Lemma 5.4. *The $\leq_i$-connected components correspond to upsets of the form $\uparrow_i\alpha_0$, where α_0 has no $\leq_i$-predecessors other than itself.*

We have given $\mathcal{T}$ the structure of a topological-partitional space and by defining $V^{\mathcal{T}}(p) = \{\alpha \in \mathcal{T} : p \in \mathsf{last}\,\alpha\}$ we have a topological-partitional model and we can prove the following:

Lemma 5.5 (Truth lemma). *For every $\alpha \in \mathcal{T}$ and ϕ in the language, $\alpha \vDash \phi$ if and only if $\phi \in \mathsf{last}\,\alpha$.*

Proof. This is again an induction on formulas in which the base case for the propositional variables follows from the definition of $V^{\mathcal{T}}$ and the induction steps for the Boolean connectives are routine.

Now, suppose the result holds for ϕ and $K_i\phi \in \mathsf{last}\,\alpha$. We need to define a locally dense open set U_i such that $\alpha \in U_i \subseteq [\alpha]_{\sim_i}$ and with the property that, for every $\beta \in U_i$, $\phi \in \mathsf{last}\,\beta$, which will give us, by induction hypothesis, that $U_i \subseteq \|\phi\|$. By Lemma 5.4, we have that $[\alpha]_{\sim_i} = \uparrow_i\alpha_i$ for some $\alpha_i \in \mathcal{T}$. In other words, every $\beta \in [\alpha]_{\sim_i}$ is of the form

$$\beta = \alpha_i \xrightarrow{R_i \text{ or } R_D} T_i \xrightarrow{R_i \text{ or } R_D} \dots \xrightarrow{R_i \text{ or } R_D} T_n.$$

Let us now partition $[\alpha]_{\sim_i}$ in two sets:

$$V_D^{[i]} := \{\beta \in [\alpha]_{\sim_i} : \alpha_i \leq_D \beta\};$$
$$V_i := \{\beta \in [\alpha]_{\sim_i} : \alpha_i \leq_i \beta \,\&\, \alpha_i \not\leq_D \beta\}.$$

Note that the elements in $V_D^{[i]}$ are of the form $\beta = \alpha_i \xrightarrow{R_D} T_1 \xrightarrow{R_D} \dots \xrightarrow{R_D} T_n$, and the elements in V_i are of the form

$$\beta = \alpha_i \xrightarrow{r_1} T_1 \xrightarrow{r_2} \dots \xrightarrow{r_n} T_n \;\text{ with } r_k \in \{R_i, R_D\} \text{ and at least one } r_k = R_i,$$

and each element in $[\alpha]_{\sim_i}$ is in exactly one of $V_i, V_D^{[i]}$. Let us define U_i as follows:

$$U_i := \{\beta \in V_D^{[i]} : (\mathsf{last}\,\alpha)R_D(\mathsf{last}\,\beta)\} \cup \{\gamma \in V_i : (\mathsf{last}\,\alpha)R_i(\mathsf{last}\,\gamma)\}.$$

The following holds:

i. $\alpha \in U_i$ by construction.
ii. *U_i is an upset.* Take any $\beta \in U_i$. If $\beta \prec_i \gamma$ then $\gamma = \beta \xrightarrow{R_i} S$ for some $S \in X$ and we clearly have $\gamma \in V_i$ and $(\mathsf{last}\,\alpha)R_i(\mathsf{last}\,\beta)R_iS$, thus $(\mathsf{last}\,\alpha)R_iS$. If $\beta \prec_D \gamma$ then $\beta = \gamma \xrightarrow{R_D} S$ and, if $\beta \in V_D^{[i]}$ we then have that $\gamma \in V_D^{[i]}$ and $(\mathsf{last}\,\alpha)R_D(\mathsf{last}\,\beta)R_DS$ (thus $(\mathsf{last}\,\alpha)R_D(\mathsf{last}\,\gamma)$) whereas if $\beta \in V_i$ we have that $\gamma \in V_i$ and similarly (given that $R_D \subseteq R_i$), $(\mathsf{last}\,\alpha)R_iS$. In any case $\gamma \in U_i$.
iii. *U_i is locally dense.* Take any $\beta \in [\alpha]_{\sim_i}$. By Lemma 5.3, we have that $\mathsf{last}\,\beta$ and $\mathsf{last}\,\alpha$ are in the same R_i-connected component and, since R_i is an $\mathsf{S4.2}$ relation, part (ii) of Lemma 3.8 gives us that there exists some $S \in X$ with $(\mathsf{last}\,\alpha)R_iS$ and $(\mathsf{last}\,\beta)R_iS$ and thus we have $\beta \xrightarrow{R_i} S \in U_i \cap \uparrow_i\beta$.
iv. *$\phi \in \mathsf{last}\,\beta$ for every $\beta \in U_i$* (given that $K_i\phi \in \mathsf{last}\,\alpha$ and $(\mathsf{last}\,\alpha)R_i(\mathsf{last}\,\beta)$).

Thus $\alpha \vDash K_i\phi$, as we intended to prove.

Conversely, if $\alpha \vDash K_i\phi$, there exists some locally dense open set U_i with $\alpha \in U_i \subseteq [\alpha]_{\sim_i} \cap \|\phi\|$. Since U_i is an upset, if $(\mathsf{last}\,\alpha)R_iS$, we have $\alpha \xrightarrow{R_i} S \in U_i$, which means $\alpha \xrightarrow{R_i} S \in \|\phi\|$ and by induction hypothesis $\phi \in S$. Every R_i-successor of $\mathsf{last}\,\alpha$ includes ϕ, which gives $K_i\phi \in \mathsf{last}\,\alpha$.

Now suppose $D\phi \in \mathsf{last}\,\alpha$. Define U_1 and U_2 as above. They are locally dense open sets contained respectively in $[\alpha]_{\sim_1}$ and $[\alpha]_{\sim_2}$. Moreover, $\alpha \in U_1 \cap U_2$ by construction. We simply need to see that $U_1 \cap U_2 \subseteq \|\phi\|$. First let us note the following: if $\beta \in [\alpha]_{\sim_1} \cap [\alpha]_{\sim_2} = \uparrow_1\alpha_1 \cap \uparrow_2\alpha_2$, then β is simultaneously of the form

$$\beta = \alpha_1 \xrightarrow{R_1 \text{ or } R_D} T_1 \xrightarrow{R_1 \text{ or } R_D} \dots \xrightarrow{R_1 \text{ or } R_D} T_n$$

and of the form

$$\beta = \alpha_2 \xrightarrow{R_2 \text{ or } R_D} S_1 \xrightarrow{R_2 \text{ or } R_D} \dots \xrightarrow{R_2 \text{ or } R_D} S_m.$$

These can only be true at the same time if β is of the form

$$\beta = \alpha_j \xrightarrow{R_D} S_1 \dots \xrightarrow{R_D} S_m \text{ with } \alpha_j = \alpha_i \xrightarrow{R_i \text{ or } R_D} T_1 \dots \xrightarrow{R_i \text{ or } R_D} T_k$$

for $i \neq j \in \{1, 2\}$. Let us assume w.l.o.g. that $i = 1, j = 2$. In particular we have that, if $\beta \in U_1 \cap U_2$, then $\beta \in V_D^{[2]}$ and hence $(\mathsf{last}\,\alpha)R_D(\mathsf{last}\,\beta)$. Since $D\phi \in \mathsf{last}\,\alpha$, this entails that $\phi \in \mathsf{last}\,\beta$ and thus that $\beta \in \|\phi\|$, whence $\alpha \vDash D\phi$.

For the converse, if $\alpha \vDash D\phi$ then $\alpha \in U_1 \cap U_2 \subseteq \|\phi\|$ for some $\leq_i$-locally dense $U_i \subseteq [\alpha]_{\sim_i}$. But then if $(\mathsf{last}\,\alpha)R_D S$ we have $\alpha \leq_D \alpha \xrightarrow{R_D} S$ and since $\leq_D = \leq_1 \cap \leq_2$ and U_1 and U_2 are respectively a $\leq_1$ and a $\leq_2$-upset, we have that $\alpha \xrightarrow{R_D} S \in U_1 \cap U_2$ and thus $\alpha \xrightarrow{R_D} S \vDash \phi$ which by induction hypothesis gives $\phi \in S$. This entails $D\phi \in \mathsf{last}\,\alpha$.

Completeness follows from this: if $\phi \notin \mathsf{Logic}_{K_i D}$, then $\{\neg\phi\}$ is consistent and can be extended as per Lindenbaum's lemma to some maximal consintent set $T_0 \in X$. We then unravel the tree around T_0 as discussed above and we have ourselves a topological-partitional model rooted in $\alpha = T_0$ with $\alpha \nvDash \phi$ as per the truth lemma.

5.2 Common Knowledge

In the context of epistemic logic, one can think of *common knowledge* as that which "every fool knows". This informal definition can be formally cashed out in several intuitive ways when one is modelling an epistemic situation. [3] compares the following approaches to common knowledge:

(1) *The iterate approach.* A fact ϕ is common knowledge for a group of agents when ϕ is true, all agents know that it is true, all agents know that all agents know that it is true, etc. If $E\phi$ is an abbreviation of $K_1\phi \wedge K_2\phi$, then

$$C\phi \equiv \phi \wedge E\phi \wedge EE\phi \wedge EEE\phi \wedge \dots$$

(2) *The fixed-point approach.* This is an approach in which common knowledge refers back to itself. The idea here is that, if ϕ is the proposition which expresses "it is common knowledge for agents a and b that p", then ϕ is equivalent to "a and b know (p and ϕ)".

[3] goes on to argue that, despite the fact that early literature considered this approach equivalent to the fixed point one, (1) and (2) offer in fact distinct accounts and the fixed point approach provides "the right theoretical analysis of the pretheoretic notion of common knowledge".

Moreover, while (1) and (2) are equivalent in relational semantics, as shown in [7] this equivalence disappears once we are working in a topological setting. If one is working topologically, one has to make a choice.

Our proposal amounts to reading the common knowledge modality C as the interior in the intersection topology $\tau_1^* \cap \tau_2^*$. More explicitly:

Definition 5.6 (Common knowledge semantics). *Let $\mathfrak{X} = (X, \tau_{1,2}, \Pi_{1,2}, V)$ be a topological-partitional model. We read*

$$\mathfrak{X}, x \vDash C\phi \text{ iff there exists } U \in \tau_1 \cap \tau_2 \text{ locally dense in } \Pi_1 \text{ and in } \Pi_2 \text{ such that } x \in U \subseteq \|\phi\|.$$

This amounts to the following: there is common knowledge of ϕ at x whenever there exists a common factive justification for ϕ.

Much like our account of distributed knowledge, this notion of common knowledge corresponds directly with the relational definition when we are dealing with a topological-partitional model stemming from two S4.2 relations: if R_1 and R_2 are S4.2, τ_i is the topology of R_i-upsets and Π_i is the set of R_i-connected components, then $\tau_1^* \cap \tau_2^*$ contains exactly the upsets of $(R_1 \cup R_2)^*$.

Another observation is that, in the spirit of [3], this definition is precisely the fixed point account of common knowledge. As pointed out in [7] and expanded in [8], the fixed point approach can be expressed in the notation of mu-calculus as

$$C\phi = \nu p(\phi \wedge Ep),$$

where p is a propositional variable which does not appear in ϕ. We read

$$\|\nu p \psi\| = \bigcup\{U \in \mathcal{P}(X) : U \subseteq \|\psi\|^{V_p^U}\},$$

where V_p^U is the valuation assigning U to p and $V(q)$ to $q \neq p$.

In particular, $\|C\phi\| = \bigcup\{U \in \mathcal{P}(X) : U \subseteq \|\phi \wedge Ep\|^{V_p^U}\}$. It is straightforward to check that this last set equals

$$\bigcup\{U \in \mathcal{P}(X) : U \in \tau_1^* \cap \tau_2^* \,\&\, U \subseteq \|\phi\|\} = \mathrm{Int}_{\tau_1^* \cap \tau_2^*} \|\phi\|,$$

which is precisely our account of common knowledge.

Some theorems in the logic of topological-partitional models with common knowledge are the following:

i. The S4.2 axioms for K_i;
ii. the S4 axioms for C;
iii. the *fixed point axiom* $C\phi \to E(C\phi \wedge \phi)$;
iv. the *induction axiom* $C(\phi \to E\phi) \to (E\phi \to C\phi)$.

Proposition 5.7 (Soundness). *All the theorems above are valid on topological-partitional models with the semantics of Definition 5.6.*

Proof. That i., ii. and iii. hold for topological-partitional models is a straightforward check. Item iv. is more involved. It amounts to checking that, on any such model, and for any $P \subseteq X$,

$$C(\neg P \vee (K_1 P \cap K_2 P)) \cap P \subseteq CP.$$

Now, let $x \in C(\neg P \vee (K_1 P \cap K_2 P))$. By the semantics of 5.6 this means that there exists some $U \in \tau_1^* \cap \tau_2^*$ such that

$$x \in U \subseteq \neg P \cup (\mathrm{Int}_{\tau_1^*} P \cap \mathrm{Int}_{\tau_2^*} P).$$

Call $V := U \cap \mathrm{Int}_{\tau_1^*}$. Now, V is a τ_1^*-open set. Note that $V \subseteq U \cap \mathrm{Int}_{\tau_2^*}$ and $U \cap \mathrm{Int}_{\tau_2^*} \subseteq V$ and thus V is also a τ_2^*-open set. Moreover, V includes x and it is contained in P. Thus there exits some $V \in \tau_1^* \cap \tau_2^*$ with $x \in V \subseteq P$, hence $x \in CP$.

Whether the preceding list of formulas constitutes a complete axiomatisation of the logic of common knowledge for topological-partitional models is a question that remains open.

6 Conclusions and Future Work

This paper presents a multi-agent generalisation for the dense interior semantics defined on topological evidence models, furthering the results in [2].

This was achieved by introducing a second epistemic agent and a partition-based semantics. We showed how this semantics generalises the single agent case and we provided a complete logic for our two-agent models. Moreover, 'generic spaces' were provided with respect to which the logic is sound and complete: the quaternary tree $\mathcal{T}_{2,2}$ and the rational plane $\mathbb{Q} \times \mathbb{Q}$. Along with this, a brief conceptual and theoretical study of notions of "group knowledge" for this group of agents was developed.

Some questions remain unanswered (and some potentially interesting results were out of the scope of this investigation). Among these are the following:

- We have proven (Corollary 4.9) that there exist partitions $\equiv_1$ and $\equiv_2$ making $\mathbb{Q} \times \mathbb{Q}$ a model for the logic $\mathsf{S4.2}_{K_1} + \mathsf{S4.2}_{K_2}$. What would be an example of such partitions?
- Are $\mathcal{T}_{2,2}$ and $\mathbb{Q} \times \mathbb{Q}$ generic models for any (all) of the fragments of the language considered in Sect. 3.2? For the distributed knowledge logic defined in the last section?
- Can we more broadly characterize a class of topological - partitional spaces which are generic for the logic? For example it is shown in [1] that, for the one-agent case, any topological space which is dense-in-itself, metrizable and idempotent is a generic model for the logic. Is a similar result true in the multi-agent setting?

- Does the list of theorems presented in Sect. 5.2 constitute a complete axiomatization of the logic of common knowledge?

Acknowledgement. We wish to thank Guram Bezhanishvili for very fruitful discussions on these topics. Special thanks go to the reviewers for their very valuable comments, which helped us improve the paper.

References

1. Baltag, A., Bezhanishvili, N., Fernández González, S.: The Mckinsey-Tarski theorem for topological evidence logics. In: Iemhoff, R., Moortgat, M., de Queiroz, R. (eds.) WoLLIC 2019. LNCS, vol. 11541, pp. 177–194. Springer, Heidelberg (2019). https://doi.org/10.1007/978-3-662-59533-6_11
2. Baltag, A., Bezhanishvili, N., Özgun, A., Smets, S.: Justified belief and the topology of evidence. In: Vaananen, J., Hirvonen, A., de Queiroz, R. (eds.) WoLLIC 2016. LNCS, vol. 9803, pp. 83–103. Springer, Heidelberg (2016). https://doi.org/10.1007/978-3-662-52921-8_6
3. Barwise, J.: Three views of common knowledge. In: Proceedings of Second Conference on Theoretical Aspects of Reasoning about Knowledge, pp. 365–379 (1988)
4. van Benthem, J., Bezhanishvili, G.: Modal logics of space. In: Aiello, M., Pratt-Hartmann, I., Van Benthem, J. (eds.) Handbook of Spatial Logics, pp. 217–298. Springer, Dordrecht (2007). https://doi.org/10.1007/978-1-4020-5587-4_5
5. van Benthem, J., Bezhanishvili, G., ten Cate, B., Sarenac, D.: Multimodal logics of products of topologies. Stud. Logica **84**(3), 369–392 (2006)
6. van Benthem, J., Pacuit, E.: Dynamic logics of evidence-based beliefs. Stud. Logica **99**(1–3), 61 (2011)
7. van Bethem, J., Sarenac, D.: The geometry of knowledge. Travaux de logique **17**, 1–31 (2004)
8. Bezhanishvili, N., van der Hoek, W.: Structures for epistemic logic. In: Baltag, A., Smets, S. (eds.) Johan van Benthem on Logic and Information Dynamics. OCL, vol. 5, pp. 339–380. Springer, Cham (2014). https://doi.org/10.1007/978-3-319-06025-5_12
9. Blackburn, P., De Rijke, M., Venema, Y.: Modal Logic, vol. 53. Cambridge University Press, Cambridge (2001)
10. Fernández González, S.: Generic Models for Topological Evidence Logics. Master's thesis, University of Amsterdam (2018)
11. Gettier, E.L.: Is justified true belief knowledge? Analysis **23**(6), 121–123 (1963)
12. Halpern, J.Y., Moses, Y.: A guide to completeness and complexity for modal logics of knowledge and belief. Artif. Intell. **54**(3), 319–379 (1992)
13. Hintikka, J.: Knowledge and Belief: An Introduction to the Logic of the Two Notions. Cornell University Press, Contemporary Philosophy, Ithaca (1962)
14. McKinsey, J.C.C., Tarski, A.: The algebra of topology. Ann. Math. **1**, 141–191 (1944)
15. Munkres, J.R.: Topology. Prentice Hall, Upper Saddle River (2000)
16. Özgün, A.: Evidence in Epistemic Logic: A Topological Perspective. Ph.D. thesis, University of Amsterdam, ILLC (2017)
17. Ramírez, A.: Topological Models for Group Knowledge and Belief. Master's thesis, University of Amsterdam (2015)
18. Stalnaker, R.: On logics of knowledge and belief. Philos. Stud. **128**(1), 169–199 (2006)

Incomplete Information and Justifications

Dragan Doder[1], Zoran Ognjanović[2], Nenad Savić[3(✉)], and Thomas Studer[3]

[1] Utrecht University, Utrecht, The Netherlands
d.doder@uu.nl
[2] Mathematical Institute of Serbian Academy of Sciences and Arts, Belgrade, Serbia
zorano@mi.sanu.ac.rs
[3] Institute of Computer Science, University of Bern, Bern, Switzerland
{nenad.savic,thomas.studer}@inf.unibe.ch

Abstract. We present a logic for reasoning about higher-order upper and lower probabilities of justification formulas. We provide sound and strongly complete axiomatization for the logic. Furthermore, we show that the introduced logic generalizes the existing probabilistic justification logic PPJ.

Keywords: Justifcation logic · Probabilistic logic · Upper and lower probabilities · Strong completeness

1 Introduction

Since the seminal paper about justification logics was published, [3], a whole family of justification logics has been established, including logics with uncertain justifications, see [2,4,9,10,12,15,17]. However, justification logics in which uncertainty originates from incompleteness of information is still not provided.

The main feature of justification logic is that evidence is representable directly in the object language, i.e. the language of justification logic includes formulas of the form $t : A$ meaning that t *justifies* A. In this paper, we distinguish the following two types of incomplete information within $t : A$:

1) *"t" is incomplete.*
A friend tells me that she saw in *some* weather forecast that tomorrow is going to rain. I know which are possible forecasts she could have checked. As a consequence of an incomplete justification t (she read in *some* weather forecast and did not specify in which one) and since each forecast provides a probability for the rain, my degree of belief that "tomorrow is going to rain" is true lies in an interval $[r, s]$, where r represents the lowest probability according to the possible forecasts she checked, and s the highest.

This work was supported by the SNSF project 200021_165549 Justifications and non-classical reasoning and by the Serbian Ministry of Education and Science through Mathematical Institute of Serbian Academy of Sciences and Arts.

© The Author(s), under exclusive license to Springer Nature Switzerland AG 2022
A. Özgün and Y. Zinova (Eds.): TbiLLC 2019, LNCS 13206, pp. 258–278, 2022.
https://doi.org/10.1007/978-3-030-98479-3_13

2) *":" is incomplete.*
After taking a medication x, a patient gets a symptom S. It is known that the symptom S is a side effect of the medication x and that there exist old and new series of the medication x. The chance that the side effect occurs is smaller when taking medication of the new series.
In this case both t: "the patient took the medication" and A: "the patient got the symptom" are certain, but we do not know if t is the reason for A or there exists another reason. Also, we do not know if the patient took the medication of the old or the new series. In the former case, the chance that x caused S is bigger than in the later case. Thus, our degree of belief that $t : A$ lies in some interval.

In this paper we formalize both types of uncertainty illustrated above. To capture uncertainty about probabilities we use the lower and upper probability measures. For an arbitrary set of probability measures P, the former assigns to an event X the infimum of the probabilities assigned to X by the measures from the set P, while the later returns their supremum.

We provide a new logic, ILUPJ[1], as an extension of the justification logic J with two families of unary operators $L_{\geq s}$ and $U_{\geq s}$, for $s \in \mathbb{Q} \cap [0, 1]$. That idea comes from some of our previous papers, see e.g., [6,18]. The intended meanings of these operators are that 'the lower (upper) probability is greater or equal to s'. Therefore, saying that our degree of belief lies in an interval $[r, s]$ is represented by saying that the lower probability is equal to r and the upper probability is equal to s.

The first case, when "t" is incomplete and therefore our degree of belief that A is true belongs to an interval $[r, s]$ we can represent in the logic ILUPJ with

$$t : L_{=r}A \wedge t : U_{=s}A.$$

The second case, when ":" is incomplete, i.e., situations in which we are not sure if t is the justification for A, can be represented by

$$L_{=r}(t : A) \wedge U_{=s}(t : A).$$

The corresponding semantics of our logic consists of special types of possible world models, where every world is equipped with a space that consists of the non-empty set of accessible worlds, algebra of subsets and a set of probability measures.

We propose a sound and complete axiomatization of the logic. In order to prove the strong completeness theorem, which is the main technical result of the paper, we use a Henkin-like construction modifying our previous techniques for probabilistic and temporal logic [5,13,14,16]. Also, the proofs in our logic can be infinite although all the formulas of the logic are finite.

We also compare our logic with the probabilistic justification logic PPJ from [10], and we prove that ILUPJ properly generalizes it. The logic PPJ is obtained

[1] I stands for iterations, LUP for lower and upper probabilities and J for the justification logic J.

by extending the justification logic J by a list of standard unary operators, $P_{\geq s}$, whose intended meaning is 'the probability is greater or equal to s'. In that approach there is no uncertainty about probabilities, and a unique probability value is assigned to an event. In our example from the first case, this would correspond to the situation where only one forecast is available. In the general case, where we consider several forecasts, we need to assign sets of probabilities to events, which lead to our more general semantics and, consequently, to different probability operators. Indeed, if r is the lowest probability according to the possible forecasts, and s the highest, we cannot always assign a truth value to the sentence "t justifies that our degree of belief that tomorrow will rain with probability at least ℓ" (in the language of PPJ: $t : P_{\geq \ell}A$), where $\ell \in (r, s)$ – according to some forecasts the sentence is true, and according to others it is false. On the other hand, in ILUPJ we can distinguish two cases: $t : L_{\geq \ell}A$ is false and $t : U_{\geq \ell}A$ is true.

The content of this paper is as follows. In Sect. 2 we define the basic notions needed for defining our logic. In Sect. 3 we propose the logic ILUPJ, whereas in Sect. 4 we prove the soundness and strong completeness theorem. In Sect. 5 we prove that our logic generalizes the logic PPJ and we conclude the paper in Sect. 6.

2 Preliminaries

We start with preliminary notions that will be used in the definition of the semantics of the logic ILUPJ.

Definition 1 (Algebra Over a Set). *Let $W \neq \emptyset$ and let $\emptyset \neq H \subseteq \mathcal{P}(W)$. H is called* algebra over W *if:*

1) $W \in H$,
2) For $X, Y \in H$, $W \setminus X \in H$ and $X \cup Y \in H$.

Definition 2 (Finitely Additive Probability Measure). *For an algebra H over W, a function $\mu : H \longrightarrow [0, 1]$ is called* finitely additive probability measure, *if:*

1) $\mu(W) = 1$,
2) For $X, Y \in H$, $\mu(X \cup Y) = \mu(X) + \mu(Y)$, whenever $X \cap Y = \emptyset$.

Definition 3 (Lower and Upper Probability Measures). *Let H be an algebra over W and P be a set of finitely additive probability measures defined on H. For $X \in H$, the* lower probability measure P_* *and the* upper probability measure P^* *are defined as follows:*

1) $P_*(X) = \inf\{\mu(X) \mid \mu \in P\}$,
2) $P^*(X) = \sup\{\mu(X) \mid \mu \in P\}$.

Now we state three properties which are used in our proof of soundness and completeness theorem for ILUPJ. The proof of these basic properties of P_* and P^* follows directly from the properties of infimum and supremum.

1) $P_*(X) \leq P^*(X)$,
2) $P_*(X) = 1 - P^*(X^c)$,
3) $P^*(X \cup Y) \leq P^*(X) + P^*(Y)$, whenever $X \cap Y = \emptyset$.

Complete characterization of P_* and P^* is needed in order to axiomatize upper and lower probabilities. We use the characterization used by Anger and Lembcke [1], which was also used by Halpern and Pucella [8, Theorem 2.3]. For that characterization we need a notion of (n, k)-cover.

Definition 4 ((n, k)-cover). *A set X is* covered n times *by a multiset*

$$\{\{X_1, \ldots, X_m\}\}$$

of sets if every element of X appears in at least n sets from $X_1, \ldots, X_m$ meaning that for all $x \in X$, there exist $i_1, \ldots, i_n \in \{1, \ldots, m\}$ such that for all $j \leq n$, $x \in X_{i_j}$.

An (n, k)-cover *of (X, W) is a multiset $\{\{X_1, \ldots, X_m\}\}$ that covers the set W k times and covers the set X $n + k$ times.*

With the notion of (n, k)-cover we are ready to define the characterization theorem:

Theorem 1 (Anger and Lembcke [1]). *Let $W \neq \emptyset$, H an algebra over W, and f a function $f : H \longrightarrow [0, 1]$. There exists a set P of probability measures such that $f = P^*$ iff the function f satisfies the following three conditions:*

(1) $f(\emptyset) = 0$,
(2) $f(W) = 1$,
(3) for all $m, n, k \in \mathbb{N}$ and all $X, X_1, \ldots, X_m \in H$, if $\{\{X_1, \ldots, X_m\}\}$ is an (n, k)-cover of (X, W), then

$$k + nf(X) \leq \sum_{i=1}^{m} f(X_i).$$

3 The logic ILUPJ

In this section we describe the syntax and semantics of the logic ILUPJ and provide an axiomatization.

3.1 Syntax

We will use the following notation:

$\mathsf{Con} = \{c_0, c_1, \ldots, c_n, \ldots\}$ for a countable set of constants,
$\mathsf{Var} = \{x_0, x_1, \ldots, x_n, \ldots\}$ for a countable set of variables, and
$\mathsf{Prop} = \{p_0, p_1, \ldots, p_n, \ldots\}$ for a countable set of atomic propositions.

Definition 5 (Justification Terms). Terms *are built from the sets* Con *and* Var *with the following grammar:*

$$t ::= c \mid x \mid t \cdot t \mid t + t \mid !t,$$

where $c \in \mathsf{Con}$ *and* $x \in \mathsf{Var}$. *The set of all terms will be denoted by* Tm.

For a term t and non-negative integer n we use the following notation:

$$!^0 t := t \quad and \quad !^{n+1} t := !(!^n t).$$

Terms represent justifications for an agent's belief (or knowledge). In the original justification logic, the *Logic of Proofs* [3], terms represent formal proofs in e.g. Peano arithmetic [11]. In possible world models for justifcation logic, first developed by Fitting [7], terms may represent arbitrary justifications like direct observation, public announcements, private communication, and so on.

Let us discuss the role of a given justification term depending on its main connective [12]:

- Constants are used in situations where the justification is not further analyzed, e.g. to justify axioms, see rule (IR1).
- Variables are used to represent arbitrary justifications.
- The operation $\cdot$ represents the agent's ability to reason by modus ponens. Assume that s justifies the agent's belief in A and t justifies the agent's belief in $A \rightarrow B$, then $t \cdot s$ will justify her belief in B, see axiom (Ax2).
- The operation $+$ combines two justifcations to a justification with broader scope, see axiom (Ax3). Often this is illustrated as follows. Let s and t be two volumes of an encyclopedia and $s + t$ be the set of those two volumes. Suppose that one of the volumes, say s, contains justification for a proposition A. Then also the larger set $s + t$ contains justification for A.
- The operation ! represents the agent's ability to perform positive introspection. In our logic ILUPJ, we only include positive introspection for axioms and iterated belief of axioms, see rule (IR1). Assume an agent believes an axiom A and c is a justification for that belief. By positive introspection the agent believes that she believes A and that A is justified by c. The term $!c$ will justify the result of the positive introspection act.

Definition 6 (Formulas of the Logic ILUPJ). Formulas *of the logic* ILUPJ *are defined with the following grammar:*

$$\mathsf{For} \qquad A ::= p \mid U_{\geq s} A \mid L_{\geq s} A \mid \neg A \mid A \wedge A \mid t : A$$

where $p \in \mathsf{Prop}$ *and* $s \in \mathbb{Q} \cap [0, 1]$.

Other connectives, $\vee, \rightarrow, \leftrightarrow$, are defined as usual. The following abbreviations will be used for introducing other types of inequalities:

$$U_{<s}A \equiv \neg U_{\geq s}A$$

$$L_{<s}A \equiv \neg L_{\geq s}A$$

$$U_{\leq s}A \equiv L_{\geq 1-s}\neg A$$

$$L_{\leq s}A \equiv U_{\geq 1-s}\neg A$$

$$U_{=s}A \equiv U_{\leq s}A \wedge U_{\geq s}A$$

$$L_{=s}A \equiv L_{\leq s}A \wedge L_{\geq s}A$$

$$U_{>s}A \equiv \neg U_{\leq s}A$$

$$L_{>s}A \equiv \neg L_{\leq s}A.$$

We set $A \wedge \neg A \equiv \bot$ and $A \vee \neg A \equiv \top$.

3.2 Axiomatization

Axioms of the logic ILUPJ:

(Ax1) $\vdash A$, where A is a propositional tautology
(Ax2) $\vdash t : (A \rightarrow B) \rightarrow (s : A \rightarrow (t \cdot s) : B)$
(Ax3) $\vdash t : A \vee s : A \rightarrow (t + s) : A$
(Ax4) $\vdash U_{\leq 1}A \wedge L_{\leq 1}A$
(Ax5) $\vdash U_{\leq r}A \rightarrow U_{<s}A$, $s > r$
(Ax6) $\vdash U_{<s}A \rightarrow U_{\leq s}A$
(Ax7) $\vdash (U_{\leq r_1}A_1 \wedge \cdots \wedge U_{\leq r_m}A_m) \rightarrow U_{\leq r}A$, if $A \rightarrow \bigvee_{J \subseteq \{1,\ldots,m\},|J|=k+n} \bigwedge_{j \in J} A_j$ and $\bigvee_{J \subseteq \{1,\ldots,m\},|J|=k} \bigwedge_{j \in J} A_j$ are propositional tautologies, where $r = \frac{\sum_{i=1}^{m} r_i - k}{n}$, $n \neq 0$
(Ax8) $\vdash \neg(U_{\leq r_1}A_1 \wedge \cdots \wedge U_{\leq r_m}A_m)$, if $\bigvee_{J \subseteq \{1,\ldots,m\},|J|=k} \bigwedge_{j \in J} A_j$ is a propositional tautology and $\sum_{i=1}^{m} r_i < k$
(Ax9) $\vdash L_{=1}(A \rightarrow B) \rightarrow (U_{\geq s}A \rightarrow U_{\geq s}B)$

Before we state the inference rules of the ILUPJ logic, we define a constant specification:

Definition 7 (Constant Specification). Constant specification CS *is any set that satisfies:*

$$\mathsf{CS} \subseteq \{(c, A) \mid c \text{ is a constant and } A \text{ is an instance of some axiom of ILUPJ}\}.$$

The constant specification is used to control an agent's reasoning capabilities, i.e. to specify which axioms the agent has a justification of. So we can model agents that are not logically omniscient. Assume that the constant specification includes (c, A) for some axiom A and some constant c. Then using rule (IR1), see below, we can infer $c : A$, i.e. the agent beliefs A and c justifies that belief. However, if for no constant c we have that $(c, A) \in \mathsf{CS}$, then the agent does not have an atomic justification for A, i.e. she may not have justified belief of the axiom A.

Inference Rules of the logic ILUPJ:

(IR1) $\vdash !^n c :!^{n-1} c : \cdots :!c : c : A$ where $(c, A) \in \mathsf{CS}$ and $n \in \mathbb{N}$
(IR2) If $T \vdash A$ and $T \vdash A \rightarrow B$ then $T \vdash B$
(IR3) If $\vdash A$ then $\vdash L_{\geq 1} A$
(IR4) If $T \vdash A \rightarrow U_{\geq s - \frac{1}{k}} B$, for every $k \geq \frac{1}{s}$ and $s > 0$ then $T \vdash A \rightarrow U_{\geq s} B$
(IR5) If $T \vdash A \rightarrow L_{\geq s - \frac{1}{k}} B$, for every $k \geq \frac{1}{s}$ and $s > 0$ then $T \vdash A \rightarrow L_{\geq s} B$

Axioms (Ax7) and (Ax8) together are the logical representation of the third condition from Theorem 1. Equivalent to saying that $\{\{X_1, \ldots, X_m\}\}$ covers a set X n times is to say that:

$$X \subseteq \bigcup_{J \subseteq \{1,\ldots,m\}, |J|=n} \bigcap_{j \in J} X_j.$$

Hence, the condition that the formula

$$A \rightarrow \bigvee_{J \subseteq \{1,\ldots,m\}, |J|=k+n} \bigwedge_{j \in J} A_j$$

is a tautology states that $[A]_{M,w}$[2] is covered $n + k$ times by a multiset

$$\{\{[A_1]_{M,w}, \ldots, [A_m]_{M,w}\}\},$$

while the condition that

$$\bigvee_{J \subseteq \{1,\ldots,m\}, |J|=k} \bigwedge_{j \in J} A_j$$

is a propositional tautology states that the set W is covered k times by a multiset $\{\{[A_1]_{M,w}, \ldots, [A_m]_{M,w}\}\}$.

Formula A is deducible from a set of formulas T, denoted by $T \vdash A$, if there exists at most countable sequence of formulas $A_0, A_1, \ldots, A$, where every A_i is an axiom or a formula that belongs to the set T, or is derived from the preceding formulas by some inference rule (exception is that the Rule (IR3) can be applied on the theorems only). Formula A is a theorem, denoted by $\vdash A$, if it can be deduced from the empty set.

[2] $[A]_{M,w}$ represents the set of all worlds from $W(w)$ in a model M where A holds and will be defined later.

3.3 Semantics

For sets of formulas X and Y, we will use the following notation:

$$X \cdot Y := \{A \mid B \to A \in X \text{ and } B \in Y, \text{ for some formula } B\}.$$

In order to provide semantics for the logic ILUPJ, we start with the notion of a basic evaluation.

Definition 8 (Basic Evaluation). *Let* CS *be a constant specification. A* basic CS-evaluation *is a function* $*$, *such that*

$$* : \mathsf{Prop} \to \{true, false\} \quad and \qquad * : \mathsf{Tm} \to \mathcal{P}(\mathsf{For}),$$

and for $s, t \in \mathsf{Tm}$, $c \in \mathsf{Con}$ *and* $A \in \mathsf{For}$ *we have:*

1) $s^* \cdot t^* \subseteq (s \cdot t)^*$
2) $s^* \cup t^* \subseteq (s + t)^*$
3) if $(c, A) \in \mathsf{CS}$ *then*
  a) $A \in c^*$
  b) $!^n c :!^{n-1} c : \cdots :!c : c : A \in (!^{n+1} c)^*$, *for* $n \in \mathbb{N}$.

We will write t^* and p^* instead of $*(t)$ and $*(p)$ respectively.

Definition 9 ($\mathsf{ILUPJ_{CS}}$-Model). *Let* CS *be any constant specification. An* $\mathsf{ILUPJ_{CS}}$-model *(or simply* model*) is a tuple* $\langle W, LUP, * \rangle$, *where:*

- W *is a nonempty set of worlds.*
- LUP *assigns to every* $w \in W$ *a space, such that* $LUP(w) = \langle W(w), H(w), P(w) \rangle$, *where:*
  - $\emptyset \neq W(w) \subseteq W$,
  - $H(w)$ *is an algebra of subsets of* $W(w)$ *and*
  - $P(w)$ *is a set of finitely additive probability measures defined on* $H(w)$.
- $*$ *is a function from* W *to the set of all basic* CS*-evaluations, i.e.* $*(w)$ *is a basic* CS*-evaluation for each world* $w \in W$.

We will denote $*(w)$ by $*_w$.

Definition 10 (Truth in a Model). *Let* CS *be any constant specification. and let* $M = \langle W, LUP, * \rangle$ *be a model. We define what does it mean for a formula* $A \in \mathsf{For_{ILUPJ}}$ *to hold in* M *at the world* w *by:*

- $M, w \models p$ *iff* $p^*_w = true$, *for* $p \in \mathsf{Prop}$
- $M, w \models U_{\geq s} A$ *iff* $P^*(w)([A]_{M,w}) \geq s$,
- $M, w \models L_{\geq s} A$ *iff* $P_*(w)([A]_{M,w}) \geq s$,
- $M, w \models \neg A$ *iff* $M \not\models A$,
- $M, w \models A \wedge B$ *iff* $M \models A$ *and* $M \models B$,
- $M, w \models t : A$ *iff* $A \in t^*_w$,

where[3] $[A]_{M,w} = \{u \in W(w) \mid M, u \models A\}$ *and* $W(w)$ *and* $P(w)$ *are given by* $LUP(w)$. *The functions* $P^*(w)$ *and* $P_*(w)$ *are defined as in Definition 3.*

Definition 11 (Measurable Model). *Let* CS *be a constant specification and let* M *be a model.* M *is said to be* measurable *if* $[A]_{M,w} \in H(w)$ *for every* $A \in \mathsf{For}$. *The class of all measurable* $\mathsf{ILUPJ}_{\mathsf{CS}}$*-models will be denoted by* $\mathsf{ILUPJ}_{\mathsf{CS},Meas}$.

For a model M, we write $M \models A$ if for every $w \in W$, $M, w \models A$. For $T \subseteq \mathsf{For}$, $M \models T$ means that $M \models A$ for every $A \in T$. Finally, $T \models A$ means that $M \models T$ implies $M \models A$.

Definition 12 (Satisfiability). *Formula* A *is satisfiable if there exists a measurable model* M *and* $w \in W$ *such that* $M, w \models A$. *A set of formulas* T *is satisfiable if every formula in* T *is satisfiable.*

As usual, we have the deduction theorem.

Theorem 2 (Deduction Theorem). *Let* $A, B \in \mathsf{For}$, T *a set of formulas and* CS *be any constant specification. Then* $T \cup \{A\} \vdash B$ *iff* $T \vdash A \rightarrow B$.

Proof The proof is completely standard. We only show the case in the direction from left to right where the last rule application is an instance of (IR4). In this case $B = C \rightarrow U_{\geq s}B'$. We have:

(1) $T, A \vdash C \rightarrow U_{\geq s-\frac{1}{k}}B'$, for all $k \geq \frac{1}{s}$
(2) $T \vdash A \rightarrow (C \rightarrow U_{\geq s-\frac{1}{k}}B')$, for all $k \geq \frac{1}{s}$ by induction hypothesis
(3) $T \vdash (A \wedge C) \rightarrow U_{\geq s-\frac{1}{k}}B'$, for all $k \geq \frac{1}{s}$
(4) $T \vdash (A \wedge C) \rightarrow U_{\geq s}B'$ by (IR4)
(5) $T \vdash A \rightarrow (C \rightarrow U_{\geq s}B')$,

which is $T \vdash A \rightarrow B$. □

We also need the following technical lemma.

Lemma 1

(a) $\vdash U_{\geq s}A \rightarrow U_{>r}A,\ s > r$
(b) $\vdash U_{>s}A \rightarrow U_{\geq s}A$
(c) If $\vdash A \leftrightarrow B$ *then* $\vdash U_{\geq s}A \leftrightarrow U_{\geq s}B$

Proof From (Ax5) and (Ax6), using contraposition we obtain proofs for (a) and (b), while (c) is a direct consequence of (IR3) and (Ax9). □

4 Soundness and Completeness

The soundness theorem can be proved as usual by transfinite induction on the depth of the derivation $T \vdash A$.

Theorem 3 (Soundness). *Let* CS *be a constant specification. For* $T \subseteq \mathsf{For}$ *and* $A \in \mathsf{For}$ *we have:*

$$T \vdash A \quad \Rightarrow \quad T \models A.$$

[3] When M is clear from the context we will write $[A]_w$.

4.1 Completeness

Definition 13 ($\mathsf{ILUPJ}_{\mathsf{CS}}$-Consistent Set). *For an arbitrary constant specification* CS *and* $T \subseteq$ For *we say that:*

(a) T *is* $\mathsf{ILUPJ}_{\mathsf{CS}}$-consistent *if and only if* $T \nvdash \bot$. *Otherwise,* T *is* $\mathsf{ILUPJ}_{\mathsf{CS}}$-inconsistent.
(b) T *is* maximal *if and only if for all* $A \in$ For, *either* $A \in T$ *or* $\neg A \in T$.
(c) T *is* maximal $\mathsf{ILUPJ}_{\mathsf{CS}}$-consistent *if and only if it is maximal and* $\mathsf{ILUPJ}_{\mathsf{CS}}$-*consistent.*

Lemma 2. *Let* CS *be an arbitrary constant specification and* T *an* $\mathsf{ILUPJ}_{\mathsf{CS}}$-*consistent set of formulas.*

(1) For any $A \in$ For, *either* $T \cup \{A\}$ *is* $\mathsf{ILUPJ}_{\mathsf{CS}}$-*consistent or* $T \cup \{\neg A\}$ *is* $\mathsf{ILUPJ}_{\mathsf{CS}}$-*consistent.*
(2) If $\neg(A \to U_{\geq s}B) \in T$, *then there exists some* $n > \frac{1}{s}$ *such that*

$$T \cup \{A \to \neg U_{\geq s-\frac{1}{n}}B\}$$

is $\mathsf{ILUPJ}_{\mathsf{CS}}$-*consistent.*
(3) If $\neg(A \to L_{\geq s}B) \in T$, *then there exists some* $n > \frac{1}{s}$ *such that*

$$T \cup \{A \to \neg L_{\geq s-\frac{1}{n}}B\}$$

is $\mathsf{ILUPJ}_{\mathsf{CS}}$-*consistent.*

Proof. (1) Suppose that both $T \cup \{A\} \vdash \bot$ and $T \cup \{\neg A\} \vdash \bot$ hold. From the Deduction Theorem, we get $T \vdash \neg A$ and $T \vdash A$ which contradicts the assumption that the set T is $\mathsf{ILUPJ}_{\mathsf{CS}}$-consistent.
(2) Assume that for all $n > \frac{1}{s}$ we have:

$$T, A \to \neg U_{\geq s-\frac{1}{n}}B \vdash \bot.$$

From Deduction Theorem and propositional reasoning, we obtain

$$T \vdash A \to U_{\geq s-\frac{1}{n}}B,$$

and from Inference Rulle 4 $T \vdash A \to U_{\geq s}B$. Contradiction with the assumption that $\neg(A \to U_{\geq s}B) \in T$.
(3) Similar to the previous case. □

Theorem 4 (Lindenbaum). *Let* CS *be an arbitrary constant specification. Every* $\mathsf{ILUPJ}_{\mathsf{CS}}$-*consistent set can be extended to a maximal* $\mathsf{ILUPJ}_{\mathsf{CS}}$-*consistent set.*

Proof. Consider a $\mathsf{ILUPJ}_{\mathsf{CS}}$-consistent set T and let $A_0, A_1, A_2, \ldots$ be an enumeration of all the formulas from For. We define a sequence of sets T_i, $i = 0, 1, 2, \ldots$ in the following way:

(1) $T_0 = T$,
(2) for every $i \geq 0$,
  (a) if $T_i \cup \{A_i\}$ is $\mathsf{ILUPJ_{CS}}$-consistent, then $T_{i+1} = T_i \cup \{A_i\}$, otherwise
  (b) if A_i is of the form $B \to U_{\geq s}C$, then $T_{i+1} = T_i \cup \{\neg A_i, B \to \neg U_{\geq s-\frac{1}{n}}C\}$, for some $n > 0$, so that T_{i+1} is $\mathsf{ILUPJ_{CS}}$-consistent, otherwise
  (c) if A_i is of the form $B \to L_{\geq s}C$, then $T_{i+1} = T_i \cup \{\neg A_i, B \to \neg L_{\geq s-\frac{1}{n}}C\}$, for some $n > 0$, so that T_{i+1} is $\mathsf{ILUPJ_{CS}}$-consistent, otherwise
  (d) $T_{i+1} = T_i \cup \{\neg A_i\}$,
(3) $T^{\spadesuit} = \bigcup_{i=0}^{\infty} T_i$.

Using induction on i, we prove that for every $i \in \mathbb{N}$, T_i is $\mathsf{ILUPJ_{CS}}$-consistent.

(i) T_0 is $\mathsf{ILUPJ_{CS}}$-consistent because T is.
(ii) Suppose that T_i is $\mathsf{ILUPJ_{CS}}$-consistent. We prove that also T_{i+1} is:
  - T_{i+1} is constructed using the step (2)(a). Trivially.
  - T_{i+1} is constructed using the step (2)(b). From Lemma 2((1) and (2)).
  - T_{i+1} is constructed using the step (2)(c). From Lemma 2((1) and (3)).
  - T_{i+1} is constructed using the step (2)(d). Since $T_i \cup \{A_i\}$ is $\mathsf{ILUPJ_{CS}}$-inconsistent, we know that $T_i \cup \{\neg A_i\}$ is $\mathsf{ILUPJ_{CS}}$-consistent.

Now let us show that $T^{\spadesuit}$ is maximal $\mathsf{ILUPJ_{CS}}$-consistent set. From the construction above we know that for any $A \in \mathsf{For}$ either $A \in T^{\spadesuit}$ or $\neg A \in T^{\spadesuit}$, i.e., $T^{\spadesuit}$ is maximal.

In order to prove that $T^{\spadesuit}$ is ILUPJ-consistent, we prove that:

(*i*) It does not contain all the formulas from For;
(*ii*) It is deductively closed.

It is clear from the construction that $T^{\spadesuit}$ does not contain all the formulas from For, so the only thing left to prove is that $T^{\spadesuit}$ is deductively closed. Assume $T^{\spadesuit} \vdash A$. Using transfinite induction on a depth of derivation we prove that $A \in T^{\spadesuit}$.

1) $A \in T^{\spadesuit}$. Trivially.
2) A is an instance of some of the axioms (Ax1)–(Ax9). There exists $k \in \mathbb{N}$ with $A = A_k$. Assuming that $\neg A_k \in T_{k+1}$, we get a contradiction from:

$$T_{k+1} \vdash A_k \quad \text{and} \quad T_{k+1} \vdash \neg A_k.$$

3) A is obtained from $T^{\spadesuit}$ by an application of (IR1), i.e.,

$$A =!^n c :!^{n-1} c : \cdots :!c : c : B,$$

for some $n \in \mathbb{N}$, axiom B and $(c, B) \in \mathsf{CS}$. There exists k such that $A = A_k$ and if $\neg A \in T_{k+1}$, then

$$T_{k+1} \vdash A \quad \text{and} \quad T_{k+1} \vdash \neg A$$

which gives us a contradiction.

4) A is obtained from $T^\spadesuit$ by an application of (IR2). Induction hypothesis tells us that there exists l, such that both premises belong to T_l. Since there exists k such that $A = A_k$, if $\neg A \in T_{max(k,l)+1}$, then

$$T_{max(k,l)+1} \vdash A \quad \text{and} \quad T_{max(k,l)+1} \vdash \neg A$$

which gives us a contradiction.
5) A is obtained from $T^\spadesuit$ by an application of (IR3), i.e., $A = L_{\geq 1}B$ and $\vdash B$. Since there exists some k such that $A = A_k$, same reasoning as in 2) gives us the claim.
6) A is obtained from $T^\spadesuit$ by an application of (IR4). That means, $A = B \rightarrow U_{\geq s}C$ and for every $k \geq \frac{1}{s}$,

$$T^\spadesuit \vdash B \rightarrow U_{\geq s-\frac{1}{k}}C.$$

Assuming that $A \notin T^\spadesuit$, i.e., $\neg(B \rightarrow U_{\geq s}C) \in T^\spadesuit$, we find a number m, such that

$$\neg(B \rightarrow U_{\geq s}C) \in T_m.$$

Also, from the construction of $T^\spadesuit$ we know that for some l,

$$\neg(B \rightarrow U_{\geq s-\frac{1}{l}}C) \in T_l.$$

Further, from inductive hypothesis,

$$B \rightarrow U_{\geq s-\frac{1}{l}}C \in T^\spadesuit.$$

Hence, there exists m' with

$$B \rightarrow U_{\geq s-\frac{1}{l}}C \in T_{m'}.$$

Contradiction with a consistency of $T_{max(l,m')+1}$, since both

$$B \rightarrow U_{\geq s-\frac{1}{l}}C \in T_{max(l,m')+1}, \quad \neg(B \rightarrow U_{\geq s-\frac{1}{l}}C) \in T_{max(l,m')+1}.$$

7) The case when A is obtained from $T^\spadesuit$ by an application of (IR5) can be proved similarly to the previous case.

We conclude that $T^\spadesuit$ is deductively closed set which does not contain all formulas meaning that $T^\spadesuit$ is consistent. □

Definition 14 (Canonical Model). *Let* CS *be an arbitrary constant specification. The* canonical model *is the tuple* $M_{can} = \langle W, LUP, *\rangle$ *, where:*

1) $W = \{w \mid w$ *is a maximal* $\mathsf{ILUPJ_{CS}}$*-consistent set of formulas*$\}$,
2) $LUP(w) = \langle W(w), H(w), P(w)\rangle$ *is defined as follows:*
   $W(w) = W$,
   $H(w) = \{\{u \mid u \in W(w), A \in u\} \mid A \in \mathsf{For}\}$,

$P(w)$ is any set of probability measures such that

$$P^*(w)(\{u \mid u \in W(w), A \in u\}) = sup\{s \mid U_{\geq s}A \in w\}.$$

3) for every world $w \in W$, the basic CS*-evaluation is defined with:*
  1. For $p \in$ Prop*:*

$$p^*_w = \begin{cases} true & if\ p \in w \\ false & if\ \neg p \in w \end{cases}$$

  2. For $t \in$ Tm*:*

$$t^*_w = \{A \mid t : A \in w\}$$

Lemma 3. *Let $M_{can} = \langle W, LUP, * \rangle$ be the canonical model. For every $u \in W$ and every formula A,*

$$\{u \mid u \in W, A \in u\} = [A]_{M_{can},u}.$$

Proof. We prove the statement by proving that $A \in u$ iff $u \models A$ by induction on the length of A. If $A = p$ the claim follows by definition of the canonical model. Cases when $A = \neg B$ or $A = B \wedge C$ are trivial.

1. Let $A = U_{\geq s}B$. First, let $U_{\geq s}B \in u$. Then

$$\sup\{r \mid U_{\geq r}B \in u\} = P^*(u)\{w \mid w \in W, B \in w\} = P^*(u)([B]_u) \geq s,$$

   so $u \models U_{\geq s}B$.
   Now, suppose that $u \models U_{\geq s}B$, i.e.

$$P^*(u)([B]_u) = \sup\{r \mid U_{\geq r}B \in u\} \geq s.$$

   If $P^*(u)([B]_u) > s$, then we have (properties of a supremum and monotonicity) $U_{\geq s}B \in u$.
   If $P^*(u)([B]_u) = s$, then as a direct consequence of (IR4), we have that $U_{\geq s}B \in u$.
2. Now, let $A = L_{\geq s}B$ or equivalently $A = U_{\leq 1-s}\neg B$. Suppose $U_{\leq 1-s}\neg B \in u$. Our goal is to show that

$$\sup\{r \mid U_{\geq r}\neg B \in u\} \leq 1 - s,$$

   hence, suppose towards contradiction that

$$\sup\{r \mid U_{\geq r}\neg B \in u\} > 1 - s.$$

   Then, there exists a rational number $q \in (1 - s, 1 - s + \epsilon]$, for some $\epsilon > 0$, such that $U_{\geq q}\neg B \in u$. Therefore, $U_{>1-s}\neg B \in u$. Contradiction. That means

$$\sup\{r \mid U_{\geq r}\neg B \in u\} \leq 1 - s,$$

   i.e., $P^*(u)([\neg B]_u) \leq 1 - s$ and therefore we obtain $u \models L_{\geq s}B$.
   For the other direction, assume that $u \models U_{\leq 1-s}\neg B$, i.e.

$$\sup\{r \mid U_{\geq r}\neg B \in u\} \leq 1 - s.$$

   We distuingish the following cases:

(1) $\sup\{r \mid U_{\geq r}\neg B \in u\} < 1-s$. In this case, if $U_{>1-s}\neg B \in u$, we would have also that $U_{\geq 1-s}\neg B \in u$, so $\sup\{r \mid U_{\geq r}\neg B \in u\} \geq 1-s$. Contradiction.

(2) $\sup\{r \mid U_{\geq r}\neg B \in u\} = 1-s$. We want to show that then it must hold

$$\inf\{r \mid U_{\leq r}\neg B \in u\} = 1-s.$$

Suppose first towards contradiction that

$$\inf\{r \mid U_{\leq r}\neg B \in u\} < 1-s.$$

Then there exists a rational number $q_1 \in [1-s-\epsilon, 1-s)$ such that $U_{\leq q_1}\neg B \in u$, and so $U_{<1-s}\neg B \in u$. Contradiction with $U_{\geq 1-s}\neg B \in u$ (this follows directly from Inference Rule 4). Now, suppose that

$$\inf\{r \mid U_{\leq r}\neg B \in u\} > 1-s,$$

i.e.,

$$\inf\{r \mid U_{\leq r}\neg B \in u\} = 1-s+\varepsilon.$$

Taking an arbitrary rational number $q_2 \in (1-s, 1-s+\varepsilon)$, we obtain that both

$$U_{\leq q_2}\neg B \in u \quad \text{and} \quad U_{\geq q_2}\neg B \in u$$

which contradicts properties of an infimum and supremum. Hence

$$\inf\{r \mid U_{\leq r}\neg B \in u\} = 1-s,$$

or equivalently

$$\inf\{r \mid L_{\geq 1-r}B \in u\} = 1-s$$

and as a consequence of an Inference Rule 5, we get $L_{\geq s}B \in u$.

3. Finally let $A = t : B$. Since $\{u \mid u \in W, u \models t : B\} = [A]_{M_{can},u}$ and

$$\{u \mid u \in W, t : B \in u\} = \{u \mid u \in W, B \in t_u^*\} = \{u \mid u \in W, u \models t : B\},$$

the proof is finished. □

Theorem 5. *Let* CS *be an arbitrary constant specification.* M_{can} *is a* $\mathsf{ILUPJ}_{\mathsf{CS},Meas}$*-model.*

Proof. Since there exists a maximal $\mathsf{ILUPJ}_{\mathsf{CS}}$-consistent set, we know that $W \neq \emptyset$ and $W(w) \neq \emptyset$. Proof that $H(w)$ is an algebra is straightforward. Also note that for every $w \in W$, $*_w$ is a basic CS-evaluation by the construction of the canonical model.

Let us prove the existence of a set of probability measures $P(w)$ claimed in 2) and that $P^*(w)$ is well defined.

1) *There exists a set of finitely additive probability measures* $P(w)$ *and* $P^*(w)$ *is an upper probability measure for* $P(w)$*:*

We prove the three conditions from Theorem 1 and since the first two conditions, $P^*(w)(\emptyset) = 0$ and $P^*(w)(W) = 1$, are trivial, we prove only the third, i.e., if

$$\{\{[A_1], \dots, [A_m]\}\}$$

is an (n, k)-cover of $([A], W)$, then

$$k + nP^*(w)([A]) \leq \sum_{i=1}^{m} P^*(w)([A_i]).$$

Let $P^*(w)([A_i]) = \sup\{r \mid U_{\geq r} A_i \in w\} = a_i$, for $i = 1, \dots, m$. For an arbitrary fixed $\varepsilon > 0$, there exist rational numbers $q_i \in (a_i, a_i + \varepsilon)$ with $U_{\leq q_i} A_i \in w$. If that would not be the case, then $U_{>q_i} A_i \in w$ which contradicts with the fact that a_i is supremum. As a consequence we get

$$w \vdash U_{\leq q_1} A_1 \wedge \dots \wedge U_{\leq q_m} A_m,$$

and by (Ax7)

$$w \vdash U_{\leq q} A,$$

where $q = \frac{\sum_{i=1}^{m} q_i - k}{n}$, $n \neq 0$. Thus, $\sup\{r \mid U_{\geq r} A_i \in w\} \leq q$ or equivalently $P^*(w)([A]) \leq q$. Thus, we have

$$P^*(w)([A]) \leq \frac{\sum_{i=1}^{m} q_i - k}{n} = \frac{\sum_{i=1}^{m} a_i + m\varepsilon - k}{n}.$$

Because this holds for every $\varepsilon > 0$ we obtain $k + nP^*(w)([A]) \leq \sum_{i=1}^{m} P^*(w)([A_i])$. If $n = 0$, we have to show that $k \leq \sum_{i=1}^{m} P^*(w)([A_i])$. Reasoning as above, we obtain

$$w \vdash U_{\leq q_1} A_1 \wedge \dots \wedge U_{\leq q_m} A_m,$$

for some $q_i \in (a_i, a_i + \varepsilon)$. From (Ax8), how

$$\bigvee_{J \subseteq \{1,\dots,m\}, |J|=k} \bigwedge_{j \in J} A_j$$

is a propositional tautology, we have that $\sum_{i=1}^{m} q_i \geq k$. Again, from the fact that it holds for every $\varepsilon > 0$, we obtain $\sum_{i=1}^{m} a_i \geq k$.

2) $P^*(w)$ *is well defined*: that Lemma 1(c) tells us that a value of the supremum does not depend on a choice of an element from $[A]$. Hence $P^*(w)([A])$ is well defined.
Finally, note that as a direct consequence of the Lemma 3 we have that this model is measurable. □

Theorem 6 (Strong Completeness for ILUPJ). *For an arbitrary constant specification* CS*,* $T \subseteq$ For *and* $A \in$ For *we have:*

$$T \models A \quad \Rightarrow \quad T \vdash A.$$

Proof. Suppose that $T \nvdash A$ or equivalently $T \nvdash \neg A \rightarrow \bot$. From Deduction Theorem we get $T, \neg A \nvdash \bot$ meaning that the set $T \cup \{\neg A\}$ is $\mathsf{ILUPJ}_{\mathsf{CS}}$-consistent. From Theorem 4 we know that there exists a maximal $\mathsf{ILUPJ}_{\mathsf{CS}}$-consistent set $T^\spadesuit$ with $T \cup \{\neg A\} \subseteq T^\spadesuit$. Finally, since $T^\spadesuit$ is a world in the canonical model, we get $M_{can}, T^\spadesuit \models T$ and $M_{can}, T^\spadesuit \models \neg A$ and thus $T \not\models A$. □

5 ILUPJ as a Generalization of the Logic PPJ

In this section we prove that the logic ILUPJ generalizes the logic PPJ from [10]. The strategy we use relies heavily on the strategy used in [6]. Let us briefly recall the logic PPJ.

The language of the logic PPJ extends the language of the justification logic J with the list of operators $P_{\geq s}$, where s is a rational number from the $[0, 1]$. For example,

$$p \wedge P_{\leq \frac{1}{2}}(t : q) \quad \text{and} \quad P_{=\frac{1}{3}} P_{\geq 1}(s : (p \vee r))$$

are well defined formulas. PPJ-models are defined as triples $M = \langle W, Prob, * \rangle$, where:

- W is a non empty set of worlds
- $Prob$ is an assignment which assigns to every $w \in W$ a probability space, such that $Prob(w) = \langle W(w), H(w), \mu(w) \rangle$, where:
  $W(w)$ is a non empty subset of W,
  $H(w)$ is an algebra of subsets of $W(w)$ and
  $\mu(w) : H(w) \rightarrow [0, 1]$ is a finitely additive probability measure.
- $*_w$ is a basic CS-evaluation.

Satisfiability of a formula is defined as expected for the justification logic formulas and

$$M, w \models P_{\geq s} A \text{ iff } \mu(w)(\{v \in W(w) \mid v \models A\}) \geq s.$$

Axiomatization of the logic PPJ is the following:

(P1) $\vdash A$, where A is a propositional tautology
(P2) $\vdash t : (A \rightarrow B) \rightarrow (s : A \rightarrow (t \cdot s) : B)$
(P3) $\vdash t : A \vee s : A \rightarrow (t + s) : A$
(P4) $P_{\geq 0} A$,
(P5) $P_{\leq r} A \rightarrow P_{< s} A$, $s > r$,
(P6) $P_{< s} A \rightarrow P_{\leq s} A$,
(P7) $(P_{\geq t} A \wedge P_{\geq s} B \wedge P_{\geq 1}(\neg A \vee \neg B)) \rightarrow P_{\geq min\{1, t+s\}}(A \vee B)$,
(P8) $(P_{\leq t} A \wedge P_{< s} B) \rightarrow P_{< t+s}(A \vee B)$, $t + s \leq 1$.

Inference Rules

(1) $\vdash !^n c : !^{n-1} c : \cdots : !c : c : A$ where $(c, A) \in \mathsf{CS}$ and $n \in \mathbb{N}$
(2) If $T \vdash A$ and $T \vdash A \rightarrow B$ then $T \vdash B$
(3) If $\vdash A$ then $\vdash P_{\geq 1} A$

(4) If $T \vdash A \rightarrow P_{\geq s-\frac{1}{k}}B$, for every $k \geq \frac{1}{s}$ and $s > 0$ then $T \vdash A \rightarrow P_{\geq s}B$

Soundness and strong completeness theorems for the logic PPJ are proved (see [10], Theorems 11 and 22).

The ILUPJ logic has the similar semantical structure as the logic PPJ. Also, it is clear that the semantics of the logic ILUPJ is more general, since reasoning about upper and lower probabilities requires *sets* of probability measures, while in the logic PPJ one measure per possible world is sufficient (thus they are isomorphic to the "sets of" probability measures which are singletons).

However, the axiomatic systems are quite different. We focus on the two proof theoretical aspects of the generalization:

1. which axioms should be added to the logic ILUPJ to reduce the proposed class of models to the class of models isomorphic to the models for the logic PPJ
2. how can we use the added axioms to formally obtain the axiomatization of PPJ.

As already stated, the subclass of the ILUPJ-models that contains only those structures where the set of probability measures is a singleton set is isomorphic to the class of PPJ-models. Thus, we add the following axiom which guarantees that it is the case:

$$(Ax10) \quad U_{\geq r}A \rightarrow L_{\geq r}A. \tag{1}$$

We will denote ILUPJ+Axiom $(Ax10)$ by ILUPJ^{Ext}.

It can easily be proved that the following holds (see the proof of Proposition 1 in [18]):

$$\vdash U_{\leq r}A \rightarrow L_{\leq r}A. \tag{2}$$

From (1) and (2) follows that operators U and L have the same behavior in the sense that for every formula A and every $r \in \mathbb{Q} \cap [0,1]$

$$\vdash U_{\geq r}A \leftrightarrow L_{\geq r}A. \tag{3}$$

As a consequence we have that in ILUPJ^{Ext} one type of operators is sufficient, since changing one type of operator with other will lead to an equivalent formula. For example, if we replace all the operators for lower probability with the operators of upper probability in $A \equiv L_{\geq\frac{1}{3}}U_{\leq\frac{1}{2}}L_{=1}(t:p)$, we will obtain the formula B equivalent to A $B \equiv U_{\geq\frac{1}{3}}U_{\leq\frac{1}{2}}U_{=1}(t:p)$. It can be proved in a straightforward manner by the induction on the complexity of a formula that this holds for any formula. This fact allows us, without loss of generality, to consider only formulas with the U operators in ILUPJ^{Ext}.

Our aim is to prove that the set of theorems of the logic PPJ is a subset of the set of theorems of the logic ILUPJ^{Ext}. In order to prove that, we show that all the axioms and inference rules of the logic PPJ can be inferred in the logic ILUPJ^{Ext}, where an operator P is replaced by U.

First not that the axioms (P1)–(P6) correspond to the axioms (Ax1)–(Ax6) and inference rules coincide as well. Our goal is to prove that the appropriate counterparts of the axioms (P7) and (P8), i.e.,

(U7) $(U_{\geq t}B \wedge U_{\geq s}C \wedge U_{\geq 1}(\neg B \vee \neg C)) \rightarrow U_{\geq min\{1,t+s\}}(B \vee C)$,
(U8) $(U_{\leq t}B \wedge U_{<s}C) \rightarrow U_{<t+s}(B \vee C)$, $t+s \leq 1$,

follow from the axiomatization of ILUPJ^{Ext}, where in that inference the essential role is played by the axioms (Ax7) and (Ax8).

In order to prove that we need the following Lemma:

Lemma 4. $\mathsf{ILUPJ}^{Ext} \vdash (U_{\leq t}B \wedge U_{\leq s}C) \rightarrow U_{\leq t+s}(B \vee C)$, $t+s \leq 1$.

Proof. We will show that the claim can be inferred from the axiom (Ax7). Consider the axiom (Ax7) for:

$m = 2$; $n = 1, k = 0$; $r_1 = t$; $r_2 = s$;
$A_1 = B$; $A_2 = C$; $A = B \vee C$.

In this case we get $r = t + s$ and therefore the Axiom (Ax7) has exactly the shape of the required formula. We also have to check whether the formulas

$$A \rightarrow \bigvee_{J \subseteq \{1,2\}, |J|=1} \bigwedge_{j \in J} A_j$$

and

$$\bigvee_{J \subseteq \{1,2\}, |J|=0} \bigwedge_{j \in J} A_j$$

are tautologies. The first formula has the form $B \vee C \rightarrow B \vee C$ which is clearly a tautology, while the second formula has the form $\bigwedge_{j \in \emptyset} A_j$, and $\bigwedge_{j \in \emptyset} A_j = \top$ by definition and hence a tautology. □

Theorem 7. *The set of theorems of the logic* PPJ *is a subset of the set of theorems of the logic* ILUPJ^{Ext}.

Proof. As already mentioned, we only need to prove that:

(a) $\mathsf{ILUPP}^{Ext} \vdash (U_{\geq t}B \wedge U_{\geq s}C \wedge U_{\geq 1}(\neg B \vee \neg C)) \rightarrow U_{\geq min\{1,t+s\}}(B \vee C)$,
(b) $\mathsf{ILUPP}^{Ext} \vdash (U_{\leq t}B \wedge U_{<s}C) \rightarrow U_{<t+s}(B \vee C)$, $t+s \leq 1$.

Proof of (a). First recall that the formula

$$(U_{\geq t}B \wedge U_{\geq s}C \wedge U_{\geq 1}(\neg B \vee \neg C)) \rightarrow U_{\geq min\{1,t+s\}}(B \vee C)$$

can be written as:

$$(U_{\leq 1-t}\neg B \wedge U_{\leq 1-s}\neg C \wedge U_{\leq 0}(B \wedge C)) \rightarrow U_{\leq 1-min\{1,t+s\}}\neg(B \vee C).$$

Now, consider the axiom (Ax7) for:

$m = 3$; $n = k = 1$; $r_1 = 1 - t$; $r_2 = 1 - s$; $r_3 = 0$;
$A_1 = \neg B$; $A_2 = \neg C$; $A_3 = B \wedge C$; $A = \neg(B \vee C)$.

We obtain that $r = 1 - (t + s)$.

(i) If $t + s > 1$ then (Axiom (Ax8), $\sum_{i=1}^{m} r_i < k$)

$$\vdash \neg(U_{\leq 1-t}\neg B \wedge U_{\leq 1-s}\neg C \wedge U_{\leq 0}(B \wedge C)),$$

so $\vdash (U_{\leq 1-t}\neg B \wedge U_{\leq 1-s}\neg C \wedge U_{\leq 0}(B \wedge C)) \to U_{\leq 1-min\{1,t+s\}}\neg(B \vee C))$.
(ii) If $t+s \leq 1$, then $1 - min\{1, t+s\} = 1 - (t+s) = r$ and it is left to check if

$$A \to \bigvee_{J\subseteq\{1,2,3\},|J|=2} \bigwedge_{j\in J} A_j$$

and

$$\bigvee_{J\subseteq\{1,2,3\},|J|=1} \bigwedge_{j\in J} A_j$$

are tautologies. Namely, in this case, the first formula has the following form:

$$\neg(B \vee C) \to ((\neg B \wedge \neg C) \vee (\neg B \wedge B \wedge C) \vee (\neg C \wedge B \wedge C)),$$

and the second formula:

$$\neg B \vee \neg C \vee (B \wedge C).$$

It is obvious that both of these formulas are tautologies and therefore this part is proved.

Proof of (b). Let us show equivalently that $\mathsf{ILUPJ}^{Ext} \vdash (U_{\leq t}B \wedge U_{\geq t+s}(B \vee C)) \to U_{\geq s}C$:

$\vdash U_{\geq t+s}(B \vee C) \to U_{>t+s'}(B \vee C)$, for all $s' < s$ (contraposition (Ax5))
$U_{\leq t}B \wedge U_{\geq t+s}(B \vee C) \vdash U_{\leq t}B \wedge U_{>t+s'}(B \vee C)$, for all $s' < s$
$U_{\leq t}B \wedge U_{\geq t+s}(B \vee C) \vdash U_{\leq t}B \wedge U_{>s'}C$, for all $s' < s$ (by Lemma 4)
$U_{\leq t}B \wedge U_{\geq t+s}(B \vee C) \vdash U_{\geq s}C$ (by (IR4))
$\vdash (U_{\leq t}B \wedge U_{\geq t+s}(B \vee C)) \to U_{\geq s}C$ (by Deduction theorem) □

6 Conclusion

We present a logic which allows making statements about upper and lower probabilities of the justification formulas. In this framework, we can represent information like: "t is justification that probability of A lies in the interval..." and our formalism, the logic ILUPJ, can be used for reasoning not only about lower and upper probabilities of a certain justification formula, but also about uncertain belief about other imprecise probabilities. The language of our logic is modal language which extends justification logic language with the unary operators $U_{\geq r}$ and $L_{\geq r}$, where r ranges over the unit interval of rational numbers. The corresponding semantics consist of the measurable Kripke models with sets of finitely additive probability measures attached to each possible world, as well as a function $*$ from the set of worlds to the set of all basic CS-evaluations. We prove that the proposed axiomatic system is strongly complete with respect to the class of measurable models.

We also provided an extension of the proposed axiomatization in order to prove that the logic ILUPJ is a generalization of the logic PPJ for reasoning about sharp probabilities of justification formulas from [10].

Acknowledgement. We would like to thank the anonymous reviewers whose comments helped to improve the paper substantially.

References

1. Anger, B., Lembcke, J.: Infinitely subadditive capacities as upper envelopes of measures. Zeitschrift fur Wahrscheinlichkeitstheorie und Verwandte Gebiete **68**, 403–414 (1985)
2. Artemov, S., Fitting, S.: Justification Logic: Reasoning with Reasons, Cambridge University Press, New York, June 2019
3. Artemov, S.N.: Explicit provability and constructive semantics. Bull. Symbol. Logic **7**(1), 1–36 (2001)
4. Artemov, S.: On aggregating probabilistic evidence. In: Artemov, S., Nerode, A. (eds.) LFCS 2016. LNCS, vol. 9537, pp. 27–42. Springer, Cham (2016). https://doi.org/10.1007/978-3-319-27683-0_3
5. Doder, D., Marinković, B., Maksimović, P., Perović, A.: A logic with conditional probability operators. Publications de l'Institut Mathématique **87**(101) (2010)
6. Doder, D., Savić, N., Ognjanović, Z.: Multi-agent logics for reasoning about higher-order upper and lower probabilities. J. Logic Lang. Inf. 1–31 (2019)
7. Fitting, M.: The logic of proofs, semantically. Ann. Pure Appl. Logic **132**(1), 1–25 (2005)
8. Halpern, J.Y., Pucella, R.: A logic for reasoning about upper probabilities. J. Artif. Intell. Res. **17**, 57–81 (2002)
9. Kokkinis, I., Maksimović, P., Ognjanović, Z., Studer, T.: First steps towards probabilistic justification logic. Logic J. IGPL **23**(4), 662–687 (2015)
10. Kokkinis, I., Ognjanović, Z., Studer, T.: Probabilistic justification logic. In: Artemov, S., Nerode, A. (eds.) LFCS 2016. LNCS, vol. 9537, pp. 174–186. Springer, Cham (2016). https://doi.org/10.1007/978-3-319-27683-0_13
11. Kuznets, R., Studer, T.: Weak arithmetical interpretations for the logic of proofs. Logic J. IGP **24**(3), 424–440 (2016)
12. Kuznets, R., Studer, F.: Logics of Proofs and Justifications. College Publications, London (2019)
13. Marinkovic, B., Glavan, P., Ognjanovic, Z., Studer, T.: A temporal epistemic logic with a non-rigid set of agents for analyzing the blockchain protocol. J. Log. Comput. **29**(5), 803–830 (2019)
14. Marinkovic, B., Ognjanovic, Z., Doder, D., Perovic, A.: A propositional linear time logic with time flow isomorphic to ω^2. J. Appl. Logic **12**(2), 208–229 (2014)
15. Milnikel, R.S.: The logic of uncertain justifications. Ann. Pure Appl. Logic **165**(1), 305–315 (2014)

16. Ognjanovic, Z., Raskovic, M., Markovic, Z.: Probability Logics - Probability-Based Formalization of Uncertain Reasoning. Springer, Cham (2016). https://doi.org/10.1007/978-3-319-47012-2
17. Ognjanović, Z., Savić, N., Studer, T.: Justification logic with approximate conditional probabilities. In: Baltag, A., Seligman, J., Yamada, T. (eds.) LORI 2017. LNCS, vol. 10455, pp. 681–686. Springer, Heidelberg (2017). https://doi.org/10.1007/978-3-662-55665-8_52
18. Savić, N., Doder, D., Ognjanović, Z.: Logics with lower and upper probability operators. Int. J. Approx. Reason. **88**, 148–168 (2017)

Unranked Nominal Unification

Besik Dundua[1,3], Temur Kutsia[2], and Mikheil Rukhaia[3(✉)]

[1] Kutaisi International University, Kutaisi, Georgia
[2] RISC, Johannes Kepler University, Linz, Austria
kutsia@risc.jku.at
[3] Institute of Applied Mathematics, Tbilisi State University, Tbilisi, Georgia
mrukhaia@logic.at

Abstract. In this paper we define an unranked nominal language, an extension of the nominal language with tuple variables and term tuples. We define the unification problem for unranked nominal terms and present an algorithm solving the unranked nominal unification problem.

1 Introduction

Solving equations between logic terms is a fundamental problem with many important applications in mathematics, computer science, and artificial intelligence. It is needed to perform an inference step in reasoning systems and logic programming, to match a pattern to an expression in rule-based and functional programming, to extract information from a document, to infer types in programming languages, to compute critical pairs while completing a rewrite system, to resolve ellipsis in natural language processing, etc. Unification and matching are well-known techniques used in these tasks.

Unification (as well as matching) is a quite well-studied topic for the case when the equality between function symbols is precisely defined. This is the standard setting. There is quite some number of unification algorithms whose complexities range from exponential [34] to linear [31]. Besides, many extensions and generalizations have been proposed. Those relevant to our interests are equational unification (more precisely, associative unification with unit element) (see, e.g., [7]), word unification [10,17,29], and sequence unification [22,25,26]. There are some good surveys on unification [8,11,18,19].

Nominal logic [12,33] extends first-order logic with primitives for renaming via name-swapping, for freshness of names, and for name-binding. Such kind of constructs are important in meta-programming and meta-deduction. Nominal logic provides a simple formalism for reasoning about abstract syntax modulo α-equivalence. A nominal term $a.t$ is an example of abstraction, binding every occurrence of atom a in t. Term equality ($t \approx t'$) in nominal language is considered modulo renaming of bound variables (atoms), i.e., it is α-equivalence,

Supported by the Austrian Science Fund (FWF), project 28789-N32, and by the Shota Rustaveli National Science Foundation, grant YS-19-367.

© The Author(s), under exclusive license to Springer Nature Switzerland AG 2022
A. Özgün and Y. Zinova (Eds.): TbiLLC 2019, LNCS 13206, pp. 279–296, 2022.
https://doi.org/10.1007/978-3-030-98479-3_14

formalized inside the language itself. The α-equivalence is a meta-relation in first-order syntax, but it is formulated at the object level in nominal languages. For such formulations it is important to explicitly define which atom can be considered as a new atom for a given term. This relation ($a\#t$), called freshness relation, is also formulated on the object level in nominal languages.

Solving equations between nominal terms needs a special unification algorithm [39], which is first-order, but can be also seen from the higher-order perspective via mapping from/to higher-order pattern unification [28]. The standard nominal language contains fixed-arity symbols and one kind of variable, corresponding to individual variables from first-order syntax. In nominal languages, a unification problem, e.g. $a.x \approx^? b.y$, is solved by a pair $\langle\{b\#x\},\ \{y \mapsto (a\,b)\cdot x\}\rangle$. The first component of the solution, the freshness constraint $b\#x$, requires that b should not occur free in any possible instantiation of x. The second component, the substitution, tells us that the solution must replace the variable y with the term $(a\,b)\cdot x$. The latter means that atoms a and b are swapped in every possible instantiation of x.

As we mentioned above, the constructs provided by nominal logic are important for meta-deduction. However, this formalism, as well as many representation formats for formalized mathematics typically do not provide a structural analog for ellipses (...) which are commonly used in mathematical texts [14,15]. In the literature, the latter problem has been addressed by permitting unranked (also known as variadic, flexary, or flexible arity) symbols in the language, introducing sequences in the meta-level, and extending the language with sequence variables, see, e.g., [15,20,21,24].

In this paper we present a combination of these two approaches, extending nominal languages by unranked symbols and studying the fundamental computational mechanism for them: unification. However, unlike the above mentioned unranked languages, where sequences are introduced in the meta-level, nominal syntax allows us to introduce their analogs in the object level. This is done by generalizing already existing syntactic constructs, pairs, to arbitrary tuples. They should be flat, which is achieved by imposing a special α-equivalence rule for them.

Term pairs, which are a part of nominal syntax in some papers (e.g., [2,39]) have been extended to term tuples in [3,4], but our approach differs in that we additionally introduce variables that can be instantiated by tuples (*tuple variables*, that resemble sequence variables), and the mentioned notion of flatness.

The paper is organized as follows: In Sect. 2, we define the language. The unification rules and a strategy that guarantees soundness and tries to minimize redundant computations are discussed in Sect. 3. Some terminating fragments of unranked nominal unification are introduced in Sect. 4. In Sect. 5, we discuss related problems and explain our design choices. Section 6 concludes.

2 Unranked Nominal Language

In our signature we have pairwise disjoint sets of atoms $(a, b, \ldots)$, function symbols $(f, g, \ldots)$, individual variables $(x, y, \ldots)$, tuple variables $(X, Y, \ldots)$, and the tuple constructor $\langle\rangle$. *Permutations* are a finite (possibly empty) sequence of swappings, which are pairs of atoms $(a\, b)$. We use π to denote permutations, write id for the identity (empty) permutation and $\pi_1 \circ \pi_2$ for concatenating two permutations.

An *unranked nominal term* ($t, s, \ldots$, shortly a term) is either an individual term r, a tuple variable with a suspended permutation $\pi \cdot X$, or a possibly empty tuple of terms $\langle t_1, \ldots, t_n\rangle$:

$$t ::= r \mid \pi \cdot X \mid \langle t_1, \ldots, t_n\rangle, \quad n \geq 0,$$

where *individual terms* are defined by the grammar

$$r ::= a \mid a.t \mid \pi \cdot x \mid f\langle t_1, \ldots, t_n\rangle, \quad n \geq 0.$$

We will write $t\colon \iota$ to indicate that t is an individual term. The terms $\pi \cdot x$ and $\pi \cdot X$ are called *suspensions*. We skip π if $\pi = id$. The inverse of a permutation π, denoted by π^{-1}, is obtained by reversing the list of swappings from π.

Permutation action on terms is defined as follows:

- $id \cdot a = a$.
- $((a_1\, a_2) \circ \pi) \cdot a = \begin{cases} a_1, & \text{if } \pi \cdot a = a_2, \\ a_2, & \text{if } \pi \cdot a = a_1, \\ \pi \cdot a, & \text{otherwise.} \end{cases}$
- $\pi \cdot (a.t) = (\pi \cdot a).(\pi \cdot t)$.
- $\pi \cdot (\pi' \cdot x) = (\pi \circ \pi') \cdot x$ and $\pi \cdot (\pi' \cdot X) = (\pi \circ \pi') \cdot X$.
- $\pi \cdot (f\langle t_1, \ldots, t_n\rangle) = f(\pi \cdot \langle t_1, \ldots, t_n\rangle)$, and
- $\pi \cdot \langle t_1, \ldots, t_n\rangle = \langle \pi \cdot t_1, \ldots, \pi \cdot t_n\rangle$.

The disagreement set of two permutations π and π' is defined as $ds(\pi, \pi') ::= \{a \mid \pi \cdot a \neq \pi' \cdot a\}$. Further, we often omit expressions like $a \neq b$, assuming that atoms differ by their names.

Substitution is a mapping from individual variables to individual terms and from tuple variables to tuples such that all but finitely many individual variables are mapped to themselves, and all but finitely many tuple variables are mapped to singleton tuples consisting of that variable only (i.e., mapping X to $\langle X\rangle$). They are usually written as finite sets, e.g., $[x_1 \mapsto t_1, \ldots, x_n \mapsto t_n, X_1 \mapsto \langle t_{11}, \ldots, t_{1n_1}\rangle, \ldots, X_m \mapsto \langle t_{m1}, \ldots, t_{mn_m}\rangle]$. We use the letter σ for substitutions in general and ε for the identity substitution, i.e., $\varepsilon(x) = x$ and $\varepsilon(X) = \langle X\rangle$ for all individual and tuple variables x and X.

Application of a substitution σ to a term t is defined as follows:

- $a\sigma = a$.
- $(a.t)\sigma = a.t\sigma$.
- $(f\langle t_1, \ldots, t_n\rangle)\sigma = f\langle t_1, \ldots, t_n\rangle\sigma$.
- $\langle t_1, \ldots, t_n\rangle\sigma = \langle t_1\sigma, \ldots, t_n\sigma\rangle$, where nested tuples are flattened.
- $(\pi \cdot x)\sigma = \pi \cdot \sigma(x)$ and $(\pi \cdot X)\sigma = \pi \cdot \sigma(X)$, where π acts on $\sigma(x)$ and $\sigma(X)$ as permutation action.

For instance, for a substitution

$$\sigma = [x \mapsto f\langle b.b\rangle,\ X \mapsto \langle a, f\langle c.c, (a\,c)\cdot Z\rangle\rangle,\ Y \mapsto \langle\rangle]$$

we have

$$a.f\langle (a\,b)\cdot x,\ (a\,b)\cdot X,\ (a\,b)\cdot Y\rangle\sigma = a.f\langle f\langle a.a\rangle,\ b,\ f\langle c.c, (a\,b)(a\,c)\cdot Z\rangle\rangle.$$

A freshness environment (denoted by ∇) is a list of freshness constraints $a\#x$ and $a\#X$, meaning that the instantiations of x and X cannot contain free occurrences of a. The flatness property of tuples is formalized by the axiom $\approx$-flat. (where $n \geq 0, k \geq 0, m \geq 0$).

The equivalence ($\approx$) is defined by the following rules:

$$\frac{}{\nabla \vdash \langle\rangle \approx \langle\rangle}\approx\text{-unit} \qquad \frac{}{\nabla \vdash a \approx a}\approx\text{-atom}$$

$$\frac{}{\nabla \vdash \langle t_1,\ldots,t_n,\langle t'_1,\ldots,t'_k\rangle,t''_1,\ldots,t''_m\rangle \approx \langle t_1,\ldots,t_n,t'_1,\ldots,t'_k,t''_1,\ldots,t''_m\rangle}\approx\text{-flat}$$

$$\frac{\nabla \vdash t_1 \approx t'_1 \quad \ldots \quad \nabla \vdash t_n \approx t'_n}{\nabla \vdash \langle t_1,\ldots,t_n\rangle \approx \langle t'_1,\ldots,t'_n\rangle}\approx\text{-tuple}$$

$$\frac{\nabla \vdash \langle t_1,\ldots,t_n\rangle \approx \langle t'_1,\ldots,t'_n\rangle}{\nabla \vdash f\langle t_1,\ldots,t_n\rangle \approx f\langle t'_1,\ldots,t'_n\rangle}\approx\text{-application}$$

$$\frac{t \approx t'}{\nabla \vdash a.t \approx a.t'}\approx\text{-abst.1} \qquad \frac{\nabla \vdash t \approx (a\ b)\cdot t' \quad \nabla \vdash a\#t'}{\nabla \vdash a.t \approx b.t'}\approx\text{-abst.2}$$

$$\frac{a\#x \in \nabla \text{ for all } a \in ds(\pi,\pi')}{\nabla \vdash \pi\cdot x \approx \pi'\cdot x}\approx\text{-susp.1}$$

$$\frac{a\#X \in \nabla \text{ for all } a \in ds(\pi,\pi')}{\nabla \vdash \pi\cdot X \approx \pi'\cdot X}\approx\text{-susp.2}$$

and the freshness predicate ($\#$) is defined by:

$$\frac{}{\nabla \vdash a\#\langle\rangle}\#\text{-unit} \qquad \frac{\nabla \vdash a\#t_1 \quad \ldots \quad \nabla \vdash a\#t_n}{\nabla \vdash a\#\langle t_1,\ldots,t_n\rangle}\#\text{-tuple}$$

$$\frac{}{\nabla \vdash a\#b}\#\text{-atom} \qquad \frac{\nabla \vdash a\#\langle t_1,\ldots,t_n\rangle}{\nabla \vdash a\#f\langle t_1,\ldots,t_n\rangle}\#\text{-application}$$

$$\frac{}{\nabla \vdash a\#a.t}\#\text{-abst.1} \qquad \frac{\nabla \vdash a\#t}{\nabla \vdash a\#b.t}\#\text{-abst.2}$$

$$\frac{(\pi^{-1}\cdot a\#x) \in \nabla}{\nabla \vdash a\#\pi\cdot x}\#\text{-susp.1} \qquad \frac{(\pi^{-1}\cdot a\#X) \in \nabla}{\nabla \vdash a\#\pi\cdot X}\#\text{-susp.2}$$

Proposition 1. *Given a freshness context ∇, a permutation π, an atom a and a term t, we have:*

(1) If $\nabla \vdash a\#\pi \cdot t$ *then* $\nabla \vdash \pi^{-1} \cdot a\#t$.
(2) If $\nabla \vdash \pi \cdot a\#t$ *then* $\nabla \vdash a\#\pi^{-1} \cdot t$.
(3) If $\nabla \vdash a\#t$ *then* $\nabla \vdash \pi \cdot a\#\pi \cdot t$.

Proof. Using induction on the structure of t and the fact that $\pi \cdot a = b$ iff $a = \pi^{-1} \cdot b$. The last statement is the consequence of (2) and the fact that permutations are bijections on atoms. □

The size of a term t, denoted by $|t|$, is defined by:

$$|\pi \cdot x| = |\pi \cdot X| = |a| = |\langle\rangle| = 1, \qquad |a.t| = 1 + |t|,$$
$$|f\langle t_1, \dots, t_n\rangle| = 1 + |\langle t_1, \dots, t_n\rangle|, \qquad |\langle t_1, \dots, t_n\rangle| = 1 + |t_1| + \dots + |t_n|.$$

Further, we define the size of an equation as $|t \approx t'| = |t| + |t'|$ and the size of a freshness constraint as $|a\#t| = |t|$.

Theorem 1. $\approx$ *is an equivalence relation.*

Proof. Similar to the corresponding results from [5,39].

- Reflexivity is by a simple induction on the structure of terms.
- Transitivity is by an induction on the size of terms using the properties:
  - permutations can be moved from one side of the freshness relation to the other by forming the inverse permutation (Proposition 1).
  - the freshness relation is preserved under $\approx$ and permutation actions.
- Symmetry is by a simple induction on the structure of terms using the Proposition 1 and preservation of freshness under alpha-equivalence.

□

3 Unification

An unranked nominal unification problem P is a finite set of equational $t \approx^? t'$ or freshness problems $a\#^?t$. Tuples occurring in the unification problem are always flattened (e.g. after substitution application, etc.). A solution for P is a pair (∇, σ) such that for all problems $t \approx^? t'$ in P we have $\nabla \vdash \sigma(t) \approx \sigma(t')$ and for all problems $a\#^?t$ in P we have $\nabla \vdash a\#\sigma(t)$.

To describe the unification algorithm we use so called labeled transformation of unification problems: $P \overset{\sigma}{\Longrightarrow} P'$ and $P \overset{\nabla}{\Longrightarrow} P'$ which are given below (note that if in $\approx$-susp.1,2 $\pi = \pi'$, then $ds(\pi, \pi')$ and thus $\{a\#^?x, X \mid a \in ds(\pi, \pi')\}$ is empty; $V(t)$ denotes the set of variables occurring in t):

$(\approx^?$-atom$)$ $\quad \{a \approx^? a\} \cup P \overset{\varepsilon}{\Longrightarrow} P.$

$(\approx^?$-unit$)$ $\quad \{\langle\rangle \approx^? \langle\rangle\} \cup P \overset{\varepsilon}{\Longrightarrow} P.$

$(\approx^?$-function$)$ $\quad \{f\langle t_1, \dots, t_n\rangle \approx^? f\langle t'_1, \dots, t'_m\rangle\} \cup P \overset{\varepsilon}{\Longrightarrow}$
$\{\langle t_1, \dots, t_n\rangle \approx^? \langle t'_1, \dots, t'_m\rangle\} \cup P.$

(≈? -abst.1) $\{a.t \approx^? a.t'\} \cup P \overset{\varepsilon}{\Longrightarrow} \{t \approx^? t'\} \cup P.$

(≈? -abst.2) $\{a.t \approx^? b.t'\} \cup P \overset{\varepsilon}{\Longrightarrow} \{t \approx^? (a\ b) \cdot t', a\#^? t'\} \cup P.$

(≈? -susp.1) $\{\pi \cdot x \approx^? \pi' \cdot x\} \cup P \overset{\varepsilon}{\Longrightarrow} \{a\#^? x \mid a \in ds(\pi, \pi')\} \cup P.$

(≈? -susp.2) $\{\pi \cdot X \approx^? \pi' \cdot X\} \cup P \overset{\varepsilon}{\Longrightarrow} \{a\#^? X \mid a \in ds(\pi, \pi')\} \cup P.$

(≈? -tuple) $\{\langle t, t_1, \ldots, t_n\rangle \approx^? \langle t', t'_1, \ldots, t'_m\rangle\} \cup P \overset{\varepsilon}{\Longrightarrow}$
$\{t \approx t', \langle t_1, \ldots, t_n\rangle \approx^? \langle t'_1, \ldots, t'_m\rangle\} \cup P,$
where t and t' are not tuple variables.

(≈? -proj.1) $\{\langle \pi \cdot X, t_1, \ldots, t_n\rangle \approx^? \langle t'_1, \ldots, t'_m\rangle\} \cup P \overset{\sigma}{\Longrightarrow}$
$\{\langle t_1\sigma, \ldots, t_n\sigma\rangle \approx^? \langle t'_1\sigma, \ldots, t'_m\sigma\rangle\} \cup P\sigma,$
where $\sigma = [X \mapsto \langle\rangle]$.

(≈? -proj.2) $\{\langle t_1, \ldots, t_n\rangle \approx^? \langle \pi \cdot X, t'_1, \ldots, t'_m\rangle\} \cup P \overset{\sigma}{\Longrightarrow}$
$\{\langle t_1\sigma, \ldots, t_n\sigma\rangle \approx^? \langle t'_1\sigma, \ldots, t'_m\sigma\rangle\} \cup P\sigma,$
where $\sigma = [X \mapsto \langle\rangle]$.

(≈? -widen.1) $\{\langle \pi \cdot X, t_1, \ldots, t_n\rangle \approx^? \langle t, t'_1, \ldots, t'_m\rangle\} \cup P \overset{\sigma}{\Longrightarrow}$
$\{\langle X', t_1\sigma, \ldots, t_n\sigma\rangle \approx^? \langle t'_1\sigma, \ldots, t'_m\sigma\rangle\} \cup P\sigma,$
where $\sigma = [X \mapsto \pi^{-1} \cdot \langle t, X'\rangle], X \notin V(t)$.

(≈? -widen.2) $\{\langle t, t_1, \ldots, t_n\rangle \approx^? \langle \pi \cdot X, t'_1, \ldots, t'_m\rangle\} \cup P \overset{\sigma}{\Longrightarrow}$
$\{\langle t_1\sigma, \ldots, t_n\sigma\rangle \approx^? \langle X', t'_1\sigma, \ldots, t'_m\sigma\rangle\} \cup P\sigma,$
where $\sigma = [X \mapsto \pi^{-1} \cdot \langle t, X'\rangle], X \notin V(t)$.

(≈? -var.1) $\{\pi \cdot x \approx^? t : \iota\} \cup P \overset{\sigma}{\Longrightarrow} P\sigma,$
where $\sigma = [x \mapsto \pi^{-1} \cdot t], x \notin V(t)$.

(≈? -var.2) $\{t : \iota \approx^? \pi \cdot x\} \cup P \overset{\sigma}{\Longrightarrow} P\sigma,$
where $\sigma = [x \mapsto \pi^{-1} \cdot t], x \notin V(t)$.

(≈? -var.3) $\{\pi \cdot X \approx^? t\} \cup P \overset{\sigma}{\Longrightarrow} P\sigma,$
where $\sigma = [X \mapsto \langle \pi^{-1} \cdot t\rangle], X \notin V(t)$.

(≈? -var.4) $\{t \approx^? \pi \cdot X\} \cup P \overset{\sigma}{\Longrightarrow} P\sigma,$
where $\sigma = [X \mapsto \langle \pi^{-1} \cdot t\rangle], X \notin V(t)$.

(#?-atom) $\{a\#^? b\} \cup P \overset{\emptyset}{\Longrightarrow} P.$

(#?-unit) $\{a\#^? \langle\rangle\} \cup P \overset{\emptyset}{\Longrightarrow} P.$

(#?-tuple) $\{a\#^? \langle t_1, \ldots, t_n \rangle\} \cup P \xRightarrow{\emptyset} \{a\#^? t_1, \ldots, a\#^? t_n\} \cup P.$

(#?-function) $\{a\#^? f\langle t_1, \ldots, t_n \rangle\} \cup P \xRightarrow{\emptyset} \{a\#^? \langle t_1, \ldots, t_n \rangle\} \cup P.$

(#?-abst.1) $\{a\#^? a.t\} \cup P \xRightarrow{\emptyset} P.$

(#?-abst.2) $\{a\#^? b.t\} \cup P \xRightarrow{\emptyset} \{a\#^? t\} \cup P.$

(#?-susp.1) $\{a\#^? \pi \cdot x\} \cup P \xRightarrow{\nabla} P$, where $\nabla = \{\pi^{-1} \cdot a\#x\}$.

(#?-susp.2) $\{a\#^? \pi \cdot X\} \cup P \xRightarrow{\nabla} P$, where $\nabla = \{\pi^{-1} \cdot a\#X\}$.

The naive algorithm, as presented in [28], is divided into two phases: first apply as many $\xRightarrow{\sigma}$ transformations as possible. It might cause branching due to tuple variables. On some branches, there might be no equational problems left. We expand them by $\xRightarrow{\nabla}$ transformations as long as possible. If we do not end up with the empty problem, then halt with failure, otherwise from the sequence of transformations $P \xRightarrow{\sigma_1} \cdots \xRightarrow{\sigma_n} P' \xRightarrow{\nabla_1} \cdots \xRightarrow{\nabla_m} \emptyset$ construct the solution $(\nabla_1 \cup \cdots \cup \nabla_m, \sigma_n \circ \cdots \circ \sigma_1)$. Some branches might directly lead to failure after application of $\xRightarrow{\sigma}$ rules. Some branches might cause more and more branching, leading to infinite sets of solutions. Employing some fair strategy of search tree development, we can have a complete method to enumerate them.

Example 1. We give examples of some unification problems and their solutions:

- Problem: $\{f\langle a.\langle X, x, Y\rangle\rangle \approx^? f\langle b.\langle f\langle X\rangle, x, b, c\rangle\rangle\}$.
  Solution: $(\emptyset, [X \mapsto \langle\rangle, x \mapsto f\langle\rangle, Y \mapsto \langle f\langle\rangle, a, c\rangle])$.
- Problem: $\{a.b.f\langle X, b\rangle \approx^? b.a.f\langle a, X\rangle\}$.
  Solution: $(\emptyset, [X \mapsto \langle\rangle])$.
- Problem: $\{f\langle X, a\rangle \approx^? f\langle a, Y\rangle\}$.
  Solutions: $(\emptyset, [X \mapsto \langle\rangle, Y \mapsto \langle\rangle])$ and $(\emptyset, [X \mapsto \langle a, Z\rangle, Y \mapsto \langle Z, a\rangle])$.
  If instead of Y we had X, then there would be infinitely many solutions: $(\emptyset, [X \mapsto \langle\rangle])$, $(\emptyset, [X \mapsto \langle a\rangle])$, $(\emptyset, [X \mapsto \langle a, a\rangle])$, ...
- Problem: $\{a.f\langle X, a\rangle \approx^? b.f\langle b, X\rangle\}$.
  Solution: $(\emptyset, [X \mapsto \langle\rangle])$.
- Problem: $\{a.f\langle X, a\rangle \approx^? b.f\langle b, Y\rangle\}$.
  Solutions: $(\emptyset, [X \mapsto \langle\rangle, Y \mapsto \langle\rangle])$ and $(\{b\#Z\}, [X \mapsto \langle a, Z\rangle, Y \mapsto \langle (a\,b){\cdot}Z, b\rangle])$.
- Problem: $\{a.f\langle X, c\rangle \approx^? b.f\langle c, Y\rangle\}$.
  Solutions: $(\emptyset, [X \mapsto \langle\rangle, Y \mapsto \langle\rangle])$ and $(\{b\#Z\}, [X \mapsto \langle c, Z\rangle, Y \mapsto \langle (a\,b){\cdot}Z, c\rangle])$.
- Problem: $\{f\langle X, Y\rangle \approx^? f\langle a, b, X\rangle, b\#X\}$.
  Solutions: $(\emptyset, [X \mapsto \langle\rangle, Y \mapsto \langle a, b\rangle])$ and $(\emptyset, [X \mapsto \langle a\rangle, Y \mapsto \langle b, a\rangle])$.

Without $b\#X$, the problem $\{f\langle X, Y\rangle \approx^? f\langle a, b, X\rangle\}$ has infinitely many solutions: $(\emptyset, [X \mapsto \langle\rangle, Y \mapsto \langle a, b\rangle])$, $(\emptyset, [X \mapsto \langle a\rangle, Y \mapsto \langle b, a\rangle])$, $(\emptyset, [X \mapsto \langle a, b\rangle, Y \mapsto \langle a, b\rangle])$, $(\emptyset, [X \mapsto \langle a, b, a\rangle, Y \mapsto \langle b, a\rangle])$, $(\emptyset, [X \mapsto \langle a, b, a, b\rangle, Y \mapsto \langle a, b\rangle])$, $(\emptyset, [X \mapsto \langle a, b, a, b, a\rangle, Y \mapsto \langle b, a\rangle])$, ...

The naive algorithm, described above, can be non-terminating even when there is a finite number of solutions. This is illustrated by the following example.

Example 2. Let us consider the following unification problem:

$$\{a.f\langle X, a\rangle \approx^? b.f\langle b, X\rangle\} \Longrightarrow_{\approx^?\text{-abst.2}}$$
$$\{f\langle X, a\rangle \approx^? f\langle a, (a\,b) \cdot X\rangle, a\#^? f\langle b, X\rangle\} \Longrightarrow_{\approx^?\text{-function}}$$
$$\{\langle X, a\rangle \approx^? \langle a, (a\,b) \cdot X\rangle, a\#^? f\langle b, X\rangle\}$$

Now, we can apply $\approx^?$ -proj.1 rule (followed by $\approx^?$ -atom and several $\stackrel{\nabla}{\Longrightarrow}$ transformations) to obtain a solution $(\emptyset, [X \mapsto \langle\rangle])$.

When we apply $\approx^?$ -widen.1 rules, we get non-terminating branch:

$$\{\langle X, a\rangle \approx^? \langle a, (a\,b) \cdot X\rangle, a\#^? f\langle b, X\rangle\} \stackrel{[X \mapsto \langle a, X_1\rangle]}{\Longrightarrow}$$
$$\{\langle X_1, a\rangle \approx^? \langle b, (a\,b) \cdot X_1\rangle, a\#^? f\langle b, a, X_1\rangle\} \stackrel{[X_1 \mapsto \langle b, X_2\rangle]}{\Longrightarrow}$$
$$\{\langle X_2, a\rangle \approx^? \langle a, (a\,b) \cdot X_2\rangle, a\#^? f\langle b, a, b, X_2\rangle\} \Longrightarrow \cdots$$

But we could apply $\stackrel{\nabla}{\Longrightarrow}$ transformations on a subset $\{a\#^? f\langle b, a, X_1\rangle\}$, obtain $\{a\#^? a, a\#^? X_1\}$ and stop with failure since $a\#^? a$ has no solution.

An obvious attempt to fix the problem for such cases would be to delay application of the $\approx^?$ -widen.1 and $\approx^?$ -widen.2 rules until all possible $\stackrel{\nabla}{\Longrightarrow}$ transformations are applied. But we should be careful not to remove freshness constraints from the problems too early.

Consider the following example: $\{a \approx^? x, a\#^? x\}$. If we apply the $\#^?$-susp.1 rule first and then $\approx^?$ -var.2, we will obtain a wrong solution $(\{a\#x\}, [x \mapsto a])$. Thus, we should delay application of the $\#^?$-susp.1 and $\#^?$-susp.2 rules until all possible $\stackrel{\sigma}{\Longrightarrow}$ transformations are applied.

The discussion above leads to the following strategy S:

- first apply as many $\stackrel{\sigma}{\Longrightarrow}$ transformations as possible except the $\approx^?$ -proj.1, 2 and $\approx^?$ -widen.1, 2 rules.
- if no other $\stackrel{\sigma}{\Longrightarrow}$ transformation is possible, $\approx^?$ -proj.1, 2 and $\approx^?$ -widen.1, 2 rules can be applied in parallel. However, before the $\approx^?$ -widen.1 and $\approx^?$ -widen.2 rules, one should apply as many $\stackrel{\nabla}{\Longrightarrow}$ transformations as possible except the $\#^?$-susp.1 and $\#^?$-susp.2 rules.
- use $\#^?$-susp.1 and $\#^?$-susp.2 rules if no other rules are applicable.
- If there is at least one equational problem in P such that no transformation rule is applicable on it, immediately halt the development of that branch with failure.

Theorem 2. *Given a unification problem P, if the unranked unification algorithm with the strategy S fails on P, then P has no solution; and if it succeeds on P, then the result is a unifier.*

Proof. The idea is that mixing equational and freshness rules except $\#^?$-susp.1, 2 does not cause soundness problems, since those freshness rules do not affect variables. Parallel application of widening and projection rules together with the iteration of the strategy make sure that no solution is lost (cf. Theorem 51 in [22]).

For a unification problem P, the strategy S fails on P if there is at least one of the following pairs $a \approx^? b$, $b \approx^? a$, $a \approx^? \langle\rangle$, $\langle\rangle \approx^? a$, $a\#^? a$ in P', obtained by applying simplification rules to P, or there is an occurs check violation in $\approx^?$-var rules. Clearly, P has no solution in these cases.

If the strategy S succeeds on P, then we get a result $(\nabla_1 \cup \cdots \cup \nabla_n, \sigma_1 \circ \cdots \circ \sigma_m)$. The proof continues by simple induction on transitions with the following induction hypothesis:

- If $P \overset{\sigma}{\Longrightarrow} P'$ and (∇, σ') is a unifier for P', then $(\nabla, \sigma \circ \sigma')$ is a unifier for P.
- If $P \overset{\nabla}{\Longrightarrow} P'$ and (∇', σ) is a unifier for P', then $(\nabla \cup \nabla', \sigma)$ is a unifier for P.

□

Example 3. We demonstrate how the strategy works on a unification problem.

$$\{a.f\langle X, x, Y, f\langle y, x\rangle\rangle \approx^? b.f\langle g\langle X\rangle, x, b, Z, f\langle g\langle X\rangle, y\rangle\rangle\} \quad \Longrightarrow_{\approx^?\text{-abst.2}}$$

$$\{f\langle X, x, Y, f\langle y, x\rangle\rangle \approx^? f\langle g\langle (a\,b)\cdot X\rangle, (a\,b)\cdot x, a, (a\,b)\cdot Z, f\langle g\langle (a\,b)\cdot X\rangle, (a\,b)\cdot y\rangle\rangle, a\#^? f\langle g\langle X\rangle, x, b, Z, f\langle g\langle X\rangle, y\rangle\rangle\} \quad \Longrightarrow_{\approx^?\text{-function}}$$

$$\{\langle X, x, Y, f\langle y, x\rangle\rangle \approx^? \langle g\langle (a\,b)\cdot X\rangle, (a\,b)\cdot x, a, (a\,b)\cdot Z, f\langle g\langle (a\,b)\cdot X\rangle, (a\,b)\cdot y\rangle\rangle, a\#^? f\langle g\langle X\rangle, x, b, Z, f\langle g\langle X\rangle, y\rangle\rangle\} \quad \overset{[X\mapsto\langle\rangle]}{\Longrightarrow}_{\approx^?\text{-proj.1}}$$

Note, that at this point $\approx^?$-widen.1 is not applicable because of occurs check: $X \in V(g\langle (a\,b)\cdot X\rangle)$.

$$\{\langle x, Y, f\langle y, x\rangle\rangle \approx^? \langle g\langle\rangle, (a\,b)\cdot x, a, (a\,b)\cdot Z, f\langle g\langle\rangle, (a\,b)\cdot y\rangle\rangle, a\#^? f\langle g\langle\rangle, x, b, Z, f\langle g\langle\rangle, y\rangle\rangle\} \quad \Longrightarrow_{\approx^?\text{-tuple}}$$

$$\{x \approx^? g\langle\rangle, \langle Y, f\langle y, x\rangle\rangle \approx^? \langle (a\,b)\cdot x, a, (a\,b)\cdot Z, f\langle g\langle\rangle, (a\,b)\cdot y\rangle\rangle, a\#^? f\langle g\langle\rangle, x, b, Z, f\langle g\langle\rangle, y\rangle\rangle\} \quad \overset{[x\mapsto g\langle\rangle]}{\Longrightarrow}_{\approx^?\text{-var.1}}$$

$$\{\langle Y, f\langle y, g\langle\rangle\rangle\rangle \approx^? \langle g\langle\rangle, a, (a\,b)\cdot Z, f\langle g\langle\rangle, (a\,b)\cdot y\rangle\rangle, a\#^? f\langle g\langle\rangle, g\langle\rangle, b, Z, f\langle g\langle\rangle, y\rangle\rangle\}.$$

Now, we have two branches: (a) continue again with $\approx^?$ -proj.1, followed by $\approx^?$ -tuple and several $\overset{\nabla}{\Longrightarrow}$ transformations (given also below), leading to the failure; and (b) continue with the $\overset{\nabla}{\Longrightarrow}$ transformations followed by $\approx^?$ -widen.1:

$$\{\langle Y, f\langle y, g\langle\rangle\rangle\rangle \approx^? \langle g\langle\rangle, a, (a\,b)\cdot Z, f\langle g\langle\rangle, (a\,b)\cdot y\rangle\rangle,$$
$$a\#^? f\langle g\langle\rangle, g\langle\rangle, b, Z, f\langle g\langle\rangle, y\rangle\rangle\} \quad \Longrightarrow_{\#^?\text{-function}}$$
$$\{\langle Y, f\langle y, g\langle\rangle\rangle\rangle \approx^? \langle g\langle\rangle, a, (a\,b)\cdot Z, f\langle g\langle\rangle, (a\,b)\cdot y\rangle\rangle,$$
$$a\#^? \langle g\langle\rangle, g\langle\rangle, b, Z, f\langle g\langle\rangle, y\rangle\rangle\} \quad \Longrightarrow_{\#^?\text{-tuple}}$$
$$\{\langle Y, f\langle y, g\langle\rangle\rangle\rangle \approx^? \langle g\langle\rangle, a, (a\,b)\cdot Z, f\langle g\langle\rangle, (a\,b)\cdot y\rangle\rangle,$$
$$a\#^? g\langle\rangle, a\#^? b, a\#^? Z, a\#^? f\langle g\langle\rangle, y\rangle\} \quad \Longrightarrow_{\#^?\text{-function,tuple}}$$
$$\{\langle Y, f\langle y, g\langle\rangle\rangle\rangle \approx^? \langle g\langle\rangle, a, (a\,b)\cdot Z, f\langle g\langle\rangle, (a\,b)\cdot y\rangle\rangle,$$
$$a\#^? g\langle\rangle, a\#^? b, a\#^? Z, a\#^? y\} \quad \Longrightarrow_{\#^?\text{-function,unit,atom}}$$
$$\{\langle Y, f\langle y, g\langle\rangle\rangle\rangle \approx^? \langle g\langle\rangle, a, (a\,b)\cdot Z, f\langle g\langle\rangle, (a\,b)\cdot y\rangle\rangle,$$
$$a\#^? Z, a\#^? y\} \quad \overset{[Y\mapsto\langle g\langle\rangle, Y_1\rangle]}{\Longrightarrow}_{\approx^?\text{-widen.1}}$$
$$\{\langle Y_1, f\langle y, g\langle\rangle\rangle\rangle \approx^? \langle a, (a\,b)\cdot Z, f\langle g\langle\rangle, (a\,b)\cdot y\rangle\rangle,$$
$$a\#^? Z, a\#^? y\} \quad \overset{[Y_1\mapsto\langle a, Y_2\rangle]}{\Longrightarrow}_{\approx^?\text{-widen.1}}$$
$$\{\langle Y_2, f\langle y, g\langle\rangle\rangle\rangle \approx^? \langle (a\,b)\cdot Z, f\langle g\langle\rangle, (a\,b)\cdot y\rangle\rangle,$$
$$a\#^? Z, a\#^? y\}.$$

Note that before the last $\approx^?$ -widen.1 rule application we should have the $\approx^?$ -proj.1 branch again leading to the failure.

Now, at this point we have several options:

(1) apply $\approx^?$ -proj.1 rule

$$\{\langle Y_2, f\langle y, g\langle\rangle\rangle\rangle \approx^? \langle (a\,b)\cdot Z, f\langle g\langle\rangle, (a\,b)\cdot y\rangle\rangle,$$
$$a\#^? Z, a\#^? y\} \quad \overset{[Y_2\mapsto\langle\rangle]}{\Longrightarrow}_{\approx^?\text{-proj.1}}$$
$$\{\langle f\langle y, g\langle\rangle\rangle\rangle \approx^? \langle (a\,b)\cdot Z, f\langle g\langle\rangle, (a\,b)\cdot y\rangle\rangle, a\#^? Z, a\#^? y\} \quad \overset{[Z\mapsto\langle\rangle]}{\Longrightarrow}_{\approx^?\text{-proj.2}}$$
$$\{\langle f\langle y, g\langle\rangle\rangle\rangle \approx^? \langle f\langle g\langle\rangle, (a\,b)\cdot y\rangle\rangle, a\#^? \langle\rangle, a\#^? y\} \quad \Longrightarrow_{\approx^?\text{-tuple,function}}$$
$$\{\langle y, g\langle\rangle\rangle \approx^? \langle g\langle\rangle, (a\,b)\cdot y\rangle, a\#^? \langle\rangle, a\#^? y\} \quad \Longrightarrow_{\approx^?\text{-tuple}}$$
$$\{y \approx^? g\langle\rangle,\ g\langle\rangle \approx^? (a\,b)\cdot y,\ \langle\rangle \approx^? \langle\rangle,\ a\#^? \langle\rangle,\ a\#^? y\} \quad \overset{[y\mapsto g\langle\rangle]}{\Longrightarrow}_{\approx^?\text{-var.1}}$$
$$\{g\langle\rangle \approx^? g\langle\rangle,\ \langle\rangle \approx^? \langle\rangle,\ a\#^? \langle\rangle,\ a\#^? g\langle\rangle\} \quad \Longrightarrow_{\approx^?\text{-function,unit}}$$
$$\{a\#^? \langle\rangle, a\#^? g\langle\rangle\} \quad \Longrightarrow_{\#^?\text{-function,unit}}$$
$$\emptyset.$$

and we obtain the solution $(\emptyset, [X \mapsto \langle\rangle, x \mapsto g\langle\rangle, Y \mapsto \langle g\langle\rangle, a\rangle, Z \mapsto \langle\rangle, y \mapsto g\langle\rangle])$. Note, that application of $\approx^?$ -widen.2 instead of $\approx^?$ -proj.2 rule in this branch will lead to failure.

(2) applying $\approx^?$ -proj.2 rule first is similar to (1), obtaining the same solution.

(3) apply $\approx^?$ -widen.1 rule

$$\begin{aligned}
&\{\langle Y_2, f\langle y, g\langle\rangle\rangle\rangle \approx^? \langle (a\,b)\cdot Z, f\langle g\langle\rangle, (a\,b)\cdot y\rangle\rangle, \\
&\quad a\#^? Z, a\#^? y\} && \overset{[Y_2\mapsto\langle (a\,b)\cdot Z, Y_3\rangle]}{\Longrightarrow}_{\approx^?\text{-widen.1}} \\
&\{\langle Y_3, f\langle y, g\langle\rangle\rangle\rangle \approx^? \langle f\langle g\langle\rangle, (a\,b)\cdot y\rangle\rangle, a\#^? Z, a\#^? y\} && \overset{[Y_3\mapsto\langle\rangle]}{\Longrightarrow}_{\approx^?\text{-proj.1}} \\
&\{\langle f\langle y, g\langle\rangle\rangle\rangle \approx^? \langle f\langle g\langle\rangle, (a\,b)\cdot y\rangle\rangle, a\#^? Z, a\#^? y\} && \Longrightarrow_{\approx^?\text{-tuple,function}} \\
&\{y \approx^? g\langle\rangle, g\langle\rangle \approx^? (a\,b)\cdot y, \langle\rangle \approx^? \langle\rangle,\ a\#^? Z,\ a\#^? y\} && \overset{[y\mapsto g\langle\rangle]}{\Longrightarrow}_{\approx^?\text{-var.1}} \\
&\{g\langle\rangle \approx^? g\langle\rangle,\ \langle\rangle \approx^? \langle\rangle,\ a\#^? Z,\ a\#^? g\langle\rangle\} && \Longrightarrow_{\approx^?\text{-function,unit}} \\
&\{a\#^? Z, a\#^? g\langle\rangle\} && \Longrightarrow_{\#^?\text{-function,unit}} \\
&\{a\#^? Z\} && \overset{\{a\#Z\}}{\Longrightarrow}_{\#^?\text{-susp.2}} \\
&\emptyset
\end{aligned}$$

and we obtain the solution $(\{a\#Z\}, [X \mapsto \langle\rangle, x \mapsto g\langle\rangle, Y \mapsto \langle g\langle\rangle, a, (a\,b)\cdot Z\rangle, y \mapsto g\langle\rangle])$. Note, that application of $\approx^?$ -widen.1 again instead of $\approx^?$ -$sfproj$.1 rule in this branch will lead to failure.

(4) applying $\approx^?$ -widen.2 rule first is similar to (3), obtaining the equivalent solution $(\{a\#Y_2\}, [X \mapsto \langle\rangle, x \mapsto g\langle\rangle, Y \mapsto \langle g\langle\rangle, a, Y_2\rangle, Z \mapsto (a\,b)\cdot Y_2, y \mapsto g\langle\rangle])$ (up to renaming of the variables).

It is clear from the example above that our algorithm is not minimal in the sense that it computes the same or equivalent solutions several times. Finding restrictions to achieve minimality is a topic for further research.

4 Terminating Fragments

Strategy S helps to detect failures early, trying to avoid redundant computations. However, it can not guarantee termination, even when the solution set is finite. It is not surprising, since the strategy does not provide the decision algorithm for unranked nominal unification.

In this section we consider three special cases for which the strategy terminates. They originate from (non-nominal) unranked unification problems with finite sets of most general unifiers [27] and, hence, keep the same property for nominal unranked unification. Characterizations of termination based on freshness constraints require further investigation.

The KIF Fragment. In this fragment, every occurrence of tuple variables is in the last argument of a tuple. The name originates from Knowledge Interchange Format (KIF), a language designed for representing and sharing information between disparate computer systems [13]. In KIF, the variables that correspond to our tuple variables are allowed to occur only as the last arguments in terms. This is a so-called unitary fragment: solvable unification problems have a single most general unifier. This property makes it suitable for reasoning, see, e.g., [16,20,30]. We can simplify the widening rules for this fragment. Instead of stepwise computation of the substitution, we can at once replace a tuple variable with the entire tuple in the other side of the equation:

$$(\approx^? \textsf{-widen.KIF.1}) \quad \{\langle \pi \cdot X\rangle \approx^? \langle t'_1, \dots, t'_m\rangle\} \cup P \stackrel{\sigma}{\Longrightarrow} P\sigma, \quad \text{where } m \geq 0,$$
$$\sigma = [X \mapsto \pi^{-1} \cdot \langle t'_1, \dots, t'_m\rangle], \text{ and } X \notin V(\langle t'_1, \dots, t'_m\rangle).$$

The second widening rule is adapted analogously, and the projection rules can be dropped as they are subsumed with these KIF-specific widening rules.

Example 4. We illustrate how the KIF-adapted rules are used to solve a unification problem in this fragment.

$$\{f\langle a.f\langle a, X\rangle, g\langle x, y, X\rangle, Y\rangle \approx^? f\langle b.f\langle b, x, Y\rangle, g\langle b, Z\rangle, U\rangle\} \Longrightarrow_{\approx^?\textsf{-function,tuple}}$$

$$\{a.f\langle a, X\rangle \approx^? b.f\langle b, x, Y\rangle, \langle g\langle x, y, X\rangle, Y\rangle \approx^? \langle g\langle b, Z\rangle, U\rangle\} \Longrightarrow_{\approx^?\textsf{-abst.2,tuple}}$$

$$\{f\langle a, X\rangle \approx^? f\langle a, (a\,b)\cdot x, (a\,b)\cdot Y\rangle, \langle Y\rangle \approx^? \langle U\rangle, g\langle x, y, X\rangle \approx^? g\langle b, Z\rangle,\ a\#^? f\langle b, x, Y\rangle\} \Longrightarrow_{\approx^?\textsf{-function,tuple,atom}}$$

$$\{\langle X\rangle \approx^? \langle (a\,b)\cdot x, (a\,b)\cdot Y\rangle,\ g\langle x, y, X\rangle \approx^? g\langle b, Z\rangle, \langle Y\rangle \approx^? \langle U\rangle,\ a\#^? f\langle b, x, Y\rangle\} \Longrightarrow_{\approx^?\textsf{-function,tuple}}$$

$$\{\langle X\rangle \approx^? \langle (a\,b)\cdot x, (a\,b)\cdot Y\rangle,\ x \approx^? b,\ \langle y, X\rangle \approx^? \langle Z\rangle, \langle Y\rangle \approx^? \langle U\rangle,\ a\#^? f\langle b, x, Y\rangle\} \stackrel{[x\mapsto b]}{\Longrightarrow}_{\approx^?\textsf{-var.1}}$$

$$\{\langle X\rangle \approx^? \langle a, (a\,b)\cdot Y\rangle, \langle y, X\rangle \approx^? \langle Z\rangle, \langle Y\rangle \approx^? \langle U\rangle, a\#^? f\langle b, b, Y\rangle\} \Longrightarrow_{\#^?\textsf{-function,tuple,atom}}$$

$$\{\langle X\rangle \approx^? \langle a, (a\,b)\cdot Y\rangle, \langle y, X\rangle \approx^? \langle Z\rangle, \langle Y\rangle \approx^? \langle U\rangle, a\#^? Y\} \stackrel{[X\mapsto\langle a,(a\,b)\cdot Y\rangle]}{\Longrightarrow}_{\approx^?\textsf{-widen.KIF.1}}$$

$$\{\langle y, a, (a\,b)\cdot Y\rangle \approx^? \langle Z\rangle, \langle Y\rangle \approx^? \langle U\rangle,\ a\#^? Y\} \stackrel{[Z\mapsto\langle y,a,(a\,b)\cdot Y\rangle]}{\Longrightarrow}_{\approx^?\textsf{-widen.KIF.2}}$$

$$\{\langle Y\rangle \approx^? \langle U\rangle,\ a\#^? Y\} \stackrel{[Y\mapsto\langle U\rangle]}{\Longrightarrow}_{\approx^?\textsf{-widen.KIF.1}}$$

$$\{a\#^? U\} \stackrel{\{a\#U\}}{\Longrightarrow}_{\approx^?\textsf{-susp.2}}$$

$$\emptyset.$$

Hence, the algorithm returns a most general unifier

$$(\{a\#U\}, [X \mapsto \langle a, (a\,b)\cdot U\rangle,\ x \mapsto b,\ Z \mapsto \langle y, a, (a\,b)\cdot U\rangle,\ Y \mapsto \langle U\rangle]).$$

Linear Fragment. Unification problems in which no variable occurs more than once are called linear. Unlike the KIF fragment, here there are unification problems that have more than one, but still finitely many solutions. For the linear fragment we can simplify the unification rules. For instance, in the rules that eliminate variables (proj, widen, var), the substitution σ does not have to apply to the whole P, because the eliminated variable can not have any other occurrence in the remaining unification equations. It may occur only in the freshness constraints and it is sufficient to apply σ only to them. Besides, the occurrence check does not have to be performed. The $\approx^?$-susp.1 and $\approx^?$-susp.2 rules never apply. The following is an example of linear unranked nominal unification.

Example 5. Let the unification problem be $\{a.f\langle X, a\rangle \approx^? b.f\langle b, Y\rangle\}$. Then we have two derivations. The first one is:

$$\{a.f\langle X,a\rangle \approx^? b.f\langle b,Y\rangle\} \quad \Longrightarrow_{\approx^?\text{-abst.2,function}}$$

$$\{\langle X,a\rangle \approx^? \langle a,(a\,b)\cdot Y\rangle,\ a\#^? f\langle b,Y\rangle\} \quad \Longrightarrow_{\#^?\text{-function,tuple,atom}}$$

$$\{\langle X,a\rangle \approx^? \langle a,(a\,b)\cdot Y\rangle,\ a\#^? Y\} \quad \overset{[X\mapsto\langle\rangle]}{\Longrightarrow}_{\approx^?\text{-proj.1}}$$

$$\{\langle a\rangle \approx^? \langle a,(a\,b)\cdot Y\rangle,\ a\#^? Y\} \quad \Longrightarrow_{\approx^?\text{-tuple,atom}}$$

$$\{\langle\rangle \approx^? \langle (a\,b)\cdot Y\rangle,\ a\#^? Y\} \quad \overset{[Y\mapsto\langle\rangle]}{\Longrightarrow}_{\approx^?\text{-proj.2,unit}}$$

$$\{a\#^?\langle\rangle\} \quad \Longrightarrow_{\#^?\text{-unit}}$$

$$\emptyset.$$

It leads to the solution $(\emptyset, [X \mapsto \langle\rangle, Y \mapsto \langle\rangle])$.
The second derivation is

$$\{a.f\langle X,a\rangle \approx^? b.f\langle b,Y\rangle\} \quad \Longrightarrow_{\approx^?\text{-abst.2,function}}$$

$$\{\langle X,a\rangle \approx^? \langle a,(a\,b)\cdot Y\rangle,\ a\#^? f\langle b,Y\rangle\} \quad \Longrightarrow_{\#^?\text{-function,tuple,atom}}$$

$$\{\langle X,a\rangle \approx^? \langle a,(a\,b)\cdot Y\rangle,\ a\#^? Y\} \quad \overset{[X\mapsto\langle a,X_1\rangle]}{\Longrightarrow}_{\approx^?\text{-widen.1}}$$

$$\{\langle X_1,a\rangle \approx^? \langle (a\,b)\cdot Y\rangle,\ a\#^? Y\} \quad \overset{[Y\mapsto\langle (a\,b)\cdot X_1,(a\,b)\cdot Y_1\rangle]}{\Longrightarrow}_{\approx^?\text{-widen.2}}$$

$$\{\langle a\rangle \approx^? \langle Y_1\rangle,\ a\#^?\langle (a\,b)\cdot X_1, (a\,b)\cdot Y_1\rangle\} \quad \Longrightarrow_{\#^?\text{-tuple}}$$

$$\{\langle a\rangle \approx^? \langle Y_1\rangle,\ a\#^?(a\,b)\cdot X_1, a\#^?(a\,b)\cdot Y_1\} \quad \overset{[Y_1\mapsto\langle a,Y_2\rangle]}{\Longrightarrow}_{\approx^?\text{-widen.2}}$$

$$\{\langle\rangle \approx^? \langle Y_2\rangle,\ a\#^?(a\,b)\cdot X_1, a\#^?\langle b,(a\,b)\cdot Y_2\rangle\} \quad \overset{[Y_2\mapsto\langle\rangle]}{\Longrightarrow}_{\approx^?\text{-proj.2,unit}}$$

$$\{a\#^?(a\,b)\cdot X_1, a\#^?\langle b\rangle\} \quad \overset{\{b\#X_1\}}{\Longrightarrow}_{\#^?\text{-susp.2,tuple,atom}}$$

$$\emptyset.$$

From this derivation, we get the second unifier $(\{b\#X_1\}, \{X \mapsto \langle a, X_1\rangle, Y \mapsto \langle (a\,b)\cdot X_1, b\rangle\})$.

Matching Fragment. Matching equations are those in which variables may occur only in one side, e.g., left. In this case, we can skip the occurrence check in variable elimination rules. The $\approx^?$ -susp.1, 2 and $\approx^?$ -susp.2 rules never apply. If the given problem does not contain freshness constraints, they will not appear in the result either.

Example 6. The matching problem $\{f\langle X, x, Y, (c\,d)\cdot x, Z\rangle \approx^? f\langle a, b.b, c, a.a, b, d\rangle\}$ has two solutions $(\emptyset, \{X \mapsto \langle a\rangle,\ x \mapsto b.b,\ Y \mapsto \langle c\rangle,\ Z \mapsto \langle b, d\rangle\})$ and $(\emptyset, \{X \mapsto \langle a, b.b\rangle,\ x \mapsto c,\ Y \mapsto \langle a.a, b\rangle,\ Z \mapsto \langle\rangle\})$.

Theorem 3. *The strategy S for unranked unification algorithm is terminating in the KIF, linear, and matching fragments.*

Proof. For a unification problem P, the measure of the size of P is a tuple of natural numbers $(n_x, n_X, n_\approx, n_\#)$, where n_x is the number of different individual variables occurring in P, n_X is the number of different tuple variables occurring in P, $n_\approx$ is the total size of all equational problems in P and $n_\#$ is the total size of all freshness problems in P.

n_x and n_X values are decreased by the $\approx^?$ -var rules and all $\stackrel{\nabla}{\Longrightarrow}$ transformations are decreasing $n_\#$. Analogously, in general, all $\stackrel{\sigma}{\Longrightarrow}$ transformations are decreasing $n_\approx$, except the $\approx^?$ -tuple, $\approx^?$ -widen.1 and $\approx^?$ -widen.2 rules. Clearly, $\approx^?$ -tuple can be applied only finitely many times, since tuples are finite. Next, it is easy to see that in the specific cases $\approx^?$ -widen.1 and $\approx^?$ -widen.2 rules are also decreasing $n_\approx$:

- if P is a unification problem from KIF-fragment, then $\approx^?$ -widen.1, 2 and $\approx^?$ -proj.1, 2 rules are replaced by $\approx^?$ -widen.KIF.1, 2 which are decreasing $n_\approx$ (and n_X as well).
- if P is a unification problem from linear fragment, then every tuple variable occurs only once in P, thus $\approx^?$ -widen.1, 2 rules are decreasing $n_\approx$.
- if P is a matching problem, then there is no tuple variables on the other side, thus $\approx^?$ -widen.1, 2 rules are decreasing $n_\approx$ in this case as well.

□

5 Discussion

Both nominal and unran4ked languages are important for formalizing informal mathematical practice. In nominal languages, one can represent and reason with syntax involving explicitly named bound variables. In unranked languages, one can express and formalize variadic operators and ellipsis that are ubiquitous in mathematical practice. By bringing these two formalisms together, one can get the best of both worlds, aiming at a combination of nominal and unranked logical frameworks. Methods for solving term equations, such as unification and matching, are the core computational mechanism for deduction, rewriting, and programming in such frameworks. Our work makes a step in this direction, providing a procedure that combines nominal and unranked features.

Unranked function symbols look similar to associative function symbols with unit element (A1 symbols). The associativity axiom (for a symbol f) can be expressed as $f\langle f\langle x,y\rangle,z\rangle \approx f\langle x,f\langle y,z\rangle\rangle$ and that e is the unit element of f can be written as $f\langle x,e\rangle \approx x$ and $f\langle e,x\rangle \approx x$. Then terms with nested associative symbols can be flattened, writing, e.g., $f\langle x,y,z\rangle$ for $f\langle f\langle x,y\rangle,z\rangle$. However, in equation solving, unranked and A1 symbols behave differently. Even without nominal binders and freshness atoms, unranked unification and A1-unification are different problems, which can be illustrated with the following example:

Example 7. If f and g are unranked symbols and X is a tuple variable, then the unranked unification problem $f\langle X,g\langle X,c\rangle\rangle \approx^? f\langle a,b,g\langle a,b,c\rangle\rangle$ is solved by $\{X \mapsto \langle a,b\rangle\}$. (In this language, tuples are flat.) If we assume that f and g are A1 symbols and X is an individual variable (in A1 unification problems, tuple variables do not occur), then the same problem does not have a solution: for the left hand side, $\{X \mapsto f\langle a,b\rangle\}$ gives $f\langle a,b,g\langle f\langle a,b\rangle,c\rangle\rangle$, while $\{X \mapsto g\langle a,b\rangle\}$ leads to $f\langle g\langle a,b\rangle,g\langle a,b,c\rangle\rangle$. None of them is equal to the right hand side. Even if we assume that $\langle a,b\rangle$ is a term of this language and X can be instantiated with it, we get $f\langle\langle a,b\rangle,g\langle\langle a,b\rangle,c\rangle\rangle$ as the instance of the left hand side, which is different from $f\langle a,b,g\langle a,b,c\rangle\rangle$, since tuples are not flat in these theories.

In recent years, equational nominal unification has been investigated e.g., in [1,3,6,36,37], but associative and associative-unit theories were not among the studied ones, although α-equivalence modulo associativity has been introduced and formalized in [2] and A-matching rules were given in [9]. One can encode flat tuples via an A1 constructor in a two-sorted language as it was shown, e.g., in [22]. Combining it with nominal techniques, we would get a nominal A1-unification problem of a special kind. However, as we have already mentioned, nominal associative unification has not been investigated so far and we would still have to develop a dedicated solving procedure for this problem. (It would look very similar to our unranked nominal unification algorithm and can be easily reconstructed along the lines of the latter.) We chose not to follow that path and, instead, stick to the unranked setting. The same approach is taken, e.g., in [15], where the authors bring various reasons in favor of the unranked (thereby called flexary) representation, among them the fact that unranked representation is often more natural for implementation. As an example, they mention implementations of type theory (e.g., the Twelf logical framework [32]), where unranked representation is preferred over associative representations both internally and at the user level. Our own experience with implementing a combination of permissive nominal unification and a restricted version of sequence unification in a mathematical assistant system [23] confirms this observation.

Nominal unification problems have been extended with context variables in [38]. Relation between context and sequence unification (without nominal terms) has been studied in [25,26]. It would be interesting to find a similar connection between our work and nominal context unification from [38], but it goes beyond the scope of this paper and can be left for future investigations.

6 Conclusion

We presented an unranked nominal language as an extension of the nominal language with tuple variables and term tuples. We developed a unification procedure for solving equality and freshness problems for unranked nominal terms and proved its soundness and completeness. The soundness property is guaranteed by a specific strategy the procedure is based on. At the same time, the strategy tries to minimize redundant computations.

Some unranked nominal unification problems have an infinite set of solutions. Our procedure, as a complete method, does not terminate for them. It may also run forever for some problems with finite set of solutions, which is not surprising since the strategy is not a decision algorithm. To address this problem, we identified three practically important finitary fragments of unranked nominal unification and proved that our procedure terminates for them.

Acknowledgements. We would like to thank the anonymous referees for their useful comments that helped us to improve our work.

References

1. Ayala-Rincón, M., de Carvalho-Segundo, W., Fernández, M., Nantes-Sobrinho, D.: Nominal C-unification. In: Fioravanti, F., Gallagher, J.P. (eds.) LOPSTR 2017. LNCS, vol. 10855, pp. 235–251. Springer, Cham (2018). https://doi.org/10.1007/978-3-319-94460-9_14
2. Ayala-Rincón, M., de Carvalho-Segundo, W., Fernández, M., Nantes-Sobrinho, D., Rocha-Oliveira, A.C.: A formalisation of nominal α-equivalence with A, C, and AC function symbols. Theoret. Comput. Sci. **781**, 3–23 (2019)
3. Ayala-Rincón, M., Fernández, M., Nantes-Sobrinho, D.: Fixed-point constraints for nominal equational unification. In: Kirchner, H. (ed.), 3rd International Conference on Formal Structures for Computation and Deduction, FSCD 2018, 9–12 July 2018, Oxford, UK, vol. 108, LIPIcs, pp. 7:1–7:16. Schloss Dagstuhl - Leibniz-Zentrum fuer Informatik (2018)
4. Ayala-Rincón, M., Fernández, M., Nantes-Sobrinho, D.: On nominal syntax and permutation fixed points. Log. Methods Comput. Sci. **16**(1), 1–36 (2020)
5. Ayala-Rincón, M., Fernández, M., Rocha-Oliveira, A.C.: Completeness in PVS of a nominal unification algorithm. ENTCS **323**(3), 57–74 (2016)
6. Ayala-Rincón, M., Fernández, M., Silva, G.F., Nantes-Sobrinho, D.: A certified functional nominal C-Unification algorithm. In: Gabbrielli, M. (ed.) LOPSTR 2019. LNCS, vol. 12042, pp. 123–138. Springer, Cham (2020). https://doi.org/10.1007/978-3-030-45260-5_8

7. Baader, F., Nipkow, T.: Term Rewriting And All That. Cambridge University Press, Cambridge (1998)
8. Baader, F., Snyder, W.: Unification theory. In: Robinson, J.A., Voronkov, A. (eds) [35], pp. 445–532
9. de Carvalho Segundo, W.: Nominal Equational Problems Modulo Associativity, Commutativity and Associativity-Commutativity. Ph.D. thesis, Universidade de Brasília, Brazil (2019)
10. Diekert, V.: Makanin's algorithm. Algebraic Comb. Words **90**, 387–442 (2002)
11. Dowek, G.: Higher-order unification and matching. In: Robinson, J.A., Voronkov, A., (eds.) [35], pp. 1009–1062
12. Gabbay, M., Pitts, A.M.: A new approach to abstract syntax with variable binding. Formal ASP Comput. **13**(3–5), 341–363 (2002)
13. Genesereth, M.R., Fikes, R.E.: Knowledge Interchange Format. Version 3.0. Reference Manual. Technical report KSL-92-86, Computer Science Department, Stanford University, June 1992
14. Horozal, F.: A Framework for Defining Declarative Languages. Ph.D. thesis, Jacobs University Bremen (2014)
15. Horozal, F., Rabe, F., Kohlhase, M.: Flexary operators for formalized mathematics. In: Watt, S.M., Davenport, J.H., Sexton, A.P., Sojka, P., Urban, J. (eds.) CICM 2014. LNCS (LNAI), vol. 8543, pp. 312–327. Springer, Cham (2014). https://doi.org/10.1007/978-3-319-08434-3_23
16. Horrocks, I., Voronkov, A.: Reasoning support for expressive ontology languages using a theorem prover. In: Dix, J., Hegner, S.J. (eds.) FoIKS 2006. LNCS, vol. 3861, pp. 201–218. Springer, Heidelberg (2006). https://doi.org/10.1007/11663881_12
17. Jaffar, J.: Minimal and complete word unification. J. ACM **37**(1), 47–85 (1990)
18. Jouannaud, J., Kirchner, C.: Solving equations in abstract algebras: a rule-based survey of unification. In: Lassez, J., Plotkin, G.D. (eds.) Computational Logic - Essays in Honor of Alan Robinson, pp. 257–321. The MIT Press, Cambridge (1991)
19. Knight, K.: Unification: a multidisciplinary survey. ACM Comput. Surv. **21**(1), 93–124 (1989)
20. Kutsia, T.: Equational prover of THEOREMA. In: Nieuwenhuis, R. (ed.) RTA 2003. LNCS, vol. 2706, pp. 367–379. Springer, Heidelberg (2003). https://doi.org/10.1007/3-540-44881-0_26
21. Kutsia, T.: Solving equations involving sequence variables and sequence functions. In: Buchberger, B., Campbell, J. (eds.) AISC 2004. LNCS (LNAI), vol. 3249, pp. 157–170. Springer, Heidelberg (2004). https://doi.org/10.1007/978-3-540-30210-0_14
22. Kutsia, T.: Solving equations with sequence variables and sequence functions. J. Symb. Comput. **42**(3), 352–388 (2007)
23. Kutsia, T.: Unification modulo alpha-equivalence in a mathematical assistant system. RISC Report Series 20–01, RISC, Johannes Kepler University Linz (2020)
24. Kutsia, T., Buchberger, B.: Predicate logic with sequence variables and sequence function symbols. In: Asperti, A., Bancerek, G., Trybulec, A. (eds.) MKM 2004. LNCS, vol. 3119, pp. 205–219. Springer, Heidelberg (2004). https://doi.org/10.1007/978-3-540-27818-4_15
25. Kutsia, T., Levy, J., Villaret, M.: Sequence unification through currying. In: Baader, F. (ed.) RTA 2007. LNCS, vol. 4533, pp. 288–302. Springer, Heidelberg (2007). https://doi.org/10.1007/978-3-540-73449-9_22
26. Kutsia, T., Levy, J., Villaret, M.: On the relation between context and sequence unification. J. Symb. Comput. **45**(1), 74–95 (2010)

27. Kutsia, T., Marin, M.: Solving, reasoning, and programming in common logic. In: Voronkov, A., (eds), 14th International Symposium on Symbolic and Numeric Algorithms for Scientific Computing, SYNASC 2012, Timisoara, Romania, 26–29 September 2012, pp. 119–126. IEEE Computer Society (2012)
28. Levy, J., Villaret, M.: Nominal unification from a higher-order perspective. ACM Trans. Comput. Log. **13**(2), 10:1-10:31 (2012)
29. Makanin, G.S.: The problem of solvability of equations in a free semigroup. Matematicheskii Sbornik **145**(2), 147–236 (1977)
30. Menzel, C.: Knowledge representation, the world wide web, and the evolution of logic. Synth. **182**(2), 269–295 (2011)
31. Paterson, M., Wegman, M.N.: Linear unification. In: Chandra, A.K., Wotschke, D., Friedman, Harrison, M.A., (eds.), Proceedings of the 8th Annual ACM Symposium on Theory of Computing, pp. 181–186. ACM (1976)
32. Pfenning, Frank, Schürmann, Carsten: System description: Twelf — a meta-logical framework for deductive systems. In: CADE 1999. LNCS (LNAI), vol. 1632, pp. 202–206. Springer, Heidelberg (1999). https://doi.org/10.1007/3-540-48660-7_14
33. Pitts, A.M.: Nominal logic, a first order theory of names and binding. Inf. Comput. **186**(2), 165–193 (2003)
34. Robinson, J.A.: A machine-oriented logic based on the resolution principle. J. ACM **12**(1), 23–41 (1965)
35. Robinson, J.A., Voronkov, A., (eds.) Handbook of Automated Reasoning (in 2 volumes). Elsevier and MIT Press (2001)
36. Schmidt-Schauß, M., Kutsia, T., Levy, J., Villaret, M.: Nominal unification of higher order expressions with recursive let. In: Hermenegildo, M.V., Lopez-Garcia, P. (eds.) LOPSTR 2016. LNCS, vol. 10184, pp. 328–344. Springer, Cham (2017). https://doi.org/10.1007/978-3-319-63139-4_19
37. Schmidt-Schauß, M., Kutsia, T., Levy, J., Villaret, M., Kutz, Y.: Nominal unification of higher order expressions with recursive let. Frank report 62, Institut für Informatik, Goethe-Universität Frankfurt am Main, October 2019
38. Schmidt-Schauß, M., Sabel, D.: Nominal unification with atom and context variables. In: Kirchner, H., (ed.), 3rd International Conference on Formal Structures for Computation and Deduction, FSCD 2018, 9–12 July 2018, Oxford, UK, vol. 108, LIPIcs, pp. 28:1–28:20. Schloss Dagstuhl - Leibniz-Zentrum für Informatik (2018)
39. Urban, C., Pitts, A.M., Gabbay, M.: Nominal unification. Theor. Comput. Sci. **323**(1–3), 473–497 (2004)

Lattices of Intermediate Theories via Ruitenburg's Theorem

Gianluca Grilletti[1] and Davide Emilio Quadrellaro[2(✉)]

[1] Munich Centre for Mathematical Philosophy (MCMP), Ludwig Maximillian University, Munich, Germany
G.Grilletti@lmu.de

[2] Department of Mathematics and Statistics, University of Helsinki, Helsinki, Finland
davide.quadrellaro@helsinki.fi

Abstract. For every univariate formula χ (i.e., containing at most one atomic proposition) we introduce a lattice of intermediate theories: the lattice of χ-logics. The key idea to define χ-logics is to interpret atomic propositions as fixpoints of the formula χ^2, which can be characterised syntactically using Ruitenburg's theorem. We show that χ-logics form a lattice, dually isomorphic to a special class of varieties of Heyting algebras. This approach allows us to build and describe five distinct lattices—corresponding to the possible fixpoints of univariate formulas—among which the lattice of negative variants of intermediate logics.

1 Introduction

This paper introduces a family of lattices of intermediate theories, building on three results from the literature: the dual isomorphism between intermediate logics and varieties of Heyting algebras, a novel algebraic semantics for inquisitive logic and negative variants, and Ruitenburg's theorem.

Intermediate logics [7,14] are classes of formulas closed under uniform substitution and modus ponens, lying between the intuitionistic propositional calculus IPC and the classical propositional calculus CPC. This family of logics has been studied using several semantics, as for example Kripke semantics, Beth semantics, topological semantics and algebraic semantics (for an overview see [2]). Among these, the algebraic semantics based on Heyting algebras plays a special role: every intermediate logic is sound and complete with respect to some class of Heyting algebras.[1]

[1] Kripke semantics is known to be incomplete for some intermediate logics, and it is still an open problem whether Beth and topological semantics are complete [2,19].

We would like to thank Nick Bezhanishvili for comments and discussions on this work. Also, we would like to thank the two anonymous referees for several useful remarks and suggestions. The first author was supported by the European Research Council (ERC, grant agreement number 680220). The second author was supported by Research Funds of the University of Helsinki.

© The Author(s), under exclusive license to Springer Nature Switzerland AG 2022
A. Özgün and Y. Zinova (Eds.): TbiLLC 2019, LNCS 13206, pp. 297–322, 2022.
https://doi.org/10.1007/978-3-030-98479-3_15

This connection between intermediate logics and Heyting algebras has been studied using tools from universal algebra. As a consequence of Birkhoff's Theorem [5], the lattice of varieties of Heyting algebras **HA** is dually isomorphic to the lattice of intermediate logics **IL**. This result makes it possible to characterise properties of intermediate logics in terms of properties of the corresponding variety, and vice versa.

Inquisitive logic `InqB` [9–11,25] is an extension of classical logic that encompasses logical relation between *questions* in addition to statements. The logic was originally defined through the *support semantics*, a generalisation of the standard truth-based semantics of `CPC`. Ciardelli et al. gave an axiomatisation of the logic, showing that it lies between `IPC` and `CPC`, and highlighting connections with other intermediate logics such as Maksimova's logic `ND`, Kreisel-Putnam logic `KP` and Medvedev's logic `ML` [8]. However, `InqB` itself is not an intermediate logic, since it is not closed under uniform substitution.

An algebraic semantics for `InqB` has been defined in [3], based on the corresponding algebraic semantics for intermediate logics. The key idea of this work is to restrict the interpretation of atomic propositions to range over regular elements of a Heyting algebra, that is, over fixpoints of the operator $\neg\neg$. This restriction allows to have a sound and complete algebraic semantics, despite the failure of the uniform substitution principle. As shown in [22], this approach can be extended to the class of `DNA`-logics, also known as *negative variants of intermediate logics* [16,20]. Moreover, this leads naturally to a dual isomorphism between `DNA`-logics and a special class of varieties, analogous to the one for intermediate logics.

Ruitenburg's theorem [26] concerns sequences of formulas of the form:

$$\alpha^0 := p \qquad\qquad \alpha^{n+1} := \alpha[\alpha^n/p].$$

where α is a formula and p is a fixed atomic proposition. In particular, the theorem states that this sequence is ultimately periodic with period 2—modulo logical equivalence. For example, if we take $\alpha := \neg p$ we can see that $\neg p \equiv \neg\neg\neg p$, showing that $\neg p$ is a fixpoint of the operator $\neg\neg$. Ghilardi and Santocanale gave an alternative proof of this result in [27], studying endomorphisms of finitely generated Heyting algebras. This proof makes use of the dual isomorphism introduced above and it highlights the relevance of the algebraic interpretation of Ruitenburg's theorem.

In this paper we use Ruitenburg's theorem and its algebraic interpretation to define a lattice of intermediate theories in the same spirit as the negative variants. For a fixed a univariate formula χ (i.e., a formula containing at most one atomic proposition), we define an algebraic semantics by restricting valuations to range over fixpoints of the formula χ^2—which can be characterised using Ruitenburg's theorem. This allows us to build the lattice of χ-logics, intermediate theories characterised in terms of the fixpoint-axiom $\chi^2(p) \leftrightarrow p$. We show that the algebraic semantics is sound and complete for these logics by introducing a lattice of special varieties of Heyting algebras—the χ-varieties—dually isomorphic to the lattice of χ-logics. We also show that there are only six possible fixpoints for univariate formulas: $\top, p, \neg p, \neg\neg p, p \vee \neg p, \bot$. This allows us to characterise and describe all the possible lattices of χ-logics built using this approach.

In Sect. 2 we introduce some preliminary notions on intermediate logics and their algebraic semantics, the dual isomorphism between these logics and varieties, and Ruitenburg's theorem. In Sect. 3 we define χ-logics and give a brief overview of their main properties that can be derived in purely syntactic terms. In Sect. 4, fixed a formula χ, we introduce a novel algebraic semantics for χ-logics based on Ruitenburg's theorem and we define a notion of χ-variety of Heyting algebras suitable to study χ-logics. In Sect. 5 we study the connection between χ-logics and χ-varieties, showing that the dual isomorphism result presented in Sect. 2 can be transferred to this setting. Finally, in Sect. 6 we show there are only 5 distinct lattices of χ-logics for any univariate formula χ, we describe their properties in more detail and we study the relations between them. Conclusions and possible directions for future work are presented in Sect. 7.

2 Preliminaries

In this section we summarise the main notions from the literature employed in this manuscript.

Algebraic Semantics for Intermediate Logics
Fix a countable set $\mathtt{AT}$ of atomic propositions and consider the set of formulas $\mathcal{L}$ generated by the following grammar:

$$\phi ::= p \mid \bot \mid \phi \wedge \phi \mid \phi \vee \phi \mid \phi \rightarrow \phi.$$

where $p \in \mathtt{AT}$. As usual, we introduce the shorthands $\phi \leftrightarrow \psi := (\phi \rightarrow \psi) \wedge (\psi \rightarrow \phi)$, $\neg\phi := \phi \rightarrow \bot$ for *negation* and $\top := \neg\bot$. Henceforth we leave the sets $\mathtt{AT}$ and $\mathcal{L}$ implicit, referring to *atomic propositions from* $\mathtt{AT}$ and to *formulas from* $\mathcal{L}$ simply as *atomic propositions* and *formulas* respectively. To indicate a tuple of propositions $\langle p_1, \ldots, p_n \rangle$ we often use the notation $\overline{p}$, and similarly for tuples of formulas ($\overline{\phi} = \langle \phi_1, \ldots, \phi_n \rangle$) and tuples of other objects.

Consider formulas ϕ, ψ and an atomic proposition p. We indicate with $\phi\,[^{\psi}/_{p}]$ the formula obtained by substituting *every* occurrence of p in ϕ with the formula ψ. More in general, given $\overline{\psi} = \langle \psi_1, \ldots, \psi_n \rangle$ a tuple of formulas and $\overline{p} = \langle p_1, \ldots, p_n \rangle$ a tuple of *distinct* atomic propositions, we indicate with $\phi\,[^{\overline{\psi}}/_{\overline{p}}]$ the formula obtained by substituting *simultaneously* each p_i with ψ_i. With abuse of notation, when we take a univariate formula χ—that is, a formula with only one free variable—we indicate the tuple $\langle \chi(p_1), \ldots, \chi(p_n) \rangle$ with the notation $\chi(\overline{p})$; for example, the notations $\phi[^{\neg\overline{p}}/_{\overline{p}}]$ and $\phi[^{\langle \neg p_1, \ldots, \neg p_n \rangle}/_{\langle p_1, \ldots, p_n \rangle}]$ indicate the same formula.

We refer to the *intuitionistic propositional calculus* as $\mathtt{IPC}$ [7,14]. With slight abuse of notation, we write $\mathtt{IPC}$ also to refer to the set of validities of this calculus. We use the notation $\phi \equiv_{\mathtt{IPC}} \psi$ to indicate that $\phi \leftrightarrow \psi \in \mathtt{IPC}$. An *intermediate logic* [7,14] is a set of formulas L with the following properties:

1. $\mathtt{IPC} \subseteq L \subseteq \mathtt{CPC}$;
2. L is closed under *modus ponens*: If $\phi \in L$ and $\phi \rightarrow \psi \in L$, then $\psi \in L$;

3. L is closed under *uniform substitution*: If $\phi \in L$ and $\overline{\psi}$ is a tuple of formulas, then $\phi\left[\overline{\psi}/\overline{p}\right] \in L$.

Given Γ a set of formulas, we indicate with $\mathrm{MP}(\Gamma)$ the smallest set of formulas extending Γ and closed under modus ponens; and with $\mathrm{US}(\Gamma)$ the smallest set of formulas extending Γ and closed under uniform substitution. It is immediate to prove that such sets always exist and that, if $\Gamma \subseteq \texttt{CPC}$, then $\mathrm{MP}(\mathrm{US}(\texttt{IPC} \cup \Gamma))$ is the smallest intermediate logic extending Γ: we call it the intermediate logic *generated* by Γ and we indicate it with $\texttt{IPC} + \Gamma$.

Intermediate logics ordered by set-theoretic inclusion form a *frame*, that is, a complete lattice [7, Theorem 4.2] satisfying the infinitary distributivity law $L \wedge (\bigvee_{i\in I} L_i) = \bigvee_{i\in I}(L \wedge L_i)$ [7, Theorem 4.6]. In particular, the (infinitary) meet and join operations are $\bigwedge_{i\in I} L_i := \bigcap_{i\in I} L_i$ and $\bigvee_{i\in I} L_i := \mathrm{MP}(\bigcup_{i\in I} L_i)$, and $\texttt{IPC}$ and $\texttt{CPC}$ are respectively the minimum and maximum of this lattice. We refer to this lattice with the notation **IL**.

In the literature, several semantics have been proposed to study intermediate logics: Kripke semantics, Beth semantics and topological semantics are some famous examples (see [2] for an overview of some well-known semantics). In this paper we focus on the *algebraic semantics* based on *Heyting algebras.*

Definition 1 (Heyting algebra [23, Section 1.12]**).** *A* Heyting algebra *is a tuple* $(H, 0_H, \wedge_H, \vee_H, \rightarrow_H)$ *such that* $(H, \wedge_H, \vee_H)$ *is a bounded distributive lattice with least element* 0_H*, and* $\rightarrow_H$ *is a binary operation on* H *such that:*[2]

$$\forall a, b, c \in H. \, (c \leq a \rightarrow_H b \quad \textit{iff} \quad a \wedge c \leq b)\,.$$

A simple example of Heyting algebra is presented in Fig. 1. We refer to the class of all Heyting algebras as HA. There is a natural way to interpret formulas in $\mathcal{L}$ as elements of a Heyting algebra H: given a function $V : \texttt{AT} \rightarrow H$—which we refer to as a *valuation*—we define recursively by the following clauses the interpretation $[\![\phi]\!]^H_V$ of a formula ϕ in H under V:

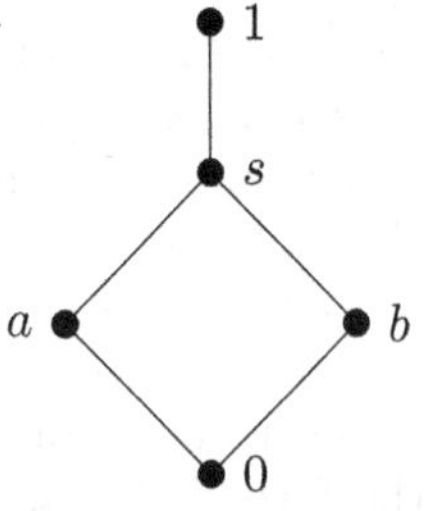

Fig. 1. An example of Heyting algebra, represented as the Hasse diagram of $\leq$.

$$[\![p]\!]^H_V = V(p) \qquad [\![\bot]\!]^H_V = 0_H$$
$$[\![\phi \wedge \psi]\!]^H_V = [\![\phi]\!]^H_V \wedge_H [\![\psi]\!]^H_V \qquad [\![\phi \vee \psi]\!]^H_V = [\![\phi]\!]^H_V \vee_H [\![\psi]\!]^H_V$$
$$[\![\phi \rightarrow \psi]\!]^H_V = [\![\phi]\!]^H_V \rightarrow_H [\![\psi]\!]^H_V.$$

We say that ϕ is *true in* H *under* V and we write $(H, V) \vDash \phi$ if $[\![\phi]\!]^H_V = 1_H$. We say that ϕ is *valid in* H and we write $H \vDash \phi$ if it is true in H under any valuation V. A first reason why this algebraic semantics is employed to study intuitionistic logic is that it provides a *correct and complete semantics* for $\texttt{IPC}$: $\phi \in \texttt{IPC}$ iff ϕ is valid in every Heyting algebra [23, Chapter 9, Sections 2 and 3].

[2] We indicate with $\leq$ the standard ordering induced by the lattice operations, that is, $a \leq b$ iff $a \wedge_H b = a$.

As a shorthand, we indicate with $[p_1 \mapsto a_1, \dots, p_n \mapsto a_n]$ an arbitrary valuation V such that $V(p_i) = a_i$—without specifying its value on the atomic formulas different from $p_1, \dots, p_n$. A function $f : H^n \to H$ is a *polynomial* if it is obtained by composing the functions $1_H, 0_H, \wedge_H, \vee_H$ and $\to_H$ (where we identify the constants 1_H and 0_H with the corresponding 0-ary functions). Given a formula $\phi(p_1, \dots, p_n)$, we can associate to it the polynomial $\boldsymbol{\phi}$ (indicated with the bold font) defined as:

$$\begin{aligned} \boldsymbol{\phi} : H^n &\to H \\ \overline{a} &\mapsto [\![\phi]\!]^H_{[p_1 \mapsto a_1, \dots, p_n \mapsto a_n]} \end{aligned}$$

Moreover, it is immediate to show that, for every polynomial f, there exists a (non-unique) formula ϕ such that $f = \boldsymbol{\phi}$.

Varieties and Dual Isomorphism

Given H, A Heyting algebras, we indicate with $H \preceq A$ that H is a *subalgebra* of A, and with $A \twoheadrightarrow H$ that H is a *homomorphic image* of A (see, e.g., [7, Sec. 7]).

Varieties are a fundamental concept which generalizes the connection between Heyting algebras and IPC to arbitrary intermediate logics. We call a class $\mathcal{V} \subseteq \mathsf{HA}$ a *variety* if $\mathcal{V}$ is closed under the operations H, S, P defined over subclasses of HA as follows:

$$\begin{aligned} \mathrm{H}(\mathcal{C}) &:= \{\, H \in \mathsf{HA} \mid \exists A \in \mathcal{C}.\, A \twoheadrightarrow H \,\} && \text{(homomorphic images)} \\ \mathrm{S}(\mathcal{C}) &:= \{\, H \in \mathsf{HA} \mid \exists A \in \mathcal{C}.\, H \preceq A \,\} && \text{(subalgebras)} \\ \mathrm{P}(\mathcal{C}) &:= \left\{ \prod_{i \in I} A_i \in \mathsf{HA} \;\middle|\; \forall i \in I.\, A_i \in \mathcal{C} \right\} && \text{(products)} \end{aligned}$$

The following are some classical results in universal algebra about algebraic varieties, which we state with reference to varieties of Heyting algebras only.

Theorem 2 (Tarski's theorem [28] and [6, Theorem 9.5]). *Given $\mathcal{C} \subseteq \mathsf{HA}$ a class of algebras,* $\mathrm{HSP}(\mathcal{C})$ *is the smallest variety containing $\mathcal{C}$.*

In light of this result, we call $\boldsymbol{\mathcal{V}}(\mathcal{C}) := \mathrm{HSP}(\mathcal{C})$ the variety *generated* by $\mathcal{C}$.

Theorem 3 (Birkhoff's theorem [5] and [6, Theorem 11.9]). *A class of algebras $\mathcal{C} \subseteq \mathsf{HA}$ is a variety iff it is equationally definable, that is, there exists a set of formulas $F \subseteq \mathcal{L}$ such that $\mathcal{C} = \{\, H \in \mathsf{HA} \mid \forall \phi \in F.\, H \vDash \phi \,\}$.*

The family of varieties of Heyting algebras has also a rather interesting structure: they form a *frame*, which we refer to as **HA**. In particular, the (infinitary) meet and join operations of this lattice are $\bigwedge_{i\in I} \mathcal{V}_i := \bigcap_{i\in I} \mathcal{V}_i$ and $\bigvee_{i\in I} \mathcal{V}_i := \boldsymbol{\mathcal{V}}(\bigcup \mathcal{V}_i)$. Notice the difference between HA (the class of all Heyting algebras) and **HA** (the lattice of *varieties* of Heyting algebras). In particular $\mathsf{HA} \in \mathbf{HA}$.

The properties of this lattice can be derived as a direct consequence of the following theorem, showing the deep connection between intermediate logics and varieties of Heyting algebras.

Theorem 4 (Dual isomorphism [7, Theorem 7.54]**).** *The lattice of intermediate logics is dually isomorphic to the lattice of varieties of Heyting algebras, that is,* $\mathbf{IL} \cong^{op} \mathbf{HA}$*. In particular,* $\mathsf{Var} : \mathbf{IL} \to \mathbf{HA}$ *and* $\mathsf{Log} : \mathbf{HA} \to \mathbf{IL}$ *defined below are isomorphisms, one inverse of the other:*

$$\mathsf{Var}(L) := \{H \in \mathsf{HA} \mid \forall \phi \in L.\, H \vDash \phi\} \qquad \mathsf{Log}(\mathcal{V}) := \{\phi \in \mathcal{L} \mid \forall H \in \mathcal{V}.\, H \vDash \phi\}.$$

We call $\mathsf{Var}(L)$ the *variety generated by* L and $\mathsf{Log}(\mathcal{V})$ the *logic of* $\mathcal{V}$.[3]

Ruitenburg's Theorem

For the remainder of this section, we indicate with p a fixed atomic proposition. Let $\phi(p, \overline{q})$ be a formula, where $p, \overline{q}$ contain all the atomic propositions appearing in ϕ. A folklore result says that the formulas

$$\phi(p, \overline{q}) \qquad \phi^3(p, \overline{q}) := \phi(\,\phi(\,\phi(p, \overline{q}),\, \overline{q}),\, \overline{q})$$

are equivalent in classical logic. Surprisingly, this result generalizes to intuitionistic logic, as Ruitenburg showed in [26].

Definition 5. *Given* $\phi(p, \overline{q})$ *a formula, define the formulas* $\{\phi^n(p, \overline{q})\}_{n \in \mathbb{N}}$ *recursively as follows:*

$$\phi^0(p, \overline{q}) := p \qquad \phi^n(p, \overline{q}) := \phi(\,\phi^{n-1}(p, \overline{q}),\, \overline{q}\,)$$

That is, ϕ^n *is obtained by substituting* ϕ^{n-1} *for* p *in* ϕ*.*

Theorem 6 (Ruitenburg's theorem [26]**).** *For every formula* $\phi(p, \overline{q})$*, the sequence* $\phi^0, \phi^1, \phi^2, \ldots$ *is—modulo logical equivalence—ultimately periodic with period* 2*. That is, there exists a natural number* n *such that:*

$$\phi^n \equiv_{\mathtt{IPC}} \phi^{n+2} \tag{1}$$

We call the smallest n for which Condition 1 holds the *Ruitenburg index* (or simply the *index*) of ϕ. Moreover, we call ϕ^n the *Ruitenburg fixpoint* (or simply the *fixpoint*) of the formula ϕ. For example the formula $\neg p$ is intuitionistically equivalent to the formula $\neg\neg\neg p = (\neg p)^3$, so in this case the index is 1 and the fixpoint is $\neg p$. Another example is the formula $p \vee \neg p$, for which we have $(p \vee \neg p)^2 = (p \vee \neg p) \vee \neg(p \vee \neg p) \equiv_{\mathtt{IPC}} p \vee \neg p$, so also in this case the index is 1 and the fixpoint is $p \vee \neg p$.

We can see Ruitenburg's result also as an algebraic fixpoint theorem. Let A be a Heyting algebra, $\overline{a}$ a tuple of elements in A and $f(x, \overline{y})$ a polynomial. Then a

[3] Admittedly, we are using an improper terminology (also adopted, e.g., in [7, Sec. 7]): varieties are proper classes, hence we cannot talk about the *set* of all varieties nor about the *lattice* of all varieties. A way to dispense of this problem is to instead consider a lattice consisting of *equational theories*, that is, sets of algebraic identities defined by a class of algebras (see, e.g., [6, Section 14] for an overview of this account). To simplify our presentation we abstract away from these issues and we maintain the terminology "lattice of varieties".

consequence of Ruitenburg's theorem is that the operator $f^2(x, \overline{a}) = f(f(x, \overline{a}), \overline{a})$ admits a fixpoint. And indeed, this is an equivalent formulation of Theorem 6, as can be easily shown by applying it to the Lindenbaum-Tarski algebra of IPC.

As proven by Ruitenburg (Example 2.5 in [26]), there is no uniform bound for the indexes of all formulas ϕ, but each formula admits an index. However, for some classes of formulas we can find a uniform bound:

Lemma 7 ([26, Proposition 2.3]). *If $\chi(p)$ is a univariate formula, then $\chi^2 \leftrightarrow \chi^4 \in$ IPC. Moreover, the fixpoint of $\chi(p)$ is equivalent to one of the following formulas: $\bot$, p, $\neg p$, $\neg\neg p$, $p \vee \neg p$, $\top$.*

We give an elementary proof of this result in Appendix A, different from the original one given by Ruitenburg in [26].

3 χ-logics

In the usual presentation of logics, there is an asymmetry between the syntactical and the semantical treatment of atomic propositions. On the one hand, atomic propositions are the basic building blocks used to construct the syntax of every formula. On the other hand, atomic propositions play the role of arbitrary formulas—which translates to the validity of the principle of uniform substitution for the logic.

However, there are several exceptions to this pattern, i.e., logics where atomic propositions are not treated as arbitrary formulas, but rather as semantical objects satisfying certain properties. For example, in *inquisitive logic* [10,11] atomic propositions are interpreted as *natural language statements*, while complex formulas are interpreted as *natural language sentences*, possibly *questions*. Another example is *dependence logic* [29], where atomic propositions are interpreted as *properties of truth-assignments*, while arbitrary formulas are interpreted as more general *relational dependencies between truth-assignments*. As it could be expected, in these logics the principle of uniform substitution fails.

Building on this general idea, we define a class of logics where atomic propositions play the special role of *fixpoints of definable operators*: the χ-logics.

Definition 8 (χ**-logic**). *Let $\chi(p)$ be a univariate formula and Γ a set of formulas. We define the* χ-logic generated by Γ *as the smallest set of formulas Γ^χ with the following properties:*

1. *IPC $\subseteq \Gamma^\chi$;*
2. *If $\phi \in \Gamma$ and σ is a substitution, then $\phi[\sigma] \in \Gamma^\chi$;*
3. *$\chi^2(p) \leftrightarrow p \in \Gamma^\chi$ for every atomic proposition p;*
4. *Γ^χ is closed under* modus ponens*: if $\phi \in \Gamma^\chi$ and $\phi \to \psi \in \Gamma^\chi$, then $\psi \in \Gamma^\chi$.*

Condition 3 requires atoms to behave like fixpoints of the operator χ^2, but we do not require this to hold for arbitrary formulas. And indeed, in general $\chi^2(\phi) \leftrightarrow \phi$ is not a valid principle, which also implies failure of uniform substitution.

An elucidating example is $\mathtt{IPC}^\top$: in this case the fixpoint axiom becomes $\top \leftrightarrow p$, and so we have $\mathtt{IPC}^\top = \mathrm{MP}(\mathtt{IPC} \cup \{\top \leftrightarrow q \,|\, q \text{ atomic proposition}\})$. In Sect. 6 we will give an alternative characterization of $\mathtt{IPC}^\top$ that will make the following a trivial observation, but for now we leave as an exercise to the reader to show that $\top \leftrightarrow \neg p \notin \mathtt{IPC}^\top$.

We could drop the requirement of χ being univariate. There are two main reasons why we adopt this additional restriction. Firstly, the presence of additional atoms requires a generalisation of the results presented in Sect. 2 to Heyting algebras with constants, which is left for future work. And secondly, although the main features of these logics can be showcased in this restricted setting, considering only univariate formulas allows us to characterize all the families of χ-logics generated as a function of χ—which are finitely many in this case, as we will prove.

Observe that we can interpret Γ^χ as the set of valid formulas of an Hilbert-style deduction system: Conditions 1 and 2 define the underlying *schematic principles*, Condition 4 specifies modus ponens as the only rule of the system, and Condition 3 imposes the fixpoint condition over atomic propositions—although it does not introduce a schematic principle. This suggests the following alternative characterisation of χ-logics.

Lemma 9. *Let L be the intermediate logic generated by Γ. Then $\Gamma^\chi = L^\chi$.*

Proof. The left-to-right containment is immediate, since the operator $(-)^\chi$ is monotone. As for the other containment, notice that Conditions 1, 2 and 4 impose that $L \subseteq \Gamma^\chi$, from which the result follows. □

Given an intermediate logic L we call L^χ the *χ-variant of L*. Notice that a direct consequence of Lemma 9 is that any χ-logic is the χ-variant of some intermediate logic L, so we can restrict ourselves to work with intermediate logics instead of arbitrary sets of formulas. And, in fact, we can show that for a fixed χ the family of χ-logics form a complete lattice, as it is the case for intermediate logics. In particular, as it is proved by the following lemma, the infinitary meet and join operations are given by set-theoretic intersection and by the closure under modus ponens of the union respectively—in complete analogy with the case of intermediate logics.

Lemma 10. *Given χ a univariate formula and $\{L_i | i \in I\}$ a family of intermediate logics we have:*

$$\bigwedge_{i\in I} L_i^\chi = \bigcap_{i\in I} L_i^\chi = \left(\bigwedge_{i\in I} L_i\right)^\chi \qquad \bigvee_{i\in I} L_i^\chi = \mathrm{MP}\left(\bigcup_{i\in I} L_i^\chi\right) = \left(\bigvee_{i\in I} L_i\right)^\chi$$

Proof. We consider only the second pair of identities, as the proof can be easily adapted to the first set. Firstly, notice that $L_i^\chi \subseteq (\bigvee_{i\in I} L_i)^\chi$. Moreover, since $L_i \subseteq L_i^\chi$, for every χ-logic Λ such that $L_i^\chi \subseteq \Lambda$ for every $i \in I$ it holds $\bigcup_{i\in I} L_i \subseteq \Lambda$; and since χ-logics are closed under modus ponens it holds $\bigvee_{i\in I} L_i = \mathrm{MP}(\bigcup_{i\in I} L_i) \subseteq \Lambda$. So in particular $(\bigvee_{i\in I} L_i)^\chi \subseteq \Lambda$. This implies that $(\bigvee_{i\in I} L_i)^\chi$ is the least upper bound of the family $\{L_i^\chi | i \in I\}$, that is, $\bigvee_{i\in I} L_i^\chi = (\bigvee_{i\in I} L_i)^\chi$.

Secondly, notice that $\mathrm{MP}(\bigcup_{i\in I} L_i^\chi)$ is the χ-logic generated by the set of formulas $\bigcup_{i\in I} L_i^\chi$. So in particular, since $L_i^\chi \subseteq \mathrm{MP}(\bigcup_{i\in I} L_i^\chi)$, we also have $\bigvee_{i\in I} L_i^\chi \subseteq \mathrm{MP}(\bigcup_{i\in I} L_i^\chi)$. Moreover, since $(\bigvee_{i\in I} L_i)^\chi$ is closed under modus ponens and $\bigcup_{i\in I} L_i^\chi \subseteq (\bigvee_{i\in I} L_i)^\chi$, it follows $\mathrm{MP}(\bigcup_{i\in I} L_i^\chi) \subseteq (\bigvee_{i\in I} L_i)^\chi = \bigvee_{i\in I} L_i^\chi$. From this we conclude that $\bigvee_{i\in I} L_i^\chi = \mathrm{MP}(\bigcup_{i\in I} L_i^\chi)$, as wanted. □

We indicate with $\mathbf{IL}^\chi$ the lattice of χ-logics. Notice that the previous proof shows that the mapping $L \mapsto L^\chi$ is a *complete lattice homomorphism*, and so χ-variants form a *frame*.

In the next sections we study the structure of these lattices for different formulas χ by employing tools from algebraic semantics. The following alternative characterization of χ-variants based on Theorem 6 will help us with our task.

Lemma 11. *Let $\chi(p)$ be a univariate formula and n be its index. Given L an intermediate logic, we have* $L^\chi = \{ \phi(\overline{p}) \mid \phi[\chi^n(\overline{p})/\overline{p}] \in L \}$.

Proof. Call the set on the right-hand side M. Firstly, we will show that M satisfies the conditions in Definition 8. Since L contains IPC and it is closed under modus ponens and uniform substitution, we easily obtain Conditions 1, 2 and 4. As for Condition 3, since n is the index of χ, we have $\chi^{n+2}(p) \leftrightarrow \chi^n(p) \in L$, from which it follows $\chi^2(p) \leftrightarrow p \in M$ for every atomic proposition p.

Secondly, we need to show that M is the smallest set satisfying these conditions, so consider a set X satisfying the conditions of Definition 8 for $\Gamma = L$. We will make extensive use of the fact that IPC proves the principle of *substitution of equivalents* (SoE for short): given formulas $\overline{\alpha} = \langle \alpha_1, \dots, \alpha_l \rangle, \overline{\beta} = \langle \beta_1, \dots, \beta_l \rangle, \gamma$ formulas and distinct atomic propositions $\overline{q} = \langle q_1, \dots, q_l \rangle$, we have

$$\bigwedge_{i\leq l} (\alpha_i \leftrightarrow \beta_i) \to (\gamma[\overline{\alpha}/\overline{q}] \leftrightarrow \gamma[\overline{\beta}/\overline{q}]) \in \mathtt{IPC} \subseteq X. \tag{2}$$

As an instance of this condition, we have $(\chi^2(q) \leftrightarrow q) \to (\chi^4(q) \leftrightarrow \chi^2(q)) \in X$. Since $\chi^2(q) \leftrightarrow q \in X$ and X is closed under modus ponens (Conditions 3 and 4 of Definition 8), we also have $\chi^4(q) \leftrightarrow \chi^2(q) \in X$. Moreover, since

$$(\alpha \leftrightarrow \beta) \to ((\beta \leftrightarrow \gamma) \to (\alpha \leftrightarrow \gamma)) \in \mathtt{IPC} \subseteq X$$

with a similar argument we obtain $\chi^4(q) \leftrightarrow q \in X$. Iterating this reasoning, we obtain that $\chi^n(q) \leftrightarrow q \in X$ for every q or $\chi^{n+1}(q) \leftrightarrow q$ for every q—depending on the parity of n. Assume the former is the case; the treatment of the other case is analogous.

Consider now an arbitrary formula $\phi(\overline{p})$ with $\overline{p} = \langle p_1, \dots, p_l \rangle$. Combining the previous facts we get:

$$\begin{array}{lll} & \chi^n(p_i) \leftrightarrow p_i & \in X \text{ for every } i \leq l \\ \text{and} & \bigwedge_{i\leq l} (\chi^n(p_i) \leftrightarrow p_i) \to (\phi[\chi^n(\overline{p})/\overline{p}] \leftrightarrow \phi(\overline{p})) & \in X \\ \text{implies} & \phi[\chi^n(\overline{p})/\overline{p}] \leftrightarrow \phi(\overline{p}) & \in X. \end{array}$$

From this we can conclude that, if $\phi[\chi^n(\overline{p})/\overline{p}] \in L$ then $\phi(\overline{p}) \in X$. As ϕ was arbitrary, we have $M \subseteq X$ and so M is the smallest set satisfying the Conditions in Definition 8, as wanted. □

Lemma 11 is especially useful to compute the χ-variant of a given intermediate logic. For example, consider the $\neg p$-variant of the logic of *weak excluded middle* $\mathtt{WEM} := \mathtt{IPC} + \{\neg p \vee \neg\neg p\}$. By Lemma 11, we have $q \vee \neg q \in \mathtt{WEM}^{\neg p}$ for every atomic proposition q. So, by a folklore result, we have

$$\mathrm{MP}(\mathtt{IPC} \cup \{q \vee \neg q \mid q \text{ atomic proposition}\}) = \mathtt{CPC} \subseteq \mathtt{WEM}^{\neg p}.$$

Finally, notice that all $\neg p$-logics are contained in $\mathtt{CPC}$ (in this particular case, the fixpoint axiom $\neg\neg p \leftrightarrow p$ is a classical tautology), and so $\mathtt{WEM}^{\neg p} = \mathtt{CPC}$.

Surprisingly, these lattices are fewer than one could expect: by Lemma 11 we have that $L^{\chi} = L^{\chi^n}$, meaning that formulas with the same Ruitenburg fixpoint determine the same lattice. By Lemma 7 there are only a finite amount of fixpoints, thus there are only finitely many lattices of χ-variants. Notice also that by Lemma 11, together with the fact that $\neg\neg\neg p \equiv_{\mathtt{IPC}} \neg p$, we have:

$$L^{\neg p} = \{ \phi(\overline{q}) \mid \phi[\neg\overline{q}/\overline{q}] \in L \} = \{ \phi(\overline{q}) \mid \phi[\neg\neg\neg\overline{q}/\overline{q}] \in L \} = L^{\neg\neg\neg p};$$

which implies that $\mathbf{IL}^{\neg p} = \mathbf{IL}^{\neg\neg\neg p}$. Therefore, we are working with 5 lattices in total. In Sect. 6 we will see that these are indeed distinct lattices:

$$\mathbf{IL}^{\bot} \qquad \mathbf{IL}^{p} = \mathbf{IL} \qquad \mathbf{IL}^{\neg p} = \mathbf{IL}^{\neg\neg p} \qquad \mathbf{IL}^{p \vee \neg p} \qquad \mathbf{IL}^{\top}.$$

4 Algebraic Semantics

As mentioned in the previous section, χ-logics treat atomic propositions as fixpoints of the operator χ^2. This intuition can be exploited to develop a sound and complete algebraic semantics by *restricting the set of admissible valuations.* This general idea has already been employed in [3] to define an algebraic semantics for inquisitive logic (see also [22] for an in-depth study of this semantics), and more recently in [15] to define a *nuclear semantics* for a novel class of inquisitive logics over an intuitionistic basis. This section adapts the methodologies employed in [22] for the class of $\mathtt{DNA}$-logics—de facto, the family of $\neg p$-logics—to the more general setting of χ-logics.

In the case of χ-logics, the key to define an algebraic semantics based on Heyting algebras lies in an algebraic interpretation of Ruitenburg's theorem. In this section we will fix a univariate formula χ with index n. As noted in Sect. 2, given a Heyting algebra H we can define a polynomial corresponding to χ:

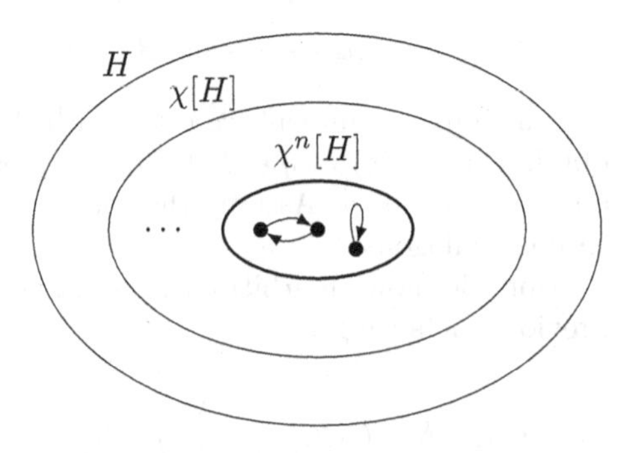

$$\chi : H \to H$$
$$a \mapsto [\![\chi(p)]\!]_{[p \mapsto a]}$$

Ruitenburg's theorem tells us that the sequence $H, \chi[H], \chi^2[H] := \chi[\chi[H]], \ldots$ is ultimately constant; and that the polynomial χ restricted to the set $\chi^n[H]$ is an *involution*, i.e. χ^2 is the identity over $\chi^n[H]$. Henceforth we will call the set $H^\chi := \chi^n[H]$ the χ-*core* (or simply *core* when χ is clear from the context) of H. Notice that the χ-core consists exactly of the fixpoints of χ^2:

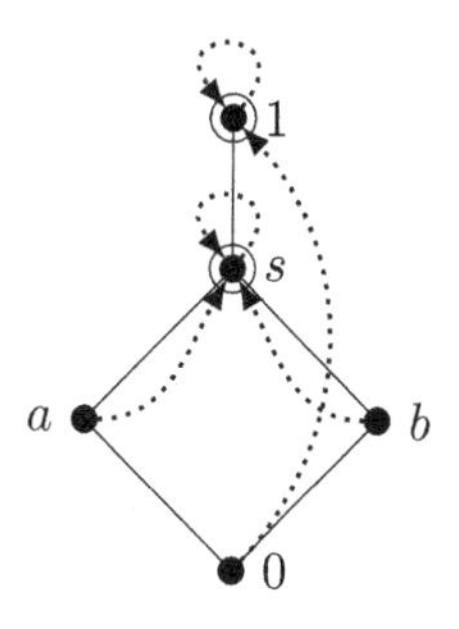

Fig. 2. An example of χ-core for $\chi = p \vee \neg p$. In this case $\chi^n(x) = x \vee \neg x$. The dotted arrows represent the function χ^n and the circles indicate the elements of the core.

Lemma 12. *H^χ is the set of fixpoints of χ^2.*

Proof. By Theorem 6, we have that $\chi^n \equiv \chi^{n+2}$. Consider an element $a \in H^\chi$, that is, an element of the form $a = \chi^n(b)$ for some $b \in H$. It follows that $\chi^2(a) = \chi^2(\chi^n(b)) = \chi^{n+2}(b) = \chi^n(b) = a$, showing that a is a fixpoint of χ^2. Conversely, let a be a fixpoint for χ^2. Then it follows that $a = \chi^2(a) = \chi^2(\chi^2(a)) = \chi^4(a) = \cdots = \chi^{2n}(a) = \chi^n\chi^n(a) \in H^\chi$.

For instance, when $\chi(p) = \neg p$ the core $H^{\neg p}$ of H consists of the *regular elements* of the algebra H, that is, fixpoints of the operator $\neg\neg$. In Fig. 2 we can also see an example of $(p \vee \neg p)$-core.

To obtain an adequate semantics for a χ-logic, we restrict the valuations of atomic propositions to the core H^χ. Let AT be an arbitrary set of atomic propositions. We say that a valuation $\sigma : \mathtt{AT} \to H$ is a χ-*valuation* if $\sigma[\mathtt{AT}] \subseteq H^\chi$. A χ-valuation over H thus sends every atomic proposition to some element of the χ-core of H. Algebraic models of χ-logics are then defined as follows.

Definition 13 (χ-Model). *A χ-model is a pair $M = (H, \sigma)$ such that H is a Heyting algebra and σ is a χ-valuation.*

The *interpretation* of a formula $\phi \in \mathcal{L}$ in the χ-model $M = (H, \sigma)$—in symbols $[\![\phi]\!]^H_\sigma$—is defined recursively exactly as in the standard algebraic semantics for intuitionistic logic. The key difference between the two semantics is that for atomic propositions we have $[\![p]\!]^H_\sigma = \sigma(p) \in H^\chi$, that is, atomic proposition are interpreted as fixpoints for χ^2. As in the standard algebraic semantics, we say that ϕ is *true in H under* $\sigma$and we write $(H, \sigma) \vDash^\chi \phi$ if $[\![\phi]\!]^H_\sigma = 1_H$. We say that ϕ is χ-*valid in* H and we write $H \vDash^\chi \phi$ if it is true in H under any χ-valuation σ. In general, we refer to this semantics as χ-*semantics* to distinguish it from the standard algebraic semantics.

Remark 14. Notice that the semantic approach we adopted reminds of the *nuclear semantics for inquisitive intuitionistic logic* employed in [15] (originally introduced for intuitionistic logic in [1]): a *nuclear algebra* is defined as a pair (H, j), where $j : H \to H$ is a *nucleus*, that is, an increasing $(a \leq ja)$, idempotent $(jja = ja)$ and multiplicative $(j(a \wedge b) = ja \wedge jb)$ map; and the corresponding semantics restricts the valuation of atomic propositions to range over *fixpoints* of the nucleus. Although the two semantics show a striking resemblance, they differ in at least two fundamental aspects. Firstly, not all the maps $\boldsymbol{\chi}^2$ are nuclei, so the fixpoint operators employed by the two logics are different. For example, this is not the case for $\chi = p \vee \neg p$, as shown in Fig. 2.[4] Secondly, in every Heyting algebra H the map $\boldsymbol{\chi}^2$ is uniquely defined by the formula χ^2, while the same does not apply in nuclear semantics. This definability condition is quite restrictive, but it allows us (as we will see in the next section) to easily transfer many results from the theory of intermediate logic to the setting of χ-logics, as for example the dual isomorphism presented in Theorem 4.

However, notice that $\neg p$-logics are examples of inquisitive intuitionistic logics ([15, Definition 3.1] with $\vee$ treated as a shorthand) and that $\neg p$-models are essentially nuclear models, so the two frameworks seem to be compatible and closely related. A comparison of the two approaches would prove to be quite fruitful and could lead to interesting results. However, in the current manuscript we will primarily focus on laying the foundations of the theory of χ-logics and we leave a thorough comparison for future works.

The interpretations under different valuations of the formula $\chi^n(p)$ range over all and only the elements of the core. This suggests the following definition: for every valuation $V : \mathtt{AT} \to H$ its χ-*variant* V^χ is the χ-valuation $V^\chi : \mathtt{AT} \to H^\chi$ such that $V^\chi(p) = \chi^n(V(p))$. The function $V \mapsto V^\chi$ maps each valuation to a χ-valuation such that $[\![\phi]\!]^H_{V^\chi} = [\![\phi\,[\chi^n(\overline{p})/\overline{p}]]\!]^H_V$. Moreover, this map is surjective: given σ a χ-valuation, since $\sigma(p) \in H^\chi = \chi^n[H]$, for any valuation V such that $V(p) \in (\chi^n)^{-1}(\sigma(p))$ we have $V^\chi(p) = \chi^n(V(p)) = \sigma(p)$. The next proposition follows directly from these observations.

Proposition 15. *For any Heyting algebra H, $H \vDash^\chi \phi$ iff $H \vDash \phi\,[\chi^n(\overline{p})/\overline{p}]$.*

Corollary 16. *Let H be a Heyting algebra and L an intermediate logic, if $H \vDash L$ then $H \vDash^\chi L^\chi$.*

Proof. Direct consequence of Lemma 11 and Proposition 15. □

The converse of Corollary 16 does not hold in general, as a formula might be true in a Heyting algebra under all χ-valuations but not under all valuations. We can however define a class of algebras for which the converse hold: let $\langle H^\chi \rangle$ be the subalgebra of H generated by the χ-core H^χ; we say that H is *core generated* if $H = \langle H^\chi \rangle$.

Lemma 17. *Let H be a Heyting algebra, $H \vDash^\chi \phi$ if and only if $\langle H^\chi \rangle \vDash^\chi \phi$.*

Proof. The algebras H and $\langle H^\chi \rangle$ share the same χ-valuations, and for any such χ-valuation σ we have $[\![\phi]\!]^{\langle H^\chi \rangle}_\sigma = [\![\phi]\!]^H_\sigma$. From this the result follows trivially. □

[4] As a consequence of Lemma 7, in the current setting this is essentially the only case for which the map $\boldsymbol{\chi}^2$ is not a nucleus. We can find other counterexamples if we lift the restriction of χ being univariate.

Proposition 18. *Let H be a Heyting algebra and L an intermediate logic. Then we have that $H \vDash^{\chi} L^{\chi}$ entails $\langle H^{\chi} \rangle \vDash L$.*

Proof. By contraposition, suppose that $(\langle H^{\chi}\rangle, V) \nvDash \phi$ for some valuation V and formula $\phi \in L$. Since $\langle H^{\chi}\rangle$ is the subalgebra generated by H^{χ}, we can express every element $x \in \langle H^{\chi}\rangle$ as a polynomial $x = \boldsymbol{\eta}(\overline{y})$ where each y_i is an element of H^{χ}. Let $\overline{p} = \langle p_1, \dots, p_n\rangle$ be the variables contained in ϕ and define $\boldsymbol{\delta}_i(\overline{y}) = V(p_i)$ for every $i = 1, \dots, n$ (without loss of generality, we can assume that $\overline{y}$ is the same tuple of elements for every i). Writing $\boldsymbol{\delta}(\overline{y})$ for the tuple $\langle \boldsymbol{\delta}_1(\overline{y}), \dots, \boldsymbol{\delta}_n(\overline{y})\rangle$, we have $\boldsymbol{\phi}(\boldsymbol{\delta}(\overline{y})) = [\![\phi(\overline{p})]\!]_V^{\langle H^{\chi}\rangle} \neq 1_H$.[5]

Since all the elements $\overline{y}$ are elements of H^{χ}, we can define a χ-valuation σ such that $\sigma(q_i) = y_i$ for every $i \leq n$. In particular, for this choice of σ we have $[\![\, \phi[^{\delta(\overline{q})}/_p] \,]\!]_{\sigma}^{H} = \boldsymbol{\phi}(\boldsymbol{\delta}(\overline{y})) \neq 1_H$ where $\delta(\overline{q})$ indicates the tuple of formulas $\langle \delta_1(\overline{q}), \dots, \delta_n(\overline{q})\rangle$. In particular, since σ is a χ-valuation we have $H \nvDash^{\chi} \phi[^{\delta(\overline{q})}/_p]$. Since L is closed under uniform substitution and $\phi \in L$, we then have $\phi[^{\delta(\overline{q})}/_p] \in L \subseteq L^{\chi}$ and so $H \nvDash^{\chi} L^{\chi}$. □

As Lemma 17 and Proposition 18 show, the subalgebra generated by the χ-core $\langle H^{\chi}\rangle$ already contains all the semantic information about H when it comes to the novel semantics. This suggests the following definition: we say that a Heyting algebra H is a *core superalgebra* of K if $K^{\chi} = H^{\chi}$ and $K \preceq H$. In particular, by Lemma 17, an algebra and its core superalgebras validate the same formulas under the χ-semantics.

In turns, the last observation leads naturally to shift our attention to *varieties* of Heyting algebras and to the following definition: given $\mathcal{V}$ a variety of Heyting algebras, its χ-closure is the class:

$$\begin{aligned}\mathcal{V}^{\chi} &= \{\, H \in \mathsf{HA} \mid \exists K \in \mathcal{V}.\ H \text{ is a core superalgebra of } K \,\}\\ &= \{\, H \in \mathsf{HA} \mid \exists K \in \mathcal{V}.\ K^{\chi} = H^{\chi} \text{ and } K \preceq H \,\}\\ &= \{\, H \in \mathsf{HA} \mid \langle H^{\chi}\rangle \in \mathcal{V} \,\}.\end{aligned}$$

We call varieties of this kind *χ-varieties.* The following theorem gives an alternative characterization of χ-varieties, more in line with the standard definition of variety.

Theorem 19. *A class of Heyting algebras $\mathcal{C}$ is a χ-variety if and only if it is closed under subalgebras, homomorphic images, products and core superalgebras.*

Proof. ($\Leftarrow$) Suppose $\mathcal{C}$ is closed under subalgebras, homomorphic images, products and core superalgebras. $\mathcal{C}$ is a variety and for any Heyting algebra H such that there is some $K \in \mathcal{C}$ with $H^{\chi} = K^{\chi}$ and $K \preceq H$, it follows by closure under core superalgebra that $H \in \mathcal{C}$. Therefore $\mathcal{C} = \mathcal{C}^{\chi}$, hence $\mathcal{C}$ is a χ-variety.

($\Rightarrow$) Suppose $\mathcal{C}$ is a χ-variety, i.e. $\mathcal{C} = \mathcal{V}^{\chi}$ for some variety $\mathcal{V}$. We need to show that $\mathcal{C}$ is closed under subalgebras, products and homomorphic images—closure under core superalgebras follows by definition of $\mathcal{V}^{\chi}$.

[5] We are omitting the line on top of $\boldsymbol{\delta}$ in favor of readability.

- *Subalgebras.* Suppose $H \in \mathcal{C}$ and $K \preceq H$. Notice that $K^\chi \subseteq H^\chi$, and consequently $\langle K^\chi \rangle \preceq \langle H^\chi \rangle$. Moreover, since $\mathcal{C} = \mathcal{V}^\chi$ we have $\langle H^\chi \rangle \in \mathcal{V}$ and so, since varieties are closed under subalgebras, $\langle K^\chi \rangle \in \mathcal{V}$. It follows that $K \in \mathcal{V}^\chi = \mathcal{C}$.
- *Homomorphic images.* Consider an algebra $H \in \mathcal{C}$ and a surjective homomorphism $f : H \twoheadrightarrow K$. We firstly prove that $f|_{H^\chi}$ is surjective over K^χ. Consider an element $k \in K^\chi$ and notice that, by Lemma 12, this amounts to $\chi^2(k) = k$. Since f is surjective, there exists an element $h \in H$ such that $f(h) = k$ (notice that h is not necessarily an element of H^χ, since nothing ensures that $\chi^2(h) = h$). Consider now the index n of χ and assume it is odd (the case n even is treated in a similar fashion). We then have that:

$$\chi^2(\chi^{n+1}(h)) = \chi^{n+3}(h) = \chi^{n+1}(h) \in H^\chi$$
$$f(\chi^{n+1}(h)) = \chi^{n+1}(f(h)) = \chi^{n+1}(k) = \chi^{n-1}(k) = \chi^{n-3}(k) = \cdots = k.$$

  And so k is the image of an element in H^χ, namely $\chi^{n+1}(h)$. Since k was arbitrary, this shows that $f|_{H^\chi}$ is surjective, as desired.
  We now show that $K \in \mathcal{C}$. Since $\mathcal{C} = \mathcal{V}^\chi$ and $H \in \mathcal{C}$ we have $\langle H^\chi \rangle \in \mathcal{V}$. As we showed, $f|_{H^\chi}$ is surjective over K^χ, which in turns implies that $f|_{\langle H^\chi \rangle}$ is a surjective homomorphism over $\langle K^\chi \rangle$. So, since $\mathcal{V}$ is closed under homomorphic images, then $\langle K^\chi \rangle \in \mathcal{V}$ and so $K \in \mathcal{X} = \mathcal{C}$.
- *Products.* Consider algebras $\{H_i | i \in I\} \subseteq \mathcal{C}$. Since $\mathcal{C} = \mathcal{V}^\chi$, it follows $\{\langle H_i^\chi \rangle | i \in I\} \subseteq \mathcal{V}$. By properties of the product we have that $\prod_{i \in I} \langle H_i^\chi \rangle = \langle (\prod_{i \in I} H_i)^\chi \rangle \preceq \prod_{i \in I} H_i$, and since $\mathcal{V}$ is a variety $\langle (\prod_{i \in I} H_i)^\chi \rangle \in \mathcal{V}$. Thus we conclude that $\prod_{i \in I} H_i \in \mathcal{X} = \mathcal{C}$.

This concludes the proof. □

A straightforward consequence of Theorem 19 is that the intersection of an arbitrary set of χ-varieties is again a χ-variety, which means that χ-varieties form a *complete sublattice of* **HA**: we denote by $\mathbf{HA}^\chi$ this lattice. If we denote by $\boldsymbol{\mathcal{X}}(\mathcal{C})$ the smallest χ-variety containing $\mathcal{C}$, then the operations of this lattice are:

$$\bigwedge_{i \in I} \mathcal{X}_i = \bigcap_{i \in I} \mathcal{X}_i \qquad\qquad \bigvee_{i \in I} \mathcal{X}_i = \boldsymbol{\mathcal{X}}\left(\bigcup_{i \in I} \mathcal{X}_i\right).$$

Together with the results of the previous sections, we have thus obtained a lattice $\mathbf{IL}^\chi$ of χ-variants of intermediate logics, and a lattice $\mathbf{HA}^\chi$ of χ-varieties. In the next section we shall see how to relate these two structures in order to prove the completeness of the algebraic semantics we introduced.

5 Dual Isomorphism

In this section we show that the lattice of χ-logics $\mathbf{IL}^\chi$ and the lattice of χ-varieties $\mathbf{HA}^\chi$ are dually isomorphic. The underlying idea behind the proof is that the isomorphisms Log and Var between the lattices **IL** and **HA** can be *transported* along the maps $(-)^\chi$ ($L \mapsto L^\chi$ and $\mathcal{V} \mapsto \mathcal{V}^\chi$), obtaining corresponding isomorphisms Log^χ and Var^χ between the lattices $\mathbf{IL}^\chi$ and $\mathbf{HA}^\chi$. Our starting point is to define explicitly the maps Log^χ and Var^χ: given Γ a set of formulas and given $\mathcal{C}$ a class of Heyting algebra, define[6]

$$\mathsf{Var}^\chi : \Gamma \mapsto \{ H \in \mathsf{HA} \mid H \vDash^\chi \Gamma \} \qquad \mathsf{Log}^\chi : \mathcal{C} \mapsto \{ \phi \in \mathcal{L} \mid \mathcal{C} \vDash^\chi \phi \}.$$

We say that a class $\mathcal{C}$ of Heyting algebras is *χ-definable* if there is a set Γ of formulas such that $\mathcal{C} = \mathsf{Var}^\chi(\Gamma)$, and we say that a χ-logic Λ is *χ-complete* with respect to a class of Heyting algebras $\mathcal{C}$ if $\Lambda = \mathsf{Log}^\chi(\mathcal{C})$. Before proceeding with the proof that Var^χ and Log^χ induce a dual isomorphism between the lattices $\mathbf{IL}^\chi$ and $\mathbf{HA}^\chi$, we first need to make sure that they are *well-defined maps* between the two lattices. We firstly show that $\mathsf{Var}^\chi(\Gamma)$ is a χ-variety, and to do so we need the following technical results.

Proposition 20. *χ-validities are preserved by taking subalgebras, products, homomorphic images and core superalgebras.*

Proof. Validity is preserved by taking subalgebras, products and homomorphic images, so by Proposition 15 also χ-validity is preserved. It remains to show that χ-validity is preserved by taking core superalgebras.

Let K be a core superalgebra of H, that is, let $K^\chi = H^\chi$ and $H \preceq K$. Consider ϕ a formula χ-valid on H and assume towards a contradiction that, for some χ-valuation σ, we have $(K, \sigma) \nvDash^\chi \phi$. Since $H^\chi = K^\chi$, σ is also a χ-valuation over H. And since $H \preceq K$ we have $[\![\phi]\!]^H_\sigma = [\![\phi]\!]^K_\sigma \neq 1$. But this contradicts our assumptions on ϕ. □

By the previous proposition we obtain the following corollary.

Corollary 21. *For every set of formulas Γ, the class of Heyting algebras $\mathsf{Var}^\chi(\Gamma)$ is a χ-variety.*

Now we focus on the map Log^χ: we want to show that, for every class $\mathcal{C}$ of Heyting algebras, the set $\mathsf{Log}^\chi(\mathcal{C})$ is a χ-logic. We show a slightly stronger result.

Proposition 22. *For every set of algebras $\mathcal{C}$, the class of formulas $\mathsf{Log}^\chi(\mathcal{C})$ is the χ-variant of $\mathsf{Log}(\mathcal{C})$.*

Proof. We have:

$$\begin{aligned}
\phi \notin \mathsf{Log}^\chi(\mathcal{C}) &\iff \exists H \in \mathcal{C} \text{ such that } H \nvDash^\chi \phi \\
&\iff \exists H \in \mathcal{C} \text{ such that } H \nvDash \phi[\chi^n(\overline{p})/\overline{p}] \quad \text{(by Proposition 15)} \\
&\iff \phi[\chi^n(\overline{p})/\overline{p}] \notin \mathsf{Log}(\mathcal{C}) \\
&\iff \phi \notin (\mathsf{Log}(\mathcal{C}))^\chi.
\end{aligned}$$

Hence $\mathsf{Log}^\chi(\mathcal{C})$ is the χ-variant of $\mathsf{Log}(\mathcal{C})$. □

[6] To lighten the notation, we use the symbol Var^χ instead of $\mathsf{Var}^\chi|_{\mathbf{IL}^\chi}$ for the function obtained by restricting the domain, omitting the explicit restriction of the domain. The same applies for the notation Log^χ, used instead of $\mathsf{Log}^\chi|_{\mathbf{HA}^\chi}$.

These two results ensure that $\mathsf{Log}^\chi : \mathbf{HA}^\chi \to \mathbf{IL}^\chi$ and $\mathsf{Var}^\chi : \mathbf{IL}^\chi \to \mathbf{HA}^\chi$ are well-defined maps. We now formalize—in Propositions 23 and 24—the intuition presented at the start of this section: Log^χ and Var^χ are obtained by *transporting* Log and Var respectively along the maps $(-)^\chi$.

Proposition 23. *For every intermediate logic L, $\mathsf{Var}^\chi(L^\chi) = \mathsf{Var}(L)^\chi$.*

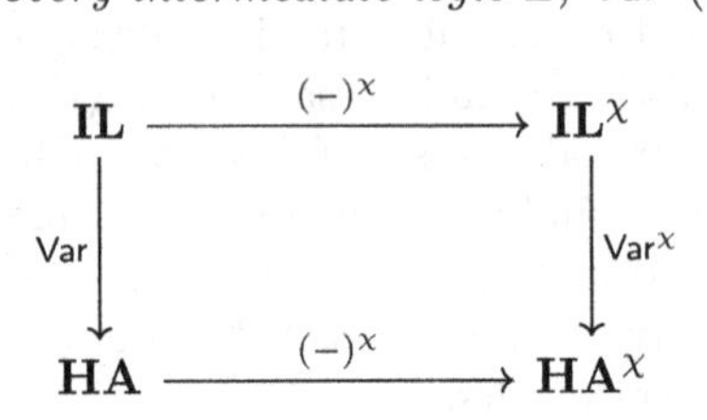

Proof. ($\subseteq$) Consider any Heyting algebra $H \in \mathsf{Var}^\chi(L^\chi)$. Then we have $H \vDash^\chi L^\chi$ and by Proposition 18 it follows $\langle H^\chi \rangle \vDash L$. So we clearly have that $\langle H^\chi \rangle \in \mathsf{Var}(L)$ and since $\langle H^\chi \rangle^\chi = H^\chi$ and $\langle H^\chi \rangle \preceq H$ also $H \in \mathsf{Var}(L)^\chi$. ($\supseteq$) Consider any Heyting algebra $H \in \mathsf{Var}(L)^\chi$, then there is some $K \in \mathsf{Var}(L)$ such that $K \preceq H$ and $H^\chi = K^\chi$. Then we have that $K \vDash L$, so by Corollary 16 above $K \vDash^\chi L^\chi$ which entails $K \in \mathsf{Var}^\chi(L^\chi)$. Finally, since χ-varieties are closed under core superalgebras, it follows that $H \in \mathsf{Var}^\chi(L^\chi)$. □

Proposition 24. *For every variety $\mathcal{V}$ of Heyting algebras $\mathsf{Log}^\chi(\mathcal{V}^\chi) = \mathsf{Log}(\mathcal{V})^\chi$.*

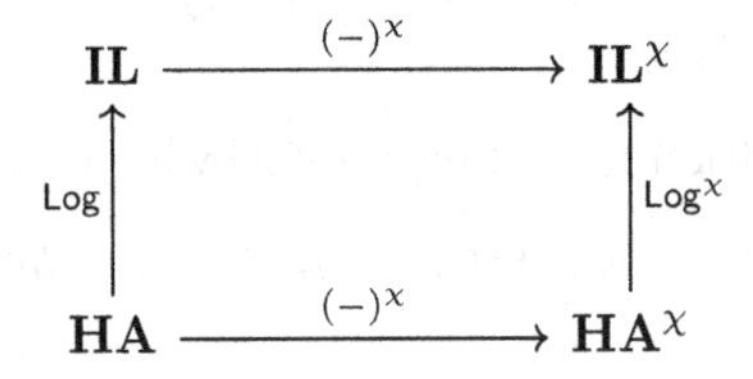

Proof. We prove both inclusions by contraposition. ($\subseteq$) Suppose $\phi \notin \mathsf{Log}(\mathcal{V})^\chi$, then $\phi[\chi^n(\overline{p})/\overline{p}] \notin \mathsf{Log}(\mathcal{V})$ and hence there is some $H \in \mathcal{V}$ such that $H \nvDash \phi[\chi^n(\overline{p})/\overline{p}]$. By Proposition 15 $H \nvDash^\chi \phi$, hence $\phi \notin \mathsf{Log}^\chi(\mathcal{V}^\chi)$. ($\supseteq$) Suppose $\phi \notin \mathsf{Log}^\chi(\mathcal{V}^\chi)$. It follows that there is some $H \in \mathcal{V}^\chi$ such that $H \nvDash^\chi \phi$, hence by Lemma 17 $\langle H^\chi \rangle \nvDash^\chi \phi$. It thus follows by Proposition 15 that $\langle H^\chi \rangle \nvDash \phi[\chi^n(\overline{p})/\overline{p}]$. Now, since $H \in \mathcal{V}^\chi$, we have for some $K \in \mathcal{V}$ that $K \preceq H$ and $K^\chi = H^\chi$. Thus it follows that $\langle H^\chi \rangle \preceq K$ and therefore $\langle H^\chi \rangle \in \mathcal{V}$. Finally, since $\langle H^\chi \rangle \nvDash \phi[\chi^n(\overline{p})/\overline{p}]$ we get that $\phi[\chi^n(\overline{p})/\overline{p}] \notin \mathsf{Log}(\mathcal{V})$ and hence $\phi \notin \mathsf{Log}(\mathcal{V})^\chi$. □

Building on the previous results, we can finally prove a definability theorem and a completeness theorem for χ-logics and χ-varieties.

Theorem 25 (Definability Theorem). *χ-varieties are defined by their set of χ-validities: $H \in \mathcal{X}$ if and only if $H \vDash^\chi \mathsf{Log}^\chi(\mathcal{X})$.*

Proof. For any χ-variety $\mathcal{V}^\chi$ we have:

$$\begin{aligned} \mathsf{Var}^\chi(\mathsf{Log}^\chi(\mathcal{V}^\chi)) &= \mathsf{Var}^\chi(\mathsf{Log}(\mathcal{V})^\chi) && \text{(by Proposition 24)} \\ &= \mathsf{Var}(\mathsf{Log}(\mathcal{V}))^\chi && \text{(by Proposition 23)} \\ &= \mathcal{V}^\chi. && \text{(by Theorem 4)} \end{aligned}$$

Hence $\mathsf{Var}^\chi \circ \mathsf{Log}^\chi = \mathbb{1}_{\mathbf{HA}^\chi}$, which proves our claim. □

Theorem 26 (Completeness Theorem). *χ-logics are complete with respect to their corresponding χ-variety: $\phi \in \Lambda$ if and only if $\mathsf{Var}^\chi(\Lambda) \vDash^\chi \phi$.*

Proof. For any χ-logic L^χ we have:

$$\begin{aligned} \mathsf{Log}^\chi(\mathsf{Var}^\chi(L^\chi)) &= \mathsf{Log}^\chi(\mathsf{Var}(L)^\chi) && \text{(by Proposition 23)} \\ &= \mathsf{Log}(\mathsf{Var}(L))^\chi && \text{(by Proposition 24)} \\ &= L^\chi. && \text{(by Theorem 4)} \end{aligned}$$

Hence $\mathsf{Log}^\chi \circ \mathsf{Var}^\chi = \mathbb{1}_{\mathbf{IL}^\chi}$, which proves our claim. □

Theorem 26 shows that the novel algebraic semantics is expressive enough to study the whole family of χ-logics. Similarly, the definability theorem for χ-varieties allows us to give a first *external*[7] characterisation of χ-varieties: they are exactly the χ-definable classes of Heyting algebras. By combining the results obtained so far, we can finally prove that the lattices $\mathbf{IL}^\chi$ and $\mathbf{HA}^\chi$ are dually isomorphic.

Theorem 27 (Dual Isomorphism). *The lattice of χ-logics is dually isomorphic to the lattice of χ-varieties of Heyting algebras, i.e. $\mathbf{IL}^\chi \cong^{op} \mathbf{HA}^\chi$. In particular, Var^χ and Log^χ are isomorphisms, one the inverse of the other.*

In addition to the external characterisation presented, we can also give *internal* characterisations of χ-varieties, employing the operations H, S and P introduced in Sect. 2 together with the novel χ-closure operation. Indeed, the first characterisation we present follows directly from the definition of χ-closure. We denote by $\mathcal{X}_{CG}$ the subclass of core generated Heyting algebras of a χ-variety $\mathcal{X}$. The following proposition is an immediate corollary of Theorem 19.

Proposition 28. *Every χ-variety is generated by its collection of core generated elements, i.e. $\mathcal{X} = \boldsymbol{\mathcal{X}}(\mathcal{X}_{CG})$.*

So now we know that every χ-variety is generated by its subdirectly irreducible elements ([6, Theorem 9.6], using that χ-varieties are also standard varieties), and by the previous proposition it is also generated by its core generated elements. We can improve this characterization result by showing that we only need

[7] We borrow this terminology from [7, Sec. 7.8]: an *external* characterization of a χ-variety $\mathcal{X}$ means a representation of $\mathcal{X}$ by means of equations, as opposed to an *internal* characterizations which "does not involve identities, [...] but uses only purely algebraic tools such as various kinds of operations on algebras".

the intersection of these two classes of generators, obtaining an analogue of [6, Theorem 9.7] for χ-varieties

To show this, we start by adapting Theorem 2 to the setting of χ-varieties. Recall that, given $\mathcal{C}$ a class of Heyting algebras, we indicate with $\boldsymbol{\mathcal{X}}(\mathcal{C})$ the least χ-variety containing $\mathcal{C}$ and with $\boldsymbol{\mathcal{V}}(\mathcal{C})$ the least variety containing $\mathcal{C}$.

Theorem 29. *Let $\mathcal{C}$ be a class of Heyting algebras, then $\boldsymbol{\mathcal{X}}(\mathcal{C}) = (\,\mathrm{HSP}(\mathcal{C})\,)^{\chi}$.*

Proof. By definition $\boldsymbol{\mathcal{X}}(\mathcal{C}) = \boldsymbol{\mathcal{V}}(\mathcal{C})^{\chi}$ and, by Theorem 2, $\boldsymbol{\mathcal{V}}(\mathcal{C}) = \mathrm{HSP}(\mathcal{C})$. □

The following proposition follows from the previous theorem.

Proposition 30. *Let $\mathcal{X}$ be a χ-variety, then $\mathcal{X} = \boldsymbol{\mathcal{X}}(\mathcal{C})$ iff $\mathsf{Log}^{\chi}(\mathcal{X}) = \mathsf{Log}^{\chi}(\mathcal{C})$.*

Proof. ($\Rightarrow$) Since $\mathcal{C} \subseteq \mathcal{X}$, the inclusion from left to right is straightforward. Suppose now that $\mathcal{X} \not\models^{\chi} \phi$ then there is some $H \in \mathcal{X}$ such that $H \not\models^{\chi} \phi$. Then since $\mathcal{X} = \boldsymbol{\mathcal{X}}(\mathcal{C})$, it follows by Theorem 29 that $H \in \mathrm{HSP}(\mathcal{C})^{\chi}$. By Proposition 20, it follows that for some $A \in \mathcal{C}$ we have $A \not\models^{\chi} \phi$. Hence $\phi \notin \mathsf{Log}^{\chi}(\mathcal{C})$.

($\Leftarrow$) Suppose $\mathsf{Log}^{\chi}(\mathcal{X}) = \mathsf{Log}^{\chi}(\mathcal{C})$. It follows that $\mathsf{Var}^{\chi}(\mathsf{Log}^{\chi}(\mathcal{X})) = \mathsf{Var}^{\chi}(\mathsf{Log}^{\chi}(\mathcal{C}))$, hence by the Duality Theorem 27, we have $\mathcal{X} = \mathsf{Var}^{\chi}(\mathsf{Log}^{\chi}(\mathcal{C}))$. Finally, since $\mathsf{Log}^{\chi}(\mathcal{C}) = \mathsf{Log}^{\chi}(\boldsymbol{\mathcal{X}}(\mathcal{C}))$ by Proposition 20 and Theorem 29, we have $\mathsf{Var}^{\chi}(\mathsf{Log}^{\chi}(\mathcal{C})) = \mathsf{Var}^{\chi}(\mathsf{Log}^{\chi}(\boldsymbol{\mathcal{X}}(\mathcal{C}))$. Finally, by Theorem 27 we have $\mathsf{Var}^{\chi}(\mathsf{Log}^{\chi}(\boldsymbol{\mathcal{X}}(\mathcal{C})) = \boldsymbol{\mathcal{X}}(\mathcal{C})$, and so it follows that $\mathcal{X} = \boldsymbol{\mathcal{X}}(\mathcal{C})$. □

We can now prove a version of Theorem 3 for χ-varieties. If $\mathcal{X}$ is a χ-variety, let $\mathcal{X}_{CGSI}$ be the subclass of core generated subdirectly irreducible Heyting algebras.

Theorem 31. *Every χ-variety is generated by its collection of core generated subdirectly irreducible elements: $\mathcal{X} = \boldsymbol{\mathcal{X}}(\mathcal{X}_{CGSI})$.*

Proof. By Theorem 27 it suffices to show that $\mathsf{Log}^{\chi}(\mathcal{X}) = \mathsf{Log}^{\chi}(\boldsymbol{\mathcal{X}}(\mathcal{X}_{CGSI}))$. By Proposition 30 this is equivalent to $\mathsf{Log}^{\chi}(\mathcal{X}) = \mathsf{Log}^{\chi}(\mathcal{X}_{CGSI})$. The direction $\mathsf{Log}^{\chi}(\mathcal{X}) \subseteq \mathsf{Log}^{\chi}(\mathcal{X}_{CGSI})$ follows from the inclusion $\mathcal{X}_{CGSI} \subseteq \mathcal{X}$. So it remains to show that $\mathsf{Log}^{\chi}(\mathcal{X}_{CGSI}) \subseteq \mathsf{Log}^{\chi}(\mathcal{X})$. To this end, we employ a classical result originally proved by Wronski [30]: For every Heyting algebra B and $x \in B\backslash\{1_B\}$, there exists a subdirectly irreducible algebra C and a surjective homomorphism $h : B \twoheadrightarrow C$ such that $f(b) \neq 1_C$.

Suppose by contraposition that $\phi \notin \mathsf{Log}^{\chi}(\mathcal{X})$. Then for some $H \in \mathcal{X}$ and some χ-valuation σ we have that $(H, \sigma) \not\models^{\chi} \phi$, and so by Lemma 17 we have that $(\langle H^{\chi}\rangle, \sigma) \not\models^{\chi} \phi$. Since $[\![\phi]\!]_{\sigma}^{\langle H^{\chi}\rangle} \neq 1_H$, by Wronski's result there exists a subdirectly irreducible algebra C and surjective homomorphism $h : \langle H^{\chi}\rangle \twoheadrightarrow C$ such that $h([\![\phi]\!]_{\sigma}^{\langle H^{\chi}\rangle}) \neq 1_C$. Consider now the valuation $\tau = h \circ \sigma$. Since h is a homomorphism, τ is again a χ-valuation—fixpoints of $\boldsymbol{\chi}^2$ are mapped on fixpoints of $\boldsymbol{\chi}^2$—and we have that $[\![\phi]\!]_{\tau}^{C} = h(\,[\![\phi]\!]_{\sigma}^{\langle H^{\chi}\rangle}\,) \neq 1_C$. In particular ϕ is not χ-valid on C.

Since $H \in \mathcal{X}$ we have that $\langle H^{\chi}\rangle \in \mathcal{X}$, and since $h : \langle H^{\chi}\rangle \twoheadrightarrow C$ we also have that $C \in \mathcal{X}$. Moreover, since $C = h[\langle H^{\chi}\rangle]$, C is core generated, which means that $C \in \mathcal{X}_{CGSI}$, showing that $\phi \notin \mathsf{Log}^{\chi}(\mathcal{X}_{CGSI})$. Since ϕ was an arbitrary formula, this proves our claim. □

6 The Lattices of χ-logics

In this section we examine each of the lattices of χ-logics for a univariate formula χ. Recall that in Lemma 7 we have shown that there are only 6 fixpoints of intuitionistic univariate formulas: $\bot, p, \neg p, \neg\neg p, p \vee \neg p, \top$. Moreover, as noted at the end of Sect. 3, $L^{\neg} = L^{\neg\neg}$ for every intermediate logic L, so there are at most five distinct lattices of χ-logics. We shall now briefly consider each of them.

p-logics: Firstly, the lattice of p-logics $\mathbf{IL}^p$ actually coincides with the lattice of intermediate logics $\mathbf{IL}$, since in this case $\chi^2(p) = p$ and so $L^p = L$. From the algebraic perspective, this means that for any Heyting algebra H its p-core is $H^p = H$, thus we are not imposing any restriction on our valuations.

$\top$-logics and $\bot$-logics: The two "limit" cases $\mathbf{IL}^{\bot}$ and $\mathbf{IL}^{\top}$ are more interesting. Notice that $H^{\bot} = \{0_H\}$ and $H^{\top} = \{1_H\}$, and so under the algebraic semantics that we have introduced $\bot$-models allow only the constant valuation with image 0_H and $\top$-models allow only the constant valuation with image 1_H. Interestingly, this means that the notion of core superalgebra collapses in both cases to that of superalgebra, as we have $\langle H^{\bot}\rangle = \langle H^{\top}\rangle = \{0_H, 1_H\}$, which is a subalgebra of every Heyting algebra.

Thus there is only one $\bot$-variety and only one $\top$-variety, in both cases the variety of all Heyting algebras. By Theorem 27 this means there are exactly one $\bot$-logic ($\mathtt{IPC}^{\bot}$) and one $\top$-logic ($\mathtt{IPC}^{\top}$), which are respectively the $\bot$-variant and $\top$-variant of every intermediate logic. These two logics are characterised by the following properties:

$$\phi(p_1, \dots, p_n) \in \mathtt{IPC}^{\bot} \quad \text{iff} \quad \phi(\bot, \dots, \bot) \in \mathtt{IPC} \quad \text{iff} \quad \phi(\bot, \dots, \bot) \in \mathtt{CPC}$$
$$\phi(p_1, \dots, p_n) \in \mathtt{IPC}^{\top} \quad \text{iff} \quad \phi(\top, \dots, \top) \in \mathtt{IPC} \quad \text{iff} \quad \phi(\top, \dots, \top) \in \mathtt{CPC}$$

Notice in particular that, although they correspond to the same variety, the two logics are distinct.

$\neg p$-logics: Apart from $\mathbf{IL}$, the lattice $\mathbf{IL}^{\neg p}$ is the only lattice of χ-logics that has already been studied in the literature, although under a different guise. In fact $\neg p$-logics have already been introduced in the literature as *negative variants* of intermediate logics [8,16,20] and they have been studied from an algebraic perspective in [4,22] under the name of DNA-logics. A well-known example of $\neg p$-logic is *inquisitive logic* InqB, which is the $\neg p$-variant of the intermediate logics KP, ND and ML—as shown in Theorem 3.4.9 of [8]. This algebraic approach to study DNA-logics has proved to be particularly useful: for instance, [22] shows that the lattice of extensions of InqB is dually isomorphic to $\omega + 1$, and also provides an axiomatisation of all such extensions by a generalisation of the method of Jankov formulas [17,18].

$\neg p$-logics have a particularly interesting feature: as mentioned before, the $\neg p$-core of a Heyting algebra is the set of its regular elements, which is a Boolean algebra in the signature $\{1, 0, \wedge, \rightarrow\}$. This easily entails the following corollary: Given an intermediate logic L and a $\vee$-free formula ϕ, $\phi \in L^{\neg p}$ iff ϕ is a classical tautology (Theorem 2.5.2 in [9]). That is, $\neg p$-logics are logics whose $\{1, 0, \wedge, \rightarrow\}$-fragment behaves classically, and which present an intuitionistic behaviour once formulas containing disjunctions are concerned. Such intuitionistic behaviour disappears once also disjunction is forced to be classical, as the following proposition shows:

Proposition 32. *Let L be an intermediate logic. Then $L^{\neg p} = \mathtt{CPC}$ iff L extends the logic of weak excluded middle $\mathtt{WEM} := \mathtt{IPC} + (\neg p \vee \neg\neg p)$.*

The original proof of this result is given in [8, Proposition 5.2.22]. Here we present an alternative proof using the machinery developed in the previous sections.

Proof. We start by claiming that $L^{\neg p} = \mathtt{CPC}$ iff $q \vee \neg q \in L^{\neg p}$ for all $q \in \mathtt{AT}$. The left-to-right implication is trivial. As for the other implication, suppose $q \vee \neg q \in L^{\neg p}$ and take an arbitrary algebra $H \in \mathsf{Var}^{\mathsf{X}}(L^{\neg p})$. Firstly notice that for an arbitrary element c the condition $c \vee \neg c = 1$ implies that c is regular:

$$c \vee \neg c = 1 \implies (c \vee \neg c) \wedge \neg\neg c = \neg\neg c \implies c \wedge \neg\neg c = \neg\neg c \implies c = \neg\neg c$$

Since the $H^{\neg p}$ is the set of the regular elements of H, for two regular elements a and b we have:

$$(a \vee b) \vee \neg(a \vee b) = a \vee b \vee (\neg a \wedge \neg b) = (a \vee \neg a \vee b) \wedge (b \vee \neg b \vee a) = 1 \wedge 1 = 1$$

Thus regular elements in $H^{\neg p}$ are closed under the operations $\wedge$, $\rightarrow$ and $\vee$ as well. It follows that $H^{\neg p}$ is a Boolean algebra. Hence by Proposition 28 and Theorem 26, it easily follows that $L^{\neg p} = \mathtt{CPC}$.

To conclude, notice that by Lemma 11, $q \vee \neg q \in L^{\neg q}$ iff $\neg q \vee \neg\neg q \in L$, which in turn is equivalent to $\mathtt{WEM} \subseteq L$. □

We refer the reader to [22] for more information on $\neg p$-logics and $\neg p$-varieties.

$(p \vee \neg p)$-logics: Finally, let us consider the lattice $\mathbf{IL}^{p \vee \neg p}$. The next proposition gives a characterisation of the $(p \vee \neg p)$-core of any Heyting algebra H. We refer the reader to [23, Ch. 4 Sec. 5] for the proof of the following result.

Proposition 33 (Rasiowa-Sikorski). *Let H be a Heyting algebra and let $x \in H$. The following are equivalent:*

1. $x = y \vee \neg y$ *for some* $y \in H$*;*
2. $\neg x = 0$*;*
3. *for every* $y \in H$*, if* $x \wedge y = 0$*, then* $y = 0$*.*

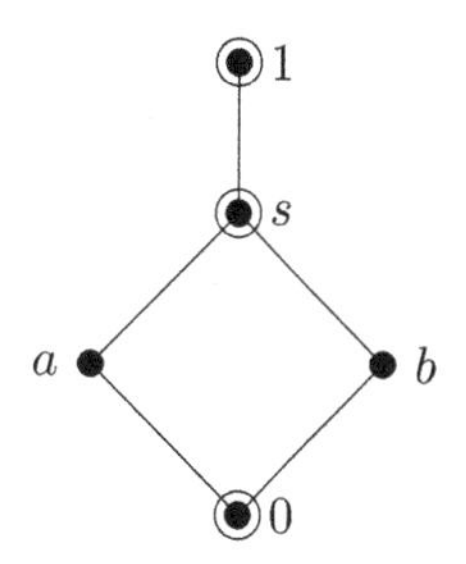

Fig. 3. The algebra H above is a member of $Var^{p\vee\neg p}(\mathtt{LC}^{p\vee\neg p})$ but not of $Var(\mathtt{LC}^{p\vee\neg p})$. The circles indicate the elements of $\langle H^{p\vee\neg p}\rangle$: notice that a and b are not in this subalgebra, so H is not core-generated.

The elements satisfying properties 1, 2 and 3 above are called *dense elements.* Notice that property 1 is exactly the condition defining the elements of the $(p\vee\neg p)$-core of H, thus $H^{p\vee\neg p}$ consists of all and only the dense elements of H.

Now, it is easy to see that the dense elements of a Heyting algebra form a *filter* and that they are closed under the operations $\wedge, \vee, \rightarrow$ and 1. [23, Ch. 4 Sec. 5] As a simple consequence of this, we have that for any Heyting algebra H its core subalgebra is $\langle H^{p\vee\neg p}\rangle = H^{p\vee\neg p} \cup \{0\}$. Therefore, the core generated algebras—which by Theorem 31 suffice to generate all the $(p \vee \neg p)$-varieties—are exactly the algebras containing only dense elements apart from 0.

We obtain an interesting example of $(p \vee \neg p)$-logic by taking the $(p \vee \neg p)$-variant of *Gödel-Dummett logic* LC. Recall that LC is the intermediate logic extending IPC with the axiom $(p \rightarrow q) \vee (q \rightarrow p)$. It can be also characterised as the logic of linear Heyting algebras [7, Example 4.15]. In the same way, the logic $\mathtt{LC}^{p\vee\neg p}$ forces a similar linearity condition, but now limited to the dense elements of a Heyting algebras. Notice that by Proposition 22, the variety $\mathsf{Var}^{p\vee\neg p}(\mathtt{LC}^{p\vee\neg p})$ is still generated by the class of linear Heyting algebras. However, the closure under core-superalgebras leads to a variety properly extending $\mathsf{Var}(\mathtt{LC})$, as shown in Fig. 3. Moreover, notice that linear algebras are core-generated since $\neg x = 0$ for every non-zero element x. Thus we found a class of core-generated algebras which generate the whole $(p \vee \neg p)$-variety.

Finally, we have seen in Proposition 32 that the intermediate logics whose $\neg p$-variant is CPC are exactly the extensions of WEM. So a natural question to ask is what intermediate logics have CPC as their $(p \vee \neg p)$-variant. The next proposition establishes that no intermediate logic has this property.

Proposition 34. CPC *is not the* $(p \vee \neg p)$*-variant of any intermediate logic.*

Proof. Suppose towards a contradiction that $L^{p\vee\neg p} = \mathtt{CPC}$ for some intermediate logic L. By Condition 3 of Definition 8, it follows that $(p \vee \neg p)^2 \leftrightarrow p \in \mathtt{CPC}$. But since $(p \vee \neg p)^2 \equiv_{\mathtt{IPC}} p \vee \neg p \in \mathtt{CPC}$, it follows that $p \in \mathtt{CPC}$, which is a contradiction. □

Now that we have described the five lattices of χ-logic more in detail, we are ready to show they are distinct. However, we need to clarify what we mean by *distinct lattices*: as we have already seen, $\mathbf{IL}^{\top}$ and $\mathbf{IL}^{\perp}$ both contain only one logic, so these lattices are isomorphic; but we also observed that the logics $\mathtt{IPC}^{\top}$ and $\mathtt{IPC}^{\bot}$ are different, and we want to consider these lattices distinct on the base of that. In general, we say that $\mathbf{IL}^{\chi}$ and $\mathbf{IL}^{\theta}$ are *the same* or *equal* iff for every intermediate logic L we have $L^{\chi} = L^{\theta}$—as in the case of $\mathbf{IL}^{\neg p}$ and $\mathbf{IL}^{\neg\neg p}$—and we say the lattices are distinct otherwise.

This suggests to study the relation between these lattices in a more systematic way: define the pointwise extension relation $\mathbf{IL}^{\chi} \preceq \mathbf{IL}^{\theta}$ to hold if for every intermediate logic L we have $L^{\chi} \subseteq L^{\theta}$. The relation $\preceq$ is a partial order between the lattices of χ-logics. In particular, $\mathbf{IL}^{\chi} \preceq \mathbf{IL}^{\theta} \preceq \mathbf{IL}^{\chi}$ if and only if the two lattices $\mathbf{IL}^{\chi}$ and $\mathbf{IL}^{\theta}$ are equal. The following theorem characterises the properties of this relation.

Theorem 35. *Let χ and θ be univariate formulas. Then the following are equivalent:*

1. $\mathbf{IL}^{\chi} \preceq \mathbf{IL}^{\theta}$*;*
2. $\mathtt{IPC}^{\chi} \subseteq \mathtt{IPC}^{\theta}$*;*
3. $(\theta^2(p) \leftrightarrow p) \rightarrow (\chi^2(p) \leftrightarrow p) \in \mathtt{IPC}$*;*
4. *For every Heyting algebra H, $H^{\theta} \subseteq H^{\chi}$.*

Proof. ($1 \Rightarrow 2$) It follows from the definition of $\preceq$. ($2 \Rightarrow 3$) Since $\mathtt{IPC}^{\chi} \subseteq \mathtt{IPC}^{\theta}$, we have in particular that $\chi^2(p) \leftrightarrow p \in \mathtt{IPC}^{\theta}$. This means that $\mathtt{IPC} + (\theta^2(p) \leftrightarrow p) \vDash \chi^2(p) \leftrightarrow p$; and so by the deduction theorem of $\mathtt{IPC}$ we have $(\theta^2(p) \leftrightarrow p) \rightarrow (\chi^2(p) \leftrightarrow p) \in \mathtt{IPC}$. ($3 \Rightarrow 4$) We prove the contrapositive of the implication: suppose that $H^{\theta} \nsubseteq H^{\chi}$ for some Heyting algebra H. Consider an element $a \in H^{\theta} \setminus H^{\chi}$. By Lemma 12 we have $\boldsymbol{\theta}^2(a) \leftrightarrow a = 1_H$ and $\boldsymbol{\chi}^2(a) \leftrightarrow a \neq 1_H$. So in particular $H \nvDash (\theta^2(p) \leftrightarrow p) \rightarrow (\chi^2(p) \leftrightarrow p)$, showing that this is not a theorem of $\mathtt{IPC}$. ($4 \Rightarrow 1$) Consider an intermediate logic L and take an arbitrary formula $\phi \notin L^{\theta}$. By Proposition 22, $L^{\theta} = \mathsf{Log}^{\theta}(\mathsf{Var}(L))$, and so there exists an algebra $H \in \mathsf{Var}(L)$ and a θ-valuation σ such that $(H, \sigma) \nvDash \phi$. Since $H^{\theta} \subseteq H^{\chi}$ by hypothesis, it follows that σ is a χ-valuation, hence $\phi \notin \mathsf{Log}^{\chi}(\mathsf{Var}(L))$ either. Again by Proposition 22, $L^{\chi} = \mathsf{Log}^{\chi}(\mathsf{Var}(L))$, and thus $\phi \notin L^{\chi}$. Since ϕ was arbitrary, it follows that $L^{\chi} \subseteq L^{\theta}$, as wanted. □

Corollary 36. *There are exactly 5 lattices of χ-logics, for χ a univariate formula:*

$$\mathbf{IL}^{p} = \mathbf{IL}, \quad \mathbf{IL}^{\neg p} = \mathbf{IL}^{\neg\neg p}, \quad \mathbf{IL}^{\bot}, \quad \mathbf{IL}^{p\vee\neg p}, \quad \mathbf{IL}^{\top}.$$

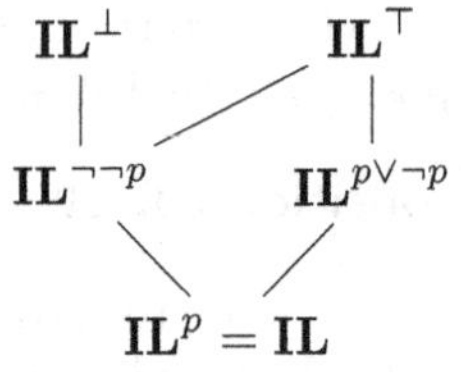

Proof. What remains to be shown is that the lattices are distinct. By Theorem 35, we can do this by exhibiting a Heyting algebra H for which the χ-cores are all distinct. Indeed, the algebra in Fig. 3 is an example of such an algebra:

$\mathbf{IL}^{\bot}$ $\mathbf{IL}^{\top}$

$\mathbf{IL}^{\neg\neg p}$ $\mathbf{IL}^{p\vee\neg p}$

$\mathbf{IL}^{p} = \mathbf{IL}$

Fig. 4. The Hasse diagram of the 5 lattices of χ-logics, ordered under the relation $\preceq$. The diagram is computed using Theorem 35.

$$H^{\bot} = \{0\} \qquad H^{\top} = \{1\}$$
$$H^{\neg p} = \{0, a, b, 1\} \qquad H^{p \vee \neg p} = \{0, s, 1\}$$
$$H^{p} = H$$

□

In Fig. 4 we give a compact representation of the results in Theorem 35 and Corollary 36.

7 Conclusion

In this article we introduced χ-logics and a sound and complete algebraic semantics for them, based on Ruitenburg's theorem. In Sect. 3 we defined the notion of χ-logics and studied them from a syntactical perspective, showing that for a fixed χ they form a complete distributive lattice, and that we have only 5 such lattices. In Sect. 4 we defined an algebraic semantics for χ-logics, by relying on an algebraic interpretation of Ruitenburg's theorem originally described in [27], and we introduced χ-varieties as the semantic counterpart of χ-logics. In Sect. 5 we initiated a more systematic study of χ-varieties providing *external* and *internal* characterisations: we showed that the lattice of χ-logics and the lattice of χ-varieties are dually isomorphic, and we showed that each χ-variety is generated by its core generated subdirectly irreducible members. Finally, in Sect. 6, we have looked more in detail at each of the 5 lattices of χ-logics and characterised explicitly the pointwise extension relation $\preceq$ between the lattices.

The results of this article provide a first approach to generate and study new logics and corresponding algebraic semantics in a systematic fashion. We adapted the approach and technical machinery of universal algebra to investigate the novel class of χ-logics. In particular, we believe this is particular interesting as it provides an algebraic perspective on a class of logics that is essentially non-standard – as χ-logics are not closed under uniform substitutions.

This work can be extended in several directions: Firstly, we limited ourselves to univariate formulas but the approach based on Ruitenburg's theorem can be generalised to the lattice produced by an arbitrary intuitionistic formula—although this would require slightly more complex algebraic structures than Heyting algebras. Secondly, even in this more general setting cores are still required to be definable, but it seems natural to consider more general notions of core (i.e., more generic fixpoint operators) and their corresponding logics. This idea was recently employed in [15] by focusing on fixpoints of *nuclei* over Heyting algebras: this naturally led to define the class of *inquisitive intuitionistic logics* and to provide a natural algebraic semantics for them. It would be of interest to develop similar techniques also for other classes of fixpoint operators. Another interesting direction of work would be to interpret the results presented in this paper in terms of topological duality, that is, Esakia duality for Heyting algebras [12]. Giving a topological interpretation to the results and constructions presented (such as the core-superalgebra operation) would give novel tools to study the structure of the lattices of χ-logics. Finally, we think that the approach of core semantics that we adopted in this work could be extended to provide an algebraic semantics to other logics without uniform substitution. In particular, we think it could be fruitfully applied to logics based on team-semantics, such as dependence logic, team logic, etc. We believe this would both provide an interesting point of study to these logics and also outline a possible generalisation of the framework of abstract algebraic logic [13], which is generally concerned with logics satisfying the principle of uniform substitution.

A Proof of Lemma 7

Proof (Proof of Lemma 7). As shown in [21,24], the following is a presentation of all the non-constant univariate intuitiornistic formulas modulo logical equivalence:

$$\beta_1 := p \qquad \beta_{n+1} := \alpha_n \vee \beta_n \qquad \alpha_1 := \neg p \qquad \alpha_{n+1} := \alpha_n \rightarrow \beta_n.$$

We consider the following two properties for a univariate formula ϕ:

1. $\neg\neg\phi \equiv \top$.
2. If ψ has property 1, then $\phi[\psi/p] \equiv \top$.

In particular, if ϕ has both properties then $\phi^2 \equiv \top$, that is, the fix-point of ϕ is $\top$.

Firstly notice that

$$\alpha_5 = ((\neg\neg p) \rightarrow p \vee \neg p) \rightarrow (\neg p \vee \neg\neg p) \qquad \beta_5 = ((\neg\neg p) \rightarrow p \vee \neg p) \rightarrow (\neg\neg p \rightarrow p)$$

have both properties.

α_5 has property 1:

$$\begin{aligned}\neg\neg\alpha_5 &\equiv \neg\neg(((\neg\neg p\rightarrow p)\rightarrow p\vee\neg p)\rightarrow(\neg p\vee\neg\neg p))\\ &\equiv \neg\neg((\neg\neg p\rightarrow p)\rightarrow p\vee\neg p)\rightarrow\neg\neg(\neg p\vee\neg\neg p)\\ &\equiv \neg\neg((\neg\neg p\rightarrow p)\rightarrow p\vee\neg p)\rightarrow\top\\ &\equiv \top\end{aligned}$$

β_5 has property 1:

$$\begin{aligned}\neg\neg\beta_5 &\equiv \neg\neg(((\neg\neg p\rightarrow p)\rightarrow p\vee\neg p)\vee(\neg\neg p\rightarrow p))\\ &\equiv \neg\neg(\neg\neg((\neg\neg p\rightarrow p)\rightarrow p\vee\neg p)\vee\neg\neg(\neg\neg p\rightarrow p))\\ &\equiv \neg\neg(\neg\neg((\neg\neg p\rightarrow p)\rightarrow p\vee\neg p)\vee\top)\\ &\equiv \neg\neg\top\\ &\equiv \top\end{aligned}$$

α_5 has property 2: for ϕ with property 1,

$$\begin{aligned}\alpha_5(\phi) &\equiv ((\neg\neg\phi\rightarrow\phi)\rightarrow\phi\vee\neg\phi)\rightarrow(\neg\phi\vee\neg\neg\phi)\\ &\equiv ((\neg\neg\phi\rightarrow\phi)\rightarrow\phi\vee\neg\phi)\rightarrow(\bot\vee\top)\\ &\equiv ((\neg\neg\phi\rightarrow\phi)\rightarrow\phi\vee\neg\phi)\rightarrow\top\\ &\equiv \top\end{aligned}$$

β_5 has property 2: for ϕ with property 1,

$$\begin{aligned}\beta_5(\phi) &\equiv ((\neg\neg\phi\rightarrow\phi)\rightarrow\phi\vee\neg\phi)\vee(\neg\neg\phi\rightarrow\phi)\\ &\equiv ((\top\rightarrow\phi)\rightarrow\phi\vee\bot)\vee(\neg\neg\phi\rightarrow\phi)\\ &\equiv (\phi\rightarrow\phi)\vee(\neg\neg\phi\rightarrow\phi)\\ &\equiv \top\vee(\neg\neg\phi\rightarrow\phi)\\ &\equiv \top\end{aligned}$$

Moreover, we can show that, if α_n and β_n have both properties, then this holds for α_{n+1} and β_{n+1} too.

α_{n+1} has property 1:

$$\begin{aligned}\neg\neg\alpha_{n+1} &\equiv \neg\neg(\alpha_n\rightarrow\beta_n)\\ &\equiv \neg\neg\alpha_n\rightarrow\neg\neg\beta_n\\ &\equiv \neg\neg\alpha_n\rightarrow\top\\ &\equiv \top\end{aligned}$$

β_{n+1} has property 1:

$$\begin{aligned}\neg\neg\beta_{n+1} &\equiv \neg\neg(\alpha_n\vee\beta_n)\\ &\equiv \neg\neg(\neg\neg\alpha_n\vee\neg\neg\beta_n)\\ &\equiv \neg\neg(\top\vee\top)\\ &\equiv \top\end{aligned}$$

α_{n+1} has property 2: for ϕ with property 1,

$$\begin{aligned}\alpha_{n+1}(\phi) &\equiv \alpha_n(\phi)\to\beta_n(\phi)\\ &\equiv \alpha_n(\phi)\to\top\\ &\equiv \top\end{aligned}$$

β_{n+1} has property 2: for ϕ with property 1,

$$\begin{aligned}\beta_{n+1}(\phi) &\equiv \beta_n(\phi)\vee\alpha_n(\phi)\\ &\equiv \top\vee\top\\ &\equiv \top\end{aligned}$$

So, by induction all the formulas α_n, β_n with $n \geq 5$ have index at most 2 and fixpoint $\top$. As for the remaining formulas, one can easily show their fix-points are as follows:

$$\begin{array}{lll}
\beta_1^2 = (p)^2 & \equiv p & \Longrightarrow (\beta_1^2)^0 \equiv (\beta_1^2)^1\\
\beta_2^2 = (p \vee \neg p)^2 & \equiv p \vee \neg p & \Longrightarrow (\beta_2^2)^1 \equiv (\beta_2^2)^3\\
\beta_3^2 \equiv (\neg p \vee \neg\neg p)^2 & \equiv \top & \Longrightarrow (\beta_3^2)^2 \equiv (\beta_3^2)^4\\
\beta_4^2 \equiv (\neg\neg p \vee (\neg\neg p \to p))^2 & \equiv \top & \Longrightarrow (\beta_4^2)^2 \equiv (\beta_4^2)^4\\
\alpha_3^2 \equiv (\neg\neg p \to p)^2 & \equiv \neg\neg p \to p & \Longrightarrow (\alpha_3^2)^1 = (\alpha_3^2)^2\\
\alpha_2^2 = (\neg\neg p)^2 & \equiv \neg\neg p & \Longrightarrow (\alpha_2^2)^1 \equiv (\alpha_2^2)^2\\
\alpha_1^3 = (\neg p)^3 & \equiv \neg p & \Longrightarrow (\alpha_1^3)^1 \equiv (\alpha_1^3)^3\\
\alpha_4^2 \equiv ((\neg\neg p \to p) \to p \vee \neg p)^2 & \equiv \top & \Longrightarrow (\alpha_4^2)^2 \equiv (\alpha_4^2)^4.
\end{array}$$

This concludes the proof. □

References

1. Bezhanishvili, G., Holliday, W.H.: Locales, nuclei, and dragalin frames. In: Beklemishev, L., Demri, S., Máté, A. (eds.) Advances in Modal Logic, vol. 11 (2016)
2. Bezhanishvili, G., Holliday, W.H.: A semantic hierarchy for intuitionistic logic. Indag. Math. **30**(3), 403–469 (2019)
3. Bezhanishvili, N., Grilletti, G., Holliday, W.H.: Algebraic and topological semantics for inquisitive logic via choice-free duality. In: Iemhoff, R., Moortgat, M., de Queiroz, R. (eds.) WoLLIC 2019. LNCS, vol. 11541, pp. 35–52. Springer, Heidelberg (2019). https://doi.org/10.1007/978-3-662-59533-6_3
4. Bezhanishvili, N., Grilletti, G., Quadrellaro, D.E.: An algebraic approach to inquisitive and DNA-logics. Rev. Symb. Log. 1–14 (2021). https://doi.org/10.1017/S175502032100054X
5. Birkhoff, G.: On the structure of abstract algebras. Math. Proc. Cambridge Philos. Soc. **31**(4), 433–454 (1935)
6. Burris, S., Sankappanavar, H.P.: A Course in Universal Algebra. Springer, New York (1981)
7. Chagrov, A., Zakharyaschev, M.: Modal Logic. Oxford University Press, Oxford (1997)
8. Ciardelli, I.: Inquisitive semantics and intermediate logics. MSc thesis, University of Amsterdam (2009)
9. Ciardelli, I.: Questions in logic. Ph.D. thesis, Institute for Logic, Language and Computation, University of Amsterdam (2016)
10. Ciardelli, I., Groenendijk, J., Roelofsen, F.: Inquisitive Semantics. Oxford University Press, Oxford (2019)
11. Ciardelli, I., Roelofsen, F.: Inquisitive logic. J. Philos. Log. **40**(1), 55–94 (2011)

12. Esakia, L.: Heyting Algebras: Duality Theory, vol. 50, 1st edn. Springer, Cham (2019)
13. Font, J.M.: Abstract Algebraic Logic. College Publication, London (2016)
14. Gabbay, D.: Semantical Investigations in Heyting's Intuitionistic Logic, vol. 148. Springer, Dordrecht (1981). https://doi.org/10.1007/978-94-017-2977-2
15. Holliday, W.H.: Inquisitive intuitionistic logic. In: Olivetti, N., Verbrugge, R., Negri, S., Sandu, G. (eds.) Advances in Modal Logic, vol. 13 (2020)
16. Iemhoff, R., Yang, F.: Structural completeness in propositional logics of dependence. Arch. Math. Log. **55**, 955–975 (2016). https://doi.org/10.1007/s00153-016-0505-8
17. Jankov, V.: On the relation between deducibility in intuitionistic propositional calculus and finite implicative structures. Soviet Mathematics Doklady **151**, 1294–1294 (1963)
18. Jankov, V.: The construction of a sequence of strongly independent superintuitionistic propositional calculi. Soviet Mathematics Doklady **9**, 806–807 (1968)
19. Kuznetsov, A.V.: Superintuitionistic logics. Mat. Issled. **10**(2(36)), 150–158, 284–285 (1975). (Russian)
20. Miglioli, P., Moscato, U., Ornaghi, M., Quazza, S., Usberti, G.: Some results on intermediate constructive logics. Notre Dame J. Form. Log. **30**(4), 543–562 (1989)
21. Nishimura, I.: On formulas of one variable in intuitionistic propositional calculus. J. Symb. Log. **25**(4), 327–331 (1960)
22. Quadrellaro, D.E.: Lattices of DNA-logics and algebraic semantics of inquisitive logic. MSc thesis, University of Amsterdam (2019)
23. Rasiowa, H., Sikorski, R.: The Mathematics of Metamathematics. Monografie matematyczne, vol. 41, 3 edn. Polish Scientific Publications (1970)
24. Rieger, L.S.: On the lattice theory of Brouwerian propositional logic. Acta Facultatis Rerum Naturalium Universitatis Carolinae (Prague). Spisy Vydávané Přiridorědeckou Fakultou University Karlovy, vol. 1949, no. 189 (1949)
25. Roelofsen, F.: Algebraic foundations for inquisitive semantics. In: van Ditmarsch, H., Lang, J., Ju, S. (eds.) LORI 2011. LNCS (LNAI), vol. 6953, pp. 233–243. Springer, Heidelberg (2011). https://doi.org/10.1007/978-3-642-24130-7_17
26. Ruitenburg, W.: On the period of sequences $(A^n(p))$ in intuitionistic propositional calculus. J. Symb. Log. **49**(3), 892–899 (1984)
27. Santocanale, L., Ghilardi, S.: Ruitenburg's theorem via duality and bounded bisimulations. In: Advances in Modal Logic, Bern, Switzerland (2018)
28. Tarski, A.: A remark on functionally free algebras. Ann. Math. **47**(1), 163–166 (1946)
29. Väänänen, J.: Dependence Logic: A New Approach to Independence Friendly Logic. Cambridge University Press, Cambridge (2007)
30. Wronski, A.: Intermediate logics and the disjunction property. Rep. Math. Log. **1**, 39–51 (1973)

Matching and Generalization Modulo Proximity and Tolerance Relations

Temur Kutsia(✉) and Cleo Pau

RISC, Johannes Kepler University, Linz, Austria
{kutsia,ipau}@risc.jku.at

Abstract. Proximity relations are fuzzy binary relations satisfying reflexivity and symmetry properties. Tolerance, which is a reflexive and symmetric (and not necessarily transitive) relation, can be also seen as a crisp version of proximity. We discuss two fundamental symbolic computation problems for proximity and tolerance relations: matching and anti-unification, present algorithms for solving them, and study properties of those algorithms.

Keywords: Fuzzy proximity relations · Matching · Anti-unification

1 Introduction

Proximity relations are reflexive and symmetric fuzzy binary relations. They generalize similarity relations, which are a fuzzy version of equivalences. Proximity relations help to represent fuzzy information in situations where similarity is not adequate.

The crisp counterpart of proximity is tolerance, which generalizes the standard equivalence relation by dropping the transitivity property. In the literature, tolerance appears under other names as well, e.g., compatibility, similarity, or proximity relation. The term 'tolerance relation' is attributed to Zeeman [17].

A tolerance relation can be expressed as an undirected graph. The vertices of the graph form the set on which the relation is defined, and two elements are related if and only if there is an edge in the graph connecting them. A similar graph but with weighted edges can be associated to a proximity relation. This graph-based view helps to easily explain two important notions related to proximity and tolerance relations: proximity/tolerance blocks and proximity/tolerance classes (of a node). Blocks correspond to maximal cliques in the graph and the class of a node corresponds to its set of adjacent nodes, together with the node itself (see, e.g., [5,8]).

Unification and anti-unification are two fundamental operations for many areas of symbolic computation. Unification aims at computing a most specific

This work was supported by the Austrian Science Fund (FWF) under project 28789-N32 and by the strategic program "Innovatives OÖ 2020" by the Upper Austrian Government.

© The Author(s), under exclusive license to Springer Nature Switzerland AG 2022
A. Özgün and Y. Zinova (Eds.): TbiLLC 2019, LNCS 13206, pp. 323–342, 2022.
https://doi.org/10.1007/978-3-030-98479-3_16

common instance of given logical expressions, while anti-unification, a technique dual to unification, computes their least general generalization. Both techniques have been studied for equivalence relations both in crisp and fuzzy settings. Syntactic and equational unification is surveyed, e.g., in [4], for syntactic and equational anti-unification see, e.g., [2,6,14,15]. Unification and anti-unification modulo similarity have been investigated, e.g., in [1,16].

On the other hand, there are very few works on unification and anti-unification modulo proximity and tolerance. In [8], the authors introduced the notion of proximity-based unification (improved later in [9]) and used it in fuzzy logic programming. It can be characterized as a block-based approach, because two terms are treated as approximate in one computation when they have the same set of positions, symbols in their corresponding positions belong to the same block, and a certain symbol is always assigned to the same block. This approach imposes the restriction that the same symbol can not be close to two symbols at the same time, when those symbols are not close to each other. One of them should be chosen as the proximal candidate to the given symbol. For matching, it means that $f(x, x)$ does not match to $f(a, c)$ when a and c are not close to each other, even if there exists a b close both to a and c. In [11,12], we reported the first results related to block-based anti-unification with proximity relations.

In this paper, we consider the class-based notion of approximation for proximity (and tolerance) relations, which helps relax the mentioned restriction. Under this approach, $f(x, x)$ matches $f(a, c)$, when there is a b that is close to both a and c, even if a and c are not close to each other. It is justified by the fact that $f(b, b)$ and $f(a, c)$ belong to the same proximity/tolerance class, and it has a natural interpretation, e.g.: for two distant points a and c on a plane, find a point x that is close to each of them. As we have already shown in [13], it is nontrivial to solve proximity constraints in this setting. Here we develop a dedicated algorithm for matching. In general, matching problems with proximity or tolerance relations might have finitely many incomparable solutions, but one can represent them in a more compact way. We show that for each matching problem there is a single answer in such a compact form, and investigate time and space complexity to compute it.

We also study class-based anti-unification for proximity/tolerance relations. This problem is closely related to matching, as generalizations (whose computation is the goal of anti-unification) are supposed to match the original terms. Also here, we aim at computing a compact representation of the solution, but unlike matching, for anti-unification there can be finitely many different solutions in compact form. If we are interested in linear generalizations (i.e., those which do not contain multiple occurrences of the same variable) then the problem has a unique compact solution. A potential application of these techniques includes, e.g., an extension of software code clone detection methods by treating certain mismatches as approximations.

The paper is organized as follows. In Sect. 2, we introduce the basic notions. The problem statement can be found in Sect. 3. In Sect. 4, we develop our matching algorithm and study its properties. Section 5 is about anti-unification. Section 6 contains concluding remarks.

2 Preliminaries

Proximity and Tolerance Relations
We define basic notions about proximity relations following [8].

A binary *fuzzy relation* on a set S is a mapping from $S \times S$ to the real interval $[0, 1]$. If $\mathcal{R}$ is a fuzzy relation on S and λ is a number $0 < \lambda \leq 1$ (called *cut value*), then the λ-*cut* of $\mathcal{R}$ on S, denoted $\mathcal{R}_\lambda$, is an ordinary (crisp) relation on S defined as $\mathcal{R}_\lambda := \{(s_1, s_2) \mid \mathcal{R}(s_1, s_1) \geq \lambda\}$.

Each fuzzy relation is characterized by a finite set of cut values, which we call *approximation levels* of the relation.

A fuzzy relation $\mathcal{R}$ on a set S is called a *proximity relation* on S iff it is reflexive ($\mathcal{R}(s, s) = 1$ for all $s \in S$) and symmetric ($\mathcal{R}(s_1, s_2) = \mathcal{R}(s_2, s_1)$ for all $s_1, s_2 \in S$). *Tolerance relations* are crisp reflexive and symmetric binary relations. A λ-cut of a proximity relation on S is a tolerance relation on S.

The *proximity class of level* $\lambda \in (0, 1]$ *of* $s \in S$ *with respect to a proximity relation* $\mathcal{R}$ (an $(\mathcal{R}, \lambda)$-*class* of s) is the set $\mathbf{pc}(s, \mathcal{R}, \lambda) := \{s' \mid \mathcal{R}(s, s') \geq \lambda\}$.

A *triangular norm* (T-norm) $\wedge$ in $[0, 1]$ is a binary operation $\wedge : [0; 1] \times [0, 1] \to [0, 1]$, which is associative, commutative, nondecreasing in both arguments, and satisfying $x \wedge 1 = 1 \wedge x = x$ for any $x \in [0, 1]$. T-norms have been studied in detail in [10]. In this paper we assume that the t-norm is minimum.

Terms and Extended Terms
Given disjoint sets of variables $\mathcal{V}$ and fixed arity function symbols $\mathcal{F}$, *terms* over $\mathcal{F}$ and $\mathcal{V}$ are defined as usual, by the grammar $t := x \mid f(t_1, \ldots, t_n)$, where $x \in \mathcal{V}$ and $f \in \mathcal{F}$ is n-ary. The set of terms over $\mathcal{V}$ and $\mathcal{F}$ is denoted by $\mathcal{T}(\mathcal{F}, \mathcal{V})$. We denote variables by x, y, z, arbitrary function symbols by f, g, h, constants by a, b, c, and terms by s, t, r.

Below we will need a notation for finite sets of function symbols, whose all elements have the same arity. They will be denoted by lower case bold face letters: $\mathbf{f}, \mathbf{g}, \mathbf{h}$. When we talk about finite sets of constants, we use $\mathbf{a}, \mathbf{b}$, and $\mathbf{c}$.

Extended terms or, shortly, *X-terms* over $\mathcal{F}$ and $\mathcal{V}$ are defined by the grammar $\mathbf{t} := x \mid \mathbf{f}(\mathbf{t}_1, \ldots, \mathbf{t}_n)$, where $\mathbf{f} \neq \emptyset$ contains finitely many function symbols of arity n. Hence, X-terms differ from the standard ones by permitting *finite non-empty sets* of n-ary function symbols in place of n-ary function symbols. Variables in X-terms are used in the standard terms. We denote the set of X-terms over $\mathcal{F}$ and $\mathcal{V}$ by $\mathcal{T}_{\mathsf{ext}}(\mathcal{F}, \mathcal{V})$, and use also bold face letters for its elements.

The set of variables for a term t and for an X-term $\mathbf{t}$ is denoted by $\mathcal{V}(t)$ and $\mathcal{V}(\mathbf{t})$, respectively. A term (resp. X-term) is called *linear* if every variable occurs in it at most once. The *head* of a term and an X-term is defined as

$$head(x) := x, \quad head(f(t_1, \ldots, t_n)) := f, \quad head(\mathbf{f}(\mathbf{t}_1, \ldots, \mathbf{t}_n)) := \mathbf{f}.$$

The *set of terms represented by an X-term* $\mathbf{t}$, denoted by $\tau(\mathbf{t})$, is defined as

$$\tau(x) := \{x\},\ \tau(\mathbf{f}(\mathbf{t}_1,\dots,\mathbf{t}_n)) := \{f(t_1,\dots,t_n) \mid f \in \mathbf{f}, t_i \in \tau(\mathbf{t}_i), 1 \le i \le n\}.$$

We also define the *intersection* operation for X-terms, denoted by $\mathbf{t} \sqcap \mathbf{s}$:

- $x \sqcap x = x$ for all $x \in \mathcal{V}$.
- $\mathbf{t} \sqcap \mathbf{s} = (\mathbf{f} \cap \mathbf{g})(\mathbf{t}_1 \sqcap \mathbf{s}_1,\dots,\mathbf{t}_n \sqcap \mathbf{s}_n)$, $n \ge 0$, if $\mathbf{f} \cap \mathbf{g} \ne \emptyset$ and $\mathbf{t}_i \sqcap \mathbf{s}_i \ne \emptyset$ for all $1 \le i \le n$, where $\mathbf{t} = \mathbf{f}(\mathbf{t}_1,\dots,\mathbf{t}_n)$ and $\mathbf{t} = \mathbf{g}(\mathbf{s}_1,\dots,\mathbf{s}_n)$.
- $\mathbf{t} \sqcap \mathbf{s} = \emptyset$ in all other cases.

Positions in terms are defined with respect to their tree representation in the standard way, as string of integers, where the empty string is denoted by ϵ. We will need another standard notion, the *subterm of* t *at position* p, denoted by $t|_p$. (See, e.g., [3] for details.) These notions straightforwardly extend to X-terms. For instance, for an X-term $\mathbf{t} = \{f\}(\{g,h\}(x,\{a,b,c\}),\{b,c,d\})$, the set of positions is $\{\epsilon, 1, 1.1, 1.2, 2\}$ and we have the X-subterms of $\mathbf{t}$ at those position $\mathbf{t}|_\epsilon = \mathbf{t}$, $\mathbf{t}|_1 = \{g,h\}(x,\{a,b,c\})$, $\mathbf{t}|_{1.1} = x$, $\mathbf{t}|_{1.2} = \{a,b,c\}$, and $\mathbf{t}|_2 = \{b,c,d\}$.

Substitutions and Extended Substitutions

Substitutions over $\mathcal{T}(\mathcal{F},\mathcal{V})$ (resp. over $\mathcal{T}_{\mathsf{ext}}(\mathcal{F},\mathcal{V})$) are mappings from variables to terms (resp. to X-terms), where all but finitely many variables are mapped to themselves. The symbols $\sigma, \vartheta, \varphi$ are used for term substitutions, and $\boldsymbol{\sigma}$, $\boldsymbol{\vartheta}$, $\boldsymbol{\varphi}$ for X-term substitutions. The identity substitution is denoted by *Id*.

The *domain* of a substitution σ is defined as $dom(\sigma) = \{x \mid \sigma(x) \ne x\}$. We use the usual set notation for substitutions, writing, e.g., σ as $\sigma = \{x \mapsto \sigma(x) \mid x \in dom(\sigma)\}$. *Substitution application* to terms is written in the postfix notation such as $t\sigma$ and is defined recursively as $x\sigma = \sigma(x)$ and $f(t_1,\dots,t_n)\sigma = f(t_1\sigma,\dots,t_n\sigma)$. In the same way, one can define the domain of an X-substitution and application of an X-substitution to an X-term.[1]

The *set of substitutions represented by an X-term substitution* $\boldsymbol{\sigma}$ is the set $\tau(\boldsymbol{\sigma}) := \{\sigma \mid \sigma(x) \in \tau(\boldsymbol{\sigma}(x))$ for all $x \in \mathcal{V}\}$.

Relations over Terms and Substitutions

Each proximity relation $\mathcal{R}$ we consider in this paper is defined on $\mathcal{F}$ so that for all $f, g \in \mathcal{F}$, we have $\mathcal{R}(f,g) = 0$ if $arity(f) \ne arity(g)$. We extend such a relation $\mathcal{R}$ from $\mathcal{F}$ to $\mathcal{F} \cup \mathcal{T}(\mathcal{F},\mathcal{V})$:

- For function symbols $\mathcal{R}$ is already defined.
- For variables: $\mathcal{R}(x,x) = 1$.
- For nonvariable terms:

$$\mathcal{R}(f(t_1,\dots,t_n), g(s_1,\dots,s_n)) = \mathcal{R}(f,g) \wedge \mathcal{R}(t_1,s_1) \wedge \dots \wedge \mathcal{R}(t_n,s_n),$$

when f and g are both n-ary.

[1] Note that notions of application of a substitution to an X-term and application of an X-substitution to a term are not defined.

– In all other cases, $\mathcal{R}(T_1, T_2) = 0$ for $T_1, T_2 \in \mathcal{F} \cup \mathcal{T}(\mathcal{F}, \mathcal{V})$.

Two terms t and s are $(\mathcal{R}, \lambda)$-*close* to each other, written $t \simeq_{\mathcal{R},\lambda} s$, if $\mathcal{R}(t, s) \geq \lambda$.

Definition 1 (Relations $\preceq_{\mathcal{R},\lambda}$ and $\leqslant$). *The relations $\preceq_{\mathcal{R},\lambda}$ and $\leqslant$ and the corresponding notions are defined as follows:*

$\preceq_{\mathcal{R},\lambda}$**:** *A term t is $(\mathcal{R}, \lambda)$-*more general than *s (or t is $(\mathcal{R}, \lambda)$-*generalization of *s, or s is an $(\mathcal{R}, \lambda)$-*instance *of t), written $t \preceq_{\mathcal{R},\lambda} s$, if there exists a substitution σ such that $t\sigma \simeq_{\mathcal{R},\lambda} s$. We say that σ is an $(\mathcal{R}, \lambda)$-*matcher *of t to s.*

$\leqslant$**:** *A term t is *syntactically more general than *s (or t is a *syntactic generalization *of s, or s is a *syntactic instance *of t), written $t \leqslant s$, if there exists a σ such that $t\sigma = s$. We say that σ is a syntactic matcher of t to s.*

*An X-term $\mathbf{t}$ is an $(\mathcal{R}, \lambda)$-*X-generalization *of a term s, if every $t \in \tau(\mathbf{t})$ is an $(\mathcal{R}, \lambda)$-*generalization *of s.*

*An X-substitution $\boldsymbol{\sigma}$ is an $(\mathcal{R}, \lambda)$-*X-matcher *of t to s, if every $\sigma \in \tau(\boldsymbol{\sigma})$ is an $(\mathcal{R}, \lambda)$-*matcher *of t to s.*

*A substitution σ that matches t to s is called a *relevant $(\mathcal{R}, \lambda)$-matcher *(resp.* relevant syntactic matcher*) of t to s if $dom(\sigma) \subseteq \mathcal{V}(t)$. A relevant $(\mathcal{R}, \lambda)$-X-matcher is defined analogously.*

The strict part of $\preceq_{\mathcal{R},\lambda}$ and $\leqslant$ are denoted respectively by $\prec_{\mathcal{R},\lambda}$ and $<$.

The relation $\preceq_{\mathcal{R},\lambda}$ is not transitive. If $a \simeq_{\mathcal{R},\lambda} b$, $b \simeq_{\mathcal{R},\lambda} c$, and $a \not\simeq_{\mathcal{R},\lambda} c$, then we have $a \preceq_{\mathcal{R},\lambda} b$, $b \preceq_{\mathcal{R},\lambda} c$, and $a \not\preceq_{\mathcal{R},\lambda} c$. Unlike $\preceq_{\mathcal{R},\lambda}$, $\leqslant$ is transitive. (In fact, $\leqslant$ is a quasi-ordering, called instantiation quasi-ordering.) We also have $\leqslant \subseteq \preceq_{\mathcal{R},\lambda}$ for any $\mathcal{R}$ and λ.

Definition 2 ($(\mathcal{R}, \lambda)$-lgg). *A term r is called an $(\mathcal{R}, \lambda)$-*least general generalization *(an $(\mathcal{R}, \lambda)$-*lgg*) of t and s iff*

- *r is $(\mathcal{R}, \lambda)$-more general than both t and s, i.e., $r \preceq_{\mathcal{R},\lambda} t$ and $r \preceq_{\mathcal{R},\lambda} s$, and*
- *there is no r' such that $r \prec_{\mathcal{R},\lambda} r'$, $r' \preceq_{\mathcal{R},\lambda} t$, and $r' \preceq_{\mathcal{R},\lambda} s$.*

*An X-term $\mathbf{r}$ is an $(\mathcal{R}, \lambda)$-*X-lgg *of t and s, if every $r \in \tau(\mathbf{r})$ is an $(\mathcal{R}, \lambda)$-*lgg *of t and s.*

Theorem 1. *If r is an $(\mathcal{R}, \lambda)$-generalization of t, then any syntactic generalization of r is also an $(\mathcal{R}, \lambda)$-generalization of t.*

Proof. From $r \preceq_{\mathcal{R},\lambda} t$, by definition of $\preceq_{\mathcal{R},\lambda}$, there exists ϑ such that $r\vartheta \simeq_{\mathcal{R},\lambda} t$. From $r' \leqslant r$, by definition of $\leqslant$, there exists φ such that $r'\varphi = r$. Then we have $r'\varphi\vartheta = r\vartheta \simeq_{\mathcal{R},\lambda} t$, which implies $r' \preceq_{\mathcal{R},\lambda} t$. □

Corollary 1. *Any syntactic generalization of an $(\mathcal{R}, \lambda)$-lgg of t and s is an $(\mathcal{R}, \lambda)$-generalization of both t and s.*

The notion of syntactic lgg can be defined analogously to $(\mathcal{R}, \lambda)$-lgg, using the relation $\leqslant$. The syntactic lgg of two terms is unique modulo variable renaming, see, e.g., [14,15]. In general, it is not difficult to show that for any terms t and s, if r and r' are their syntactic lgg and $(\mathcal{R}, \lambda)$-lgg, respectively, then $r \leqslant r'$.

Example 1. Let $\mathcal{R}$ and λ be such that $a \simeq_{\mathcal{R},\lambda} b$, $b \simeq_{\mathcal{R},\lambda} c$, and $a \not\simeq_{\mathcal{R},\lambda} c$. Then $(\mathcal{R},\lambda)$-lgg of a and c is b, while their syntactic lgg is x.

Given a term t, a proximity relation $\mathcal{R}$, and a cut value λ, the $(\mathcal{R},\lambda)$-*proximity class* of t is an X-term $\mathbf{pc}(t,\mathcal{R},\lambda)$, defined as

$$\mathbf{pc}(x,\mathcal{R},\lambda) := \{x\},$$
$$\mathbf{pc}(f(t_1,\dots,t_n),\mathcal{R},\lambda) := \mathbf{pc}(f,\mathcal{R},\lambda)(\mathbf{pc}(t_1,\mathcal{R},\lambda),\dots,\mathbf{pc}(t_n,\mathcal{R},\lambda)).$$

Theorem 2. *Given a proximity relation $\mathcal{R}$, a cut value λ, and two terms t and s, each $r \in \mathbf{pc}(t,\mathcal{R},\lambda) \sqcap \mathbf{pc}(s,\mathcal{R},\lambda)$ is $(\mathcal{R},\lambda)$-close both to t and to s.*

Proof. Follows directly from the definition of proximity class of a term. □

The examples below illustrate some of the notions introduced in this section.

Example 2. Let the proximity relation $\mathcal{R}$ be defined as

$$\mathcal{R}(g_1,g_2) = \mathcal{R}(a_1,a_2) = 0.5, \qquad \mathcal{R}(g_1,h_1) = \mathcal{R}(g_2,h_1) = 0.6,$$
$$\mathcal{R}(g_1,h_2) = \mathcal{R}(a_1,b) = 0.7, \qquad \mathcal{R}(g_2,h_2) = \mathcal{R}(a_2,b) = 0.8.$$

The set of approximation levels of $\mathcal{R}$ is $\{0.5, 0.6, 0.7, 0.8\}$.

Let t be the term $f(g_1(a_1), g_2(a_2))$. Then the proximity class $\mathbf{pc}(t,\mathcal{R},\lambda)$ for different values of λ is:

$$\begin{aligned}
0 < \lambda \le 0.5 &: \{f\}(\{g_1,g_2,h_1,h_2\}(\{a_1,a_2,b\}), \{g_1,g_2,h_1,h_2\}(\{a_1,a_2,b\})).\\
0.5 < \lambda \le 0.6 &: \{f\}(\{g_1,h_1,h_2\}(\{a_1,b\}), \{g_2,h_1,h_2\}(\{a_2,b\})).\\
0.6 < \lambda \le 0.7 &: \{f\}(\{g_1,h_2\}(\{a_1,b\}), \{g_2,h_2\}(\{a_2,b\})).\\
0.7 < \lambda \le 0.8 &: \{f\}(\{g_1\}(\{a_1\}), \{g_2,h_2\}(\{a_2,b\})).\\
0.8 < \lambda \le 1 &: \{f\}(\{g_1\}(\{a_1\}), \{g_2\}(\{a_2\})).
\end{aligned}$$

Example 3. Let $\mathcal{R}$ be defined as in Example 2. Let $t = f(x,x)$ and $s = f(g_1(a_1), g_2(a_2))$. Then for each of the following X-substitution σ, the set $\tau(\sigma)$ contains all relevant $(\mathcal{R},\lambda)$-matchers of t to s for different values of λ:

$$\begin{aligned}
0 < \lambda \le 0.5 &: \quad \sigma = \{x \mapsto \{g_1,g_2,h_1,h_2\}(\{a_1,a_2,b\})\}.\\
&\quad\ \tau(\sigma) \text{ contains 12 substitutions.}\\
0.5 < \lambda \le 0.6 &: \quad \sigma = \{x \mapsto \{h_1,h_2\}(\{b\})\}.\\
&\quad\ \tau(\sigma) = \{\{x \mapsto h_1(b)\},\ \{x \mapsto h_2(b)\}\}.\\
0.6 < \lambda \le 0.7 &: \quad \sigma = \{x \mapsto \{h_2\}(\{b\})\}. \qquad \tau(\sigma) = \{\{x \mapsto h_2(b)\}\}.\\
0.7 < \lambda \le 1 &: \quad \text{No substitution matches } t \text{ to } s.
\end{aligned}$$

Example 4. Let $\mathcal{R}$ be a proximity relation defined as

$$\mathcal{R}(a_1,a) = \mathcal{R}(a_2,a) = \mathcal{R}(b_1,b) = \mathcal{R}(b_2,b) = 0.5,$$
$$\mathcal{R}(a_2,a') = \mathcal{R}(a_3,a') = \mathcal{R}(b_2,b') = \mathcal{R}(b_3,b') = 0.6, \qquad \mathcal{R}(f,g) = 0.7.$$

Its set of approximation levels is $\{0.5, 0.6, 0.7\}$.

Let $t = f(a_1, a_2, a_3)$ and $s = g(b_1, b_2, b_3)$. Then x is the syntactic lgg of t and s. As for proximity-based generalizations, for each of the following X-term $\mathbf{r}$, the set $\tau(\mathbf{r})$ contains all $(\mathcal{R}, \lambda)$-lggs of t and s for different values of λ:

$0 < \lambda \leq 0.5:$

$$\mathbf{r}_1 = \{f, g\}(x_1, x_1, x_3). \quad \tau(\mathbf{r}_1) = \{f(x_1, x_1, x_3),\ g(x_1, x_1, x_3)\}.$$
$$\mathbf{r}_2 = \{f, g\}(x_1, x_2, x_2). \quad \tau(\mathbf{r}_2) = \{f(x_1, x_2, x_2),\ g(x_1, x_2, x_2)\}.$$

$0.5 < \lambda \leq 0.6:$

$$\mathbf{r} = \{f, g\}(x_1, x_2, x_2). \quad \tau(\mathbf{r}) = \{f(x_1, x_2, x_2),\ g(x_1, x_2, x_2)\}.$$

$0.6 < \lambda \leq 0.7:$

$$\mathbf{r} = \{f, g\}(x_1, x_2, x_3). \quad \tau(\mathbf{r}) = \{f(x_1, x_2, x_3),\ g(x_1, x_2, x_3)\}.$$

$0.7 < \lambda \leq 1: \quad \mathbf{r} = x. \quad \tau(\mathbf{r}) = \{x\}.$

If we are interested only in linear generalizations, we will get a single X-term $(\mathcal{R}, \lambda)$-lgg for each fixed λ:

$$0 < \lambda \leq 0.7: \qquad \mathbf{r} = \{f, g\}(x_1, x_2, x_3). \quad \tau(\mathbf{r}) = \{f(x_1, x_2, x_3),\ g(x_1, x_2, x_3)\}.$$

$$0.7 < \lambda \leq 1: \qquad \mathbf{r} = x. \quad \tau(\mathbf{r}) = \{x\}.$$

3 Matching and Anti-unification: Problem Statement

Matching and anti-unification problems for terms are formulated as follows: Given a proximity relation $\mathcal{R}$, a cut value λ, and two terms t and s, find

- an $(\mathcal{R}, \lambda)$-matcher of t to s (the matching problem) or
- an $(\mathcal{R}, \lambda)$-lgg of t and s (the anti-unification problem).

Below we develop algorithms to solve these problems. as we will see, each of them has finitely many solutions. It is important to mention that instead of computing all the solutions to the problems, we will be aiming at computing their compact representations in the form of X-substitutions (for matching) and X-terms (for generalization). Hence, our algorithms will solve the following reformulated version of the problems:

Matching problem
Given: a proximity relation $\mathcal{R}$, a cut value λ, and two terms t and s.
Find: an X-substitution $\boldsymbol{\sigma}$ s.t. each $\sigma \in \tau(\boldsymbol{\sigma})$ is an $(\mathcal{R}, \lambda)$-matcher of t to s.

Anti-unification problem
Given: a proximity relation $\mathcal{R}$, a cut value λ, and two terms t and s.
Find: an X-term $\mathbf{r}$ such that each $r \in \tau(\mathbf{r})$ is an $(\mathcal{R}, \lambda)$-lgg of t and s.

Such a reformulation will help us compute a single X-substitution instead of multiple matchers, and fewer X-lggs than lggs. Moreover, if we restrict ourselves to linear lggs (i.e., those with a single occurrence of generalization variables), then also here we get a single answer.

4 Matching

Given $\mathcal{R}$, λ, t, and s (where s does not contain variables), to solve an $(\mathcal{R},\lambda)$-matching problem $t \ll s$, we create the initial pair $\{t \ll s\}; \emptyset$ and apply the rules given below. They work on pairs $M; S$, where M is a set of matching problems, and S is the set of equations of the form $x \approx \mathbf{s}$. The rules are as follows ($\uplus$ stands for disjoint union):

Dec-M: **Decomposition**

$\{f(t_1,\dots,t_n) \ll g(s_1,\dots,s_n)\} \uplus M; S \Longrightarrow M \cup \{t_i \ll s_i \mid 1 \le i \le n\}; S,$
if $n \ge 0$, $\mathcal{R}(f,g) \ge \lambda$.

VE-M: **Variable elimination**

$\{x \ll s\} \uplus M;\ S \Longrightarrow M;\ S \cup \{x \approx \mathbf{pc}(s,\mathcal{R},\lambda)\}.$

Mer-M: **Merging**

$M;\ \{x \approx \mathbf{s}_1, x \approx \mathbf{s}_2\} \uplus S \Longrightarrow M;\ S \cup \{x \approx \mathbf{s}_1 \sqcap \mathbf{s}_2\}, \qquad \text{if } \mathbf{s}_1 \sqcap \mathbf{s}_2 \ne \emptyset.$

Cla-M: **Clash**

$\{f(t_1,\dots,t_n) \ll g(s_1,\dots,s_m)\} \uplus M;\ S \Longrightarrow \bot, \qquad \text{where } \mathcal{R}(f,g) < \lambda.$

Inc-M: **Inconsistency**

$M;\ \{x \approx \mathbf{s}_1,\ x \approx \mathbf{s}_2\} \uplus S \Longrightarrow \bot, \qquad \text{if } \mathbf{s}_1 \sqcap \mathbf{s}_2 = \emptyset.$

The matching algorithm $\mathfrak{M}$ uses the rules to transform pairs as long as possible, returning either $\bot$ (indicating failure), or the pair $\emptyset; S$ (indicating success). In the latter case, each variable occurs in S at most once and from S one can obtain an X-substitution $\{x \mapsto \mathbf{s} \mid x \approx \mathbf{s} \in S\}$. We call it the *computed X-substitution.*

We call a substitution σ an $(\mathcal{R},\lambda)$-solution of an $M; S$ pair, iff σ is an $(\mathcal{R},\lambda)$-matcher of M and for all $x \approx \mathbf{t} \in S$, we have $x\sigma \in \mathfrak{r}(\mathbf{t})$. We also assume that $\bot$ has no solution.

Lemma 1. *If $M_1; S_1 \Longrightarrow M_2; S_2$ is a step made by $\mathfrak{M}$, then σ is an $(\mathcal{R},\lambda)$-solution of $M_1; S_1$ iff it is an $(\mathcal{R},\lambda)$-solution of $M_2; S_2$.*

Proof. For the rules Dec-M and Cla-M, the lemma follows by definition of matcher. For Mer-M and Inc-M it is implied by definition of $\sqcap$. For VE-M, by definition of **pc**, we have $x\sigma \in \mathbf{pc}(s,\mathcal{R},\lambda)$ iff $\mathcal{R}(x\sigma,s) \ge \lambda$, which is equivalent to the fact that σ is an $(\mathcal{R},\lambda)$-matcher of $x \ll s$. $\square$

In the theorems below the *size* of a syntactic object (term, matching problem, set of matching problems, a set of equations) is the number of alphabet symbols in it: $size(x) = 1$, $size(f(t_1,\dots,t_n)) = 1 + \sum_{i=1}^{n} size(t_i)$, $size(t \ll s) = size(t \approx s) = size(t) + size(s)$, and $size(S) = \sum_{p \in S} size(p)$, where S is a set of matching problems or equations.

Theorem 3. *Given an $(\mathcal{R}, \lambda)$-matching problem $t \ll s$, the matching algorithm $\mathfrak{M}$ terminates and computes an X-substitution $\boldsymbol{\sigma}$ such that $\tau(\boldsymbol{\sigma})$ consists of all relevant $(\mathcal{R}, \lambda)$-matchers of t to s.*

Proof. The theorem consists of three parts: termination, soundness, and completeness. We prove each of them separately.

Termination. The rules Dec-M and VE-M strictly reduce the number of symbols in M. The rule Mer-M does the same for S, without changing M. Cla-M and Inc-M stop the algorithm immediately. Hence, the algorithm strictly reduces the lexicographic combination $\langle size(M), size(S)\rangle$ of sizes of M and S, which implies termination.

Soundness. If $\sigma \in \tau(\boldsymbol{\sigma})$, then σ is a relevant $(\mathcal{R}, \lambda)$-matcher of t to s.
Let $\{t \ll s\}; \emptyset \Longrightarrow^{+} \emptyset; S$ be the derivation in $\mathfrak{M}$ that computes $\boldsymbol{\sigma}$. By definition of computed X-substitution, we can conclude that $\sigma \in \tau(\boldsymbol{\sigma})$ iff σ is a solution of $\emptyset; S$. By induction on the length of the given derivation, using Lemma 1, we can prove that σ is an $(\mathcal{R}, \lambda)$-matcher of t to s. In $\mathfrak{M}$, no new variables are created and put in S. All variables there come from the original problem. It implies that σ is a relevant matcher of t to s.

Completeness. If σ is a relevant $(\mathcal{R}, \lambda)$-matcher of t to s, then $\sigma \in \tau(\boldsymbol{\sigma})$.
Since $t \ll s$ is solvable, we can construct a derivation $\{t \ll s\}; \emptyset \Longrightarrow^{+} \emptyset; S$ in $\mathfrak{M}$. This follows from the fact that for each form of matching equation we have a rule in $\mathfrak{M}$, and if we have two equations with the same variable in S we can also transform it. Moreover, by Lemma 1, we would never apply Cla-M and Inc-M rules, because it would contradict the solvability of $t \ll s$. Hence, we can construct the mentioned derivation, for which, again by Lemma 1, we have that σ is a $(\mathcal{R}, \lambda)$-solution of $\emptyset; S$. By definitions of computed X-substitution $\boldsymbol{\sigma}$ and τ, it implies that $\sigma \in \tau(\boldsymbol{\sigma})$. □

Hence, $\mathfrak{M}$ computes all relevant $(\mathcal{R}, \lambda)$-X-matchers for matching problems.

Example 5. We illustrate the steps of the algorithm $\mathfrak{M}$ for the matching problem in Example 3 for $\lambda = 0.6$ and $\lambda = 0.8$.

$\lambda = 0.6:$

$$\{f(x,x) \ll f(g_1(a_1), g_2(a_2))\}; \emptyset \Longrightarrow_{\text{Dec-M}}$$
$$\{x \ll g_1(a_1),\ x \ll g_2(a_2)\}; \emptyset \Longrightarrow_{\text{VE-M}}$$
$$\{x \ll g_2(a_2)\}; \{x \approx \{g_1, h_1, h_2\}(\{a_1, b\})\} \Longrightarrow_{\text{VE-M}}$$
$$\emptyset; \{x \approx \{g_1, h_1, h_2\}(\{a_1, b\}),\ x \approx \{g_2, h_1, h_2\}(\{a_2, b\})\} \Longrightarrow_{\text{Mer-M}}$$
$$\emptyset; \{x \approx \{h_1, h_2\}(\{b\})\}.$$

$\lambda = 0.8:$

$$\{f(x,x) \ll f(g_1(a_1), g_2(a_2))\}; \emptyset \Longrightarrow_{\text{Dec-M}}$$
$$\{x \ll g_1(a_1),\ x \ll g_2(a_2)\}; \emptyset \Longrightarrow_{\text{VE-M}}$$
$$\{x \ll g_2(a_2)\}; \{x \approx \{g_1\}(\{a_1\})\}; \Longrightarrow_{\text{VE-M}}$$
$$\emptyset; \{x \approx \{g_1\}(\{a_1\}),\ x \approx \{g_2, h_2\}(\{a_2, b\})\} \Longrightarrow_{\text{Inc-M}} \bot.$$

The proximity relation $\mathcal{R}$ can be represented as a weighted undirected graph, whose vertices form a (finite) subset of $\mathcal{F}$ and if $\mathcal{R}(f, g) = \mathfrak{d} > 0$ for two vertices f and g, then there is an edge of weight $\mathfrak{d}$ between them. When we consider $\mathcal{R}$ as a graph, we represent it as a pair $(V_{\mathcal{R}}, E_{\mathcal{R}})$ of the sets of vertices $V_{\mathcal{R}}$ and edges $E_{\mathcal{R}}$. We denote by $|S|$ the number of elements in the (finite) set S.

Graphs induced by proximity relations are sparse, since symbols of different arities are not close to each other. Therefore, in the proofs of complexity results below, we choose to represent the graphs by adjacency lists rather than by adjacency matrices.

Theorem 4. *Let $\mathcal{R} = (V_{\mathcal{R}}, E_{\mathcal{R}})$ be a proximity relation and M be a matching problem with size$(M) = n$. Then the algorithm $\mathfrak{M}$ needs $O(n|V_{\mathcal{R}}| + n|E_{\mathcal{R}}|)$ time and $O(n|V_{\mathcal{R}}| + |E_{\mathcal{R}}|)$ space to compute the $(\mathcal{R}, \lambda)$-solution to M for a given λ.*

Proof. We represent the graph for $\mathcal{R}$ as adjacency lists, in which proximity degrees are weights of edges. Such a weight of an edge (v_1, v_2) is stored at the vertex v_2 in the adjacency list of v_1 and vice versa [7]. Further, from the given matching problem $t \ll s$ we can construct its directed acyclic graph (dag) representation with shared variables (see, e.g., [4]). At each node g of s, we add a pointer to the entry in the adjacency list of $\mathcal{R}$ for the symbol g. The nodes in the representation of t are labeled by function symbols and variables occurring in t. In fact, we have a graph representation $dag(t)$ of t and a tree representation $tree(s)$ of s, since there are no variables to share in s.

During the run of the algorithm, we follow the structures top-down both in $dag(t)$ and $tree(s)$, comparing the node labels pairwise. Assume the label of a nonvariable node f in $dag(t)$ is an element of the adjacency list of a node g in $tree(s)$, and $\mathfrak{d} \geq \lambda$ for the degree $\mathfrak{d}$ stored together with f in the adjacency list. Then the DEC-M rule is applied and we proceed to the successor nodes of f and g (pairwise), as usual. Otherwise we stop (CLA-M rule).

When we reach a variable node x in $dag(t)$ and a node g in $tree(s)$, we check whether there already exists a pointer from x to the root h of some tree $tree_h$. If not, we make a copy $tree_g$ of the subtree $subtree(s, g)$ of $tree(s)$ rooted at g. It means that the adjacency lists of the nodes of this subtree are also copied, not shared. We call the copies of those lists the class labels. After that, we make a pointer from x to g in $tree_g$, and continue with the next unvisited node-pairs in $dag(t)$ and $tree(s)$ (VE-M rule). If $tree_h$ to which x points already exists, we go top-down to the trees $tree_h$ and $subtree(s, g)$, updating the class label at each node of $tree_h$: if the class label at some node in this tree is L_1, and the adjacency list of the corresponding node in $subtree(s, g)$ is L_2, we replace L_1 in $tree_h$ by $L_1 \cap L_2$, provided that $L_1 \cap L_2 \neq \emptyset$, and continue with the next unvisited node-pair. This process corresponds to the MER-M rule, eagerly applied immediately after VE-M. If either the intersection is empty, or one tree is deeper than the other, then we stop with failure (the INC-M rule).

First we make the space analysis. The adjacency list representation of $\mathcal{R}$ needs $O(|V_{\mathcal{R}}| + |E_{\mathcal{R}}|)$ space [7]. The graph/tree representation of the matching problem requires $O(n)$ space. All the copies of trees generated by the VE-M rule may contain in total at most n nodes, each labeled with at most $|V_{\mathcal{R}}|$ symbols, i.e., to store them we need $O(n|V_{\mathcal{R}}|)$ space. Hence, the total amount of required memory is $O(n|V_{\mathcal{R}}| + |E_{\mathcal{R}}|)$.

For the runtime complexity, constructing the adjacency list needs $O(|V_{\mathcal{R}}| + |E_{\mathcal{R}}|)$ time. Construction of the dag/tree representation of the matching problem can be done in $O(n)$ time [4]. Each node in $dag(t)$ and $tree(s)$ is visited once. Hence, the structure traversal is done in linear time with respect to n. Checking the membership of some vertex f from $dag(t)$ in the adjacency list of some vertex g in $tree(s)$, needed in the DEC-M rule, requires $O(degree(g))$ time. Since this check is performed $O(n)$ times, and a (rough) upper bound for vertex degrees is $|E_{\mathcal{R}}|$, we can say that the total time needed for the adjacency list membership operation during the run of $\mathfrak{M}$ is $O(n|E_{\mathcal{R}}|)$. Creating the copies of trees by the VE-M rule is constant for each symbol, thus needing $O(n|V_{\mathcal{R}}|)$ time. Computation of intersections between two proximity classes needs $O(|V_{\mathcal{R}}|)$ time. We may need to perform $O(n)$ such intersections, hence for them we need $O(n|V_{\mathcal{R}}|)$ time. It implies that the runtime complexity of the matching algorithm is $O(n|V_{\mathcal{R}}| + n|E_{\mathcal{R}}|)$. □

4.1 Computing Approximation Degrees for Matching

The algorithm above does not compute approximation degrees for the returned matchers. We can add this feature with a small modification of the notions.

A *graded set of function symbols* is a finite set of pairs $\{\langle f_1, \alpha_1\rangle, \ldots, \langle f_n, \alpha_n\rangle\}$, where $\alpha_i \in (0, 1]$ and all f_i's have the same arity. *Graded X-terms* are constructed from graded function symbol sets and variables in the same way X-terms were constructed from non-graded symbol sets and variables. We reuse the bold face letters $\mathbf{f}, \mathbf{a}, \mathbf{t}$, etc. that denote the non-graded counterparts of these notions.

The intersection of graded function symbol sets is defined as $\mathbf{f}_1 \cap \mathbf{f}_2 := \{\langle f, \alpha_1 \wedge \alpha_2\rangle \mid \langle f, \alpha_1\rangle \in \mathbf{f}_1, \langle f, \alpha_2\rangle \in \mathbf{f}_2\}$. Then the intersection of graded X-terms $\mathbf{t}_1 \sqcap \mathbf{t}_2$ is defined as it was done for non-graded X-terms, using the intersection of graded sets of functions symbols.

The *graded* $(\mathcal{R}, \lambda)$*-proximity class* for a symbol f is a set $\{\langle g, \alpha\rangle \mid \mathcal{R}(f, g) = \alpha \geq \lambda\}$. Also here, we reuse the notation from its non-graded version: $\mathbf{pc}(f, \mathcal{R}, \lambda)$. The proximity class for a term is defined and denoted similarly.

Example 6. Let $\mathcal{R}$ be the proximity relation defined in Example 2.

Let t be the term $f(g_1(a_1), g_2(a_2))$. Then the graded proximity class for it, $\mathbf{pc}(t, \mathcal{R}, \lambda)$, for different values of λ is:

$$
\begin{aligned}
0 < \lambda \leq 0.5 :\ & \\
& \{\langle f, 1\rangle\}(\{\langle g_1, 1\rangle, \langle g_2, 0.5\rangle, \langle h_1, 0.6\rangle, \langle h_2, 0.7\rangle\}(\{\langle a_1, 1\rangle, \langle a_2, 0.5\rangle, \langle b, 0.7\rangle\}), \\
& \qquad \{\langle g_1, 0.5\rangle, \langle g_2, 1\rangle, \langle h_1, 0.6\rangle, \langle h_2, 0.8\rangle\}(\{\langle a_1, 0.5\rangle, \langle a_2, 1\rangle, \langle b, 0.8\rangle\})).
\end{aligned}
$$

$$
\begin{aligned}
0.5 < \lambda \leq 0.6 :\ & \{\langle f, 1\rangle\}(\{\langle g_1, 1\rangle, \langle h_1, 0.6\rangle, \langle h_2, 0.7\rangle\}(\{\langle a_1, 1\rangle, \langle b, 0.7\rangle\}), \\
& \qquad \{\langle g_2, 1\rangle, \langle h_1, 0.6\rangle, \langle h_2, 0.8\rangle\}(\{\langle a_2, 1\rangle, \langle b, 0.8\rangle\})).
\end{aligned}
$$

$$
0.6 < \lambda \leq 0.7 : \{\langle f, 1\rangle\}(\{\langle g_1, 1\rangle, \langle h_2, 0.7\rangle\}(\{\langle a_1, 1\rangle, \langle b, 0.7\rangle\}),
$$

$$\{\langle g_2, 1\rangle, \langle h_2, 0.8\rangle\}(\{\langle a_2, 1\rangle, \langle b, 0.8\rangle\})).$$

$$0.7 < \lambda \leq 0.8 : \{\langle f, 1\rangle\}(\{\langle g_1, 1\rangle\}(\{\langle a_1, 1\rangle\}),$$
$$\{\langle g_2, 1\rangle, \langle h_2, 0.8\rangle\}(\{\langle a_2, 1\rangle, \langle b, 0.8\rangle\})).$$

$$0.8 < \lambda \leq 1 \quad : \{\langle f, 1\rangle\}(\{\langle g_1, 1\rangle\}(\{\langle a_1, 1\rangle\}), \{\langle g_2, 1\rangle\}(\{\langle a_2, 1\rangle\})).$$

The modified version of the matching algorithm works on triples $M; S; \alpha$, where S is a set of equations between variables and graded X-terms, and α is the approximation degree between function symbols, initialized with 1. The only rule we need to change is Dec-M, which should update the approximation degree α:

DEC-M: **Decomposition**
$\{f(t_1, \dots, t_n) \ll g(s_1, \dots, s_n)\} \uplus M;\ S;\ \alpha \Longrightarrow$
$\quad M \cup \{t_i \ll s_i \mid 1 \leq i \leq n\};\ S;\ \alpha \wedge \mathcal{R}(f, g),$
if $n \geq 0$, $\mathcal{R}(f, g) \geq \lambda$.

The graded counterpart of τ, denoted by τ_{gr}, takes a graded X-term and returns a set of pairs $\langle t, \alpha\rangle$ where t is a term and $\alpha \in (0, 1]$. It is defined as

$$\tau_{\mathrm{gr}}(x) := \{\langle x, 1\rangle\},$$
$$\tau_{\mathrm{gr}}(\mathbf{f}(\mathbf{t}_1, \dots, \mathbf{t}_n) := \{\langle f(t_1, \dots, t_n), \alpha\rangle \mid \langle f, \alpha_0\rangle \in \mathbf{f}, \langle t_i, \alpha_i\rangle \in \tau_{\mathrm{gr}}(\mathbf{t}_i),$$
$$1 \leq i \leq n,\ \alpha = \alpha_0 \wedge \dots \wedge \alpha_n\}.$$

Similarly, we define τ_{gr} for graded X-substitutions: $\tau_{\mathrm{gr}}(\boldsymbol{\sigma})$ is a set of pairs $\langle \sigma, \alpha\rangle$, where σ is a substitution and $\alpha \in (0, 1]$, defined as

$$\tau_{\mathrm{gr}}(\{x_1 \mapsto \mathbf{t}_1, \dots, x_n \mapsto \mathbf{t}_n\}) :=$$
$$\{\langle \{x_1 \mapsto t_1, \dots, x_n \mapsto t_n\}, \alpha_1 \wedge \dots \wedge \alpha_n\rangle \mid \langle t_i, \alpha_i\rangle \in \tau_{\mathrm{gr}}(\mathbf{t}_i),\ 1 \leq i \leq n\}.$$

When it succeeds, the matching algorithm stops with the triple of the form $\emptyset$; $\{x_1 \approx \mathbf{s}_1, \dots, x_n \approx \mathbf{s}_n\}$; α. From this representation, we take $\boldsymbol{\sigma} = \{x_1 \mapsto \mathbf{s}_1, \dots, x_n \mapsto \mathbf{s}_n\}$ as the computed graded X-substitution and α as the upper bound of approximation degrees of all matchers. We can use $\boldsymbol{\sigma}$ and α as the basis from which various concrete matchers of the original $(\mathcal{R}, \lambda)$-matching problem M and their approximation degrees can be obtained. For instance:

- Extract each solution and its approximation degree from $\tau_{\mathrm{gr}}(\boldsymbol{\sigma})$: For each $\langle \sigma, \alpha_\sigma\rangle \in \tau_{\mathrm{gr}}(\boldsymbol{\sigma})$ we get $\langle \sigma, \alpha_\sigma \wedge \alpha\rangle$.
- If we are interested only in matchers with the maximal approximation degree, we select $\langle \sigma, \alpha_\sigma\rangle \in \tau_{\mathrm{gr}}(\boldsymbol{\sigma})$ so that $\alpha_\sigma \wedge \alpha$ is maximal (there can be several of them), without unpacking the whole set of solutions.

Example 7. Let the proximity relation $\mathcal{R}$ be obtained by adding $\mathcal{R}(f_1, f_2) = 0.8$ to the one defined in Example 2. Let $t = f_1(x, x)$ and $s = f_2(g_1(a_1), g_2(a_2))$.

We illustrate the steps of the algorithm $\mathfrak{M}$ for the matching problem $f_1(x,x) \ll f_2(g_1(a_1), g_2(a_2))$ for $\lambda = 0.6$.

$$\{f_1(x,x) \ll f_2(g_1(a_1), g_2(a_2))\}; \emptyset; 1 \Longrightarrow_{\text{DEC-M}}$$
$$\{x \ll g_1(a_1),\ x \ll g_2(a_2)\}; \emptyset; 0.8 \Longrightarrow_{\text{VE-M}}$$
$$\{x \ll g_2(a_2)\}; \{x \approx \{\langle g_1, 1\rangle, \langle h_1, 0.6\rangle, \langle h_2, 0.7\rangle\}(\{\langle a_1, 1\rangle, \langle b, 0.7\rangle\})\}; 0.8$$
$$\Longrightarrow_{\text{VE-M}}$$
$$\emptyset; \{x \approx \{\langle g_1, 1\rangle, \langle h_1, 0.6\rangle, \langle h_2, 0.7\rangle\}(\{\langle a_1, 1\rangle, \langle b, 0.7\rangle\}),$$
$$x \approx \{\langle g_2, 1\rangle, \langle h_1, 0.6\rangle, \langle h_2, 0.8\rangle\}(\{\langle a_2, 1\rangle, \langle b, 0.8\rangle\})\}; 0.8 \Longrightarrow_{\text{MER-M}}$$
$$\emptyset; \{x \approx \{\langle h_1, 0.6\rangle, \langle h_2, 0.7\rangle\}(\{\langle b, 0.7\rangle\})\}; 0.8.$$

If we want to extract all $(\mathcal{R}, \lambda)$-matchers, we would return $\langle\{x \mapsto h_1(b)\}, 0.6\rangle$ and $\langle\{x \mapsto h_2(b)\}, 0.7\rangle$. The maximal solution would be only $\langle\{x \mapsto h_2(b)\}, 0.7\rangle$.

If we had $\mathcal{R}(f_1, f_2) = 0.6$, then the sets of all $(\mathcal{R}, \lambda)$-matchers and all maximal $(\mathcal{R}, \lambda)$-matchers would coincide. They both would be $\{\langle\{x \mapsto h_1(b)\}, 0.6\rangle, \langle\{x \mapsto h_2(b)\}, 0.6\rangle\}$.

5 Anti-unification

Given $\mathcal{R}$ and λ, for solving an $(\mathcal{R}, \lambda)$-anti-unification problem between two terms t and s, we create the anti-unification triple (AUT) $x : \mathbf{pc}(t, \mathcal{R}, \lambda) \triangleq \mathbf{pc}(s, \mathcal{R}, \lambda)$ where x is a fresh variable. Then we put it in the initial tuple $\{x : \mathbf{pc}(t, \mathcal{R}, \lambda) \triangleq \mathbf{pc}(s, \mathcal{R}, \lambda)\}; \emptyset; x$, and apply the rules given below. They work on tuples $A; S; \mathbf{r}$, where A is a set of AUTs to be solved (called the AU-problem set), S is the set consisting of AUTs already solved (called the store), and $\mathbf{r}$ is the generalization X-term computed so far. The rules transform such tuples in all possible ways as long as possible, returning $\emptyset; S; \mathbf{r}$. In this case, we call $\mathbf{r}$ the *computed X-term.* We denote the algorithm by $\mathfrak{G}$. The rules are as follows:

DEC-AU: **Decomposition**

$\{x : \mathbf{f}(\mathbf{t}_1, \dots, \mathbf{t}_n) \triangleq \mathbf{g}(\mathbf{s}_1, \dots, \mathbf{s}_n)\} \uplus A;\ S;\ \mathbf{r} \Longrightarrow$
$\{y_1 : \mathbf{t}_1 \triangleq \mathbf{s}_1, \dots, y_n : \mathbf{t}_n \triangleq \mathbf{s}_n\} \cup A;\ S;\ \mathbf{r}\{x \mapsto (\mathbf{f} \cap \mathbf{g})(y_1, \dots, y_n)\}$,
where $n \geq 0$, $\mathbf{f} \cap \mathbf{g} \neq \emptyset$.

SOL-AU: **Solving**

$\{x : \mathbf{t} \triangleq \mathbf{s}\} \uplus A;\ S;\ \mathbf{r} \Longrightarrow A;\ \{x : \mathbf{t} \triangleq \mathbf{s}\} \cup S;\ \mathbf{r}$,
if $head(\mathbf{t}) \cap head(\mathbf{s}) = \emptyset$.

MER-AU: **Merging**

$\emptyset;\ \{x_1 : \mathbf{t}_1 \triangleq \mathbf{s}_1,\ x_2 : \mathbf{t}_2 \triangleq \mathbf{s}_2\} \uplus S;\ \mathbf{r} \Longrightarrow \emptyset;\ \{x_1 : \mathbf{t} \triangleq \mathbf{s}\} \cup S;\ \mathbf{r}\{x_2 \mapsto x_1\}$,
if $\mathbf{t} = \mathbf{t}_1 \sqcap \mathbf{t}_2 \neq \emptyset$ and $\mathbf{s} = \mathbf{s}_1 \sqcap \mathbf{s}_2 \neq \emptyset$.

MER-AU can be applied in different ways, which might lead to multiple X-lggs. One may notice that we do not have a rule for AUTs containing variables.

This is because one can treat the input variables as constants. Then AUTs such as $x : y \triangleq y$ are dealt by the DEC-AU rule, and AUTs of the form $x : y \triangleq z$ with $y \neq z$ are processed by SOL-AU.

Theorem 5. *Given a proximity relation $\mathcal{R}$, a cut value λ, and two terms t and s, the anti-unification algorithm $\mathfrak{G}$ terminates and computes X-terms $\mathbf{r}_1, \dots, \mathbf{r}_n$, $n \geq 1$, such that $\cup_{i=1}^n \tau(\mathbf{r}_i)$ contains all $(\mathcal{R}, \lambda)$-least general generalizations of t and s (modulo variable renaming).*

Proof. Like Theorem 3, here also we have three parts: termination, soundness and completeness.

Termination. The algorithm obviously terminates, since the rules DEC-AU and SOL-AU strictly reduce the number of symbols in the AU-problem set A, and MER-AU strictly reduces the number of symbols in the store S.

Soundness. We will prove that if $r \in \cup_{i=1}^n \tau(\mathbf{r}_i)$, then r is an $(\mathcal{R}, \lambda)$-generalization of t and s.

If $r \in \cup_{i=1}^n \tau(\mathbf{r}_i)$, then $r \in \tau(\mathbf{r}_j)$ for some $1 \leq j \leq n$. It means that there exists a derivation

$$\{x : \mathbf{pc}(t, \mathcal{R}, \lambda) \triangleq \mathbf{pc}(s, \mathcal{R}, \lambda)\}; \emptyset; x\boldsymbol{\vartheta}_0 \Longrightarrow^k \emptyset; S; x\boldsymbol{\vartheta}_0\boldsymbol{\vartheta}_1 \cdots \boldsymbol{\vartheta}_k, \qquad (1)$$

where $\boldsymbol{\vartheta}_0 = Id$, $k \geq 1$ and $\mathbf{r}_j = x\boldsymbol{\vartheta}_0\boldsymbol{\vartheta}_1 \cdots \boldsymbol{\vartheta}_k$. For the reference, we denote the tuple at step l in this derivation by $A_l; S_l; \mathbf{r}_l$. Observe that:

- by DEC-AU rule, whenever an AUT $x' : \mathbf{t}' \triangleq \mathbf{s}'$ appears in some A_l in this derivation $(0 \leq l \leq k)$, then we have $x' \in \mathcal{V}(x\boldsymbol{\vartheta}_0 \cdots \boldsymbol{\vartheta}_l)$, $\mathbf{t}' = \mathbf{pc}(t, \mathcal{R}, \lambda)|_{p'}$ for some position p' in $\mathbf{pc}(t, \mathcal{R}, \lambda)$, and $\mathbf{s}' = \mathbf{pc}(s, \mathcal{R}, \lambda)|_{p'}$ for the same position p' in $\mathbf{pc}(s, \mathcal{R}, \lambda)$;
- by SOL-AU rule, the same is true for any $x' : \mathbf{t}' \triangleq \mathbf{s}'$, which appears in some S_l in this derivation $(0 \leq l \leq k)$ with $A_l \neq \emptyset$;
- by MER-AU rule, for any AUT $x' : \mathbf{t}' \triangleq \mathbf{s}'$, which appears in some S_l in this derivation $(0 \leq l \leq k)$ with $A_l = \emptyset$, we have $x' \in \mathcal{V}(x\boldsymbol{\vartheta}_0 \cdots \boldsymbol{\vartheta}_l)$, $\tau(\mathbf{t}') \subseteq \tau(\mathbf{pc}(t, \mathcal{R}, \lambda)|_{p'})$ for some position p' in $\mathbf{pc}(t, \mathcal{R}, \lambda)$, and $\tau(\mathbf{s}') \subseteq \tau(\mathbf{pc}(s, \mathcal{R}, \lambda)|_{p'})$ for the same position p' in $\mathbf{pc}(s, \mathcal{R}, \lambda)$.

Coming back to the derivation in (1), we prove that for all $0 \leq i < k$, if $x\boldsymbol{\vartheta}_0\boldsymbol{\vartheta}_1 \cdots \boldsymbol{\vartheta}_i$ is an $(\mathcal{R}, \lambda)$-X-generalization of t and s, then $x\boldsymbol{\vartheta}_0\boldsymbol{\vartheta}_1 \cdots \boldsymbol{\vartheta}_{i+1}$ is an $(\mathcal{R}, \lambda)$-X-generalization of t and s. For $i = 0$ it is obvious. We assume that this statement is true for some $0 \leq i < k$ and show it for $i + 1$. We should look at all possible ways to make the step

$$A_i; S_i; x\boldsymbol{\vartheta}_0\boldsymbol{\vartheta}_1 \cdots \boldsymbol{\vartheta}_i \Longrightarrow A_{i+1}; S_{i+1}; x\boldsymbol{\vartheta}_0\boldsymbol{\vartheta}_1 \cdots \boldsymbol{\vartheta}_{i+1}.$$

- The step is made by DEC-AU. It means that the problem set A_i contains an AUT of the form $x_i : \mathbf{f}_i(\mathbf{t}_{i_1}, \dots, \mathbf{t}_{i_{n_i}}) \triangleq \mathbf{g}_i(\mathbf{s}_{i_1}, \dots, \mathbf{s}_{i_{n_i}})$ with $\mathbf{f}_i \cap \mathbf{g}_i \neq \emptyset$, which is replaced in A_{i+1} by new AUTs $y_1 : \mathbf{t}_{i_1} \triangleq \mathbf{s}_{i_1}, \dots, y_{n_i} : \mathbf{t}_{i_{n_i}} \triangleq \mathbf{s}_{i_{n_i}}$, and $\boldsymbol{\vartheta}_{i+1} = \{x_i \mapsto (\mathbf{f}_i \cap \mathbf{g}_i)(y_1, \dots, y_{n_i})\}$. There is a position p in both $\mathbf{pc}(t, \mathcal{R}, \lambda)$ and $\mathbf{pc}(s, \mathcal{R}, \lambda)$ such that $\mathbf{pc}(t, \mathcal{R}, \lambda)|_p = \mathbf{f}_i(\mathbf{t}_{i_1}, \dots, \mathbf{t}_{i_{n_i}})$

and $\mathbf{pc}(s, \mathcal{R}, \lambda)|_p = \mathbf{g}_i(\mathbf{s}_{i_1}, \ldots, \mathbf{s}_{i_{n_i}})$. In the same position p in t and s, we have respectively $t_p = f_i(t_{i_1}, \ldots, t_{i_{n_i}}) \in \tau(\mathbf{f}_i(\mathbf{t}_{i_1}, \ldots, \mathbf{t}_{i_{n_i}}))$ and $s_p = g_i(s_{i_1}, \ldots, s_{i_{n_i}}) \in \tau(\mathbf{s}_i(\mathbf{s}_{i_1}, \ldots, \mathbf{s}_{i_{n_i}}))$. Moreover, in the same p in the X-term $x\boldsymbol{\vartheta}_0\boldsymbol{\vartheta}_1 \cdots \boldsymbol{\vartheta}_i$ we have x_i and we know (by the assumption) that $x\boldsymbol{\vartheta}_0\boldsymbol{\vartheta}_1 \cdots \boldsymbol{\vartheta}_i$ is an $(\mathcal{R}, \lambda)$-generalization of t and s. Besides, by definition of X-generalization, its is obvious that $x_i\boldsymbol{\vartheta}_{i+1} = (\mathbf{f}_i \cap \mathbf{g}_i)(y_1, \ldots, y_{n_i})$ is an $(\mathcal{R}, \lambda)$-generalization of $f_i(t_{i_1}, \ldots, t_{i_{n_i}})$ and $g_i(s_{i_1}, \ldots, s_{i_{n_i}})$. By replacing x_i with $x_i\boldsymbol{\vartheta}_{i+1}$, we obtain that $x\boldsymbol{\vartheta}_0\boldsymbol{\vartheta}_1 \cdots \boldsymbol{\vartheta}_i\boldsymbol{\vartheta}_{i+1}$ is an $(\mathcal{R}, \lambda)$-generalization of t and s.

- The step is made by SOL-AU. In this case, $x\boldsymbol{\vartheta}_0\boldsymbol{\vartheta}_1 \cdots \boldsymbol{\vartheta}_{i+1} = x\boldsymbol{\vartheta}_0\boldsymbol{\vartheta}_1 \cdots \boldsymbol{\vartheta}_i$ and the statement holds.
- The step is made by MER-AU. In this case, S_i contains two AUTs $x_{i_1} : \mathbf{t}_{i_1} \triangleq \mathbf{s}_{i_1}$, $x_{i_2} : \mathbf{t}_{i_2} \triangleq \mathbf{s}_{i_2}$ with $\mathbf{t}_{i_1} \sqcap \mathbf{t}_{i_2} \neq \emptyset$ and $\mathbf{s}_{i_1} \sqcap \mathbf{s}_{i_2} \neq \emptyset$. In S_{i+1} these AUTs are replaced by a single AUT $x_{i_1} : \mathbf{t}_{i_1} \sqcap \mathbf{t}_{i_2} \triangleq \mathbf{s}_{i_1} \sqcap \mathbf{s}_{i_2}$, and $\boldsymbol{\vartheta}_{i+1} = \{x_{i_2} \mapsto x_{i_1}\}$. There are two positions p_1 and p_2 in $\mathbf{pc}(t, \mathcal{R}, \lambda)$ such that $\tau(\mathbf{t}_{i_j}) \subseteq \tau(\mathbf{pc}(t, \mathcal{R}, \lambda)|_{p_j})$, $j = 1, 2$. From $\mathbf{t}_{i_1} \sqcap \mathbf{t}_{i_2} \neq \emptyset$ we have $\tau(\mathbf{t}_{i_1}) \cap \tau(\mathbf{t}_{i_2}) \neq \emptyset$ and, as a consequence, $\tau(\mathbf{pc}(t, \mathcal{R}, \lambda)|_{p_1}) \cap \tau(\mathbf{pc}(t, \mathcal{R}, \lambda)|_{p_2}) \neq \emptyset$. Similarly, we get $\tau(\mathbf{s}_{i_1}) \cap \tau(\mathbf{s}_{i_2}) \neq \emptyset$.
  Since $x\boldsymbol{\vartheta}_0\boldsymbol{\vartheta}_1 \cdots \boldsymbol{\vartheta}_i$ is an $(\mathcal{R}, \lambda)$-X-generalization of t and s, for any $q \in \tau(x\boldsymbol{\vartheta}_0\boldsymbol{\vartheta}_1 \cdots \boldsymbol{\vartheta}_i)$ there exist substitutions σ_t and σ_s such that $q\sigma_t \simeq_{\mathcal{R},\lambda} t$ and $q\sigma_s \simeq_{\mathcal{R},\lambda} s$. For σ_t, we have $x_{i_1}\sigma_t \simeq_{\mathcal{R},\lambda} t|_{p_1}$ and $x_{i_2}\sigma_t \simeq_{\mathcal{R},\lambda} t|_{p_2}$. For σ_s, we have $x_{i_1}\sigma_s \simeq_{\mathcal{R},\lambda} s|_{p_1}$ and $x_{i_2}\sigma_s \simeq_{\mathcal{R},\lambda} s|_{p_2}$.
  In $x\boldsymbol{\vartheta}_0\boldsymbol{\vartheta}_1 \cdots \boldsymbol{\vartheta}_{i+1}$, we have x_{i_1} both in position p_1 and in position p_2. Let φ_t be a substitution such that $dom(\varphi_t) = dom(\sigma_t) \setminus \{x_{i_2}\}$, $x_{i_1}\varphi_t \in \tau(\mathbf{pc}(t, \mathcal{R}, \lambda)|_{p_1}) \cap \tau(\mathbf{pc}(t, \mathcal{R}, \lambda)|_{p_2})$, and $y\varphi_t = y\sigma_t$ for all $y \in dom(\varphi_t) \setminus \{x_{i_1}\}$. Such a φ_t exists, since we have shown that $\tau(\mathbf{pc}(t, \mathcal{R}, \lambda)|_{p_1}) \cap \tau(\mathbf{pc}(t, \mathcal{R}, \lambda)|_{p_2}) \neq \emptyset$. By definition, we know that every element of the set $\tau(\mathbf{pc}(t, \mathcal{R}, \lambda)|_{p_1}) \cap \tau(\mathbf{pc}(t, \mathcal{R}, \lambda)|_{p_2})$ is $(\mathcal{R}, \lambda)$-close to both $t|_{p_1}$ and $t|_{p_2}$. Hence, $x_{i_1}\varphi_t \simeq_{\mathcal{R},\lambda} t|_{p_1}$ and $x_{i_1}\varphi_t \simeq_{\mathcal{R},\lambda} t|_{p_2}$. We can define φ_s analogously, and by a similar reasoning conclude that $x_{i_1}\varphi_s \simeq_{\mathcal{R},\lambda} s|_{p_1}$ and $x_{i_1}\varphi_s \simeq_{\mathcal{R},\lambda} s|_{p_2}$. For every position other than those where x_{i_2} appeared in $x\boldsymbol{\vartheta}_0\boldsymbol{\vartheta}_1 \cdots \boldsymbol{\vartheta}_i$, the X-terms $x\boldsymbol{\vartheta}_0\boldsymbol{\vartheta}_1 \cdots \boldsymbol{\vartheta}_{i+1}$ and $x\boldsymbol{\vartheta}_0\boldsymbol{\vartheta}_1 \cdots \boldsymbol{\vartheta}_i$ coincide. Hence, for every $q \in \tau(x\boldsymbol{\vartheta}_0\boldsymbol{\vartheta}_1 \cdots \boldsymbol{\vartheta}_{i+1})$, we get $q\varphi_t \simeq_{\mathcal{R},\lambda} t$ and $q\varphi_s \simeq_{\mathcal{R},\lambda} s$, implying that $x\boldsymbol{\vartheta}_0\boldsymbol{\vartheta}_1 \cdots \boldsymbol{\vartheta}_{i+1}$ is an $(\mathcal{R}, \lambda)$-X-generalization of t and s.

Hence, we proved that in derivation (1), $x\boldsymbol{\vartheta}_0\boldsymbol{\vartheta}_1 \cdots \boldsymbol{\vartheta}_k$ is an $(\mathcal{R}, \lambda)$-X-generalization of t and s. Since $x\boldsymbol{\vartheta}_0\boldsymbol{\vartheta}_1 \cdots \boldsymbol{\vartheta}_k = \mathbf{r}_j$ with $r \in \tau(\mathbf{r}_j)$, we get that r is an $(\mathcal{R}, \lambda)$-generalization of t and s, which proves soundness.

Completeness. If r is an $(\mathcal{R}, \lambda)$-lgg of t and s, then there exists $r' \in \cup_{i=1}^{n} \tau(\mathbf{r}_i)$ such that r and r' are equal modulo variable renaming.

We prove completeness by structural induction on r. First, assume r is a variable. Since it is an $(\mathcal{R}, \lambda)$-lgg of t and s, we have $\tau(head(\mathbf{pc}(t, \mathcal{R}, \lambda))) \cap \tau(head(\mathbf{pc}(s, \mathcal{R}, \lambda))) = \emptyset$. But in this case we apply the rule SOL-AU and get also a variable as a computed X-generalization, which may differ from r only by the name.

Now assume $r = h(r_1, \ldots, r_m)$. Then we have that $t = f(t_1, \ldots, t_m)$, $s = g(s_1, \ldots, s_m)$, and $h \in \mathbf{f} \cap \mathbf{g}$, where $\mathbf{f} = \mathbf{pc}(f, \mathcal{R}, \lambda)$ and $\mathbf{g} = \mathbf{pc}(g, \mathcal{R}, \lambda)$. We apply the rule DEC-AU to $x : \mathbf{pc}(t, \mathcal{R}, \lambda) \triangleq \mathbf{pc}(s, \mathcal{R}, \lambda)$ and obtain new AUTs $y_k : \mathbf{pc}(t_k, \mathcal{R}, \lambda) \triangleq \mathbf{pc}(s_k, \mathcal{R}, \lambda)$, $1 \leq k \leq m$. Note that each r_k, $1 \leq k \leq m$, is an $(\mathcal{R}, \lambda)$-lgg of t_k and s_k. Then by the induction hypothesis, for each $1 \leq k \leq m$ we compute $\mathbf{r}'_k$ so that there exists $r'_k \in \tau(\mathbf{r}'_k)$ which is a renamed copy of r_k. We combine the initial step DEC-AU with the derivations that compute $\mathbf{r}'_i$ to obtain a derivation computing $(\mathbf{f} \cap \mathbf{g})(\mathbf{r}'_1, \ldots, \mathbf{r}'_m)$. However, this does not yet guarantee that $(\mathbf{f} \cap \mathbf{g})(\mathbf{r}'_1, \ldots, \mathbf{r}'_m)$ contains a renamed copy of r, since by being an $(\mathcal{R}, \lambda)$-lgg, r might contain the same variable in multiple positions (in different r_i and r_j), which we have not captured yet. Let p_i and p_j be such positions in r, containing the same variable y, but having different variables y_i and y_j in $(\mathbf{f} \cap \mathbf{g})(\mathbf{r}'_1, \ldots, \mathbf{r}'_m)$. Since r is a generalization of t and s, having the same variable in p_i and p_j implies that $\tau(\mathbf{pc}(t, \mathcal{R}, \lambda)|_{p_i}) \cap \tau(\mathbf{pc}(t, \mathcal{R}, \lambda)|_{p_j}) \neq \emptyset$. Therefore, $\mathbf{pc}(t, \mathcal{R}, \lambda)|_{p_i} \sqcap \mathbf{pc}(t, \mathcal{R}, \lambda)|_{p_j} \neq \emptyset$. Similarly, we have $\mathbf{pc}(s, \mathcal{R}, \lambda)|_{p_i} \sqcap \mathbf{pc}(s, \mathcal{R}, \lambda)|_{p_j} \neq \emptyset$. Having different y_i and y_j in positions p_i and p_j in $(\mathbf{f} \cap \mathbf{g})(\mathbf{r}'_1, \ldots, \mathbf{r}'_m)$ implies that we have $y_i : \mathbf{pc}(t, \mathcal{R}, \lambda)|_{p_i} \triangleq \mathbf{pc}(s, \mathcal{R}, \lambda)|_{p_i}$ and $y_j : \mathbf{pc}(t, \mathcal{R}, \lambda)|_{p_j} \triangleq \mathbf{pc}(s, \mathcal{R}, \lambda)|_{p_j}$ in the store in the derivation we just constructed. But then we can extend this derivation by applying MER-AU rule for y_i and y_j obtaining $(\mathbf{f} \cap \mathbf{g})(\mathbf{r}'_1, \ldots, \mathbf{r}'_i, \ldots, \mathbf{r}''_j, \ldots, \mathbf{r}'_m)$ which reduces the difference with r in distinct variables. We can repeat these steps as long as there are positions which contain different variables in the generalization computed by us, and the same variable in r. In this way, we obtain an X-generalization $\mathbf{r}'$ of t and s such that there exists $r' \in \tau(\mathbf{r}')$ which is a renamed copy of r. □

Hence, the algorithm computes $(\mathcal{R}, \lambda)$-X-lggs of the given terms. To compute linear generalizations, we do not need the MER-AU rule. In this case the anti-unification algorithm returns a single X-term $\mathbf{r}$ such that $\tau(\mathbf{r})$ contains all linear lggs of s and t (modulo variable renaming).

Example 8. Let $\mathcal{R}$ be a proximity relation defined as

$$\mathcal{R}(a_1, a) = \mathcal{R}(a_2, a) = \mathcal{R}(b_1, b) = \mathcal{R}(b_2, b) = 0.5,$$
$$\mathcal{R}(a_2, a') = \mathcal{R}(a_3, a') = \mathcal{R}(b_2, b') = \mathcal{R}(b_3, b') = 0.6, \qquad \mathcal{R}(f, g) = 0.7.$$

Let $t = f(a_1, a_2, a_3)$ and $s = g(b_1, b_2, b_3)$. Then the anti-unification algorithm run ends with the following pairs consisting of the store and an $(\mathcal{R}, \lambda)$-lgg, for different values of λ:

$0 < \lambda \leq 0.5$:

$$\text{store}_1 = \{x_1 : \{a\} \triangleq \{b\},\ x_3 : \{a_3, a'\} \triangleq \{b_3, b'\}\},$$
$$\text{X-lgg}_1 = \{f, g\}(x_1, x_1, x_3).$$

$$\text{store}_2 = \{x_1 : \{a_1, a\} \triangleq \{b_1, b\},\ x_2 : \{a'\} \triangleq \{b'\}\},$$
$$\text{X-lgg}_2 = \{f, g\}(x_1, x_2, x_2).$$

$0.5 < \lambda \leq 0.6$:

$$\text{store} = \{x_1 : \{a_1\} \triangleq \{b_1\},\ x_2 : \{a'\} \triangleq \{b'\}\},$$
$$\text{X-lgg} = \{f, g\}(x_1, x_2, x_2).$$

$$0.6 < \lambda \leq 0.7: \qquad \text{store} = \{x_1 : \{a_1\} \triangleq \{b_1\}, x_2 : \{a_2\} \triangleq \{b_2\}, \\ x_3 : \{a_3\} \triangleq \{b_3\}\}, \\ \text{X-lgg} = \{f, g\}(x_1, x_2, x_3).$$

$$0.7 < \lambda \leq 1: \qquad \text{store} = \{x : \{f(a_1, a_2, a_3)\} \triangleq g(b_1, b_2, b_3)\}, \quad \text{X-lgg} = x.$$

The store shows how to obtain terms which are $(\mathcal{R}, \lambda)$-close to the original terms. For instance, when $0 < \lambda \leq 0.5$, store$_1$ tells us that for any substitution σ from the set $\tau(\{x_1 \mapsto \{a\}, x_3 \mapsto \{a_3, a'\}\})$, the instances of the generalizations, $f(x_1, x_1, x_3)\sigma$ and $g(x_1, x_1, x_3)\sigma$ are $(\mathcal{R}, \lambda)$-close to the original term t. We have two such σ's, $\sigma_1 = \{x_1 \mapsto a, x_3 \mapsto a_3\}$ and $\sigma_2 = \{x_1 \mapsto a, x_3 \mapsto a'\}$. They give, respectively, $f(x_1, x_1, x_3)\sigma_1 = f(a, a, a_3) \simeq_{\mathcal{R},\lambda} f(a_1, a_2, a_3)$, $g(x_1, x_1, x_3)\sigma_1 = g(a, a, a_3) \simeq_{\mathcal{R},\lambda} f(a_1, a_2, a_3)$, and $f(x_1, x_1, x_3)\sigma_2 = f(a, a, a') \simeq_{\mathcal{R},\lambda} f(a_1, a_2, a_3)$, $g(x_1, x_1, x_3)\sigma_2 = g(a, a, a') \simeq_{\mathcal{R},\lambda} f(a_1, a_2, a_3)$.

Similarly, for any substitution ϑ from the set $\tau(\{x_1 \mapsto \{b\}, x_3 \mapsto \{b_3, b'\}\})$ (which is also extracted from store$_1$), the instances of the generalizations $f(x_1, x_1, x_3)\vartheta$ and $g(x_1, x_1, x_3)\vartheta$ are $(\mathcal{R}, \lambda)$-close to the original term s.

Now we illustrate how the first two X-lggs have been computed. Let $\lambda = 0.5$. For the initial problem we take $\mathbf{pc}(t, \mathcal{R}, \lambda) = \{f, g\}(\{a_1, a\}, \{a_2, a, a'\}, \{a_3, a'\})$ and $\mathbf{pc}(s, \mathcal{R}, \lambda) = \{g, f\}(\{b_1, b\}, \{b_2, b, b'\}, \{b_3, b'\})$ and proceed as follows:

$$\{x : \{f, g\}(\{a_1, a\}, \{a_2, a, a'\}, \{a_3, a'\}) \triangleq \{g, f\}(\{b_1, b\}, \{b_2, b, b'\}, \{b_3, b'\})\}; \\ \emptyset; x \Longrightarrow_{\text{DEC-AU}}$$
$$\{x_1 : \{a_1, a\} \triangleq \{b_1, b\}, x_2 : \{a_2, a, a'\} \triangleq \{b_2, b, b'\}, x_3 : \{a_3, a'\} \triangleq \{b_3, b'\}\}; \\ \emptyset; \{f, g\}(x_1, x_2, x_3) \Longrightarrow_{\text{SOL-AU}\times 3}$$
$$\emptyset; \{x_1 : \{a_1, a\} \triangleq \{b_1, b\}, x_2 : \{a_2, a, a'\} \triangleq \{b_2, b, b'\}, x_3 : \{a_3, a'\} \triangleq \{b_3, b'\}\}; \\ \{f, g\}(x_1, x_2, x_3).$$

Now there are two alternatives: to merge x_1 and x_2, or x_2 and x_3. They give:

$$\emptyset;\ \{x_1 : \{a\} \triangleq \{b\}, x_3 : \{a_3, a'\} \triangleq \{b_3, b'\}\};\ \{f, g\}(x_1, x_1, x_3), \quad \text{or} \\ \emptyset;\ \{x_1 : \{a_1, a\} \triangleq \{b_1, b\}, x_2 : \{a'\} \triangleq \{b'\}\};\ \{f, g\}(x_1, x_2, x_2).$$

These are exactly the stores and X-lggs we saw at the beginning of this example.

Example 9. Consider again the proximity relation and the terms from Example 8, but this times assume we are interested in linear generalizations. Then the stores and X-lggs are the following:

$$0 < \lambda \leq 0.5: \\ \text{store} = \{x_1 : \{a_1, a\} \triangleq \{b_1, b\}, x_2 : \{a_2, a, a'\} \triangleq \{b_2, b, b'\}, \\ x_3 : \{a_3, a'\} \triangleq \{b_3, b'\}\}. \\ \text{X-lgg} = \{f, g\}(x_1, x_2, x_3).$$

$$0.5 < \lambda \leq 0.6:$$

$$\text{store} = \{x_1 : \{a_1\} \triangleq \{b_1\}, x_2 : \{a_2, a'\} \triangleq \{b_2, b'\},$$
$$x_3 : \{a_3, a'\} \triangleq \{b_3, b'\}\}.$$
$$\text{X-lgg} = \{f, g\}(x_1, x_2, x_3).$$

$0.6 < \lambda \leq 0.7$:

$$\text{store} = \{x_1 : \{a_1\} \triangleq \{b_1\},\ x_2 : \{a_2\} \triangleq \{b_2\}, x_3 : \{a_3\} \triangleq \{b_3\}\}.$$
$$\text{X-lgg} = \{f, g\}(x_1, x_2, x_3).$$

$0.7 < \lambda \leq 1$: store $= \{x : \{f(a_1, a_2, a_3)\} \triangleq g(b_1, b_2, b_3)\}$. X-lgg $= x$.

Theorem 6. *Let $\mathcal{R} = (V_{\mathcal{R}}, E_{\mathcal{R}})$ be a proximity relation and λ be a cut value. Assume t and s are terms with size(s) + size$(t) = n$. Then*

- *$\mathfrak{G}$ needs $O(n^2|V_{\mathcal{R}}|+|E_{\mathcal{R}}|)$ time and $O(n|V_{\mathcal{R}}|+|E_{\mathcal{R}}|)$ space to compute a single $(\mathcal{R}, \lambda)$-X-lgg of t and s;*
- *$\mathfrak{G}$ needs $O(n|V_{\mathcal{R}}| + |E_{\mathcal{R}}|)$ time and space to compute a linear $(\mathcal{R}, \lambda)$-X-lgg of t and s.*

Proof. To represent the relation $\mathcal{R}$, we use adjacency lists in the same way as we did for the matching algorithm (see the proof of Theorem 4). For adjacency lists, the required amount of memory is $O(|V_{\mathcal{R}}|+|E_{\mathcal{R}}|)$. The input can be represented as trees in $O(n)$ space. The same amount is needed for the store. The generalization X-term contains $O(n)$ nodes, each labeled with at most $|V_{\mathcal{R}}|$ symbols. Hence, the total space requirement is $O(n|V_{\mathcal{R}}| + |E_{\mathcal{R}}|)$, and it is independent whether we compute a single $(\mathcal{R}, \lambda)$-X-lgg or a linear $(\mathcal{R}, \lambda)$-X-lgg.

As for the runtime complexity, constructing the adjacency list representation is done in $O(|V_{\mathcal{R}}| + |E_{\mathcal{R}}|)$ time. Besides, whenever DEC-AU or SOL-AU is applied, we need to compute the intersection between proximity classes of two function symbols, which needs $O(|V_{\mathcal{R}}|)$ time. Hence, applying these two rules as long as possible requires $O(n|V_{\mathcal{R}}|)$ time. It implies that the runtime complexity for computing linear $(\mathcal{R}, \lambda)$-X-lgg of t and s is $O(n|V_{\mathcal{R}}| + |E_{\mathcal{R}}|)$.

To compute an unrestricted $(\mathcal{R}, \lambda)$-X-lgg, we should further apply MER-AU as long as possible. This may require $O(n^2)$ steps. At each step we perform the intersection of proximity classes which is done in $O(|V_{\mathcal{R}}|)$ time. Therefore, exhaustive application of MER-AU for computing one $(\mathcal{R}, \lambda)$-X-lgg of t and s needs $O(n^2|V_{\mathcal{R}}|)$. Together with the complexity of maximal applications of the DEC-AU or SOL-AU rules considered above, it gives the $O(n^2|V_{\mathcal{R}}| + |E_{\mathcal{R}}|)$ runtime bound for computing a single $(\mathcal{R}, \lambda)$-X-lgg of t and s. □

5.1 Computing Approximation Degrees for Anti-unification

We can incorporate the approximation degree computation in anti-unification easier than we did for matching. To $(\mathcal{R}, \lambda)$-anti-unify t and s, we just take their graded proximity classes $\mathbf{pc}(t, \mathcal{R}, \lambda)$ and $\mathbf{pc}(s, \mathcal{R}, \lambda)$ and run the algorithm as described above. The operations $\cap$ and $\sqcap$ will be performed on graded sets of functions symbols and graded X-terms, respectively.

Example 10. Let $\mathcal{R}$ be a proximity relation from Example 8 with $\mathcal{R}(f,g) = 0.7$ replaced by $\mathcal{R}(f,h) = 0.7$ and $\mathcal{R}(h,g) = 0.8$. Let $t = f(a_1,a_2,a_3)$ and $s = g(b_1,b_2,b_3)$. Then for $0 < \lambda \leq 0.5$ we get

$$\begin{aligned}
\text{store}_1 &= \{x_1 : \{\langle a, 0.5\rangle\} \triangleq \{\langle b, 0.5\rangle\},\\
&\qquad x_3 : \{\langle a_3, 1\rangle, \langle a', 0.6\rangle\} \triangleq \{\langle b_3, 1\rangle, \langle b', 0.6\rangle\}\},\\
\text{X-lgg}_1 &= \{\langle h, 0.7\rangle\}(x_1, x_1, x_3).\\
\text{store}_2 &= \{x_1 : \{\langle a_1, 1\rangle, \langle a, 0.5\rangle\} \triangleq \{\langle b_1, 1\rangle, \langle b, 0.5\rangle\},\\
&\qquad x_2 : \{\langle a', 0.6\rangle\} \triangleq \{\langle b', 0.6\rangle\}\},\\
\text{X-lgg}_2 &= \{\langle h, 0.7\rangle\}(x_1, x_2, x_2).
\end{aligned}$$

From the X-lgg's we get the actual generalizations. For instance, X-lgg$_2$ gives $r = h(x_1, x_2, x_2)$. From the generalizations, we can "get close" to the original terms by applying the substitutions composed from the store: $\mathcal{R}(r\{x_1 \mapsto a_1, x_2 \mapsto a'\}, t) = \mathcal{R}(h(a_1, a', a'), t) = 0.6$ and $\mathcal{R}(r\{x_1 \mapsto b_1, x_2 \mapsto b'\}, s) = \mathcal{R}(h(b_1, b', b'), s) = 0.6$. Another instance would be $\mathcal{R}(r\{x_1 \mapsto a, x_2 \mapsto a'\}, t) = \mathcal{R}(h(a, a', a'), t) = 0.5$ and $\mathcal{R}(r\{x_1 \mapsto b, x_2 \mapsto b'\}, s) = \mathcal{R}(h(b, b', b'), s) = 0.5$.

6 Conclusion

In this paper, we investigated two fundamental matching and anti-unification problems with fuzzy proximity relations. Fuzzy proximity (and its crisp counterpart, tolerance) is not a transitive relation, which makes these problems challenging. In general, there is no single solution to them.

We developed algorithms that solve the mentioned problems, aiming at computing a compact representation of solution sets. We use extended terms (X-terms) to represent term sets. In X-terms, instead of function symbols, finite sets of function symbols are permitted. X-substitutions map variables to X-terms.

Our matching algorithm computes a single X-substitution solution for solvable proximity (and tolerance) matching problems. We prove that it is sound and complete: every standard substitution obtained from the computed X-matcher is a matcher, and any relevant solution of the matching problem is contained in the set of substitutions induced by the computed X-matcher. Time and space complexities of the algorithm are analyzed.

Unlike matching, proximity/tolerance anti-unification problems, in general, do not have a single solution even if we restrict computed least-general generalizations to X-terms. Our anti-unification algorithm computes a finite complete set of X-lggs. If we consider the linear variant (i.e., if generalizations are not permitted to contain more than one occurrence of each generalization variable), then there exists a single linear X-lgg (which still represents a finite set of lggs as standard terms), and our algorithm computes it. We also analyze time and space complexities of our anti-unification algorithm and its linear variant.

References

1. Aït-Kaci, H., Pasi, G.: Fuzzy unification and generalization of first-order terms over similar signatures. In: Fioravanti, F., Gallagher, J.P. (eds.) LOPSTR 2017. LNCS, vol. 10855, pp. 218–234. Springer, Cham (2018). https://doi.org/10.1007/978-3-319-94460-9_13
2. Alpuente, M., Escobar, S., Espert, J., Meseguer, J.: A modular order-sorted equational generalization algorithm. Inf. Comput. **235**, 98–136 (2014)
3. Baader, F., Nipkow, T.: Term Rewriting and All That. Cambridge University Press, Cambridge (1998)
4. Baader, F.: Unification theory. In: Schulz, K.U. (ed.) IWWERT 1990. LNCS, vol. 572, pp. 151–170. Springer, Heidelberg (1992). https://doi.org/10.1007/3-540-55124-7_5
5. Bartol, W., Miró, J., Pióro, K., Rosselló, F.: On the coverings by tolerance classes. Inf. Sci. **166**(1–4), 193–211 (2004)
6. Baumgartner, A., Kutsia, T.: A library of anti-unification algorithms. In: Fermé, E., Leite, J. (eds.) JELIA 2014. LNCS (LNAI), vol. 8761, pp. 543–557. Springer, Cham (2014). https://doi.org/10.1007/978-3-319-11558-0_38
7. Cormen, T.H., Leiserson, C.E., Rivest, R.L., Stein, C.: Introduction to Algorithms, 3rd edn. MIT Press, Cambridge (2009)
8. Julián-Iranzo, P., Rubio-Manzano, C.: Proximity-based unification theory. Fuzzy Sets Syst. **262**, 21–43 (2015)
9. Julián-Iranzo, P., Sáenz-Pérez, F.: An efficient proximity-based unification algorithm. In: 2018 IEEE International Conference on Fuzzy Systems, FUZZ-IEEE 2018, Rio de Janeiro, Brazil, 8–13 July 2018, pp. 1–8. IEEE (2018)
10. Klement, E., Mesiar, R., Pap, E.: Triangular Norms. Trends in Logic, vol. 8. Springer, Cham (2000). https://doi.org/10.1007/978-94-015-9540-7
11. Kutsia, T., Pau, C.: Proximity-based generalization. In: Ayala Rincón, M., Balbiani, P. (eds.), 32nd International Workshop on Unification, UNIF 2018, Proceedings (2018)
12. Kutsia, T., Pau, C.: Computing all maximal clique partitions in a graph. RISC Report 19–04, RISC, Johannes Kepler University Linz (2019)
13. Kutsia, T., Pau, C.: Solving proximity constraints. In: Gabbrielli, M. (ed.) LOPSTR 2019. LNCS, vol. 12042, pp. 107–122. Springer, Cham (2020). https://doi.org/10.1007/978-3-030-45260-5_7
14. Plotkin, G.D.: A note on inductive generalization. Mach. Intel. **5**(1), 153–163 (1970)
15. Reynolds, J.C.: Transformational systems and the algebraic structure of atomic formulas. Mach. Intel. **5**(1), 135–151 (1970)
16. Sessa, M.I.: Approximate reasoning by similarity-based SLD resolution. Theor. Comput. Sci. **275**(1–2), 389–426 (2002)
17. Zeeman, E.C.: Topology of the brain. In: Waddington, C.H. (ed.) Towards a Theoretical Biology, vol. 1, pp. 140–151. Edinburgh University Press, Edinburgh (1965)

From Paradox to Truth

An Introduction to Self-reference in Formal Language

Graham E. Leigh(✉)

Department of Philosophy, Linguistics and Theory of Science,
University of Gothenburg, Gothenburg, Sweden
graham.leigh@gu.se

Abstract. We present a short introduction to the logical analysis of truth and related concepts. We examine which assumptions are implicit in the paradoxes of truth and self-reference, and present some of the important formal theories of truth that have arisen out of these considerations.

Keywords: Paradox · Truth · Self-reference · First-order logic

1 Introduction

Consider the following statement which we name *.

(*) *If the sentence named $*$ is true then all Georgians drink chacha.*

We give a short argument about the sentence:

1. a. Assume the statement named * is true.
   b. That is, if the statement named * is true then all Georgians drink chacha.
   c. Combining a and b, we conclude all Georgians drink chacha.
2. So, if the statement named * is true then all Georgians drink chacha.
3. That is, the statement named * is true.
4. By 2 and 3 we conclude all Georgians drink chacha.

There is something puzzling about the above argument. It claims to establish as fact a statement ('all Georgians drink chacha') without depending in any way on what the statement asserts. As such, we could replace throughout the argument this statement by any other statement, for example 'no Georgian drinks chacha', and the argument is as valid as before. This is an example of a **logical paradox**:

This paper was presented by the author in three lectures at the Thirteenth International Tbilisi Symposium on Language, Logic and Computation held in Batumi, Georgia in September 2019, and is based on work by the author and Volker Halbach [10]. The author would like to thank the anonymous referee for their comments and improvements. The work was supported by the Knut and Alice Wallenberg Foundation (grant no. 2015.0179) and Swedish Research Council VR (grant no. 2017-05111).

© The Author(s), under exclusive license to Springer Nature Switzerland AG 2022
A. Özgün and Y. Zinova (Eds.): TbiLLC 2019, LNCS 13206, pp. 343–353, 2022.
https://doi.org/10.1007/978-3-030-98479-3_17

a sequence of apparently unproblematic steps in logical reasoning resulting in an absurd conclusion. The particular argument above (for an arbitrary conclusion) is known as **Curry's paradox**.

The analysis of paradoxes arising from self-referential statements has a long history. Variants of the liar paradox 'this statement is false' have been examined since at least the 4th century BCE. But it is only relatively recently that the paradoxes have been given formal treatment using the techniques and systems of mathematical logic. The idea of such an analysis is to isolate the logical and ontological assumptions necessary and sufficient for the paradoxes, and to develop formal (and consistent) theories of complex concepts such as truth.

This survey is organised as follows. In the next section we present a formal version of Curry's paradox and identify the basic assumptions about truth and syntax on which it depends. Section 3 applies these ideas to paradoxes arising from other truth-theoretic assumptions. In Sect. 4 we present some natural collections of truth assumptions which we can show are consistent. The article concludes with further philosophical and mathematical considerations.

2 Formalising Paradox

We present a more formal account of the paradoxical argument above using the framework of first-order logic.[1] In what follows, given a set U, a **U-expression** means a finite string of non-zero length of elements of U.

Let $\mathscr{L}$ be a formal language which we assume to consist of (at least) the logical connectives implication $\rightarrow$ and conjunction $\wedge$, a unary predicate $\mathsf{T}(\cdot)$ and a constant symbol $\overline{e}$ for every $\mathscr{L}$-expression e. Symbols of the final kind provide names for $\mathscr{L}$-expressions and are called **quotation constants**. We will say $\overline{e}$ **names** the $\mathscr{L}$-expression e, or that $\overline{e}$ is the **quotation** of e. For example, $\overline{\wedge}$ is the quotation of the symbol $\wedge$ whereas $\overline{\rightarrow\overline{\wedge}}$ is the quotation of the two-symbol expression $\rightarrow\overline{\wedge}$. Notice that quotation constants are symbols of $\mathscr{L}$, so $\overline{\rightarrow\overline{\wedge}}$ is itself an $\mathscr{L}$-expression of length 1 but names an $\mathscr{L}$-expression of length 2.

The role of the unary predicate symbol $\mathsf{T}(\cdot)$ is to render references to truth in the informal argument. Thus, the phrase *the sentence (named by) x is true* is represented as the $\mathscr{L}$-formula $\mathsf{T}(x)$. The direct translation of our sentence * is therefore the $\mathscr{L}$-formula $\mathsf{T}(\overline{\alpha}) \rightarrow \beta$ where β expresses *all Georgians drink chacha* and $\overline{\alpha}$ is the constant naming this very sentence. But it is easily verified that a quotation constant can never occur within the $\mathscr{L}$-expression it names, so there can be no such α. Fortunately, such a strong assumption on self-reference is unnecessary, and it suffices to determine a formula α which is merely equivalent to $\mathsf{T}(\overline{\alpha}) \rightarrow \beta$. It remains to explain how such sentences can be constructed in a formal language, but first let us confirm that the informal argument from before can be given formal treatment. Thus, let β be any formula and suppose we have to hand a theory S such that

$$(\dagger) \qquad S \vdash \alpha \leftrightarrow (\mathsf{T}(\overline{\alpha}) \rightarrow \beta)$$

[1] Throughout this article we assume familiarity with the basic concepts of first-order logic, such as in [6].

where $\alpha_1 \leftrightarrow \alpha_2$ abbreviates $(\alpha_1 \to \alpha_2) \wedge (\alpha_2 \to \alpha_1)$ and $S \vdash \gamma$ expresses that γ is a theorem of S.

We can now carry out the paradoxical argument within the theory S. As we do, we can isolate the specific additional assumptions required for the argument to go through.

1. a. Assume $\mathsf{T}(\overline{\alpha})$
   b. α — if $S \vdash \mathsf{T}(\overline{\alpha}) \to \alpha$
   c. $\mathsf{T}(\overline{\alpha}) \to \beta$ — b and †
   d. β — a and c
2. So $\mathsf{T}(\overline{\alpha}) \to \beta$ — from 1(a–d)
3. α — 2 and †
4. $\mathsf{T}(\overline{\alpha})$ — if $S \vdash \alpha \to \mathsf{T}(\overline{\alpha})$
5. β — 2 and 4

In other words, we have established the following theorem.

Theorem 1. *Suppose S is a first-order theory in the language $\mathscr{L}$. If S satisfies the following two requirements, then S is inconsistent.*[2]

1. *For every choice of β a formula α can be found such that † holds.*
2. *$S \vdash \gamma \leftrightarrow \mathsf{T}(\overline{\gamma})$ for every γ.*

Condition 2 of Theorem 1 states that S believes, for each sentence α, that α is true iff α. The theorem was proved by Alfred Tarski in 1935 [15], who considered a formal counterpart to the infamous liar sentence, *this sentence is not true*:

$$(\lambda) \qquad S \vdash \lambda \leftrightarrow \neg\mathsf{T}(\overline{\lambda}).$$

The proof we outlined earlier is due to Haskell Curry [2] and generalises Tarski's result by requiring only that S has a logical connective for material implication.

As it stands, however, Theorem 1 is not particularly informative because we have not yet shown the existence of any theory S satisfying condition †. Our next task, therefore, is to present natural examples of such theories. In fact, it turns out that † must hold for any theory in which the following three basic operations on $\mathscr{L}$-expressions are available:

Concatenation Given two $\mathscr{L}$-expressions a, b we can form the expression ab.
Substitution Given expressions a and b, and a symbol u, we can form the result of substituting throughout a every occurrence of u by b.
Quotation Given an $\mathscr{L}$-expression a we can form its quotation, the symbol $\overline{a}$.

In what follows, we assume, in addition to the requirements laid out at the start of the section, that $\mathscr{L}$ contains function symbols corresponding to these three operations: a binary function symbol $*$ expressing concatenation (written as $s * t$), a ternary function symbol sub for substitution, and a unary function

[2] For the present work we consider a theory S to be **inconsistent** if every formula of the language is derivable in S.

symbol quot for quotation. Moreover, $\mathscr{L}$ should contain sufficient symbols to formulate classical first-order logic: a binary predicate symbol $=$, connectives $\wedge$, $\rightarrow$, and a quantifier $\forall$. Other standard connectives are assumed defined. Note that $\mathscr{L}$ may contain other function, constant or predicate symbols beyond the ones listed above. Henceforth, $\mathscr{L}^*$ denotes the set of $\mathscr{L}$-expressions.

Definition. *A* ***syntax theory*** *is any theory in the language $\mathscr{L}$ in which the following three formulas are derivable for all $a, b, c \in \mathscr{L}^*$ and $u \in \mathscr{L}$.*

S1 $\overline{a} * \overline{b} = \overline{ab}$.
S2 $\mathsf{sub}(\overline{a}, \overline{u}, \overline{b}) = \overline{c}$ *if c is the result of substituting b for the symbol u in a.*
S3 $\mathsf{quot}(\overline{a}) = \overline{\overline{a}}$.

The ***minimal syntax theory****, S_{min}, is the theory axiomatised by all instances of the above formulas only.*

To better separate our logical and meta-logical vocabulary we introduce a relation $\doteq$ for expressing identity between $\mathscr{L}$-expressions. Thus, $a \doteq b$ holds iff a and b are identical $\mathscr{L}$-expressions. In contrast, $a = b$ represents the $\mathscr{L}$-expression which happens to be a formula if a and b are two $\mathscr{L}$-terms. In particular, writing $\alpha \doteq \mathsf{T}(x) \rightarrow \beta$ means that α is the formula $\mathsf{T}(x) \rightarrow \beta$.

It is important to confirm the axioms outlined above are not contradictory. We do this by presenting a model of the minimal syntax theory. Let $\mathbb{E} = (\mathscr{L}^*, I)$ be an interpretation of the language $\mathscr{L}$ where $\mathscr{L}^*$ is the domain of the interpretation and I interprets the symbols of $\mathscr{L}$ as follows.

1. The function symbols $*$, quot and sub of $\mathscr{L}$ are interpreted as the functions concatenation, quotation and substitution on $\mathscr{L}^*$ respectively. For example, quot is interpreted as the function $\mathsf{quot}^I \colon \mathscr{L}^* \rightarrow \mathscr{L}^*$ given by $\mathsf{quot}^I(e) \doteq \overline{e}$.
2. The quotation constant $\overline{e}$ is interpreted as the $\mathscr{L}^*$-expression $\overline{e}^I \doteq e \in \mathscr{L}^*$.
3. The interpretation of other predicate and function symbols of $\mathscr{L}$ is arbitrary.

An interpretation of the kind above is called a **standard model**. Note, $\mathbb{E}$ is not uniquely determined since the interpretation of not all function and relation symbols (for instance $\mathsf{T}(\cdot)$) was specified. We leave it to the reader to verify that every standard model validates the axioms of the minimal syntax theory S_{min}.

It turns out that the self-referential formulas required by Theorem 1 can be constructed in any syntax theory. This is the task of the **diagonal lemma**.

Diagonal Lemma. *For every formula $\beta(v_0)$ there exists a term t such that $S \vdash t = \overline{\beta(t)}$ and a sentence α such that $S \vdash \alpha \leftrightarrow \beta(\overline{\alpha})$.*

The second claim is what we require. The first claim is a stronger, more explicit, form of diagonalisation that also holds in any syntax theory. That the first claim implies the second is immediate when we consider the sentence $\alpha \doteq \beta(t)$.

Proof. Fix $\beta(v_0)$. Renaming bound variables if necessary, we may assume v_0 does not occur bound in β. Consider the term $d(v_0) \doteq \mathsf{sub}(v_0, \overline{v_0}, \mathsf{quot}(v_0))$. Since $\overline{v_0}$ is a constant symbol, given any term s we have $d(s) \doteq \mathsf{sub}(s, \overline{v_0}, \mathsf{quot}(s))$. Writing s for the term $\overline{\beta(d(v_0))}$, the following equations are derivable in S.

$$\begin{aligned} S \vdash d(\overline{s}) &= \mathsf{sub}(\overline{s}, \overline{v_0}, \overline{\overline{s}}\,) && \text{(axiom S3)} \\ &= \overline{\beta(d(\overline{s}))} && \text{(axiom S2)} \end{aligned}$$

Selecting $t \doteq d(\overline{s})$, we have $S \vdash t = \overline{\beta(t)}$ as desired. □

3 Paradoxes

The diagonal lemma confirms that every syntax theory satisfies the first condition of Theorem 1. Hence no consistent extension of the minimal syntax theory S_{min} can prove the truth bi-conditional $\alpha \leftrightarrow \mathsf{T}(\overline{\alpha})$ for every $\mathscr{L}$-formula α. We have seen that only a few logical and syntactic assumptions are required to derive the paradoxes. In the present section we observe that even under weaker truth-theoretic assumptions paradoxes arise.

One natural weakening of the truth bi-conditionals is to conceive them as rules instead of implications:

$$\begin{aligned} \text{Necessitation } (\mathsf{NEC}){:}\quad & \textit{If } S \vdash \alpha \textit{ then } S \vdash \mathsf{T}(\overline{\alpha}) \\ \text{Co-Necessitation } (\mathsf{CoNEC}){:}\quad & \textit{If } S \vdash \mathsf{T}(\overline{\alpha}) \textit{ then } S \vdash \alpha \end{aligned}$$

The first rule, necessitation, is the predicate logic formulation of the inference rule from modal logic bearing the same name; the latter states the converse direction. It is immediately clear that the two rules are weakened forms of the two implications making up the truth bi-conditional $\alpha \leftrightarrow \mathsf{T}(\overline{\alpha})$. That they are strictly weaker than the corresponding implications will follow from later observations. For now, we simply observe that from the implication $\alpha \rightarrow \mathsf{T}(\overline{\alpha})$ the 'contraposition' $\neg\mathsf{T}(\overline{\alpha}) \rightarrow \neg\alpha$ immediately follows, yet necessitation need not entail the rule '*if* $S \vdash \neg\mathsf{T}(\overline{\alpha})$ *then* $S \vdash \neg\alpha$'.

The first strengthening of Theorem 1 we present is due to Richard Montague and concerns the use of necessitation in place of one direction of the bi-conditional.

Theorem 2 (Montague's Paradox). *Suppose* $S \vdash \mathsf{T}(\overline{\alpha}) \rightarrow \alpha$ *for every sentence* α*, and* S *satisfies the rule of necessitation. Then* S *is inconsistent.*

Proof. The theorem results from a closer inspection of Curry's paradox. Steps 1–3 of the argument still hold because $S \vdash \mathsf{T}(\overline{\alpha}) \rightarrow \alpha$. So $S \vdash \alpha$. If S contains NEC, this leads to $S \vdash \mathsf{T}(\overline{\alpha})$ and so $S \vdash \beta$. As β was arbitrary we are done.

Montague's theorem appears to lay the blame on the implication $\mathsf{T}(\overline{\alpha}) \rightarrow \alpha$ since the other direction of the equivalence can be weakened and the paradox remains. However, the same can be said for the converse implication $\alpha \rightarrow \mathsf{T}(\overline{\alpha})$:

Theorem 3. *Suppose* $S \vdash \alpha \rightarrow \mathsf{T}(\overline{\alpha})$ *for all* α*. Then*

1. *if* S *satisfies the rule of co-necessitation,* S *is inconsistent.*

2. if $S \vdash \mathsf{T}(\overline{\alpha \rightarrow \beta}) \wedge \mathsf{T}(\overline{\alpha}) \rightarrow \mathsf{T}(\overline{\beta})$ *for every* α *and* β *then* $S \vdash \mathsf{T}(\overline{\bot})$.

Proof. We consider the liar sentence λ for both parts. The diagonal lemma gives $S \vdash \lambda \rightarrow \neg\mathsf{T}(\overline{\lambda})$ whereas it is also the case that $S \vdash \lambda \rightarrow \mathsf{T}(\overline{\lambda})$ by assumption. So $S \vdash \neg\lambda$ and, because $S \vdash \mathsf{T}(\overline{\lambda}) \rightarrow \lambda$, we deduce $S \vdash \mathsf{T}(\overline{\lambda})$. So $S \vdash \mathsf{T}(\overline{\lambda}) \wedge \mathsf{T}(\overline{\neg\lambda})$. From this final observation, both parts of the theorem are immediate. □

Part 2 of this theorem, although not deriving an inconsistency, is also paradoxical. It concludes that two natural assumptions about truth, the implication $\alpha \rightarrow \mathsf{T}(\overline{\alpha})$ and that truth is closed under basic logical reasoning, trivialise truth.

We conclude this section with one more 'paradox' of truth, due to Vann McGee [13]. Unlike the previous examples, the undesirability of this set of truth assumptions is more subtle. It states that working with the minimal syntax theory S_{min} (which, recall, is true in all standard models) and adding three natural principles of truth yields a theory which, although perhaps consistent, admits no standard model.

In order to state McGee's theorem it is necessary to introduce some notation. Let $y(\dot{x})$ abbreviate the term $\mathsf{sub}(y, \overline{v_0}, \mathsf{quot}(x))$ which represents the result of substituting, in place of the variable v_0 in the expression y, the quotation of x. When utilising this notation in the form $\overline{\alpha}(\dot{x})$ we assume v_0 does not occur bound in α. We must also assume a syntax theory S that derives a further property of $\mathscr{L}$-expressions, namely that every $\mathscr{L}$-expression has a unique decomposition as a sequence of $\mathscr{L}$-symbols.

Theorem 4 (McGee's Theorem). *Suppose* S *satisfies the following: i)* NEC*; ii)* $S \vdash \mathsf{T}(\overline{\alpha \rightarrow \beta}) \wedge \mathsf{T}(\overline{\alpha}) \rightarrow \mathsf{T}(\overline{\beta})$ *and* $S \vdash \mathsf{T}(\overline{\neg\alpha}) \rightarrow \neg\mathsf{T}(\overline{\alpha})$ *for every of* α *and* β*; and iii)* $S \vdash \forall x \mathsf{T}(\overline{\alpha}(\dot{x})) \rightarrow \mathsf{T}(\overline{\forall v_0 \alpha})$ *for every* α*. Then* S *has no standard model.*

Assumption (iii) of the theorem states that a formula $\forall v_0 \alpha(v_0)$ is true if every instantiation $\alpha(\overline{e})$ is true. The proof, which we omit due to limitations of space, analyses a formula γ, constructed by the diagonal lemma, expressing *for some natural number, that many iterations of truth over* $\overline{\gamma}$ *is not true.*[3]

4 Models of Truth

In the previous section we saw that even assumptions strictly weaker than the truth bi-conditionals give rise to undesirable, or inconsistent, notions of truth. We now turn to the task of identifying consistent collections of principles, starting with the observation that the paradoxes depend on self-reference.

Definition (Typed Truth). *Let* $\mathscr{L}_0 = \mathscr{L} \setminus \{\mathsf{T}\}$ *be the sub-language of* $\mathscr{L}$ *in which the predicate* T *has been removed. Let TB be the expansion of the minimal syntax theory* S_{min} *by the collection of sentences*

(TB) $\quad \mathsf{T}(\overline{\alpha}) \leftrightarrow \alpha \quad$ *for* α *a sentence of* $\mathscr{L}_0$

[3] For a proof of McGee's theorem in the context of syntax theories, including clarification of additional syntactic assumptions, we refer the reader to [10].

The notation *TB* stands for typed bi-conditionals. For example, the equivalence $\mathsf{T}(\overline{\overline{\wedge} = \overline{\wedge}}) \leftrightarrow \overline{\wedge} = \overline{\wedge}$ is an instance of TB, but $\mathsf{T}(\overline{\lambda}) \leftrightarrow \lambda$, where λ is the liar sentence, is not because λ is not in the language $\mathscr{L}_0$. This theory was introduced by Tarski [15] in his seminal work on formal truth.

Theorem 5. *TB is consistent.*

We introduce some notation useful for this and later proofs. Given a standard model $\mathbb{E}$ and a set $T \subseteq \mathscr{L}^*$ of $\mathscr{L}$-expressions, $\mathbb{E}_T$ denotes the standard model which agrees with $\mathbb{E}$ in all respects except that it interprets the predicate T by the set T. That is,

$$\mathbb{E}_T \models \mathsf{T}(\overline{e}) \iff e \in T$$

Observe that if α does not contain the predicate T then $\mathbb{E}_T \models \alpha$ iff $\mathbb{E} \models \alpha$.

Proof. Fix a standard model $\mathbb{E}$ and consider the standard model $\mathbb{E}_T$ where $T = \{\alpha \mid \mathbb{E} \models \alpha\}$ is the set of sentences validated by $\mathbb{E}$. Then, for α in $\mathscr{L}_0$, we have $\mathbb{E}_T \models \alpha$ iff $\mathbb{E}_T \models \mathsf{T}(\overline{\alpha})$. □

The model $\mathbb{E}_T$ constructed in the proof satisfies more than just the typed truth bi-conditionals. It also validates the bi-conditionals

$$\forall x(\alpha(x) \leftrightarrow \mathsf{T}(\overline{\alpha}(\underset{\cdot}{x})))$$

for $\alpha(v_0)$ in $\mathscr{L}_0$ with at most v_0 free, as well as its generalisation to more than one free variable. This schema is known as the **uniform (typed) bi-conditionals**. Notice that by the definition of the set T, a conjunction $\alpha \wedge \beta \in T$ iff both $\alpha \in T$ and $\beta \in T$, and a quantified formula $\forall x \gamma(x) \in T$ iff for every $e \in \mathscr{L}^*$, $\gamma(\overline{e}) \in T$. From these observations it immediately follows that

$$\mathbb{E}_T \models \forall x \forall y \big(\mathsf{T}(x \underset{\cdot}{\wedge} y) \leftrightarrow (\mathsf{T}(x) \wedge \mathsf{T}(y))\big) \qquad \mathbb{E}_T \models \forall x \big(\mathsf{T}(\overline{\forall v_0} x) \leftrightarrow \forall z\, \mathsf{T}(x(\underset{\cdot}{z}))\big)$$

where $x \underset{\cdot}{\wedge} y \doteq \overline{(} * x * \overline{\wedge} * y * \overline{)}$ is the term forming a conjunction out of x and y, and $x(\underset{\cdot}{z})$ is the abbreviation for substitution introduced earlier. Equivalences for the other connectives and quantifiers hold also. These principles, combined with the uniform bi-conditionals for atomic formulas from $\mathscr{L}_0$, are collectively known as the **compositional clauses**; the theory axiomatised by the clauses is denoted *CT* for **compositional truth** and is also consistent by our model $\mathbb{E}_T$.

The consistency of *TB* (indeed *CT*) confirms the suspicion that it is self-application of truth, i.e. the truth predicate applied to formulas that refer to truth, which is to blame for the paradoxes. The result invites us to explore whether there are natural classes of untyped bi-conditionals which are still consistent. Such sets clearly cannot contain the liar sentence, yet there are untyped formulas that are not inherently paradoxical, such as $\mathsf{T}(\overline{\neg \mathsf{T}(\overline{\wedge} = \overline{\vee})})$ or even the **truth-teller**, the diagonal formula $\tau \leftrightarrow \mathsf{T}(\overline{\tau})$ expressing *this sentence is true.* Judging collections of truth bi-conditionals on the basis of consistency alone is not possible however, due to another negative result by McGee who showed there are as many consistent sets of bi-conditionals as there are consistent extensions of S_{min} (i.e., uncountably many) [14]. Instead, when designing interesting and consistent theories of truth we should look for new methods of model construction or for collections of truth principles that express some 'natural' conception of truth. In the following we consider examples of both kind.

The first construction we present is Saul Kripke's semantic theory of truth introduced in [11]. The underlying motivation is that statements whose truth or falsity can be inferred ultimately from truth-free statements, so-called 'grounded' statements, are unproblematic. Formulas $\mathsf{T}(\overline{\overline{v_0} = \overline{v_1}})$ and $\mathsf{T}(\overline{\neg\mathsf{T}(\overline{\overline{v_0} = \overline{v_1}})})$ are grounded in the truth-value of the truth-free statement $\overline{v_0} = \overline{v_1}$. The liar and truth-teller are examples of ungrounded statements: repeatedly stripping away the truth predicate will never result in a truth-free statement.

To present Kripke's model it is necessary to examine a non-classical conception of truth. Let $\mathbb{E}$ be a standard model (in the usual sense) and suppose $T, F \subseteq \mathscr{L}^*$ are two sets of $\mathscr{L}$-expressions. By $\mathbb{E}_{T,F}$ we denote the **partial model** which agrees with $\mathbb{E}$ on the interpretation of all symbols in $\mathscr{L}_0 = \mathscr{L} \setminus \{\mathsf{T}\}$, but makes true the formula $\mathsf{T}(\overline{e})$ if $e \in T$ and makes $\neg\mathsf{T}(\overline{e})$ true if $e \in F$. Thus, T collects the sentences marked as true by the model $\mathbb{E}_{T,F}$ and F the sentences marked as false. The classical models hitherto considered are simply partial models where $F = \mathscr{L}^* \setminus T$. But if $\alpha \notin T \cup F$ then $\mathsf{T}(\overline{\alpha}) \vee \neg\mathsf{T}(\overline{\alpha})$ does not hold in $\mathbb{E}_{T,F}$, whereas $\mathsf{T}(\overline{\alpha}) \wedge \neg\mathsf{T}(\overline{\alpha})$ holds if $\alpha \in T \cap F$. So some formulas, for example the liar sentence, may be designated as neither true nor false or both true and false, which is impossible for classical models. Rules for connectives and quantifiers are the same as before but we assume negation appears only in front of atomic formulas. Partial models provide a natural semantics for the non-classical logic known as Strong Kleene logic.

We begin the construction at the trivial model $\mathbb{E}_{\emptyset,\emptyset}$ in which neither formula $\mathsf{T}(\cdot)$ nor $\neg\mathsf{T}(\cdot)$ is assumed to hold. Setting $T_0 = F_0 = \emptyset$ to name the first approximations to the two concepts, we now construct a transfinite hierarchy of refinements. Given T_κ and F_κ have been defined, set

$$T_{\kappa+1} = \{\alpha \mid \mathbb{E}_{T_\kappa,F_\kappa} \models \alpha\} \qquad F_{\kappa+1} = \{\alpha \mid \mathbb{E}_{T_\kappa,F_\kappa} \models \neg\alpha\}$$

For a limit ordinal κ, define $T_\kappa = \bigcup_{\kappa_0<\kappa} T_{\kappa_0}$ and F_κ similarly. By induction along ordinals, if $\kappa_0 < \kappa_1$ then $T_{\kappa_0} \subseteq T_{\kappa_1}$ and $F_{\kappa_0} \subseteq F_{\kappa_1}$. Cardinality considerations show these hierarchies stabilise, i.e., there exists an ordinal κ such that $(T_{\kappa+1}, F_{\kappa+1}) = (T_\kappa, F_\kappa)$. Considering the partial model $\mathbb{E}_{T_\kappa,F_\kappa}$ we observe

$$\mathbb{E}_{T_\kappa,F_\kappa} \models \alpha \iff \alpha \in T_{\kappa+1} \iff \alpha \in T_\kappa \iff \mathbb{E}_{T_\kappa,F_\kappa} \models \mathsf{T}(\overline{\alpha}).$$

This model can be used to show that a certain collection of untyped truth bi-conditionals is consistent over classical logic. Let us call a formula α **positive** if every occurrence of the predicate T in α is under the scope of an even number of negation symbols.

Theorem 6. *The extension of S_{min} by the collection of truth bi-conditionals for positive formulas is a consistent theory.*

Proof. We form a classical model $\mathbb{E}_*$ from the partial model by setting $\mathbb{E}_* = \mathbb{E}_{T_\kappa}$. For α a positive formula we have $\mathbb{E}_* \models \alpha$ iff $\mathbb{E}_{T_\kappa, F_\kappa} \models \alpha$. Thus it follows that positive bi-conditionals are validated in $\mathbb{E}_*$. □

Like our model of *TB*, the compositional axioms for connectives $\wedge, \vee$ and two quantifiers are also validated in $\mathbb{E}_*$. Note though that the clause for negation, $\mathsf{T}(\overline{\neg\alpha}) \leftrightarrow \neg\mathsf{T}(\overline{\alpha})$, does not hold in $\mathbb{E}_*$ except if $\alpha \in \mathscr{L}_0^*$. Instead, we observe

$$\mathbb{E}_* \models \forall x \forall y (\mathsf{T}(\dot{\neg}(x \dot{\wedge} y)) \leftrightarrow \mathsf{T}(\dot{\neg} x) \vee \mathsf{T}(\dot{\neg} y))$$

expressing that a conjunction is false if, and only if, at least one of the conjuncts is false, as well as analogous versions of the other clauses.

Selecting the above compositional clauses as axioms together with the positive bi-conditionals yields an expressive theory of truth and falsity known as the **Kripke–Feferman theory** *KF*.[4] This theory was introduced and analysed by Solomon Feferman [3] as an axiomatic counterpart to Kripke's semantic theory. Both the axiomatic and semantic theories are attractive theories of truth because they validate a large class of untyped truth bi-conditionals as well as satisfying many untyped compositional clauses. However, the theories represent conceptions of truth that do not validate classical reasoning. If α is liar sentence or truth teller (or one many other ungrounded sentences) then $\mathsf{T}(\overline{\alpha \vee \neg\alpha})$ does not hold in $\mathbb{E}_*$.

The next, and final, theory we present is based on a thoroughly classical conception of truth. It shifts the attention away from truth bi-conditionals to extensions of the compositional axioms. The theory was proposed by Harvey Friedman and Michael Sheard [5] and commonly denoted *FS*.

Definition. *FS is the axiomatic theory extending the compositional theory of truth CT by the two rules* NEC *and* CoNEC.

In particular, $FS \vdash \forall x(\mathsf{T}(\dot{\neg} x) \leftrightarrow \neg\mathsf{T}(x))$. By necessitation, also $FS \vdash \mathsf{T}(\overline{\alpha \vee \neg\alpha})$ for every α and indeed $FS \vdash \mathsf{T}(\overline{\forall x(\mathsf{T}(\dot{\neg} x) \leftrightarrow \neg\mathsf{T}(x))})$. Neither formula holds in Kripke's model $\mathbb{E}_*$. On the other hand, $\mathbb{E}_*$ validates the implication $\mathsf{T}(\overline{\alpha}) \to \alpha$ which cannot, by Montague's theorem, be consistently added to *FS*. Consistency of *FS* is deduced using a different construction, known as **revision semantics**.

Theorem 7. *FS is consistent.*

We utilise a construction due to Anil Gupta and Nuel Belnap [1] which, like Kripke's model, builds a hierarchy of 'approximations' to truth $\mathbb{R}_n$. Unlike the construction of $\mathbb{E}_*$ though, the interpretation of truth at each level will remove, as well as add, statements from the previous. Moreover, the hierarchy can begin from any starting assignment. Thus, let $\mathbb{R}_0 = \mathbb{E}$ be any standard model. Assuming the model $\mathbb{R}_n = \mathbb{E}_{R_n}$ has been defined, set

$$\mathbb{R}_{n+1} = \mathbb{E}_{R_{n+1}} \text{ where } R_{n+1} = \{\alpha \mid \mathbb{R}_n \models \alpha\}.$$

[4] As with *CT*, our definition of *KF* brushes over some important details. We refer the interested reader to [9] for an axiomatisation of *KF* and other theories of truth; the monograph [7] provides a detailed introduction to axiomatic theories of truth.

This hierarchy is not increasing, for $\lambda \in R_n$ iff $\lambda \notin R_{n+1}$. For the consistency of *FS* it is not necessary to extend the construction into the transfinite. Instead we observe that the theorems of *FS* hold co-finally in the hierarchy.

Proof. By induction on the length of derivations in *FS*, it can be shown that if $FS \vdash \alpha$ then for some positive n and every $k \geq n$, $\mathbb{R}_k \models \alpha$. □

As a theory of truth, *FS* has one significant drawback compared to, say, Kripke–Feferman truth. Given that it contains the compositional clauses and is closed under NEC, by McGee's paradox (Theorem 4) the theory has no standard model. Even so, some authors have argued that *FS* provides an acceptable formal theory of truth, since it embodies a notion of truth that respects classical reasoning and the symmetry of truth creation (NEC) and truth destruction (CoNEC).

5 Conclusion

We surveyed some common paradoxes of self-reference and natural (and consistent) theories of truth. What we presented is far from exhaustive. Much more can be said about the interplay of logic, language and truth. So we conclude this article by pointing to some important considerations glossed over in our account.

The starting point of our analysis was the simple syntax theory S_{min}. Almost any formal theory which can encode its own syntax and meta-theory will suffice in place of S_{min}. Common choices are theories of arithmetic or set theory. For most purposes, the choice of background theory is immaterial because, as we have seen, mild syntactic assumptions are sufficient to present both theories of truth and the paradoxes. Nevertheless, alternative background theories provide different frameworks for analysing the interplay of truth and (formal) language, and can alter the deductive strength of theories of truth.

A second point to be elaborated is our logical assumptions concerning truth. We assumed classical logic throughout and placed fault on our truth-theoretic assumptions. Some authors have argued in the other direction, that the biconditional $\alpha \leftrightarrow \mathsf{T}(\overline{\alpha})$ is not, inherently, at fault, but rather the assumption that truth respects classical reasoning. This naturally leads to considering non-classical logics, for instance intuitionistic [12], partial [8], and even paraconsistent logics [4].

Finally, even within the informal constraints of this article, there are naturally arising theories of truth which we did not present. The Gupta–Belnap revision hierarchy can be extended into the transfinite providing more refined notions of stable truth. Another prominent 'classical' theory of truth appeals to van Fraassen's supervaluation schema. The interested reader can consult [7,9] for more details and references.

References

1. Belnap, N., Gupta, A.: The Revision Theory of Truth. MIT Press, Cambridge (1993)
2. Curry, H.B.: The inconsistency of certain formal logics. J. Symb. Log. **7**, 115–117 (1942)
3. Feferman, S.: Reflecting on incompleteness. J. Symb. Log. **56**, 1–49 (1991)
4. Field, H.: Saving Truth from Paradox. Oxford University Press, Oxford (2008)
5. Friedman, H., Sheard, M.: An axiomatic approach to self-referential truth. Ann. Pure Appl. Logic **33**, 1–21 (1987)
6. Halbach, V.: The Logic Manual. Oxford University Press, Oxford (2010)
7. Halbach, V.: Axiomatic Theories of Truth, Revised edn. Cambridge University Press, Cambridge (2014). First edition 2011
8. Halbach, V., Horsten, L.: Axiomatizing Kripke's theory of truth. J. Symb. Log. **71**, 677–712 (2006)
9. Halbach, V., Leigh, G.E.: Axiomatic theories of truth. In: Zalta, E.N. (ed.) The Stanford Encyclopedia of Philosophy, Spring 2020 Edition (2020)
10. Halbach, V., Leigh, G.E.: The Road to Paradox: A Guide to Syntax, Truth, and Modality. Cambridge University Press (forthcoming)
11. Kripke, S.: Outline of a theory of truth. J. Philos. **72**, 690–712 (1975)
12. Leigh, G.E., Rathjen, M.: The Friedman-Sheard programme in intuitionistic logic. J. Symb. Log. **77**, 777–806 (2012)
13. McGee, V.: How truthlike can a predicate be? A negative result. J. Philos. Log. **14**, 399–410 (1985)
14. McGee, V.: Maximal consistent sets of instances of Tarski's schema (T). J. Philos. Log. **21**, 235–241 (1992)
15. Tarski, A.: The concept of truth in formalized languages. In: Logic, Semantics, Metamathematics, pp. 152–278. Clarendon Press, Oxford (1956)

Author Index

Author Index

Printed in the United States
by Baker & Taylor Publisher Services